EVOLUTION
PRINCIPLES AND PROCESSES

D1297138

JONES AND BARTLETT TOPICS IN BIOLOGY

The Jones & Bartlett Topics in Biology Series

We are pleased to offer a series of full-length textbooks designed specifically for your special topics courses in biology. Our goal is to supply comprehensive texts that will introduce non-science majors to the wonders of biology. With coverage of topics in the news, emerging diseases, and important advances in biotechnology, students will enjoy learning and relating science to current events.

AIDS: The Biological Basis, Fifth Edition
Benjamin S. Weeks, Adelphi University; I. Edward Alcamo, PhD,
formerly of State University of New York at Farmingdale

Human Embryonic Stem Cells, Second Edition
Ann Kiessling, PhD, Harvard Medical School, Boston, Massachusetts; Scott C. Anderson

The Microbial Challenge: Science, Disease, and Public Health, Second Edition
Robert Krasner, PhD, Emeritus, Providence College

Microbes and Society, Second Edition
Benjamin S. Weeks, PhD, Adelphi University; I. Edward Alcamo, PhD,
formerly of State University of New York at Farmingdale

Related Titles in Evolution and Genetics

Defending Evolution: A Guide to the Evolution/Creation Controversy
Brian J. Alters

Essential Genetics: A Genomics Perspective, Fifth Edition
Daniel L. Hartl

Genetics: Analysis of Genes and Genomes, Seventh Edition
Daniel L. Hartl & Elizabeth W. Jones (late)

Genetics of Populations, Fourth Edition
Philip W. Hedrick

Lewin's Essential GENES, Second Edition
Jocelyn E. Krebs, Elliott S. Goldstein, & Stephen T. Kilpatrick

Lewin's GENES X
Jocelyn E. Krebs, Elliott S. Goldstein, & Stephen T. Kilpatrick

Microbial Genetics, Second Edition
Stanley R. Maloy, John E. Cronan, Jr., & David Freifelder

Strickberger's Evolution, Fourth Edition
Brian K. Hall & Benedikt Hallgrímsson

EVOLUTION
PRINCIPLES AND PROCESSES

BRIAN K. HALL
Department of Biology
Dalhousie University
Halifax, Nova Scotia
Canada

JONES AND BARTLETT PUBLISHERS
Sudbury, Massachusetts
BOSTON TORONTO LONDON SINGAPORE

WORLD HEADQUARTERS
Jones and Bartlett Publishers
40 Tall Pine Drive
Sudbury, MA 01776
978-443-5000
info@jbpub.com
www.jbpub.com

Jones and Bartlett Publishers Canada
6339 Ormindale Way
Mississauga, Ontario L5V 1J2
Canada

Jones and Bartlett Publishers
 International
Barb House, Barb Mews
London W6 7PA
United Kingdom

Jones and Bartlett's books and products are available through most bookstores and online booksellers. To contact Jones and Bartlett Publishers directly, call 800-832-0034, fax 978-443-8000, or visit our website, www.jbpub.com.

Substantial discounts on bulk quantities of Jones and Bartlett's publications are available to corporations, professional associations, and other qualified organizations. For details and specific discount information, contact the special sales department at Jones and Bartlett via the above contact information or send an email to specialsales@jbpub.com.

Copyright © 2011 by Jones and Bartlett Publishers, LLC

All rights reserved. No part of the material protected by this copyright may be reproduced or utilized in any form, electronic or mechanical, including photocopying, recording, or by any information storage and retrieval system, without written permission from the copyright owner.

PRODUCTION CREDITS
Chief Executive Officer: Ty Field
Chief Operating Officer: Don W. Jones, Jr.
President, Higher Education and Professional Publishing: James Homer
V.P., Sales: William J. Kane
V.P., Design and Production: Anne Spencer
V.P., Manufacturing and Inventory Control: Therese Connell
Publisher, Higher Education: Cathleen Sether
Acquisitions Editor: Molly Steinbach
Senior Editorial Assistant: Jessica S. Acox
Editorial Assistant: Caroline Perry
Production Manager: Louis C. Bruno, Jr.
Senior Marketing Manager: Andrea DeFronzo
Text Design: Anne Spencer
Cover Design: Kristin E. Parker
Illustrations: Elizabeth Morales
Senior Photo Researcher: Christine Myaskovsky
Photo and Permissions Associate: Emily Howard
Composition Services: Circle Graphics
Cover Image: Photo by Tom Rutledge, www.photosbytomandpolly.com
Printing and Binding: Courier Kendallville
Cover Printing: Courier Kendallville

Library of Congress Cataloging-in-Publication Data

Hall, Brian Keith, 1941–
 Evolution : principles and processes / Brian K. Hall. — 1st ed.
 p. cm.
 Includes bibliographical references and index.
 ISBN 978-0-7637-6039-7 (alk. paper)
 1. Evolution (Biology) I. Title.

QH366.2.H346 2011
576.8—dc2 2010008085

6048

Printed in the United States of America
14 13 12 11 10 10 9 8 7 6 5 4 3 2 1

About the Cover
A zeedonk (zonkey) is a hybrid of a male zebra and female donkey, two species within a horse family. This zeedonk has well-developed stripes of the zebra parent on the legs and muzzle but less obvious stripes on the flank.

Brief Contents

Contents

■ PART VII HUMAN ORIGINS AND EVOLUTION374

Preface

If you want to know about the science of evolution, how life began, what Charles Darwin really said about evolution, why a fungus is more closely related to a human being than to a plant, how experiments in evolution are carried out, why birds are flying dinosaurs, how we manipulate the evolution of other species, and if you want a clear treatment of the processes that result in evolution, then this is the book for you. It is written for those who want to learn about evolution but have minimal background in sciences such as chemistry or mathematics. The approach and level of treatment makes the book suitable for biology majors in an introductory class in evolution, non-biology majors seeking an evolution class, students in Arts, Liberal Arts, Social Science or Health professions in which a science class is required, business schools or departments of economics in which students need to understand the importance of natural selection, and community colleges offering a one-term or full-year class in evolution for students or for the public.

Evolution is both a science and the science of life. Cultures, societies, populations, organisms, cell and tissue types, genes and other molecules, can be compared because similarities and differences can be explained on the basis of common inheritance from earlier forms. The processes that explain evolution can be examined in nature. Evolutionary processes can be tested experimentally in the field or in the laboratory. Evolution is both real and ongoing.

Theories that life evolved have been proposed since the early 18th century. Not until 1859, however, when Charles Darwin's book *On the Origin of Species* appeared in print, was the evidence for evolution gathered together on such a massive scale in support of a particular mechanism of evolution: evolution by natural selection. Darwin's theory revolutionized human existence and caused enormous changes in how we view the world, understand our place in it, and explain life and living things.

The central aim of evolutionary biology, which is to explain the origins and diversity of life, treats such issues as:

- How whales evolved from ancestors that walked on land
- How birds acquired flight as they evolved from dinosaurs
- Why genes that cause human diseases have not disappeared as humans evolved
- Whether human cultural evolution has overtaken biological evolution
- Whether the Universal Tree of Life (UToL) has multiple trunks

These and many other questions are asked and answered by the study of evolution.

The aim of this book is twofold: (1) To bring together the major **principles and processes** that explain evolutionary change; (2) To discuss the **patterns** of life that

have resulted from the operation of evolution over the past 3.5 billion years. To achieve these aims, the first third of the book discusses evolution in relation to:

- the scientific method and evolution as a science in Chapter 1;
- the origins of matter, of the universe, and of Earth in Chapter 2;
- the origin of molecules, cells and the first organisms, and the nature of life in Chapter 3 and 4;
- the role played by Charles Darwin and Alfred Wallace in independently proposing a theory of evolution by natural selection in Chapters 5 through 7.

The next four chapters, Chapters 8 through 11, outline the transformation of single-celled organisms into multicellular organisms — animals, plants and fungi — that can be grouped into kingdoms in a universal tree of life. Chapters 12 to 15 discuss:

- the presence and importance of variation at the level of genes, gene regulation and in populations; and
- how populations as reproductively isolated units provide the raw material upon which natural selection can act.

In Chapters 16 through 18, you will see that organisms are members of species, which arise and usually become extinct. Chapters 19 to 21 examine the principles, processes and patterns of evolution as seen in the origination and evolution of humans, in the evolution of culture and society, and in the evolution of belief systems and of religion.

Although the 7 parts and 21 chapters are logically arranged, the structure of the book allows you to begin with any part — for example, Part VII, which contains chapters 19–21 on Human Origins and Evolution — or to read individual parts as units independent of the others. Neither your reading nor the curriculum need follow the order of topics.

The book is extensively illustrated with four-color images and includes the following pedogogical tools:

Key concepts are listed at the beginning of each chapter. Checking these both before and after you read a chapter will alert you to the content and enable you to be sure you have mastered it. Key concepts in the text and in boxes are highlighted. **Boxes** are used to draw attention to particular topics, often ones that relate to more than one chapter.

Marginal notes are used to provide comments on or to expand upon the text. A list of **recommended reading** for each chapter allows you to explore particular topics in more depth by looking to primary literature and other relevant resources.

A **glossary** of definitions is provided at the end of the book, along with a separate **cross-listing (lexicon)** of key concepts, principles, processes, and groups of organisms. This cross-tool allows you to access related (or opposite) terms and concepts. For example, "Abiotic" links to "Biotic" and to "Life." "Adaptive radiation" links to "Biogeography," "Speciation," and "Zoogeography." This cross-listing, which may alert you to relationships you otherwise might not have thought of, will be especially useful for those readers who want to examine or study related topics and for those who need to assign projects or related topics in examination formats.

Ancillary Materials
To assist the instructor with teaching the course and to provide students with the best in learning aids, Jones and Bartlett has prepared an ancillary package.

For the Instructor

The *Instructor's Media CD-ROM* provides instructors who have adopted the text with the following traditional ancillaries:

The **PowerPoint® Image Bank** provides a library of all of the illustrations, photographs and tables, to which Jones and Bartlett Publishers holds the copyright or has permission to reprint digitally, inserted into PowerPoint slides. Instructors can easily copy individual images into existing PowerPoint slides or add more images to the PowerPoint Lecture Outline Slides.

The **PowerPoint Lecture Outline Slides**, prepared by Jeff Meldrum of Idaho State University, combine text and images into a presentation that is both educational and engaging. Using the Microsoft® PowerPoint program, instructors can easily modify and edit the slide show to fit individual presentation needs.

An online *Test Bank*, prepared by Dalton Gossett of Louisiana State University–Shreveport, features questions in plain text files that can be easily integrated with your existing course management software.

For the Student

Jones and Bartlett has established a **Student Companion Web Site**, at http://biology.jbpub.com/hall/evolutionprinciples, to host a wide array of additional resources and activities for independent study, review and research.

An **Interactive Glossary**, on-screen **Flashcards** and electronic **Crosswords** offer fun and easy ways for you to review key terms, concepts and processes.

To help with reading comprehension and exam preparation, **Study Quizzes** for each chapter, prepared by James K. Dooley of Adelphi University, provide you with short Web-based tests to assess your knowledge of the content.

Web Links recommend external Web sites that provide further information about topics covered in the text. James K. Dooley compiled this list of helpful outside resources.

Acknowledgments

Special thanks to the following reviewers who provided comments on chapters of the manuscript:

Jeffrey Meldrum, Idaho State University
Robert A. Martin, Murray State University
Nicholas J. Cheper, East Central University
Robert P. Benard, American International College
Bill Brindley, Utah State University
Robert Loiselle, Université du Québec à Chicoutimi
Mike Toliver, Eureka College
Moira E. Royston, St. Joseph's College
Dorothy B. Payne, Alabama State University
Jonathan Frye, McPherson College
Lee Christianson, University of the Pacific
Elizabeth L. Rich, St. Joseph's University
Callie A. Vanderbilt, San Juan College
Jay P. Clymer III, Marywood University
Jeffrey D. Sack, Valley Regional High School
Raymond Pierotti, University of Kansas

Christine Andrews, The University of Chicago
Dalton R. Gossett, Louisiana State University–Shreveport
Keith W. Pecor, The College of New Jersey
Jeremy Moynihan, Florida International University
Ken Gobalet, California State University, Bakersfield
James K. Dooley, Adelphi University
Elka T. Porter, University of Maryland
Ralph L. Holloway, Columbia University

Brian K. Hall
Dalhousie University and Arizona State University

About the Author

The author relaxing at Milford House, South Milford, Nova Scotia.

Brian Hall, born, raised and educated in Australia, has been associated with Dalhousie University in Halifax, Nova Scotia since 1968, most recently as University Research Professor and George S. Campbell Professor of Biology, and University Research Professor Emeritus since July 2007.

Trained as an experimental embryologist, his laboratory researches the development and evolution of cartilage, bone and vertebrate skeletal systems. He played a major role in integrating evolutionary and developmental biology into the discipline now known as *Evolutionary Developmental Biology (evo-devo)*, including writing the first evo-devo textbook, *Evolutionary Developmental Biology* (Kluwer Academic Publishers, Netherlands), published in 1990 and in a second edition in 1999. He and Benedikt Hallgrímsson co-authored the fourth edition of *Strickberger's Evolution: The Integration of Genes, Organisms and Populations* (Jones and Bartlett Publishers, Sudbury, MA), from which the present book evolved.

A fellow of the Royal Society of Canada and Foreign Honorary Member of the American Academy of Arts and Sciences, Dr. Hall has earned many awards for his research, writing and teaching, including the 2005 Killam Prize in Natural Sciences (one of the top scientific prizes in Canada) and induction in 2009 into the Discovery Centre Hall of Fame (http://www.loudandclear.ca/video/Hall_HQ.mov). He is widely recognized for his scholarship, excellence as a teacher and mentor, and ability to communicate complex material and concepts to specialist and general audiences alike. To this end, he has written 11 books, edited or co-edited 30 others, and given many lectures and led many workshops around the world.

Science and Evolution as Science

I

1

The Nature of Science and Evolution, and Evolution as Science

KEY CONCEPTS

- Science can investigate events that occurred in the past.
- Science can investigate evolution.
- The scientific method can be applied to evolution.
- Evolution is a science.
- The word *evolution* has had different meaning over time.
- Evolution acts on individuals and on populations but in different ways.
- Individuals in the same species can adapt differently to different parts of their geographical range.
- Organisms modify the environment in which they live.
- Genes are defined and studied in different ways.

Overview

The scientific method is a universal means of proposing a hypothesis, designing experiments, collecting data to test that hypothesis, interpreting the data in the context of past knowledge, and accepting or rejecting the hypothesis. New conclusions may

Above: The wonders created by the relentless action of wind and water.

add evidence to support existing scientific knowledge or they may overthrow an existing theory and replace it with a new theory that changes the way we view the world. Isaac Newton's discovery of **gravitation** and Charles Darwin's discovery of **evolution by natural selection** are two examples of theories that changed our understanding of how the world and its organisms function.

Much scientific investigation relates to events that occurred in the past. The origin of the universe and how volcanoes form are two examples. Evolution as a science is no different. The science of evolution investigates past and ongoing processes concerning how organisms arise and change over time. The origin of stars and of glaciers and the process of evolution were all revealed in the same way — using the scientific method.

In addition to examining the scientific method and evolution as a science, this chapter lays out a brief history of evolution as a term and as a concept, and introduces **natural selection, variation,** and **inheritance** as three fundamental principles and processes underlying evolution. The different ways in which evolution "sees" individuals and populations are outlined. The nature of the unit of inheritance — the **gene** — is introduced, as is the special way in which organisms relate to their environment — the **niche.**

The Scientific Method

The essential nature of science is discovery through application of a method — the **scientific method** — that allows discoveries to be made. The discovery may be a previously unknown object — a previously unknown type of organisms from the ocean deeps — or a new explanation — how such organisms survived and evolved at depth.

The scientific method consists of

1. producing a hypothesis;
2. designing and performing controlled experiments or making observations that allow data relevant to the hypothesis to be collected;
3. analyzing the data in an objective way against the background of existing knowledge; and,
4. drawing conclusions that support or refute the hypothesis.

Through the repeated application of this method, science progresses by accumulating evidence consistent with one interpretation and inconsistent with others.

When possible, experimentation is an important way to test hypotheses. However, when experimentation is not possible, data can be collected and hypotheses accepted or rejected without experimental verification. The scientific method thus is applied to astronomy, geology, and past evolutionary events because the scientific method is sufficiently precise to allow explanations of past events. If those explanations contradicted present events, new hypotheses would be generated and new data obtained. As a consequence, astronomy, geology and evolution are sciences.

Careful observation, recording and systematic analysis are important ways of applying the scientific method. Comparative study of systems that are alike in many respects but differ in others is another important application of the scientific method. We thus can compare ants and termites as two types of social insects, or

Our quest for discovery, whether by the scientific method or through exploration, invention, or just plain thirst to understand the unknown, has never been better treated than by Daniel J. Boorstein in *The Discoverers,* first published in 1983.

we can compare chimpanzees and modern humans as closely related primates. We can go further and compare ants and humans as two groups of social organisms. Through systematic application of the scientific method, unsupported hypotheses are eliminated and a single interpretation emerges as the one best explaining the data. This process may take years or centuries, as it did for the discovery of the relationship between the orbits of Earth, other planets and the Sun (FIGURE 1.1), and for the discovery of deoxyribonucleic acid (DNA) as the molecule of inheritance carrying the genetic code from generation to generation.

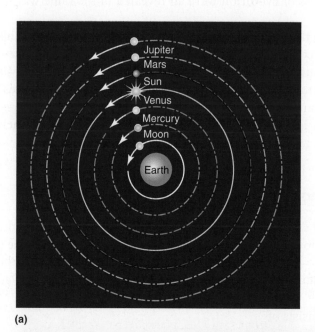

(a)

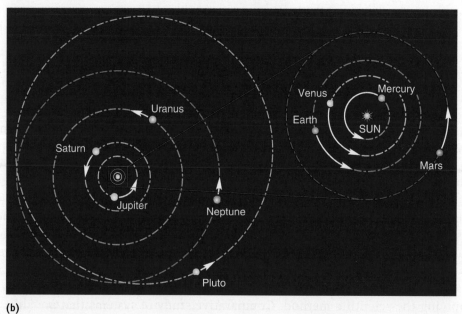

(b)

FIGURE 1.1 Two hypotheses reflecting theories of the orbits of the planets, one in which the planets orbit Earth (**a**), the other in which the planets orbit the Sun (**b**).

Other discoveries may be, or appear to be, instantaneous: The apple falling on Newton's head led him to the discovery of the law of universal gravitation; the rising level of the water in his bath led Archimedes to the discovery of the principle of buoyancy. Even such "Eureka moments" cannot be isolated from past knowledge or the way of thinking of the society in which the discoverer resides. This is so even when the discovery totally changes how we view natural phenomena, as when Albert Einstein discovered that matter and energy are not separate but interconvertible, expressed in the formula $E = mc^2$ (energy equals mass times the speed of light squared [multiplied by itself]). Even though the formula revolutionized our thinking, it did not arise in a vacuum. Earlier theories of matter and energy existed.

The social context in which science is conducted — indeed, whether science is conducted at all — what is considered an appropriate topic for analysis using the scientific method, how scientific knowledge is used, and the proportion of the population engaged in or benefiting from scientific discovery, vary among and within modern human societies. A century ago, psychology was regarded as "no more than" a social science (and some objected to the word science in the phrase "social science"). Today, psychologists utilize the scientific method, expect variation around the norm, propose and test hypotheses, and use sophisticated statistical methods to inform their conclusions.

■ Archimedes' principle is that, "Any body fully or partially submerged in a fluid is buoyed up by a force equal to the weight of the fluid displaced."

Newton's law of universal gravitation: "I deduced that the forces which keep the planets in their orbs must be reciprocally as the squares of their distances from the centers about which they revolve" (Newton, 1687).

Science and Events in the Past

Can we apply the scientific method to understand and explain events that happened in the past? Can we, for example, design experiments and/or collect data to investigate how Earth arose, how gravity originated, how cells evolved? Yes, we can. All sciences concerned with the past — astronomy, cosmology, geology, paleontology and evolution — make use of observations to refute or support proposed hypotheses. This entire book is devoted to obtaining scientific evidence for various aspects of evolution, not only biological evolution but evolution of the Universe, the Solar System and Earth.

Furthermore, we can gather data to explain events that occurred in the past, even if they occurred only once. The latter is of special interest in historical sciences where the emphasis may be on understanding a particular sequence of historical events rather than discovering general laws such as those of physics and chemistry. The evolutionary biologist and popularizer of science Stephen J. Gould (1941–2002) was especially prominent in promoting the view that past events condition present and future change. Topics of historical science include events that led to our solar system, to the separation of South America from Africa, and to the origin of humans, recognizing that these events are singular and may not apply to all stars, all continental separations, or the evolution of all species. Sciences that deal with the past make use of general laws such as those of gravity, mechanics and biochemistry with the aim of discovering the causes of diversity and uniqueness as well as the principles and laws that apply uniformly to all matter or all life. The ability to extrapolate from processes or events that affect all organisms — the existence of variation, mutation, changing environments — and to explain the role of past history, is central to the dynamic interdisciplinary nature of the study of evolution, and makes it a robust and testable field of inquiry. Does this make evolution a science?

Evolution as a Science

A criticism sometimes raised against evolution as a science that can be studied and understood using the scientific method, is that evolutionary explanations (hypotheses) cannot be tested and supported (or falsified) in the same way as hypotheses in physics and chemistry.

The claim is sometimes made that because evolution deals with events that occurred in the past — events that are generally impossible to repeat in a laboratory — evolutionary biology can never reach the status of the sciences of physics and chemistry. As discussed in the previous section, we can explain historical events as rationally as we explain other scientific events. In particular, evolution that occurred in the past can be observed, documented, studied and tested. Evidence to test evolutionary hypotheses exists in the fossil record, and evolution can be tested experimentally.

- The sequence of primate-like hominins (humans and their closest relatives) in the fossil record supports the hypothesis that humans have a primate origin (see Chapter 19).
- Correspondence in the basic chemical sequences of myoglobin and hemoglobin, two classes of iron-containing molecules that bind and transport oxygen, supports an evolutionary relationship between them (see Chapter 8).

Because either hypothesis could be disproved by finding, for example, frog- or reptile-like hominid fossilized ancestors or by discovering a species that lacks chemical sequence similarities between myoglobin and hemoglobin, such hypotheses are scientific. The fact of past evolutionary events can be tested scientifically.

The ever-present influence of past evolutionary history is far more than a theoretical postulate. Evolution can be demonstrated experimentally. One example is when replicated populations of the fruit fly, *Drosophila melanogaster*, are returned to a common ancestral environment and allowed to breed for 50 generations. These populations undergo changes to an ancestral state, but do not all revert to the same universal state. The changes we see depend upon the particular **character** analyzed, a reflection of **mosaic evolution** in which characters in the same species evolve at different rates. This important experiment illustrates that we can demonstrate past evolutionary history on a character-by-character basis.

In a second example described by Jeffrey Barrick and his colleagues in the journal *Nature* on October 29, 2009, all mutations occurring over 20 years (40,000 generations) in the bacterium *Escherichia coli* were recorded. After 20,000 generations, a mutation in a gene involved in repairing DNA resulted in an acceleration of the rate at which mutations became established in the populations — 45 mutations in the first 20,000 generations and 600 in the second 20,000 generations.

The properties of different hydrogen atoms can be explained on a common physical basis. The properties of different organisms — the organization and function of their component parts — only can be understood in the context of their history, which includes adjustments to specific lifestyles at particular times that influence (and have influenced) the type of change possible in the future (BOX 1.1). The adjustment of an organism to its environment (including adjustment to other organisms in that environment) is measured by the biological attribute fitness, which differs from organism to organism, population to population, and species to species. Nevertheless, when historical conditions are repeated, and different organisms are subjected to similar selective evolutionary forces, some common responses can be predicted; geographi-

Ways of Knowing. A New History of Science, Technology and Medicine, by J. V. Pickstone, 2001, contains a unified argument against such criticisms in an analysis of the development of the scientific method in science, technology and medicine — an analysis described by one of the ablest historians of science and medicine, the late Roy Porter (1946–2002) as, "the most exciting synthesis we now possess."

The terms **character(s)** and **feature(s)** are used interchangeably for any distinguishable attribute of an organism, whether morphological (blue eyes), behavioral (burrows), functional (breathes oxygen) or molecular (has the gene for hemoglobin).

Fitness can be defined as the measure of the ability of an individual to survive and transmit its genes (genotype) to the next generation, relative to the ability of other individuals in the same population (see Chapter 15).

BOX 1.1

Environment-Organism Interactions

Climate, terrain, prey and/or predators are often reflected in the features of organisms. If a species is distributed over a wide north–south range, adaptations in northern populations often allow us to distinguish northern from southern individuals. Adjustment to similar environments can bring forth parallel evolutionary changes in different species, as seen on a local scale, as in cave-dwelling animals (see Figure 1.2) or on a continental scale, as seen in American and Australian mammals (see Figure 6.6).

A well-known pattern of inherited response to a specific environment is the relationship in warm-blooded vertebrates between body size and average temperature, a relationship known as **Bergmann's rule.** Individual members of a species in cooler climates tend to be larger than individuals of the same species in warmer climates. This relationship exists primarily because bodies with larger volumes have proportionately less exposed surface area than do bodies with smaller volumes. Because heat loss relates to surface area, larger bodies can retain heat more efficiently in cooler climates. Smaller bodies can eliminate heat more efficiently in warmer climates.

■ Karl Georg Lucas Christian Bergmann (1814–1865), a German anatomist, physiologist and biologist developed this rule in 1847.

Patterns of adaptation also can be detected in the fossil record. Body size of bushy-tailed woodrats (*Neotoma cinerea*) closely follows climatic fluctuations from the time of the last glacial period about 25,000 years ago to the present with larger body size in colder periods and smaller body size in warmer periods. Similarly, among populations of modern humans, we can compare thick-chested, short-limbed Inuit in the north with slender Africans in equatorial regions.

cally widely separated populations and species of fish, amphibians, spiders and insects adapted to cave conditions, consistently show rudimentary eyes and loss of pigmentation (FIGURE 1.2). Predictability is an essential element of a scientific theory.

■ Evolution as Only a Theory

It is sometimes proposed that because we cannot see evolution happening, evolution will always be a theory and not a "fact."

Any of the foundational theories in science — the atomic theory, the theory of relativity, the universal theory of gravitation — account for events we cannot see in everyday life. Nevertheless, we experience the effects of these forces of nature daily. The atomic theory is accepted by scientists because it explains chemical reactions, although atoms are not visible to the naked eye. The universal theory of gravitation is accepted as explaining why objects remain tethered to Earth even though we cannot see the gravitational waves that cause the effect.

Similarly, the theory of evolution accounts for the historical sequence of past and present organisms by explaining their existence in terms of factors that cause changes in the genetic inheritance of organisms over time.

The facts of evolution come from the anatomical similarities and differences among organisms, the places where they live, the metabolic pathways they use, the embryological stages through which they develop, the fossil forms they leave behind, and the genetic, chromosomal, and molecular features that relate them. The theory of evolution (inherited changes over time) is a coherent explanation of the historical course of biology (facts) in terms of natural processes, such as mutation, selection, genetic drift, migration and altered gene regulation (see Chapters 12–15). These

FIGURE 1.2 Adaptations to life in the dark. (**a**) Surface-dwelling (eyed, pigmented) and cave-dwelling (blind, non-pigmented) forms of Mexican cave fish. (**b, c**) Blind and non-pigmented (albino) cave-dwelling axolotls shown from the top (**b**) and from the front (**c**). Note prominent gills in (**c**).

explanations are consistent with all observations made so far. Such an approach is not different from the *fact* that an apple dropped on Newton's head, an event Newton explained by developing a *theory* of gravity.

For our purposes, an appropriate way to present the science of evolution is to document the facts of evolution — the historical record available to us in fossils, molecules, and living organisms — and to seek to understand, through application of the scientific method, the evolutionary processes by which the historical record of evolutionary change came about.

A good place to begin is to consider the meaning of the word *evolution*.

Evolution: an Overview of the Term and the Concept

The word *evolution* had different meanings in the past to the way we use the word today.

Evolution as the Development of an Individual

The first meaning — evolution as the development of an organism — reflects the original 17th century definition, when the word evolution (from the Latin *evolutio*, unrolling) was used for the unfolding of the parts and organs of an embryo to reveal

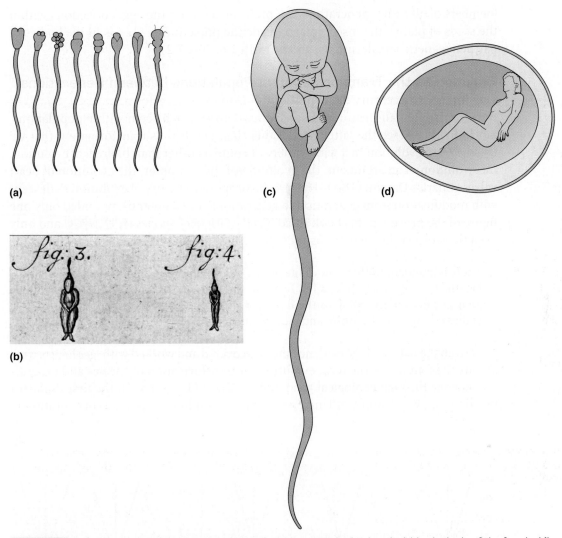

FIGURE 1.3 Four views of adult humans as preformed within sperm (**a, b, c**) and within the body of the female (**d**). (**a** Adapted from Antoni van Leeuwenhoek. *Philosophical Transactions*, 1678; **c** adapted from Nicholas Hartsoeker, 1694; **d** adapted from Needham, J. *History of Embryology*. Cambridge University Press, 1959.)

a preformed body plan (**FIGURE 1.3**, and see Box 7.2). Here evolution applies to individuals. Evolution as development can be traced to the Swiss botanist, physiologist, lawyer, and poet, Albrecht von Haller (1708–1777), who in 1774 used evolution to describe the development of the individual in the egg:

> But the 17th century theory of evolution proposed by Jan Swammerdam and Marcello Malpighi prevails almost everywhere: all human bodies were created fully formed and folded up in the ovary of Eve and that these bodies are gradually distended by alimentary humor until they grow to the form and size of animals (Haller, 1774, cited from Adelmann, 1966, pp. 893–894).

Another Swiss lawyer, Charles Bonnet (1720–1793), further solidified evolution as preformation in his theory of encapsulation (*emboîtment*). Bonnet wrote that all

members of all future generations are preformed within the egg: cotyledons within the seeds of plants; the insect imago inside the pupa; future aphids in the bodies of parthenogenetic female aphids, and so forth (see Box 7.2).

Evolution as the Transformation of Populations between Generations

Only in the 19th century did evolution come to mean transformation of a species or transformation of the features of organisms, both of which occur within populations, not individuals (see the latter part of this chapter). It is as transformation that we use the term evolution in a wide variety of contexts other than biological evolution: the evolution of an argument, the evolution of the computer, the evolution of heart valves. Charles Darwin (1809–1882), who proposed a theory of evolution as descent with modification from generation to generation (see Chapter 6), provided only one figure of the process in his book titled, *On the Origin of Species* (FIGURE 1.4) and only used the word *evolution* once, as the last word of the book.

> There is grandeur in this view of life, with its several powers, having been originally breathed by the Creator into a few forms or into one; and that, whilst this planet has gone cycling on according to the fixed law of gravity, from so simple a beginning endless forms most beautiful and most wonderful have been, and are being evolved.

Given the nature of the evidence they discovered and worked with, geologists were among the first to use the term evolution for transformation of species and progressive change through geological time. Robert Grant (1793–1874), the first Professor of Comparative Anatomy at University College London, used the term evolution in

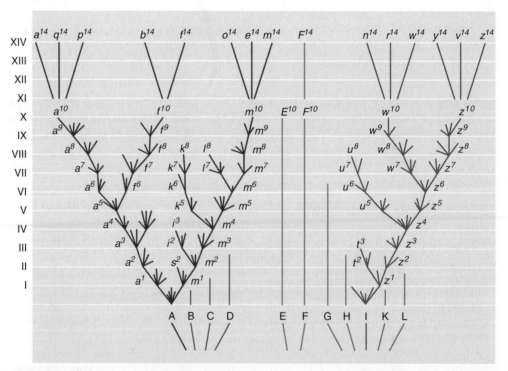

FIGURE 1.4 The illustration Charles Darwin used in his book, *On the Origin of Species* to represent progressive divergence within individual species (a^1–a^{10}; m^1–m^{10}; z^1–z^{10}) and the splitting of species into multiple lineages (S^2, i^2, i^3 from m^1 and m^2, for example).

1826 for the gradual origin of invertebrate groups. Charles Lyell (1797–1875), one of the founders of geology and a close friend of Charles Darwin, used evolution in 1832 for gradual improvement associated with the transformation of aquatic to land-dwelling organisms: "the testacea [shelled animals] of the ocean existed first, until some of them, by gradual evolution, were improved into those inhabiting the land." Even here, an argument can be made that both Grant and Lyell were using evolution in the sense of change during development.

Transformation and Descent with Modification: 1859–Now

This section contains a very brief overview of major changes in the way evolution has been studied, changes that reflect advances in understanding biology at deeper and deeper levels.

From the publication of *On the Origin of Species* in 1859 until 1900, evolution was studied as

- the origination and transformation of species (one species of horse → another species of horse);
- the transformation of major groups/lineages of organisms and the search for ancestors (invertebrates → vertebrates; fish → amphibians); and
- the transformation of features such as jaws, limbs, kidneys, nervous systems within lineages of organisms.

■ **Origination** is used throughout the text in the same sense as in the *Oxford English Dictionary*'s definition of evolution, that is, for the first appearance or origin of.

A new genetic approach to evolution followed the rediscovery in 1900 of Gregor Mendel's experiments breeding pea plants and the development of Mendelian genetics (see Chapter 7). Geneticists began to work with inbred lines of organisms, with animals maintained in laboratories or with plants in greenhouses. The discovery of mutations — changes in a sequence of DNA — further focused attention on genes as fundamental to the evolutionary process.

Two discoveries in the early 20th century placed the emphasis on mechanisms of evolutionary change within populations. In 1908, Geoffrey Hardy (1877–1947) in England and Wilhelm Weinberg (1863–1937) in Germany independently derived a formula for calculating gene frequencies in populations under natural selection (soon named the *Hardy-Weinberg law* or *principle*; see Chapter 14). In 1918 the brilliant English mathematician R. A. Fisher (1890–1962) fused Mendelian inheritance with population genetics (see Chapter 7). By the 1930s, speciation was seen as resulting from genetic changes within a lineage as reflected in changes in gene frequency (see Chapter 14). In the 1940s, the synthesis of population genetics, systematics, and adaptive change forged what we know as the **modern synthesis** or the **modern evolutionary synthesis**.

■ Fisher, along with his fellow Englishman J. B. S. Haldane (1892–1964) and the American geneticist Sewall Wright (1889–1988) are the acknowledged founders of the modern synthesis of evolution based on population genetics (see Chapter 14).

Population genetics does not provide a complete theory of evolution, however. Evolution now is recognized as hierarchical; genes, structures, populations, species and ecosystems all evolve. To a considerable extent hierarchical evolution is a reflection of the hierarchical organization of life itself, a concept outlined in Box 4.1.

Perhaps the most well recognized hierarchy of life is that from

genes → molecules → organelles → cells → tissues → organs → organisms → populations → species

Hierarchical systems aid and stabilize evolution, enabling organisms to evolve, incorporate and maintain new functional properties. A hierarchical approach to understanding evolution has emerged and consolidated over the past 40 years, replacing

or running parallel with the reductionist approach, which proposes that explanations for events on one level of complexity can and should be reduced to (deduced from) explanations on a more basic level. The biological hierarchy extends across many levels, from atoms to molecules to cells to tissues to organs to individuals to populations to species to cultures, each with specific functional properties.

Why does evolutionary biology not reduce to a reductionist approach? In no small part we require a hierarchy of mechanisms to explain emerging properties and increasing complexity because properties at one level are insufficient to explain properties at higher level(s). Evolution operates on at least four levels.

1. The **genetic** level, through mutations, changes in gene number and regulation, and changes in gene networks.
2. The **organismal** level, seen as individual variation and differential survival through adaptation and the evolution of new structures, functions and/or behaviors.
3. Changes in populations of organisms, operating through mechanisms that limit gene flow between populations.
4. The subsequent origin, radiation and adaptation of **species**.

From this list you may have inferred that the term *gene* (*genes*) must be being used differently at levels 1 and in 3: a unit that mutates (1) and a unit that can flow through a population (3). Current concepts of what we mean by a gene are outlined in BOX 1.2.

Because evolution acts at genetic, organismal and population levels, an ideal definition should reflect evolution at all three levels. In many respects, Darwin's phrase descent with modification, illustrated in Figure 1.4, remains an inclusive definition of biological evolution. Evolution as descent with modification encompasses evolutionary change at genetic, organismal and population levels, although integration of all three levels is required to fully comprehend evolution.

BOX 1.2
What Is a Gene?

It may surprise you to find that there are several definitions for the basis unit of inheritance, the *gene*. In large part, this is because different specialists study the gene in different ways and at different levels. Reflecting these different approaches, a gene can be defined as one or more of the following (an example of each is provided).

- a region of DNA, the activation of which leads to the formation of a feature or character (the gene for blue eyes);
- a region of DNA, the activation of which leads to the formation of a protein or RNA (the gene for the protein collagen);
- a region of DNA encompassing coding and non-coding segments (exons and introns) (a definition that applies only to eukaryotes);
- a unit of inheritance located on a chromosome (the gene for muscular dystrophy in humans mapped to chromosome 9);
- a set of nucleotides reliably copied and transferred from generation to generation (the unit of heredity).

How our views of evolution originated and have changed, and how evolution operates at the three levels of genes, organisms and populations are the topics of this book. Some themes that will emerge are

- We cannot predict the continued evolutionary success of a particular group on the basis of its dominance at a particular earlier time.
- Crucial evolutionary advantages may accrue to groups that already have characteristics adaptable to new circumstances.
- Long-term evolutionary replacement among groups is not predictable.
- New modes of biological organization can enhance group survival.
- New levels of organization occur because of coordinated changes in many traits over long intervals of time.
- Once such new levels have been attained, widespread radiation often begins.
- Extinction is common if not inevitable because of constraints on the potential of a species to adapt to large or rapid environmental changes.

Individuals, Populations and Evolution

Natural selection reflects the differential survival and reproduction of individual organisms with particular features. Differential survival and reproduction are mechanisms of evolution. Natural selection is the outcome. Although selection reflects the fate of individuals, the response to selection lies in the information content of the genomes of all individuals in the population, information that can change because of mutation or random processes (see Chapters 6, 12, and 15).

It is important to have a clear view of individuals and populations with respect to evolution (TABLE 1.1), not merely because use of the word evolution has shifted

TABLE

1.1 Comparison of Characteristics of Individuals and Populations

Characteristic	Individual	Population
Life span	One generation	Many generations
Spatial continuity	Limited	Extensive
Genetic characteristics	Genotype	Gene frequencies
Genetic variation	Expressed in one lifespan	Expressed in evolutionary change
Evolutionary characteristics	No changes, because an individual has only one genotype and is limited to a single generation	Can evolve (change in gene frequency), because evolution occurs between generations
Selection	Operates on phenotype in one generation	With mutation, leads to change in genotype from generation to generation
Mutation	Somatic influence individual	Gametic inheritance/gametic mutation transferred through reproduction
Variation	Phenotypic not inherited Genotypic transferred through reproduction	Genotype inherited

from individuals to populations, but because individuals respond to selection but populations evolve. Because this is important, the essential differences between individuals and populations as far as evolution is concerned are outlined as 12 points below. Each is presented as a statement of fact. You can regard them as conclusions, the evidence for which is provided in the remainder of the book. You also can read them as a summary of the evolutionary process.

1. Organisms exist as individuals. Individual multicellular organisms (animals, plants, fungi) develop, grow, mature, reproduce, senesce (in most cases) age and die. Individual unicellular organisms (bacteria in everyday language) reproduce and die.

2. Natural selection acts on individuals but individuals do not evolve. Individuals pass on their genes (see Box 1.2) along with mutations in those genes to individuals of the next generation.

3. In most species of uni- and multicellular organisms, individuals exist in populations that inhabit discrete ecological niches (**BOX 1.3**). Populations of multicellular organisms have a structure that usually includes different age classes of a single generation and often includes overlapping generations, especially in species with short reproductive cycles and long lives.

4. Populations of a sexually reproducing organism consist of individuals that are **not identical** to one another — they are not clones. In a population of a species in which individuals reproduce asexually, for example by budding or fission, all individuals in a population may be identical.

5. Resources are often limited with the consequence that not all individuals in a population will survive to reproduce and contribute offspring to the next generation.

6. Populations do not reproduce, individuals reproduce. Populations pass to the next generation(s) a gene pool from some of the individuals who reproduced.

7. Variation is an essential prerequisite for evolution to act: natural selection allows some variants to survive and others not.

8. Because resources are limited, natural selection results in survival to the next generation through reproduction of the individuals (variants) that are best suited to the conditions of their existence.

9. Because the genetic background of individual sexually reproducing organisms differs, those that are selected on the basis of their fit to the environment will preferentially pass their genes to the next generation. Those individuals that are less well fitted to the environment will tend not to pass their genes to the next generation.

10. Because of differential reproduction the genetic composition of a population will change gradually from generation to generation. Genetic changes also accumulate because of random **drift** of genes from generation to generation and/or because spontaneous **mutations** change the genetic composition of populations.

11. Populations may **subdivide** into smaller groups. Differences that emerge in the subgroups can provide the basis for speciation.

12. Populations or subsets or populations may "**crash**" or become **extinct** because of environmental catastrophes.

■ In one species of frog, the adults die soon after producing their offspring whose embryonic life is prolonged. Consequently, for much of the year this species exists only as embryos.

BOX 1.3
What Is a Niche?

Niche or **ecological niche** is defined in various ways in the literature of ecology and evolutionary biology. Most definitions emphasize the role of the organism (population, species) in its environment, that is, the niche is not defined as a place but by a set of interactions. Each has a particular emphasis, illustrated by the italicized words in the following definitions.

1. Where a living thing *is found* and what *it does there* (FIGURE B1.1)
2. The *role* of an organism in an ecosystem
3. The *role* of a species within a community
4. All the *functional roles* of an organism in a biological community
5. A *unique ecological role* of an organism in a community
6. The environmental *habitat* of a population or species, the *resources* it uses and its *interactions* with other organisms
7. The *status* of an organism within its environment and community as it affects the survival of the species
8. The *role* or functional position of a species *within the community* of an ecosystem
9. The physical and functional *role* of an organism *within an ecosystem*

FIGURE B1.1 Several species of lichens (distinguishable by color, shape, size and habit) occupy and utilize different niches on the rock outcrop.

As you can see, most definitions consider the niche in relation to organisms, although definitions 3 and 8 emphasize the species. All except 1, 6 and 7 use *role* as the primary aspect of the definition. Number 4 emphasizes functional role, number 5 ecological role, and number 9 physical and functional roles. Numbers 1 and 6 are the most general or all encompassing; number 7 is the most specific (it ties niche to species survival). Four uses the word *community* without defining what a community is. Number 6 uses *habitat* without a definition. However defined, ecological niches are now being reconstructed in phylogenetic analyses and shown to be conserved over long periods of evolutionary time (Donoghue, 2008).

Recommended Reading

Allen, G., and J. Baker, 2003. *Biology: Scientific Process and Social Issues*. Wiley Blackwell, Hoboken, NJ.

Appleman, P. (ed.), 2001. *A Norton Critical Edition. Darwin: Texts, Commentary,* 3d ed. Philip Appleman (ed.). W. W. Norton, New York.

Barrick, J. E., S. Y. Dong, S. H. Yoon, et al., 2009. Genome evolution and adaptation in a long-term experiment with *Escherichia coli. Nature*, 481, 1243–1247. (See pp. 1219–1221 for commentary.)

Biological Sciences Curriculum Study (BSCS) contains much on evolution, including a virtual tour activity of the Galapagos Islands that allows computer-based collection and analysis of organisms. See http://www.bscs.org/.

Boorstein, D. J., 1983. *The Discoverers*. Random House, New York.

Bowler, P. J., 1996. *Life's Splendid Drama: Evolutionary Biology and the Reconstruction of Life's Ancestry, 1860–1940*. The University of Chicago Press, Chicago.

Bowler, P. J., 2003. *Evolution: The History of an Idea*, 3d ed. University of California Press, Berkeley, CA.

Donoghue, M. J., 2008. A phylogenetic perspective on the distribution of plant diversity. *Proc. Natl. Acad. U.S.A.,* **105**, 11549–11555.

Mayr, E., 2001. *What Evolution Is. With a Foreword by Jared Diamond*. Basic Books, New York.

Pickstone, J. V., 2001. *Ways of Knowing. A New History of Science, Technology and Medicine*. The University of Chicago Press, Chicago.

Origins

II

2

The Origins of Matter, the Universe and Earth

KEY CONCEPTS

- The universe came into existence 13.73 billion years ago (Bya) with the Big Bang.
- The Solar System formed from condensations within a huge cloud of dust and gas.
- Our Sun is about 4.6 By old.
- Earth began forming 4.5 Bya.
- Abiotic molecules exist in meteorites that are 4.6 By old.
- If life is defined as based on organic molecules, then life likely had a non-biological origin.
- Continents form, drift and break apart through the movement of plates of crustal rocks, and the sea floors spread.
- Rocks can be classified into geological strata and their ages determined.
- Darwin calculated that Earth was hundreds of millions of years old.
- Rocks contain fossils, which are the physical evidence of past life.
- Fossils record the transformation of life over time — evolution.

Above: Energy and majesty combined in a display of the power of nature.

18

Overview

This chapter sets the stage for all that follows. It starts by asking the question: What makes our planet so special — what makes it suitable for life? The rest of the chapter recounts the story of how Earth got this way, beginning with the **Big Bang**, 13.73 billion years ago, and ending with the forces that have shaped (and are still shaping) our home, planet Earth. Without that most fleeting of beginnings, and the unimaginable events that have since occurred, there would be no atoms, no molecules, no solar system, no Earth, and thus no life as we know it.

Humans have long speculated about the origin of the universe, but only in recent decades have we had the tools to investigate our theories. Today, however, *cosmology* — the study of the universe — is an exciting whirlwind of research and discovery. Aided by the Hubble Space Telescope, satellites, space probes, and research on the ground, we are "seeing" further and further into the past, discovering planets orbiting far-distant stars, and much more. Yet the more we learn, the more we realize how little we know. For instance, is ours the only universe that has ever existed (or indeed, exists now), and is there life on other planets?

What Makes Earth So Special?

Our planet provides what may be a unique set of features that allowed life to evolve and flourish over billions of years.

- A Sun of moderate size providing even radiant energy over hundreds of millions of years
- An orbit that is just the right distance from the Sun, providing just the right amount of heat and light, and thus a moderately stable climate
- The right mix of atomic elements, especially carbon, hydrogen, nitrogen, oxygen, phosphorus and sulfur
- The presence of liquid water
- An ozone layer that protects life from harmful ultraviolet rays, and
- An iron core that casts a gigantic magnetic shield far into space, protecting us from harmful solar and cosmic radiation.

But is Earth unique? Over the last 20 or so years, scientists have located a large number of planets orbiting stars other than our own — 374 as of June 2009. Almost all, however, are gigantic balls of gas; only two of those discovered so far appear to resemble Earth in any way. The search goes on, however. Given that there are billions of galaxies in the universe, we can only guess at the likelihood that there is life elsewhere. Closer to home, we are still searching for clues that life might once have existed on Mars.

The First 380,000 Years

Although debate continues, almost all cosmologists agree that the universe began 13.73 billion years ago in a sequence of events known as the *Big Bang*. Despite its name, the Big Bang did not involve an explosion, nor did it begin at a particular place or expand into anything. Amazingly, space appeared everywhere, all at once.

The term *Big Bang* was coined in 1949 by an early opponent of the theory, British astronomer Fred Hoyle (1915–2001).

■ The National Aeronautics and Space Administration (NASA) provides a fascinating introduction to the history of the universe at http://map.gsfc.nasa.gov/universe/.

■ 10^{-35} seconds (sec) is an unimaginably short period of time. Recollect that 10^{-1} sec = one tenth of a second, 10^{-2} = a hundredth of a second, and so on. At the other extreme, 10^{24} meters is a huge, huge distance: 10 followed by 23 zeroes or a trillion trillion (or septillion) meters, according to U.S. terminology.

■ The nucleus of the most common form (isotope) of hydrogen consists of a single proton. Hydrogen dominated the early universe, and still does.

From this beginning evolved everything that exists today: hundreds of billions of stars in more than 100 billion galaxies — every atom of which owes its origin to the universe's early days — and much else besides.

TABLE 2.1 provides a recent view of the first 100 seconds of our universe. Look particularly at the left column to gain a sense of the speed of the processes involved and at the right column for the extraordinarily high temperatures (up to 100,000 times hotter than the Sun's core) and rapid rate of cooling. Then factor in size: the universe is thought to have increased from virtually nothing to about 10^{24} meters or one hundred million light years, all in an instant at around 10^{-35} of the first second. Since then, the universe has continued to expand, albeit at a much slower pace. **Inflation**, as this expansion is known, has been called one of the most important scientific discoveries of the 20th century (see below for more on this topic).

Within the hot, super-dense brew of these first moments, a great deal of complicated physics was going on, physics that is well beyond the scope of this book. Simplifying greatly, once the initial inflation ended, the universe entered a phase during which a variety of elementary particles first appeared, then electrons, protons and neutrons, and then the atomic nuclei of helium and a few other low-molecular-weight elements (Table 2.1). The latter stage involved nuclear fusion and ended about 20 minutes after the Big Bang, by which time it was too cool (though still extremely hot) for fusion to continue. Among other important early events, the four fundamental forces of the universe, including gravity and the electromagnetic force, came into being. Matter, energy, space and time all now existed.

But still no atoms, and nor would there be for a long time. It was just too hot. Instead, **photons** — the elementary particles responsible for light and electromagnetic radiation — dominated the energy of the universe. Because free electrons scatter photons quite efficiently, an observer at the time, if such were possible, would have been immersed in a dense fog, like the inside of a cloud. This is important to our story, for once it was cool enough, *electrons* (which are negatively charged) could combine with positively charged protons to form neutral hydrogen atoms, and combine with helium nuclei to form neutral helium atoms. Finally — around 380,000 years after the Big Bang — the universe became transparent, and the photons streamed forth. Today, the universe

TABLE

2.1 **The First 100 Seconds of the Evolution of the Universe**

Time (seconds)	Event	Temperature (°C)
0	Birth of the universe	
10^{-43}	Era of quantum gravity and exotic physics	10^{32}
10^{-35}	Universe expands exponentially	10^{28}
10^{-11}	Electromagnetic and weak forces differentiate; quarks and gluons emerge	10 quadrillion
0.1 microsecond		20 trillion
1 microsecond		6 trillion
10 microseconds	Quarks bound into protons and neutrons	2 trillion
100 seconds	Beginning of formation of the nuclei of helium and a few other elements from hydrogen	1 billion

is bathed in a remarkably uniform sea of those same photons — the afterglow of radiation from the initial Big Bang. Discovery of minute fluctuations in this **cosmic microwave background (CMB)** has proved invaluable, for it has given us an exact snapshot of the universe as it was back then.

The Universe Evolves

FIGURE 2.1 provides a time line of the evolution of the universe, shown as a bell to simulate change in three dimensions over time (the universe is actually flat, not bell-shaped; FIGURE 2.2a). On the left, you can see the instantaneous inflation of the Big Bang and the afterglow of the CMB. Next comes a band labeled Dark Ages. For 200 million years, the universe would have appeared dark because there were no stars to emit visible light. Gradually, however, gravity had its way, and matter began to clump together, making larger and larger bodies, until finally the first generation of stars blinked on.

Stars are crucial to this story because they are the nuclear fusion reactors where chemical elements are manufactured, and from which they are dispersed in a star's dying days. It's a complicated story involving several types of stars and a lot of time. Those first generation stars — massive, hot and short-lived — produced a few heavy elements, especially oxygen, neon, silicon, magnesium and sulfur. As a result, all later stars, including those we see today, contain heavy elements. Eventually, through a process of accumulation, all the naturally occurring elements were formed. Our own

■ The word *evolved* is used in a different sense in astronomy and cosmology than in biology. In astronomy, evolution means changes with time of individual objects and systems, changes that can be explained by the laws of physics. In biological evolution, contingency (chance) plays a major role; in cosmology, it almost certainly does not.

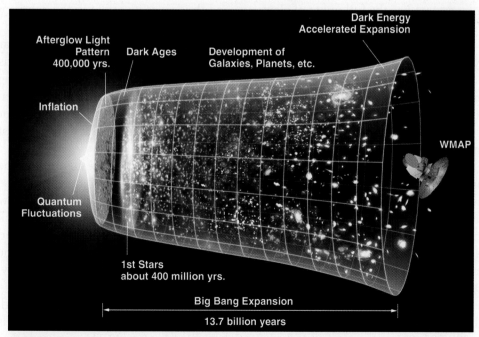

FIGURE 2.1 Time line of the universe, from the Big Bang to now, shown as a bell to simulate change in three dimensions over time. The initial inflation at 10^{-35} of the first second (left) was replaced by a far smaller rate of inflation that is currently accelerating as a result of dark energy (hence the bell's flare), implying that the universe will expand forever. WMAP — the Wilkinson Microwave Anisotropy Probe operated by the National Aeronautics and Space Administration (NASA) — here represents the present.

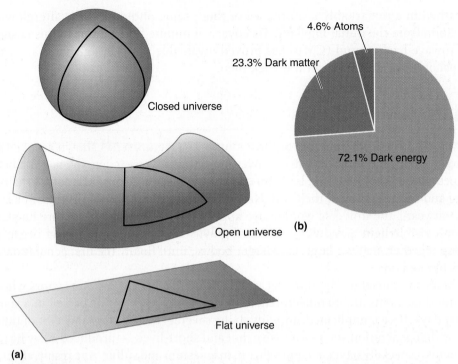

FIGURE 2.2 The universe, which is flat (**a**) rather than spherical (closed) or saddle-shaped (open) is made up of atoms, dark matter and dark energy in the proportions shown in (**b**), although it is important to realize that most of the universe is virtually empty.

solar system has the full quota of elements, and thus owes its origin to the debris of many earlier generations of stars.

Today the universe is highly structured, with planets and moons within solar systems in galaxies, some of which form clusters and superclusters that appear as occasional bumps (more properly, nodes) in sheets and filaments of galaxies separating immense, bubble-like voids. Within this unimaginable amount of material, hydrogen still rules. In our own galaxy, hydrogen accounts for 74 percent by mass, helium for a further 24 percent. Add in oxygen at 1 percent and carbon at 0.46 percent, and not much is left for the other elements.

> Missing from this account are all sorts of things: black holes, quasars, supernovae, and much, much more.

But that's not all. Ordinary atoms make up just 4.6 percent of the contents of the universe (Figure 2.2b). Mysterious (and hypothetical) dark matter and dark energy, neither of which emits or absorbs light, account for the rest. Dark matter we know through its gravitational effects on regular matter. Dark energy brings us back to Figure 2.1, and the topic of inflation. Notice that the bell continues to expand (inflate) as time passes, and that it is flared near its mouth, reflecting the fact that the rate of expansion of the universe is picking up speed — another recent discovery. Dark energy is thought to mediate this increase.

But what does inflation mean? How was the universe, in that first split second, able to expand at speeds far, far in excess of the speed of light (which itself travels at 299,792,458 meters per second, about 9.46 trillion km/year)? The answer is that space–time itself was inflating; any matter inside it was (and is, since the universe is still expanding) merely carried along for the ride. A difficult concept, no doubt, but remember, gravity continues to operate. Things held together by gravity — say, a galaxy or a supercluster — are not affected.

The Future of the Universe

Despite the enormous amount of matter involved, the density of the universe is extraordinarily low, equivalent to only six atoms of hydrogen per cubic meter on average. In effect, then, most of the universe is empty. Yet research tells us that if the universe had contained much more matter, it would have collapsed back on itself; much less and it would have expanded forever, but probably never formed stars. Scientists are using this and other information to work out both the shape of the universe and its future.

We now have strong evidence, from a variety of sources, that the universe is flat, like a sheet of paper (Figure 2.2a), not curved, as had been hypothesized before. It is also — as a result of the negative pressure exerted by dark energy — likely to expand forever. But there is a limit to the amount of fuel (hydrogen, etc.) available for stars to burn. In the end, under this scenario, all galaxies will run out of fuel and we will return to total darkness, with just the cold remnants of celestial bodies (but not any time soon).

Origin of the Solar System

On February 14, 1990, from a distance of 6.4 billion km, cameras onboard the spacecraft *Voyager I* gave us the first view of our solar system. Had they been present 5 to 5.6 Bya they could have witnessed its very beginnings, for it was then that the huge cloud of dust and gas (the **solar nebula**) out of which our system developed began to collapse under its own weight.

By 4.6 Bya, the large condensing mass at the center of the cloud had become a star: the Sun. Around it, the remaining material formed a whirling "accretion disk" (FIGURE 2.3) in which smaller condensations grew as the material cooled. At first, grains of dust merely stuck together, but once clumps of matter reached about a kilometer in diameter, gravity kicked in, and vast numbers of these bodies collided and merged, eventually forming **protoplanets** and finally planets — eight in all — a process occupying some 100 million years.

As the planets formed, the distribution of elements changed dramatically. The result? Four small, terrestrial inner planets (Mercury, Venus, Earth and Mars) that are composed mainly of rock and metal and have only three satellites (**moons**) between them, and four giant, gaseous outer planets (Jupiter, Saturn, Uranus and Neptune) that are primarily made of helium and hydrogen and are surrounded by rings and many moons. Completing the picture are asteroids, comets, interplanetary dust, and **dwarf planets** such as Pluto, long considered a planet but voted out of the club at a meeting of the International Astronomical Union on August 24, 2006 — an extraordinary way to come to a scientific conclusion.

The Sun has also continued to evolve, becoming progressively warmer over the past 4.6 By. It was thus cooler (and dimmer) in our neighborhood of the solar system during Earth's early days.

Although Pluto is a celestial object that orbits the Sun and is large enough for the force of its gravity to have formed it into a sphere — two of the criteria for recognizing a planet — it does not meet the third criterion — that it dominate the neighborhood around its orbit, sweeping up and incorporating unto itself asteroids, comets and other small celestial bodies.

Earth's Structure and the Origin of the Moon

Earth owes its origin to a series of cataclysmic collisions, collisions that would have generated a great deal of heat, causing Earth to become at least partially molten

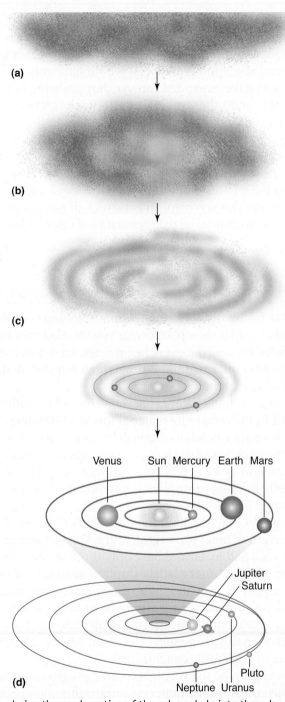

FIGURE 2.3 Stages during the condensation of the solar nebula into the solar planetary system.
(**a**) Fragmentation of an interstellar cloud. (**b**) Contraction and flattening of the solar nebula.
(**c**) Condensation of nebular material into protoplanets and smaller bodies such as asteroids.
(**d**) Solidification of planets, with an indication of present orbits. The four inner planets are shown
at a larger scale than are the outer planets.

and allowing material to move around. Influenced by gravity, the heavier elements (especially iron) sank inward and lighter material rose toward the surface.

Today our planet is layered, its concentric layers nested within one another (FIGURE 2.4). Starting at the center, we have:

- an extremely hot **inner core**, a solid mass of iron (with some nickel)
- an **outer core**, a molten envelope of iron mixed with sulfur or silicon
- a partly plastic or ductile **mantle**, a layer of rock that comprises about four-fifths of Earth's volume and has experienced repeated melting and recrystallization
- a crust, of which there are two types: the thicker (and mostly older) continental crust, consisting of several rigid plates, and the thinner, younger oceanic crust. Together they form the **lithosphere**, which varies considerably in thickness from place to place.

Development of Earth's layers must have been well under way around 4.53 billion years ago, when, according to the currently favored hypothesis, a Mars-sized object crashed into Earth at an oblique angle, blasting vast quantities of material from both bodies into orbit. So catastrophic was the impact (often called the *Big Whack*) that both Earth and the intruder from space would have melted. Some of the debris cast into space clumped together to form the moon, which remained tied to Earth's orbit. The bulk of the intruder would have merged with Earth.

Not surprisingly, no rocks remain from Earth's earliest days (the oldest rock discovered thus far is 4.03 billion years old), but the moon itself can tell us much about the timing and nature of this event. First, because we've analyzed rock samples brought back to Earth by the American (manned) and Soviet Union (unmanned) missions of the late 1960s and early to mid 1970s, we know that the moon (and thus Earth) is at

■ The heat released by relatively short-lived radioactive elements already present would have helped raise the temperature of Earth at this time.

■ The lithosphere varies in thickness from about 150 km or more under mountains to a few kilometers in mid-oceanic ridges, though in general the oceanic crust is of the order of several tens of kilometers thick.

■ The moon rocks sampled are between 3.16 and 4.5 By old. The youngest parts of the moon, the result of volcanic eruption, are believed to be about 1.2 By old, but no rocks of that age were brought back to Earth.

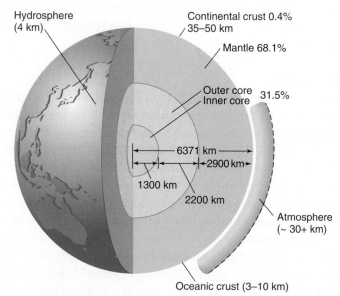

Hydrosphere (4 km)

Continental crust 0.4% 35–50 km

Mantle 68.1%

Outer core
Inner core 31.5%

6371 km
2900 km
1300 km
2200 km

Atmosphere (~ 30+ km)

Oceanic crust (3–10 km)

FIGURE 2.4 A section through Earth. The lithosphere consists of relatively rigid plates composed of the rocklike crust plus a portion of the underlying mantle, reaching a depth of about 96 km at the oceanic basins and 96 to 145 km at the continents. Below the lithosphere is a more plastic, deformable layer that allows the lithospheric plates to move about.

least 4.5 billion years old. And because the moon has far less iron than does Earth, the impact must have happened after much of Earth's iron had sunk into its middle.

Earth Evolves

Little is known about Earth's history following the moon's birth. Until recently it was assumed that the planet was a place of roiling magma for the next 500 million or so years, a time that would have richly deserved its name, the **Hadean Era** (from Hades, hell). But since the early 2000s, when zircon crystals 4.4 By old were discovered in Western Australia, scientists have had to rethink. Zircon crystals require liquid water and low temperatures to form, so the planet may well have cooled, and landmasses formed, far earlier than we had thought. Because intense bombardment by **meteorites** and comets between 4.1 and 3.8 Bya would have sterilized the planet's surface, we do know that present life could have begun no earlier than the end of the Hadean Era.

Meteorites are fragments of asteroids that have fallen to Earth, the rocky bodies familiar to us as shooting stars. Meteorites come in several varieties, the most significant for the story of life being **chondrites**. These meteorites have remained essentially unchanged since their condensation and origin from the solar nebula 4.56 to 4.6 Bya. Significantly, and as discussed in Box 3.1, some of them contain an assortment of carbon compounds. Meteoric impacts almost certainly brought to Earth huge amounts of water, nitrogen and other elements as well as organic molecules that could be used for biological purposes once life got under way.

The term *organic* is applied to a wide range of molecules containing carbon, and does not necessarily mean "synthesized by life," as was believed of all such molecules in the 19th century, when the term was first applied to them. The organic molecules found in meteorites are abiotic; there is no need to conjure up alien life forms to explain their origin. Organic molecules preceded the origin of life — a simple statement that has enormous consequences for how likely it is that life would arise in the universe (see Chapter 3).

Earth's Atmospheres

As Earth cooled, a crust began to form. Eventually, water condensed to form rain (a lot of it!) and finally oceans between large landmasses. Over time, Earth's atmosphere changed radically from the hydrogen and helium briefly present at the outset, through a second atmosphere that accumulated between 4.2 and 3.8 Bya, to a third and final atmosphere, the one we have today (Figure 2.4).

The second atmosphere, which mostly consisted of water vapor and carbon dioxide (CO_2), owed its origin to massive out-gassing from volcanoes and to material imported by meteorites and comets. One of the many gases present was methane, which, like CO_2, is a **greenhouse gas**. Together, these two gases may well have acted as a warming blanket, preventing the icy conditions that were otherwise likely, given that the Sun then gave off only about 80% of its current heat. At least some of the methane, it has been suggested, was a byproduct of a group of single-celled microbes known as **methanogens** (see Chapter 4).

Once liquid water formed, CO_2 levels fell as the gas was absorbed by the oceans and reacted with silicates to produce carbonates. We see evidence of this in thick

■ We know this because a large number of impact craters were formed on the moon at this time, and thus by inference on Earth and the other inner planets.

■ Greenhouse gases such as methane and CO_2 are much in the news today. By trapping radiant energy from the Sun, thereby preventing it from returning to space, they keep the planet warm, and therefore suitable for life. But through the burning of fossil fuels, land clearing and other means, humans are raising the atmospheric levels of these gases, resulting in global climate change.

river deposits laid down in South Africa and other places around 3 Bya, deposits that include compounds that are highly unstable in the presence of oxygen.

The third — and present — atmosphere mostly consists of nitrogen and oxygen. Geochemists generally agree that the proportion of free oxygen in the atmosphere began to increase about 2.3 Bya. Driving this increase were **cyanobacteria** (see Chapter 4), the first organisms to leave behind fossils, beginning more than 3.5 Bya. Known to some as "the architects of the atmosphere," cyanobacteria are aquatic and employ the process known as photosynthesis (see Chapters 4 and 8) to obtain energy from the Sun's rays. Through a series of steps, and after hundreds of millions of years, oxygen began accumulating in the atmosphere.

Two other results of this process were a further decline in levels of CO_2 — consumed during photosynthesis — and the creation of an **ozone layer** through ultraviolet (UV) irradiation of oxygen high in the atmosphere. Life on land was not possible without the protection against UV light afforded by this layer. In addition, oxygen had enormous impacts on life on Earth (see Chapters 3 and 4).

Dating Earth's Rocks and Fossils

An ability to date Earth and its past inhabitants is plainly necessary before a time line of life on Earth can be constructed. Before the mid-20th century, scientists used a variety of methods to arrive at their conclusions, most of which involved comparison of rock layers. None of their calculations, some of which are set out in BOX 2.1, produced ages even vaguely approaching the real situation.

Essential to Charles Darwin's theory of evolution through natural selection (see Chapter 6) was his belief that Earth's age extended beyond anything proposed before. As Darwin pointed out in an 1844 essay:

BOX 2.1
Some Early Attempts to Determine Earth's Age

Isaac Newton (1643–1727) calculated that a sphere the size of Earth would take about 50,000 years to cool down to its present temperature. Because even such a short period contradicted the 5,000 or so years of history allowed in the Judeo-Christian Bible, Newton piously rejected his calculations. The French naturalist, Georges-Louis LeClerc (Compte de Buffon, 1707–1788, usually known as Buffon), calculated approximately 75,000 years — 74,832 to be precise — for the age of Earth. "For 35,000 years our globe has only been a mass of heat and fire which no sensible being could get close. Then, for 15,000 or 20,000 years, its surface was only a universal sea."

In Darwin's time, William Thomson (Lord Kelvin, 1824–1907) reassessed the temperature gradients observed in mine shafts, the conductivity of rocks, and the presumed temperature and cooling rate of the Sun, to calculate the total age of Earth's crust at about 100 My, but suggested that only the last 20 to 40 My could have been sufficiently cool for life to exist. Although similar to a calculation of 96 My made by John Phillips in 1860 (see footnote a to Box 2.2), in Kelvin's view, this number was still too small to account for the Darwinian evolution of organisms. What Kelvin did not know was that radioactive decay (discovered at the end of the 19th century) provides a source of heat that significantly reduces the rate of Earth's cooling.

The mind cannot grasp the full meaning of the term of a million or hundred million years, and cannot consequently add up and perceive the full effects of small successive variations accumulated during almost infinitely many generations.

Darwin thus recognized that evolution needs time — a great deal of time, far more time than any previous scientific calculation of Earth's age would allow (BOX 2.2) — and used his own calculations (hundreds of millions of years) of the age of Earth in the first and second editions of *On the Origin of Species*.

Before discussing how to determine the age of rocks, we need to describe the three major classes of rock making up Earth's crust.

■ The first five editions of Darwin's book were titled *On the Origin of Species by Means of Natural Selection, or the Preservation of Favoured Races in the Struggle for Life*. Darwin dropped "On" from the title, beginning with the 6th edition published in 1872. Consequently, the book became known as *The Origin of Species*. . . .

1. **Igneous rocks** (65% of the crust) form when molten rock (magma) within the mantle cools and solidifies. Slow cooling deep within Earth results in the crystallization of minerals that form coarse-grained rocks such as granite. Magma deposited on the surface as lava from volcanoes cools quickly and forms fine-grained rocks such as basalt.

2. **Sedimentary rocks** (8% of the crust) form by the weathering, transportation and deposition of existing rocks by water, wind, or glaciers or from the debris emitted by volcanoes. Sandstone and limestone will be familiar examples of sedimentary rocks, the oldest of which were deposited 2.5 Bya.

BOX 2.2
Darwin and the Age of Earth

Although in the generally accepted view, Darwin could not deal with the problem of the long periods of time needed for life to evolve, in fact he did make calculations based on rates of erosion that demonstrated the enormity of geological time; not the 100 million years estimated by Kelvin (see Box 2.1) but three times longer. Darwin came to regret these numbers.

It is an interesting story. Darwin included his calculations in the first and second editions of *The Origin*, but removed them from later editions (over objections from a number of his scientific friends). The appropriate section from the second edition (*emphases added*) is long but worth reproducing in full.

I am tempted to give one other case, the well-known one of the denudation [erosion] of the Weald [the rolling countryside between the South and North Downs in Kent and adjacent counties in southern England]. Though it must be admitted that the denudation of the Weald has been a mere trifle, in comparison with that which has removed masses of our Palaeozoic strata, in parts ten thousand feet in thickness, as shown in Prof. Ramsay's masterly memoir on this subject. Yet it is an admirable lesson to stand on the North Downs and to look at the distant South Downs; for, remembering that at no great distance to the west the northern and southern escarpments meet and close, *one can safely picture to oneself the great dome of rocks which must have covered up the Weald within so limited a period as since the latter part of the Chalk formation*. The distance from the northern to the southern Downs is about 22 miles, and the thickness of the several formations is on an average about 1100 feet, as I am informed by Prof. Ramsay. But if, as some geologists suppose, a range of older rocks underlies the Weald, on the flanks of which the overlying sedimentary deposits might have accumulated in thinner masses than elsewhere, the above estimate would be erroneous; but this source of doubt probably would not greatly affect the estimate as applied to the western extremity of the district. If, then, we knew the rate at which the sea commonly wears away a line of cliff of any given height, we could measure the time requisite to have denuded the Weald. This, of course, cannot be done; *but we may, in order to form some crude notion on the subject, assume that the sea would eat into cliffs 500 feet [152 m] in height at the rate of one inch [25.4 mm] in a century*. This will at first appear much too small an

BOX 2.2

Darwin and the Age of Earth (*continued*)

allowance; but it is the same as if we were to assume a cliff one yard in height to be eaten back along a whole line of coast at the rate of one yard in nearly every twenty-two years. I doubt whether any rock, even as soft as chalk, would yield at this rate excepting on the most exposed coasts; though no doubt the degradation of a lofty cliff would be more rapid from the breakage of the fallen fragments. On the other hand, I do not believe that any line of coast, ten or twenty miles in length, ever suffers degradation at the same time along its whole indented length; and we must remember that almost all strata contain harder layers or nodules, which from long resisting attrition form a breakwater at the base. Hence, under ordinary circumstances, I conclude that for a cliff 500 feet in height, a *denudation of one inch per century for the whole length would be an ample allowance. At this rate, on the above data, the denudation of the Weald must have required 306,662,400 years; or say three hundred million years.* But perhaps it would be safer to allow two or three inches per century, and this would reduce the number of years to 150 or 100 million years. So that it is not improbable that a longer period than 300 million years has elapsed since the latter part of the Secondary period (Darwin, *On the Origin of Species*, 2d ed., p. 285).

In 1860, in his presidential address to the Geological Society of London, the geologist, paleontologist and Reader in Geology at Oxford, John Phillips (1800–1874) provided his own estimate of Earth's age on the basis of rates of sedimentation by the Ganges River (Darwin had used rates of erosion) and the thickness of the known geological strata. On this basis, Phillips came up with an estimate of 96 My.

Darwin's estimates of geological times caused him considerable anguish. How could a piece of the English countryside be three times older than Kelvin's estimate of the age of Earth itself, especially when Kelvin's calculations were based on physical rates (see Box 2.1) and Darwin's on "back of the envelope" calculations? We see Darwin's anguish in two letters written to his close friend and confidant, Charles Lyell. On January 10, 1860, responding to suggestions for changes to the first edition made by Lyell, Darwin wrote (*emphasis added*).

. . . It is perfectly true that I owe nearly all the corrections to you, and several verbal ones to you and others; I am heartily glad you approve of them, *as yet only two things have annoyed me*; those confounded millions of years (not that I think it is probably wrong), and my not having (by inadvertence) mentioned Wallace towards the close of the book in the summary, not that any one has noticed this to me.

On November 20th of the same year, in a footnote to a letter to Lyell concerning changes to the next edition of his book, Darwin wrote, "The confounded Wealden Calculation to be struck out, and a note to be inserted to the effect that I am convinced of its inaccuracy from a review in the *Saturday Review,* and from Phillips, as I see in his Table of Contents[a] that he alludes to it."

Darwin never referred to the age of Earth again, which is a real shame; his would have stood as the best estimates of the age of Earth for quite some time.

[a]The Table of Contents referred to by Darwin is the printed version of Phillips' lecture published in 1860, entitled, *Life on the Earth: Its Origin and Succession* (Phillips, 1860).

3. **Metamorphic rocks** (25% of the crust) form when existing rocks are subjected to heat, pressure, and/or chemical interactions. For example, limestone metamorphoses to marble, shale to slate.

Over time, these three types of rock are transformed from one to the other (FIGURE 2.5) and undergo substantial movement, especially through the process of

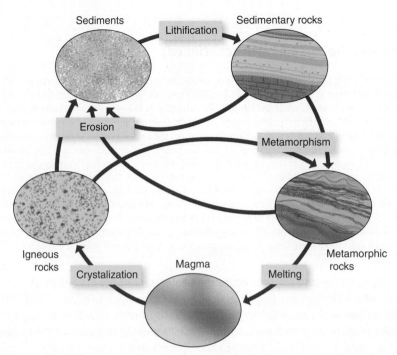

FIGURE 2.5 Diagrammatic representation of the rock cycle. Some crustal rocks have been recycled many times; others have persisted with little change from their initial formation. Geologists estimate that about half of all crustal rocks have formed during the last 600 My. (Adapted from Hawkesworth, C. J. and A. I. A. Kemp, *Nature* **443**, 2006: 811–817.)

plate tectonics (see below). Indeed, Earth's dynamic nature means that nowhere on Earth is there a continuous record of what has happened. Despite this, three clear principles can be drawn, principles that have enormous significance for those interested in the history of life.

1. For a given series of sedimentary rocks, the oldest layers (strata) lie at the bottom and the youngest at the top, observations canonized as the law of superposition by the Danish "Father of Geology," Nicholas Steno (1638–1686).

2. Strata can be identified on the basis of the fossils within them, arranged into eons, eras, periods and epochs analogous to centuries, years, months and days, and geological maps produced, the first by the "Father of English Geology," William Smith (1769–1839).

3. Fossils from the uppermost (younger) strata are more like living organisms than are fossils from lower (older) strata, as clearly laid out by "the Father of Paleontology," Georges Cuvier (1769–1832).

The method, which is still in use today, relies on comparisons with strata in other places (sometimes other continents) to build a picture of the geological record. It was mostly on this basis that Darwin and his contemporaries based their estimates of Earth's age and that of its fossils (Box 2.2). It would take the tools of twentieth century chemistry and physics, however, to deliver truly accurate ages through the method known as **radiometric dating**, which relies on the decay of radioactive isotopes (BOX 2.3).

BOX 2.3
Radiometric Dating

The most accurate method of dating rocks, fossils and other ancient things is **radiometric dating**. The atoms of a particular element all have the same number of protons, but the number of neutrons (and so the atomic weight of the element) can vary. The different forms of an element are known as isotopes, some of which decay naturally over time, producing new isotopes and releasing energy. The decay of radioactive isotopes is orderly, allowing geologists to date tiny fragments of rocks even billions of years old with considerable precision.

For instance, the radioactive isotope uranium-238 (U-238) is present in the mineral zircon, found in most igneous rocks. U-238 disintegrates to form the lead isotope Pb-206 at a rate that transforms half of the uranium into lead over a period (the half-life) of about 4.5 billion years. The relative amounts of the two isotopes in a rock provide a fairly accurate way to date mineral crystals older than about 100 million years.

Geologists have mostly applied this method to igneous rocks, but they can extend the dates to sedimentary rocks (and thus the fossils they contain) by comparing the relative positions of the two rocks and by dating layers of volcanic rock and ash between sedimentary layers. The method can also be applied to the remains of living organisms. Carbon-14 (C-14) is especially useful; it allows us to date organic remains up to 50,000 years old.

When it comes to dating Earth itself, the oldest rocks identified so far come from near Slave Lake in northwestern Canada (4.03 billion years) and from western Greenland (3.7–3.8 billion years), but rocks older than 3.5 billion years are found on all continents. Even older are zircon crystals found in Western Australia.

For the age of the solar system, and of Earth, we must rely on the analysis of meteorites, fragments of asteroids that have fallen to Earth. Meteorites are some of the oldest and most primitive objects in our system (*primitive* meaning little altered since the origin of the solar system).

TABLE 2.2 provides a detailed time line of Earth from its earliest days, listing major events in the evolution of life. This table is referred to often in future chapters.

Fossils and the Nature of Geological Change

Because of the way they are laid down (see Box 5.1), **fossils** — the physical evidence of once-living organisms — are usually found in sedimentary rocks, occasionally (although often distorted) in metamorphic rocks such as marble but never in igneous rocks. When arranged by stratigraphic age, with deeper strata signifying older age than superimposed strata, fossils in older strata show greater differences from extant species than do fossils in younger strata, indicative of changes over time.

It had long been known that the fossilized bones of animals do not resemble *extant* (currently living) species, and that strange seashells can be found in the most unlikely places, such as mountaintops (see Chapter 5). The ancient Greeks were aware of such fossils, and a number of ancient writers, including Herodotus (484–425 BC), suggested that their distribution could be explained by changes in the positions of sea and land. To Aristotle, there was no question that these changes occurred over considerable periods of time.

TABLE

2.2 Geological Ages and Associated Organic Events

Eon	Era	Period	Epoch	Millions of Years Before Present (approx.)	Duration in Millions of Years (approx.)	Some Major Organic Events
Phanerozoic	Cenozoic	Quaternary	Recent (last 5,000 years)	0.01	1.8	Appearance of humans
			Pleistocene	1.8		
		Tertiary	Pliocene	5.3	3.5	Dominance of mammals and birds
			Miocene	23.8	18.5	Proliferation of bony fishes (teleosts)
			Oligocene	34	10.2	Rise of modern groups of mammals and invertebrates
			Eocene	55	21	Dominance of flowering plants
			Paleocene	65	10	Radiation of primitive mammals
	Mesozoic	Cretaceous		142	77	First flowering plants / Extinction of dinosaurs
		Jurassic		206	64	Rise of giant dinosaurs / Appearance of first birds
		Triassic		248	42	Development of conifer plants
	Paleozoic	Permian		290	42	Proliferation of reptiles / Extinction of many early forms (invertebrates)
		Carboniferous	Pennsylvanian	320	30	Appearance of early reptiles
			Mississippian	354	34	Development of amphibians and insects
		Devonian		417	63	Rise of fishes / First land vertebrates
		Silurian		443	26	First land plants and land invertebrates
		Ordovician		495	52	Dominance of invertebrates / First vertebrates
		Cambrian		545	40	Sharp increase in fossils of invertebrate phyla
Precambrian	Proterozoic	Upper		900	355	Appearance of multicellular organisms
		Middle		1,600	700	Appearance of eukaryotic cells
		Lower		2,500	900	Appearance of planktonic prokaryotes
	Archean			4,000–4,400	1,400	Appearance of sedimentary rocks, stromatolites and benthic prokaryotes
	Hadean			4,560	160–560	From the formation of Earth until first appearance of sedimentary rocks; no observable fossil organisms

Dates derived mostly from Gradstein et al. *A Geological Time Scale*. Cambridge University Press, 2004, and from *Geologic Time Scale*, e from http://www.stratigraphy.org, accessed January 2010.

The whole vital process of the earth takes place so gradually and in periods of time which are so immense compared with the length of our life, that these changes are not observed; and before their course can be recorded from the beginning to end, whole nations perish and are destroyed (*Treatise on Meteorology*).

Once the reality of fossils and of extinction was accepted, it was possible to conceive of a "law of succession" in which one form replaced another. However, gaps in the geological and fossil records presented major difficulties for this hypothesis.

During the late 1700s and early 1800s, a commonly held theory to explain such gaps was **catastrophism**, popularized largely by followers of Georges Cuvier (1769–1832), a gifted French comparative anatomist and the founder of paleontology. In this view, such gaps indicated sudden upheavals caused by catastrophes such as glaciation, floods, volcanic activity and so on. Fossils were recognized as extinct species "whose place those which exist today have filled, perhaps to be themselves destroyed and replaced by others."

In contrast to this position, the French naturalist Jean-Baptiste Lamarck (1744–1829) proposed that geological discontinuities represented **gradual changes in the environment and climate** to which species were exposed. Through their effects on organisms these changes led to species transformation. This **uniformitarian** concept that the steady uniform action of the forces of nature could account for Earth's features was foreshadowed by Buffon and others (Box 2.1). Later, in *Principles of Geology* published in 1830, Charles Lyell, a geologist and contemporary of Charles Darwin, offered the uniformitarian reply to catastrophism using the following arguments.

1. Sharp, catastrophic discontinuities are absent if geological strata are examined over widespread geographical areas. Any widely distributed stratum often shows considerable regularity in its structure and composition (Box 2.2). Only in specific localities do rapid shifts seem to appear and then because of local changes.

2. Changes in the geological record arise from the action of erosive natural forces such as rain and wind as well as from volcanic upthrusts and flood deposits (Box 2.2). The laws of motion and gravity that govern natural events are constant through time. Past events are caused by the same forces that produce phenomena today, although the extent of activities such as volcanism may have fluctuated in the past. This means that all natural causes for phenomena should be investigated before supernatural causes are used to explain them.

3. Earth must be very old for its many geological changes to have taken place by such gradual processes.

Lyell united geology, botany and zoology; the full title of his treatise is *Principles of Geology, or the Modern Changes of the Earth and Its Inhabitants, Considered as Illustrative of Geology*. He was the first to demonstrate the importance of comparing the geographical distribution of animals and plants with the geological history of the region(s) in which they were found, a correlation that had considerable influence upon Charles Darwin, who read the first volume soon after leaving England on HMS Beagle (see Chapter 5). The frontispiece of Lyell's book is a portrait of the three remaining columns of the ruined "Temple of Serapis" in Pozzuoli, Italy, showing that they had been subjected to past changes in sea level. A three-meter section of the columns contains holes bored by marine organisms, showing that the columns were once partially submerged (FIGURE 2.6). Lyell used an image of these columns through 12 editions of his book as an example of gradual geological change.

FIGURE 2.6 A contemporary photograph of the three remaining columns of the ruined "Temple of Serapis" in Pozzuoli, Italy, showing that they had been historically subjected to rise and fall in sea level. A dark, three-meter section of the columns is filled with holes bored by marine organisms, evidence that the columns were once partly submerged.

Although uniformitarianism did not exclude sudden geological changes such as floods, volcanic eruptions and meteorite impacts — events that were of common or recorded knowledge — it led to the view that even such "catastrophes" could be naturally caused and rationally explained. The transition from catastrophism to uniformitarianism had profound effects because it helped liberate scientific thinking from the concept of a static Earth to one that is perpetually dynamic and more historically understandable. Nowhere is that dynamism more astounding than in the discovery that the continents have moved over vast distances during the course of Earth's evolution.

Continental Drift

The **continents** have not always been where they are today. Continents break apart, and continents move. Not until the 1960s, however, was this fact widely accepted and a mechanism discovered.

But scientists had long been puzzled, for the continental margins of South America and Africa appear to fit together like hand and glove (**FIGURE 2.7**), and other parts of the globe also can be matched, though less closely. Further, geologists had discovered remarkable similarities in some geological formations and fossils found across southern continents. Attempts were made to explain these phenomena, but it took

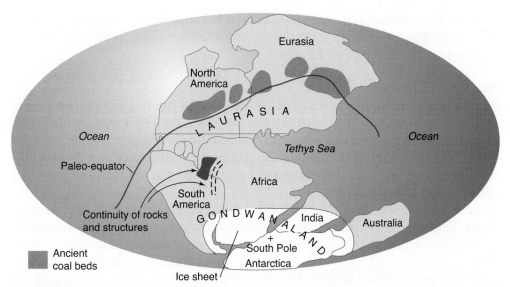

FIGURE 2.7 Fit of the continental shelves at a depth of 500 fathoms (915 meters) on opposite sides of the Atlantic Ocean. (Adapted from Eichler, D. L. and A. L. McAlester. *History of the Earth*. Prentice Hall, 1980.)

a meteorologist with wide-ranging interests, Alfred Wegener (1880–1930), to come up with a coherent hypothesis: **continental drift**. In 1912, Wegener proposed that there was once just one continent — **Pangaea** — that broke into pieces that have drifted apart. His hypothesis was controversial, to put it mildly, but over his lifetime he amassed a wide range of evidence to support it. What he could not do, however, was come up with a mechanism.

We now know that 250 Mya there was indeed a single continent, Pangaea, and that over many tens of millions of years its parts have slowly drifted to where they are today, in the process undergoing many changes. Thanks to a proposal by Alexander L. Du Toit (1878–1948), a South African geologist and strong supporter of Wegener's hypothesis, we now know that a section of Pangaea, **Gondwana**, broke away as two massive blocks: one represented today as South America and Africa, the other as Antarctica, Australia and India (Figure 2.7). The piece left behind — Laurasia — eventually became North America and Eurasia. In fact, continents have probably existed and moved from quite early in Earth's existence.

In the mid-20th century, discovery of two previously unknown natural phenomena provided evidence that helped prove the hypothesis of continental drift and point the way to an overarching theory to explain it. The two are *magnetization of iron* in rocks when they formed and *spreading of the sea floor*.

Paleomagnetism

As new rocks arise from the cooling of magma, iron compounds within them become magnetized, the direction of magnetization depending on the location and strength of the magnetic field present when the minerals form. If the location and strength of the magnetic field remain stable, the minerals deposited subsequently should show the same direction of magnetization as the first-formed rocks. The discovery that this is not so provided evidence that Earth's magnetic poles have

changed over time, wandering thousands of kilometers, and even reversing by 180° (FIGURE 2.8) — something that is best explained by movement of the continents relative to each other and to the poles.

Sea-Floor-Spreading

Running down the middle of some oceans (the Atlantic, for instance) is a mountainous ridge on either side of which, it was discovered in the 1960s, run parallel bands of oceanic crust that have alternating magnetic orientation (Figure 2.8). Because the further you move away from the ridge, the older the rocks, it was hypothesized that this echoed past reversals of the magnetic poles, and that new ocean floor was being added along the ridges.

We have known since the 1950s that the mid-oceanic ridges form a gigantic, interconnecting system through the world's oceans. In fact, the ridges are giant fissures from which molten magma emerges, the makings of new ocean floor. In time, as more material is added, the ocean floor grows in width, spreading away from the ridge in either direction (Figure 2.8) — currently at a rate of 1 cm a year from the

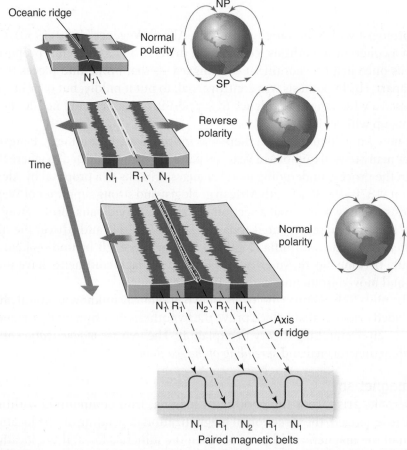

FIGURE 2.8 Diagrammatic sections through an oceanic ridge showing how sea-floor spreading produces differently magnetized belts, with normal (N) or reversed (R) polarity. Hot molten material adds to the ridge from the mantle, falls away on both sides and, as it cools, magnetizes in the orientation of the prevailing magnetic field. With time, the magnetic field changes in strength and/or direction, and new material added to the ridge forms a pair of belts that differ from adjacent belts.

North Atlantic Ridge, 3 cm a year from the South Atlantic Ridge, and 9 cm a year in parts of the Eastern Pacific Ridge.

Plate Tectonics

One very obvious feature of our planet is the "Ring of Fire," a jagged, almost horseshoe-shaped band circling much of the Pacific Ocean (FIGURE 2.9). Known for its earthquakes and volcanic eruptions, it is aptly named, containing as it does three quarters of the world's volcanoes. The Ring of Fire, along with similar zones of high geological activity, completed the picture of how continents move. The mechanism by which this happens is known as **plate tectonics**. Earth's crust — the *lithosphere* — consists of eight gigantic (and several smaller) plates that can separate, slide past, or converge upon each other (Figure 2.9). These massive but slow movements are the motor for continental drift and explain why earthquake zones mostly occur at the margins between plates. The three major movements are

- **sea-floor spreading**
- **sliding** of one plate past another at a geological fault. The most well-known example of this is movement of the Pacific Plate past the North American Plate along the San Andreas Fault in California. Although these two plates pass each other at a rate of only 3.5 cm/year, it all adds up. Ten million years from now Los Angeles will lie at the latitude currently occupied by San Francisco.

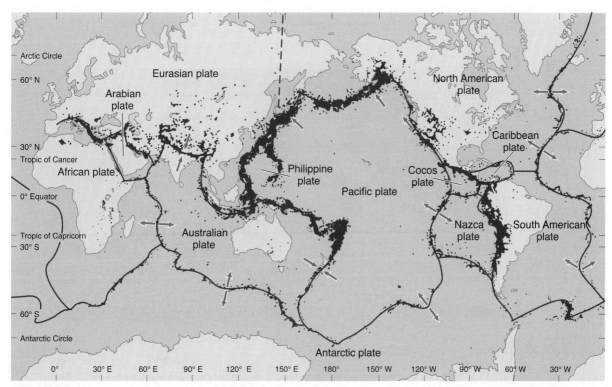

FIGURE 2.9 The boundaries (dark brown lines) between tectonic plates (for example between the Australian and Pacific plates), the distribution of active volcanoes (shown by the density and amount of dark brown along the boundaries), and the direction of movement of the plates (shown as double light brown arrows) illustrate the dynamics of plate tectonics. (Adapted from Cloud, P. *Cosmos, Earth and Man: A Short History of the Universe.* Yale University Press, 1978.)

- **subduction** (sliding under) of one plate beneath another. At such sites, crust is essentially destroyed, being turned back into mantle (Figure 2.9).

Plate tectonics subjects moving landmasses to new climatic conditions and geographical relationships, enabling their inhabitants to be selected for different evolutionary adaptations. The movement, separation and rejoining of landmasses over the past 2.5 By have had important influences on the distribution of all organisms, especially land plants and terrestrial animals (FIGURE 2.10), a topic discussed in Chapter 18.

As landmasses move away from each other, groups of organisms that were once associated are set on different evolutionary paths. The joining of landmasses, on the other hand, initiates competition among previously separated plants and animals. These organisms can then interact, leading to increased complexity for some groups and extinction for others (see Chapter 18). The evolution of the continents thus had major impacts on the evolution of life, especially plant and animal life. One of the most prominent examples of the effect of plate tectonics on the distribution (biogeography) and evolution of organisms is the unique collection of mammals found in Australia and South America (see Chapter 18).

To recap this chapter and to look ahead, here are some origin milestones and estimated times of occurrence:

- 13.73 Bya — origin of the universe
- 4.6–4.7 Bya — origin of the solar system
- 3.5–3.8 Bya — origin of single-celled organisms
- 2 Bya — origin of multicellular organisms

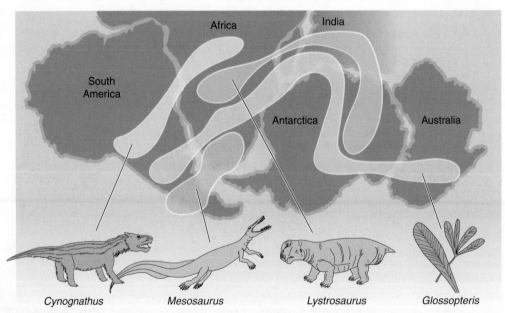

FIGURE 2.10 Distribution of several fossil animals on and between Gondwanan continents. The presumed fit of the continental margins during the Permian-Triassic periods is also shown. Fossil leaves of *Glossopteris*, a group of plants that lived in the early Permian, are found on all Gondwana formations. (Adapted from Colbert, C. H. *Wandering Lands and Animals*. Hutchinson, 1973.)

Recommended Reading

Davies, P., 2006. *The Goldilocks Enigma: Why Is the Universe Just Right for Life?* Allen Lane, London.

Hawking, S. W., 2001. *The Universe in a Nutshell*. Bantam Books, Toronto.

Hogan, J., 2007. Unseen universe: Welcome to the dark side. *Nature* **448**, 240–245.

Lineweaver, C. H. and T. M. Davis, 2005. Misconceptions about the Big Bang. *Sci. Amer.* **292**(3), 24–33.

Loeb, A., 2006. The dark ages of the universe. *Sci. Amer.* **292**(1), 46–53.

National Aeronautics and Space Administration (NASA), 2008. *Cosmology; The Study of the Universe* (Web site devoted to the Wilkinson Microwave Anisotropy Probe.) Available at http://map.gsfc.nasa.giv/m_uni.html.

Phillips, J., 1860. *Life on the Earth: Its Origin and Succession*. Macmillan, London. (Reprinted 1980, Arno Press, New York.)

Riordan, M., and W. A. Zajc, 2006. The first few microseconds. *Sci. Amer.* **294**(5), 24–31.

Rudwick, M. J. S., 2005. *Bursting the Limits of Time. The Reconstruction of Geohistory in the Age of Revolution*. The University of Chicago Press, Chicago.

Saal, A. E., E. H. Hauri, M. L. Cascio, et al., 2008. Volatile content of lunar volcanic glasses and the presence of water in the Moon's interior. *Nature* **454**, 192–195.

Tarbuck, E. J., and F. K. Lutgens, 2006. *Earth Science*, 9th ed. Prentice Hall, Englewood Cliffs, NJ.

3

The Origin of Molecules and the Nature of Life

KEY CONCEPTS

- Early Earth contained the necessary preconditions for carbon-based (organic) molecules to arise.
- Water early in Earth's history provided the necessary solvent for molecules to arise.
- Hydrothermal vents and volcanoes are possible sites for the origin of molecules.
- The origin of molecules can be replicated in the laboratory.
- Molecules in meteorites could have contributed to molecules on Earth.
- DNA, RNA, proteins, lipids and sugars are molecules of life.
- The first molecular world was an RNA world.
- Proteins originated after RNA; DNA originated after proteins.
- The genetic code arose after molecules already existed.
- Elements such as iron can act as catalysts to speed up chemical reactions.
- Enzymes co-opted iron; therefore, these organic molecules have a nonbiological component.
- Four conclusions concerning molecules and the nature of life can be drawn from the evidence presented in this chapter.

Above: A modern-day sulfur pool in which new molecules form.

Overview

Four topics are relevant to an understanding of the origin of molecules. One is the necessary preconditions for molecules to arise at all. Another is whether we can reproduce in the laboratory conditions under which molecules could have arisen. The third is whether molecules arose only on Earth or in other parts of the solar system. The fourth is whether molecules arose only in specific locations on Earth, for example, in the oceans or in volcanoes. These four topics/questions are addressed in this chapter before discussing the origin of the molecules most associated with life: DNA, RNA and proteins. As these are organic molecules — molecules produced by the action of cells — you might think that discussing their origin means discussing the origin of cells and or of organisms; discussing the origin of life itself. However, you will find that "organic" molecules arose in the absence of life (*abiotically*) before organisms appeared, perhaps as much as a billion years before organisms arose. This discovery causes us to consider our definition of life. Did life arise with cells 3.5 to 3.8 Bya (see Chapter 4) or with organic molecules around 4.5 Bya (this chapter)? A related question is whether life equates with organisms. Conclusions concerning the origin of molecules and the nature of life are gathered at the end of this chapter.

Prerequisites for the Origin of Molecules

The classes of carbon-containing molecules that arose on Earth can be interpreted as responses to a set of conditions, without any one of which molecules would not have arisen at all, in which case life as we know it would not exist. Alternatively, the molecules that arose could have been very different — based on the element silicon found in many minerals, for example — in which case life and life forms would have been very different from those that did evolve and exist today. Because the number of types of minerals has increased since the origin of the Solar System, some Earth Scientists such as Robert Hazen and colleagues have provided evidence for the evolution of minerals and for the role that life played in mineral evolution.

The evolution of our solar system offered a number of **essential conditions** that enabled the origin and persistence of life. Minimally these are the presence of a sun, chemical diversity, water, and the nature of Earth's orbit. Two of these conditions — the size of the **Sun** and Earth's **orbit** — are consequences of the type of solar system to which Earth belongs. The others — the formation of chemical elements, and the availability of water — are consequences of how Earth arose and evolved.

Our Sun and Earth's Orbit

Our Sun is of moderate size as suns go and is long-lived. Consequently, the Sun has emitted a steady stream of radiation for 4.56 By without major events in its history that could (would?) have destroyed **carbon**-containing molecules present (see Chapter 2).

Since its origin, Earth has maintained a uniform distance from the Sun and an orbit that is almost perfectly circular. Earth circles the Sun once every 365.25 days — hence one in four years is a leap year — at a speed of 108,000 km per hour. The consequences of a uniform orbit are that temperature extremes or wide temperature fluctuations, in which newly formed molecules might not have survived, are minimized but were not nonexistent (see "Snowball Earth," Chapter 11).

The Sun is 1,392,000 km in diameter — 1,391,980 km if you want to be precise — and 110 times the diameter of Earth. One million three hundred thousand Earth-size planets would fit inside the Sun. The biggest known star, VY Canis Majoris in the Canis Major constellation, is 21,000 times larger than our sun.

Chemical Elements

At or near its origin, Earth was chemically diverse, with hydrogen, nitrogen, sulfur, carbon and phosphorus, all of which are molecules that interact in varied simple and complex ways.

With increasing oxygen levels (see Chapter 2), new classes of molecules arose, many of which were based on **carbon**. If DNA is the molecule of life — defining *life* as the transfer of hereditable information from generation to generation — then carbon is the molecule of life as far as the maintenance of structure, physiology and metabolism are concerned.

Molecules produced by organisms are organic (biotic) and so contrast with anorganic or abiotic molecules. Organic molecules based on carbon are not restricted to organisms, or indeed to Earth; they are present in the interstellar clouds that serve as the starting point for stars and planets (FIGURE 3.1). Organic molecules, therefore, need not be organismal in origin but can be (and indeed were) formed abiotically in the absence of life, a concept elaborated further below.

Water

The fourth prerequisite for molecular evolution, especially for biotic molecular evolution is the existence of large volumes of a **solvent**, a liquid that can dissolve molecules or compounds to form a solution. Water (H_2O) is that solvent. Geochemical evidence is consistent with water having been present from the outset in the cooling planetesimal from which earth arose. Volcanic activity ejected more water into the early atmosphere (see Chapter 2).

Water is often referred to as the universal solvent. Many molecules are soluble in water. Water can remain liquid over a wide range of temperatures — down to 0°C

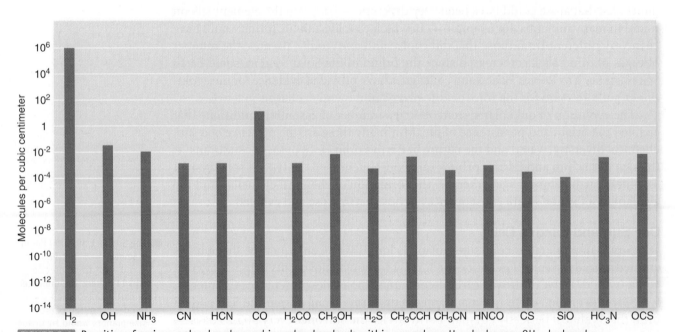

FIGURE 3.1 Densities of various molecules observed in molecular clouds within our galaxy. H_2 = hydrogen, OH = hydroxyl, NH_3 = ammonia, CN = cyanogen, HCN = hydrogen cyanide, CO = carbon monoxide, H_2CO = formaldehyde, CH_3OH = methyl alcohol, H_2S = hydrogen sulfide, CH_3CCH = methylacetylene, CH_3CN = methyl cyanide, HNCO = isocyanic acid, CS = carbon monosulfide, SiO = silicon monoxide, HC_3N = cyanoacetylene, and OCS = carbonyl sulfide. (Adapted from Buhl, D., *Origins of Life,* **5,** 1974:29–40.)

or even lower if the water contains dissolved salts, that is, if it is saline. An important property of salt water is that although it freezes, the temperature can go below 0°C before water freezes.

On average, the oceans today are about 3.5% salt. At this salinity water freezes at −2°C. This is why we strew salts on our sidewalks and roads in cold climates; ice will form at a lower temperature than in the absence of the salt. Make water as saline as you can, which means 23.3% salt, and water does not freeze until the temperature drops to −21.1°C. Not only is water the universal solvent, it was present at the origin of the solar system. The presence of water on Mars, suspected from results obtained by the NASA Mars Odyssey satellite in 2001, came closer to being proven in June 2008 when the NASA Phoenix Mars Lander detected small pieces of ice away from the Martian polar ice caps. Located just beneath the surface, these pieces of ice "melted" (changed from solid to gas) over several days after being exposed by the Mars Lander.

Replicating the Production of the First Molecules in the Laboratory

An active field of research in the middle of the last century was designing experiments to test whether molecules could be generated under conditions that mimic those projected to have existed on Earth when molecules arose. One successful approach involved the application of heat or an electric spark to mixtures of different molecules in either a gaseous or a liquid state.

The most well-publicized experiment performed by Stanley Miller in 1953 involved placing mixtures of methane, ammonia, and hydrogen gases in 5-liter flasks, generating an electric spark via an electrode and circulating water vapor through the mixture for a week or more. Any volatile compounds formed would circulate through what was, in effect a chemical condensation system. At least 20 compounds formed, including many amino acids — amino acids linked by chemical bonds (peptide linkages) are the basic units from which proteins are assembled (see below) — and urea, which is the breakdown product of proteins. So, we have two major classes of compounds produced by organisms and characteristic of metabolism being produced "in a test tube" in the absence of any organisms.

Ultraviolet light, ß-, γ- or x-rays were equally efficient as sources of energy in initiating the reactions. From such results we began to understand that the evolution of molecules, including carbon-based molecules was not as unlikely an event as had previously been thought. Laboratory studies on the origin of molecules entered a new phase this century with the application of lightning to mixes of ammonia, methane and other gases along with buffers that prevented degradation of newly formed amino acids; many more amino acids formed in the early experiments but were destroyed before they could be detected.

We saw in Chapter 2 that the presence of biotic and abiotic methane in ancient rocks complicates interpreting when organisms first arose; also see Chapter 4.

Possible Sites for the Origin of the First Molecules on Earth

The unusual living conditions of many organisms discovered in the past two decades — oceanic sea vents, boiling hot springs — show that life can be maintained, thrive, and that molecules and life could have arisen under the most stringent conditions.

Hydrothermal Vents

Thermophilic (heat-loving) organisms have been found at temperatures well above the boiling point of water. Some live a kilometer and a half below the ocean surface in association with **hydrothermal vents**, which are openings in the ocean floor from which water flows at temperatures up to 350°C. The superheated plumes arising from hydrothermal vents produce large amounts of ammonia and can produce polymers, which are large molecules comprised of many similar linked units and regarded as products of organisms. Discovered in 1977, hydrothermal vents are now regarded as likely sites for the origin of molecules early in Earth's evolution, and for the origin of cell-based life at a later stage. Lipids (see Chapter 4) typical of those found in unicellular organisms (Archaea and bacteria; see Chapter 8) have been isolated from 2.7 By-old hydrothermal environments.

Hydrothermal vents are a rich source of metallic ions, especially the mineral iron pyrite ("fool's gold") and iron ions (ferric ions, Fe^{3+}). As discussed in more detail under **Enzymes and Energy Transfer** below, when in aqueous solution, ferric ions function as catalysts and speed up chemical reactions. The catalytic activity of Fe^{3+} is limited, however. Organisms took advantage of this property, enhanced it, and made iron an essential component of the activity of many enzymes; as we will see below, enzymes are enormously effective catalysts. This cooption of iron gives us an indication of how organic molecules/life emerged from existing nonorganic processes.

Acetic acid (the acid most commonly associated with vinegar), peptides and ammonia (normally a by-product of animal waste) formed in experiments conducted in the late 1990s under temperatures and pressures that mimicked conditions in hydrothermal vents. Ammonia was produced when nitrogen or oxides of nitrogen were combined with iron or other mineral catalysts. Combining pyruvic acid (which is produced as an intermediate in metabolism or fermentation) with the amino acid alanine resulted in the production of short peptides previously thought only to be formed by the action of biological organisms, that is, to be organic in origin.

Volcanoes

Surfaces near some volcanic regions, or upwelling from shallow marine hydrothermal plumes, may have maintained appropriate temperatures for the condensation of amino acids. Volcanic sediments can produce sponge-like minerals (*zeolites*) that can retain and catalyze organic compounds. Proposals that much of Earth's early landmass was composed of volcanic islands, and that high global "greenhouse" temperatures persisted for more than a billion years because of high levels of CO_2 in the atmosphere are consistent with molecules forming in association with volcanic activity.

Clays

Other researchers have proposed that layered clays such as montmorillonite served as polymerizing templates for polypeptide or nucleotide chains. The early availability of heat, clays, and other condensing agents makes it highly probable that polypeptides, polysaccharides, lipids, and perhaps even polynucleotides were present early in Earth's history and could have been used for early organism-like reactions and structures (see Chapter 4).

■ Molecules in Meteorites

Formation of carbon-based (organic) molecules under abiotic conditions makes sense of the otherwise paradoxical discovery of molecules in meteorites and in interstellar clouds (Figure 3.1). As they are 4.6 By-old, meteorites provide evidence for the existence of

■ Recognition of the three kingdoms of life discussed in Chapter 8 would not have come about had we not begun to investigate organisms living in extreme conditions such as in association with hydrothermal vents.

■ Montmorillonite is a soft mineral (an aluminum silicate) named after the 12th century town of Montmorillon in France. Montmorillonite forms as one-micrometer (1 μm) crystals that produce a clay used in agriculture to hold soil moisture, in drilling for oil to keep the drilling mud viscous, and in earthen dams and foundry sand.

■ A polypeptide is a chain of amino acids (typically between 10 and 100), linked together. Polypeptides are the building blocks of proteins. Nucleotides are the building blocks of the nucleic acids DNA RNA, each nucleotide consisting of a chemical (purine or pyrimidine) base, a phosphate group and a sugar molecule.

BOX 3.1
Molecules in Meteorites

As discussed in the text, meteorites contain molecules previously thought only to be produced by organismal activity. Given the bombardment of Earth with meteorites, it has been claimed that the fundamental molecules from which life arose could have come from a meteoric source, specifically the carbon-containing meteorites known as **carbonaceous chondrites.**

However — and there is always a "however" when dealing with extraterrestrial structures/processes — although carbon-based molecules are present in carbonaceous chondrites, a fundamental property of those molecules differs from molecules found in organisms today. The property is the way molecules deflect light, a process known as chirality or handedness. Molecules either deflect light to the left (are levorotary) or to the right (are dextrorotary). A mixture of equal forms averages out the optical activity and the molecule exhibits no chirality.

The amino acids in carbonaceous chondrites possess both right- and left-handedness in approximately equal amounts. Therefore, they exhibit little if any optical activity. In contrast, amino acids produced by organisms are generally levorotary. Furthermore, as a number of the amino acids found in meteorites do not appear in organismal proteins, any contribution to life from meteoric molecules must have been minimal and may have been nonexistent. A terrestrial origin of life is considered much more likely.

molecules in the original solar condensation. Further, these are **carbonaceous meteorites** (*chondrites*), meaning that the molecules within them contain carbon. Organic molecules therefore preceded life on Earth by at least a billion years, leading to hypotheses that life had an extraterrestrial origin (BOX 3.1).

A well-studied carbonaceous meteorite, the Murchison meteorite, fell to Earth in Australia in 1969. Eighty kg of carbonaceous material have been isolated from the Murchison meteorite, including a range of amino acids similar to those generated in the laboratory experiments discussed above. Eight carbon-long fatty acids also have been isolated; fatty acids are part of the phospholipid that forms cell membranes. As discussed above high temperatures and pressures found in hydrothermal vents, which are common along mid-oceanic ridges, provided conditions appropriate for fatty acid synthesis on Earth.

DNA, RNA and Proteins

DNA, RNA and proteins are the three fundamental classes of molecules associated with life.

Although life can be defined in a number of ways, the **continuity of life** is based on the presence, function and replication of deoxyribonucleic acid (DNA) within cells. Both DNA and cells are required in any definition of life or living. Why? Because

- DNA does not enable a cell to function (to be alive) unless the DNA is transcribed into another nucleic acid, ribonucleic acid (RNA).
- RNA does not enable a cell to function (to be alive) unless it in turn is translated into a protein or functions to trigger the activity of other molecules of DNA, RNA or protein (FIGURE 3.2).

■ Concepts pertaining to the definition of life are summarized at the end of this and the following chapter.

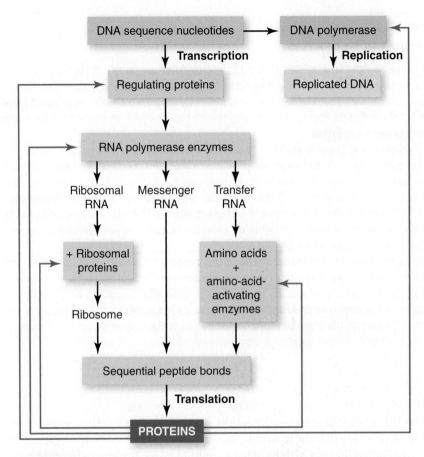

FIGURE 3.2 Schematic diagram showing the mutual dependence of information carried by nucleotide sequences and function governed by proteins. Black lines indicate the general directions of information transfer, and blue lines point to proteins synthesized by this process. Clearly, the nucleotide sequence information determines the amino acid sequences of proteins, and proteins in turn regulate and catalyze the transfer of nucleotide information; one process could not have developed without the other.

The discovery in the 1960s that DNA → RNA → protein in a one-way and nonreversible pathway became the **central dogma of biology**, although we now know that some RNA exert functional roles without being translated into protein.

The essential functions and nature of the interactions between DNA, RNA and proteins are threefold.

- **Replication**, the process of duplication by which DNA makes another copy of itself, a process we can represent as DNA → DNA (FIGURE 3.3).
- **Transcription**, the process by which the message coded in DNA is transferred to RNA, a process we can represent as DNA → RNA.
- **Translation**, the process by which the RNA message is translated into the peptides from which proteins are made, a process we can represent as RNA → protein (Figure 3.3).

(a) Replicating DNA

Replicating DNA

Template strands

Daughter strands

REPLICATION of each complementary DNA strand

(b) DNA

TRANSCRIPTION of a DNA template strand

(c) Messenger RNA

Messenger RNA

G U C U U U A G G C A U U G C G U U G A G G C A

Codon

TRANSLATION of messenger RNA

(d) Polypeptide

Amino acid

Valine Leucine Serine Histidine Cysteine Glutamine Leucine Alanine

FIGURE 3.3 Illustration of how DNA replicates and how information is transferred from DNA to RNA to protein. In DNA replication (**b** to **a**) proteins break the hydrogen bonds between paired bases, allowing the two template strands to unwind. Each unwound daughter strand then acts as a template producing a new complementary strand. As a result, two double-stranded DNA molecules form, each an exact replica of the original parental double helix. In transcription (**b** to **c**), one of the two DNA strands serves as a template upon which a molecule of messenger RNA is transcribed (green). This messenger then serves in turn as a template on which a molecule of protein is translated (**c** to **d**). The messenger RNA is exactly complementary to its DNA template and that a sequence of three nucleotides (a triplet codon) on the messenger specifies one amino acid (**c**).

Which of These Molecules Evolved First: DNA, RNA or Protein?

DNA First

Because — with few but fascinating exceptions — protein synthesis depends on the prior existence of RNA and DNA, for many years the origin of life was investigated on the very reasonable assumption that nucleic acids (DNA and RNA) originated before proteins. Logic coupled with the conservation of life's processes over time would appear to dictate the hypothesis that DNA evolved before RNA, which would have evolved before protein. Life on Earth initially would have been a "**DNA World.**"

Proteins First

We saw above that amino acids and urea can be produced in the laboratory in the absence of DNA or RNA. Therefore, a reasonable hypothesis is that proteins could have evolved first, with a genetic code evolving later. In this scenario, the first world would have been a "**Protein World,**" or at least an "Amino Acid or Polypeptide World." If we look at the history of our knowledge of the molecules associated with life, the proposal for an initial protein world makes considerable sense.

In the first half of the last century DNA and proteins were both contenders for the title "the molecule of life," DNA having been discovered in 1869 and named *nuclein*, a name preserved in the modern term *nucleic acids*. Only in 1952 were proteins eliminated as the genetic material. James Watson and Francis Crick deduced the double helical structure of DNA in 1953 (FIGURE 3.4). Four years later, the year that the first satellite Sputnik was launched by the USSR, the first enzyme involved in the replication of DNA was isolated. Watson and Crick realized that complementary pairing between purine and pyrimidine bases on opposite strands of DNA (FIGURE 3.5) would allow copying the base sequences of DNA providing the mechanism that could enable DNA to copy genetic material; see Table 7.1 for key events in our understanding of the nature of the gene. (DNA can reproduce [replicate] because one of the two strands serves as a template for the formation of a new strand. Base pairing [and, therefore, the nucleotide sequence] of the new strand is complementary to the template and to the double-stranded molecule that provided the template.)

RNA First

We now have sufficient evidence to state that the first world was neither DNA- nor protein-based but was an "**RNA World.**" Furthermore, it is hypothesized that in this early world, RNA functioned both as an enzyme (as proteins do today) and could reproduce itself, as DNA does today (but as RNA does not, except in some special situations). DNA and RNA have shown quite remarkable evolutionary change, not in their structures, but in their interactions and functions (BOX 3.2).

First RNA, Then DNA

It is surprisingly commonplace in science to find the same discovery made simultaneously by two or more individuals or groups. This applies to the now accepted hypothesis that the first chemical world was an RNA World, proposed independently in the early 1982 by Leslie Orgel, Francis Crick and Carl Woese.

■ For those of you who relish coincidences, the last paper published by Charles Darwin appeared in the journal *Nature* on April 6, 1882, only 13 days before Darwin died. The topic was the potential transport (dispersal) from one pond to another of a freshwater clam on the leg of a water beetle, transport that could initiate a new population. The specimen on which the paper was based was sent to Darwin by one W. D. Crick, Francis Crick's grandfather.

FIGURE 3.4 Helical structure of DNA.

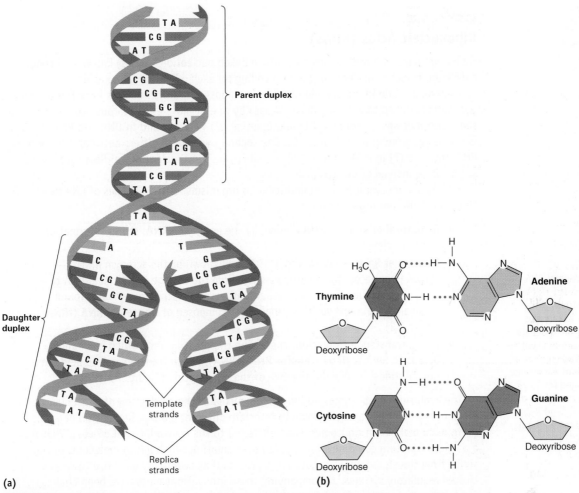

FIGURE 3.5 (a) The Watson-Crick model of the standard DNA double helix. (b) Examples of hydrogen bond pairing between bases (Adenine-Thymine and Guanine-Cytosine). The two strands on the double helix are bridged by parallel rows of such paired nucleotide bases stacked at regular 3.4 Å intervals, where 1 Å = 0.0000001 millimeters (mm).

The hypothesis is that when RNA originated it could both catalyze chemical reactions as enzymes (proteins) do today, and replicate, as DNA does today. Over time interactions between RNAs and evolving amino acids and proteins set up the classic conditions for selection to operate. The two functions — enzymatic activity and replication — became segregated into proteins and RNA, respectively. In a 1993 paper, Joyce and Orgel argued that fewer RNA molecules (40–60 nucleotides) are required to carry our replication and catalytic activity than if the two functions are distributed between nucleic acids and proteins. Compare this with what may be the smallest amount of information in a living organism: 40 to 60 RNA nucleotides, 30 DNA molecules and 90 different proteins in one of the smallest bacteria, *Mycoplasma genitalium*.

Joyce and Orgel's hypothesis of the dual catalytic and replication function of early RNA was given a substantial boost in the late 1980s and 1990s by several discoveries, including that:

1. RNA can replicate. Enzymes made of short sequences of RNA (named *ribozymes*) can cut and rejoin nucleotides to form templates that combine with complementary RNA sequences.

■ Selection originated, not with the first organisms, but with the first appearance of nonidentical molecules in solution (see Chapter 4).

■ *Mycoplasma genitalium* lives parasitically in the respiratory and genital tracts of primates and is the bacterium with the smallest genome: 521 genes. As outlined in Box 1.3, defining a gene is not as easy as you might imagine.

BOX 3.2
Ribonucleic Acids (RNAs)

Each functional unit of RNA consists of three linked nucleotides (bases) known as a codon, a different triplet of nucleotides (codon) coding for each of the 20 amino acids.

Each nucleotide in a molecule of RNA is composed of a nitrogenous base (purine or pyrimidine) linked to a sugar (deoxyribose) by one or more phosphorous groups. The same purine bases — adenine (A) and guanine (G) — are found in DNA and RNA. So is the same pyrimidine, cytosine (C). The second pyrimidine differs between DNA and RNA: thymine (T) in DNA (Figure 3.5); uracil (U) in RNA. Nucleotides of RNA contain the sugar ribose instead of deoxyribose.

RNA exists in several forms in multicellular organisms, different types of RNA functioning in different organelles.

- **Structural** or storage **RNA**, known by the acronym **sRNA**, stores information in the nucleus.
- **Messenger RNA** — discovered in 1960 and known by the acronym **mRNA** — carries (translates) the genetic message from the storage nucleic acid DNA to amino acid sequences in the protein synthesizing machinery, which is found on small cytoplasmic vesicles (ribosomes) composed of **ribosomal RNA** (**rRNA**; Figure 3.2).
- mRNA transfers information, specifying the position into which each amino acid is to be placed by **transfer RNA** (**tRNA**), which transfers amino acids to polypeptides, increasing the size of the polypeptide (Figure 3.2).

The evolution of ribosomes no longer committed to the production of particular proteins enabled the same ribosome to translate different mRNAs. Regulation of which proteins to make from ribosomes was transferred to the transcriptional process. That is, a particular protein could now be selected by regulating which mRNA molecules to transcribe from the stored genetic material, a process that eventually gave rise to sophisticated regulatory systems. Accompanying these innovations must have been changes from depending on only few enzymes and proteins to employing greater numbers with restricted functions and higher binding specificity.

■ Classes of ribosomal RNA (rRNA) include 16*S* rRNA molecules in prokaryotes and 18*S* rRNA eukaryotes, both of which are essential for ribosomal protein synthesis. *S* is the Sverberg unit of sedimentation of a molecule when centrifuged for 10^{-13} seconds. It is named for a Swedish chemist and Nobel Laureate Theodor Sverberg (1884–1971).

■ You may be thinking, "Isn't this evidence for the spontaneous generation of life?" Read on, and see Chapter 6 for spontaneous generation.

2. Nucleotides can act as catalysts. Nucleotides that can catalyze the joining of other nucleotides were isolated and shown to obtain the necessary energy from triphosphates, specifically adenosine triphosphate (ATP), which is the major energy transfer molecule in organisms (see Chapter 8).
3. The catalytic activity of RNAs can be selected for and improved. Molecules with much greater catalytic activity formed when RNAs were selected for in the laboratory.

If ancient RNA stored information in nucleotide sequences and could make polypeptides, how did DNA arise? In organisms today, DNAs (deoxyribonucleotides) like any molecules are synthesized from ribonucleotides (RNAs), so we can see how RNAs could have been a substrate for DNA. We also can see that, once RNA arose, locating the two functions of information storage/replication and protein manufacture in different but linked molecules would be of selective advantage. Without this selection, naked replicating DNA and polypeptide-producing RNA would likely have remained as separate classes of molecular activity, and life would have been very different. Once

DNA appeared, selection pressure would have established (or maintained) a tie with RNA as DNA evolved as a **genetic code**.

The Genetic Code

In all extant organisms information transfer between nucleic acids and proteins follows a genetic code (BOX 3.3) that determines the placement of a particular type of amino acid in a protein on the basis of the placement of a particular sequence of three nucleotides in messenger RNA (mRNA). The code is **universal** because it specifies the same 20 amino acids in all organisms with only few exceptions. Although no ancestral codes have been discovered, we presume that as for any other biological trait, the genetic code evolved from an earlier and presumably simpler form. The universality of the genetic code (with a few rare codon exceptions) indicates that only the ancestral bearers of a code specifying 20 amino acids survived.

Enzymes and Energy Transfer

It is not an overstatement to affirm that transfer and utilization of energy and regulation of biochemical activity is a function of sequences of amino acids. The specific structure (including the 3-D configuration), properties, and often the function of any protein arises from the specific linear sequence of amino acids that define it. This is so whether the protein functions as an **enzyme** (catalyst) or whether it serves another function.

One of the components of a definition of life, perhaps even a sufficient definition of life, is transformation of external sources of material and energy into metabolism, growth, and development. The origination of enzymes — molecules that catalyze chemical reactions — and their sequestration into specialized compartments within

BOX 3.3
The Genetic Code

Five of the major features of the genetic code as it operates in extant life are:

1. Messenger RNA molecules consist of only four kinds of nucleotide bases — adenine (A), guanine (G), uracil (U), and cytosine (C) — from which are built chains of varying lengths and varying sequences.
2. Each amino acid is specified by a triplet of nucleotides of mRNA known as a codon, one codon for each amino acid, just as words are made up of sequences of letters.
3. The code is universal, reflecting its ancient origin. All organisms share the same codon for the same amino acid with a few exceptions; some mitochondrial genes use different codons than those used by nuclear genes for the same amino acid, reflecting the independent evolutionary history of mitochondrial and nuclear genes (see Chapter 9).
4. No two amino acids are specified by the same codon, although 18 of the 20 amino acids are specified by more than one codon.
5. Each codon is translated as a continuous, uninterrupted sequence in a specific direction.

■ Mitochondrial DNA (mtDNA) has proven to be exceedingly useful in analyses of DNA ("ancient DNA") from museum specimens, from extinct organisms and even from fossils; see Box 9.2.

■ Because each triplet of nucleotides is read as if reading three words without intervening commas (AGU rather than A, G, U) the code is said to be comma-less.

evolving cells (see Chapter 9) allowed early cellular life to increase the efficiency of material and energy intake, transformation and utilization.

The property of functioning as an enzyme is not limited to biotic organic molecules. Organisms did not invent enzymes but harnessed the existing but weak catalytic actions of ions. As we saw when discussing hydrothermal vents as a possible site for the origination of molecules, iron was a common element in early rocks. Iron ions (ferric ions, Fe^{3+}) increase the efficiency of, and indeed are required for, many biological reactions. As an example, Fe^{3+} is required for the protein hemoglobin to transport oxygen, and for the function of many hormones. Having appropriate levels of iron is critical for most living systems; the amount of iron in an organism is regulated by many different genes. The first organisms took advantage of the catalytic properties of Fe^{3+}, coupled it to evolving enzymes, made it more efficient, and in the process made Fe^{3+} essential for many enzymatic reactions. Do you think we can regard Fe^{3+} as an essential component of life?

Molecules and Life

With regard to the pursuit of the question "**What Is Life?**" the following conclusions can be drawn from evidence presented in this chapter.

1. The molecules most associated with life are DNA, RNA and proteins. All have maintained stable structures but shown remarkable evolutionary change in their interactions and functions.
2. DNA is the molecule of life where life is defined as the transfer of hereditable information from generation to generation. Carbon is the molecule of life when life is defined on structural, physiological and metabolic bases.
3. The same molecules produced by organisms (organic molecules) can be formed abiotically in the absence of life and arose before the origin of organismal life. We therefore can identify a continuum between the abiotic production of molecules and first life.
4. Consequently, we will discuss in the next chapter whether life should be considered as having begun with molecules around 4.5 Bya, with cells 3.5 to 3.8 Bya, or with the first organisms.

A set of conclusion on the nature of life in relation to cellular organization is listed at the end of the next chapter.

Recommended Reading

Hazen, R., D. Papineau, W. Bleeker, et al., 2008. Mineral Evolution, *American Mineralogist* **93**, 1693–1720. (See also *Nature*, 2008 **456**, 456–458.)

Joyce, G. F., and L. E. Orgel, 1993. Prospects for understanding the origin of the RNA world. In *The RNA World*. R. F. Gesteland and J. F. Atkins (eds.). Cold Spring Harbor Laboratory Press, Cold Spring Harbor, New York, pp. 1–25.

Loomis, W. F., 1988. *Four Billion Years: An Essay on the Evolution of Genes and Organisms*. Sinauer Associates, Sunderland, MA.

Maynard Smith, J., and E. Szathmáry, 1995. *The Major Transitions in Evolution*. Freeman, Oxford, England.

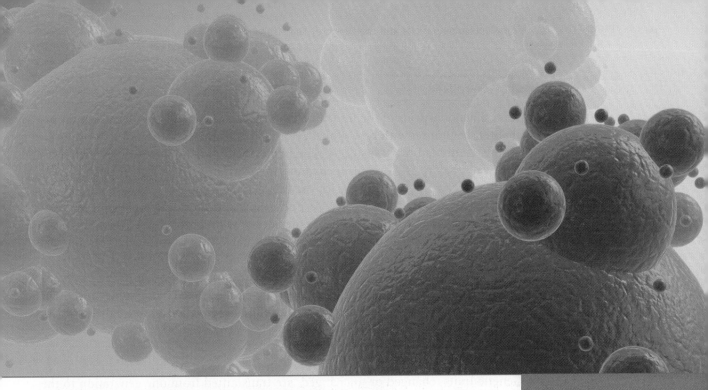

From Molecules to Cells and the Origin of Selection

4

KEY CONCEPTS

- Earth has had two atmospheres, the first of which, the primary atmosphere, arose 4.6 to 4.2 Bya and lacked oxygen.
- The second atmosphere, which arose 4.2 to 3.8 Bya, had little oxygen until cells arose.
- The first molecules were short chains and linear.
- Linear molecules rounded up as membrane-bound vesicles → the first protocells.
- Membranous structures (vesicles, protocells) can be formed under laboratory conditions thought to mimic early condition on Earth.
- Variation arose as soon as different types of vesicles formed in a population.
- Selection originated once different types of vesicles formed in a population.
- The first cells, which originated 3.5 to 3.8 Bya, lacked a nuclear membrane or organelles.
- Cell division originated in these cells.
- Life may most accurately be defined as based on cells.

Above: Molecules form microspheres, from which protocells arose.

Overview

Several major aspects of the transition from molecules to cells are discussed in this chapter, beginning with how global events, such as the evolution of Earth's atmosphere, facilitated the transitions. The first transition is how molecules that arose abiotically before any organisms had originated, transformed into sheets or droplets that became the prototypes for the first cell membranes. Selection would have operated on these molecules. The second is how concentrating molecules within a membrane-bound droplet could have produced the first protocells. Selection would have operated on these protocells. Between 3.5 and 3.8 Bya, protocells became organized into a central region containing DNA and a peripheral region containing proteins and RNA. This is the fundamental organization of those unicellular organisms (prokaryotes) that lack a nuclear membrane. The origin among some unicellular organisms of nuclear membranes and intracellular organelles some 2.5 Bya gave rise to eukaryotic cells.

Life depends on the accuracy/constancy with which organisms transmit genetic information to their offspring, but evolution cannot occur without genetic variation. Genetic traits, changed or unchanged, are transmitted from one generation to the next via cell division. Life became based in cells with an inheritance system that could pass their properties to future generations.

Earth's Atmosphere(s) and the First Cells/Organisms

Earth has been enveloped in least two different atmospheres during its history. The **first or primary atmosphere** composed of hydrogen and helium persisted for less than half a billion years after the origin of Earth 4.6 Bya. Possible reasons for the loss of the primary atmosphere are the low gravitational force of Earth itself, extremely strong winds from the Sun, and/or heat generated by Earth and by the Sun. Organisms had nothing to do with the origin or loss of this primary atmosphere.

Four billion years ago, the first volcanoes and hot springs spewed gases into the atmosphere. Comets colliding with Earth also may have contributed gases, especially water vapor. These combined events resulted in the formation of a **secondary atmosphere** composed largely of water vapor and carbon dioxide, with smaller amounts of hydrogen, carbon monoxide, nitrogen, ammonia, methane, hydrochloric acid, and hydrogen sulfide. This atmosphere, which arose between 4.2 and 3.8 Bya, contained no or very little oxygen.

Both the first and second atmospheres arose before the origin of organisms. By 3.5 Bya, methane (CH_4) levels in the atmosphere began to increase because of the release of methane by the first organisms, which may have originated as early as 3.8 Bya. These were single celled **methanogens** (methane generators), organisms that only survive in the absence of oxygen.

Methanogens are known, not from fossilized cells, but from molecular fossils — the presence of methane of biological origin in rocks dated at 3.5 Mya; methanogens produce methane as a metabolic by-product. (Along with cows, we produce methane as the result of the activity of methanogens in our intestines.) Whether all the methane in these rocks is the result of methanogen activity (that is, is biotic), or whether some was deposited by geological-chemical (abiotic) processes is not easy to determine.

Methane is now recognized as a **greenhouse gas**. In the early oxygen-free atmosphere, methane would have accumulated and formed an insulating layer, raising the surface temperature and so would have prevented ice accumulation. Between 3.5 and 3.0 Bya, Earth's surface temperature cooled, liquid water formed and the level of CO_2 in the atmosphere began to fall. The presence in 3 By-old river deposits of grains of sand composed of sulfides of iron, lead and zinc is consistent with an absence of atmospheric oxygen 3 Bya; sulfides do not form in the presence of oxygen.

> ■ Greenhouse gases such as carbon dioxide (CO_2) and methane (CH_4) absorb infrared radiation, reduce the loss of heat from Earth's surface and so raise the global temperature.

evidence?

Oxygen and Cyanobacteria

Before organisms requiring oxygen for their metabolism arose, any O_2 released would have reacted with other elements in the atmosphere and with minerals on the surfaces of rocks. This chemical activity initiated the first weathering of rocks whose inexorable breakdown produced the soils of today.

The current level of oxygen in the atmosphere is 21% O_2 (FIGURE 4.1). Geochemists have determined that a gradual accumulation of oxygen in the secondary atmosphere began about 2.3 Bya. Much of this accumulation of O_2 is considered to be a consequence of the metabolic activity of newly evolving life. Organisms released oxygen during the synthesis and production of organic compounds from carbon sources and water, a process known as **photosynthesis**. During photosynthesis, free O_2 is produced as hydrogen ions are removed from water molecules. The major photosynthetic organisms on Earth today are flowering plants.

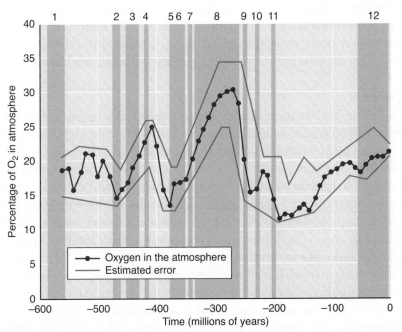

FIGURE 4.1 Changes in the percentage of oxygen in the atmosphere over the past 600 million years along with upper and lower limits of the estimated error in predicting oxygen levels. Numbers 1 to 12 correspond to significant evolutionary events, for example: origin of animal body plans (1) and increase in body size of mammals (12) correlate with high concentrations of oxygen; major extinction events (5, 9, 11) correlate with low levels of oxygen. (Reproduced from Berner, Robert, A., et al., *Science* **316**[2007]: 557 [http://www.sciencemag.org]. Reprinted with permission from AAAS.)

Chloroplasts, which are energy-producing organelles in plant cells, arose when the ancestors of plants ingested and transformed cyanobacteria into chloroplasts (see Chapter 10).

It is important to realize that life based on something other than carbon could have arisen and evolved had Earth taken a different course in its evolution. Whether we would recognize such forms of life is another issue.

Global warming, increase in levels of atmospheric CO_2 and depletion of the ozone layer, are the three primary atmospheric forces being impacted by human activity.

The first organisms preserved as fossils are photosynthetic, single-celled aquatic prokaryotes known as **cyanobacteria**. The earliest cyanobacterial fossils are in deposits laid down 3.5 Bya. We can extrapolate from the presence of cyanobacteria 3.5 Bya and the rise in atmospheric O_2 2.3 Bya (Figure 4.1) to conclude that it took cyanobacteria 1.2 By to accumulate to the level where they began to affect the composition of the atmosphere. Because photosynthesis utilizes CO_2, the increase in cyanobacterial mass lowered atmospheric CO_2 as O_2 was rising. As photosynthetic organisms expanded over time and as atmospheric O_2 increased, atmospheric CO_2 declined to the current level of 0.04%.

Free O_2 penetrated as far as the upper levels of the atmosphere. A further critical event required before life as we know it could arise on Earth was the UV irradiation of free O_2 in the upper levels of the atmosphere to form ozone (O_3). The formation of ozone gradually produced an ozone layer that protected Earth's surface and the lower atmosphere from the very UV radiation that caused its formation. The ozone layer remains to this day.

With increasing levels of O_2 and declining levels of CO_2, the stage was being set for new types of molecules and new forms of molecular organization to arise. Their origin and the **transition to the origin of life**, is taken up in the remainder of this chapter.

Membranes → Protocells

Before the origin of cells — or before any evidence for the presence of cells — carbon-based molecules such as amino acids, fatty acids and small peptides existed (see Chapter 3). These molecules are **organic** according to the definition of organic used in chemistry — they are molecules based on or containing the element carbon (see Chapter 3). They are not organic in the sense of having being produced by organisms.

Membranes

Sometime before 3.5 Bya these **abiotic organic molecules** began to form structural aggregates that would have included chains of fatty acids and polypeptides comprised of linked amino acids. Such long-chain molecules are unlikely to form aggregates or spheres — their 3-D configuration is more consistent with organization into linear molecules. This simple chemical fact may explain why life when it arose was based on membranous structures, cells. How might such membranes have originated?

The chemical properties of these molecules reflect features at the atomic level. Those fundamental features — the formation of bonds between molecules of hydrogen, the ability of these molecules to dissolve in water and to release hydrogen ions — are not likely to have changed much if at all since molecules originated. End-to-end attachment and elongation of chains of molecules — like stringing together a chain of paper clips — stabilized intramolecular and supramolecular structural organization.

A phospholipid is a lipid (fat) with a phosphorus-containing polar group at one end and a non-polar fatty acid group at the other (FIGURE 4.2). A **cell membrane** is constructed as a phospholipid bilayer. Place a phospholipid in water and the polar end orients toward water while the non-polar end orients away from water. Consequently, phospholipid molecules can join and align to form membranes, which if they round up, become vesicles (Figure 4.2).

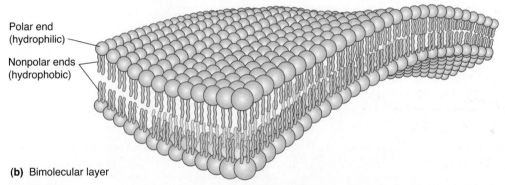

(a) Lecithin

Polar end
(hydrophilic)

Nonpolar ends
(hydrophobic)

(b) Bimolecular layer

FIGURE 4.2 **(a)** A phospholipid molecule, lecithin. **(b)** A diagrammatic view of a bimolecular sheet like double layer of phospholipid molecules that have self-assembled with their hydrophilic phosphate heads (colored circles) facing the water solvent, and their hydrophobic hydrocarbon tails facing each other. Such polar–nonpolar molecules characterize cell membranes.

Protocells

Any molecules attached to a lipid vesicle, even molecules in solution, will be trapped inside the vesicle, forming what have been called **membranous droplets**. Droplet formation would have been an important step in the evolution of cells.

Droplets can grow by adding more fatty acids, and will do so in response to mechanical agitation (**FIGURE 4.3**), as would have occurred in ancient pools or ponds. As they grow, vesicles become large enough for concentration gradients to be established across the membrane. The gradients would be molecular — H_2 ions (measured as pH), CO_2 or O_2 — allowing differences to develop between the inside and the outside of the vesicles. In essence, with such vesicles, we would have had the first protocells, life forms that could maintain their internal environments in the face of a different external environment.

The rise in concentration of molecules within protocells facilitated the next stage of cellular evolution, which would have occurred much more slowly, if at all, had the molecules remained in solution. Membranes → protocells therefore established

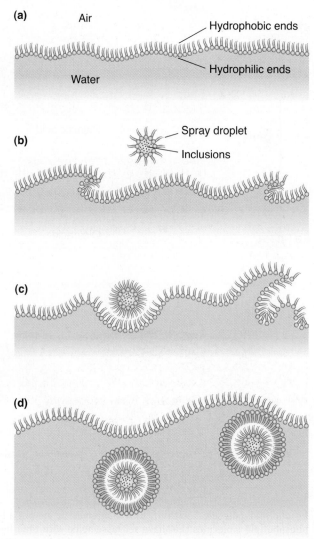

should be easy to prove
or disprove

FIGURE 4.3 Effect of wave action on a surface film containing (a) molecules oriented with one end pointed away from water (hydrophobic) and the other toward the water (hydrophilic). Wave action (b and c) causes the formation of droplets and the bilayered vesicles shown in (d). Double-membrane structures may have formed through incorporation of one bilayered vesicle within another.

a major feature of cellular life, which is acquiring energy from outside and using that energy to drive internal chemical processes that enhanced protocell survival.

Laboratory Studies

As with the search to understand the origin of molecules (see Chapter 3), laboratory experiments began in the 1920s to determine whether membrane-bound vesicles could be produced, and if they could, whether they could grow. Both were found to occur. Dehydrating solutions of molecules produced droplets that aggregated into what were named **coacervates** (FIGURE 4.4). Importantly in these experiments, both droplet and coacervates increased in size with time. The formation of droplet organization would have been a crucial step in the evolution of life on anything other than a two-dimensional pavement.

■ A coacervate is a spherical aggregation of colloidal particles in liquid suspension.

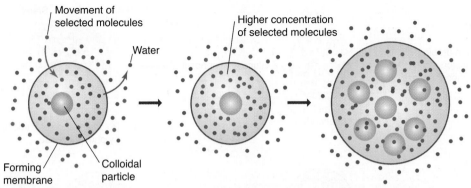

FIGURE 4.4 Formation of coacervates by the exclusion of water molecules (blue dots) from associated colloidal particles (tan circles). The intervening water molecules can be removed through dehydration (for example, increased salt concentration), or when colloidal particles are attracted to each other because they have opposite charges or because some colloids are basic (for example, histones) and others are acidic (for example, nucleic acids). (Adapted from Kenyon, D. H. and G. Steinman. *Biochemical Predestination.* McGraw-Hill, 1969.)

The pioneer in these studies was the Russian biochemist Aleksandr Oparin (1894–1980). In 1924, Oparin published a paper that set out what we now regard as the modern theory of the origin of life. Oparin's theory brought a radically new way of thinking about the origin of life. In his words: "There is no fundamental difference between a living organism and lifeless matter. The complex combination of manifestations and properties so characteristic of life must have arisen in the process of the evolution of matter."

Oparin produced the first coacervates and showed that they could grow. He also demonstrated that incorporating enzymes into the droplets enabled them to perform metabolic functions — synthesis and hydrolysis of starch, for example — previously known only as a result of cellular activity (FIGURE 4.5). This is important. Enzymatic functions were produced that previously were held to be the properties of proteins produced in organisms. The reality of hypotheses of the origination of cells from abiotic molecules and membranes had been shown. Again, in Oparin's words:

> At first there were the simple solutions of organic substances . . . governed by the properties of their component atoms . . . But gradually, as the result of growth and increased complexity of the molecules, new properties have come into being and a new colloidal-chemical order was imposed . . . In this process biological orderliness already comes into prominence. Competition, speed of growth, struggle for existence and, finally, natural selection determined such a form of material organization which is characteristic of living things of the present time.

Several decades later an American biochemist, Sidney Fox (1912–1998), showed that a gram of polymers when heated and then cooled produced hundreds of millions of tiny **microspheres**, each with a double bounding membrane. Some of these microspheres underwent fission and/or budding as cells do (FIGURE 4.6). As with membrane-bound vesicles, protocells and coacervates, these microspheres could absorb chemicals selectively from the medium, taking in some and excluding others. As with coacervates, microspheres increased in size, providing a further experimental means of exploring the possible origin of life.

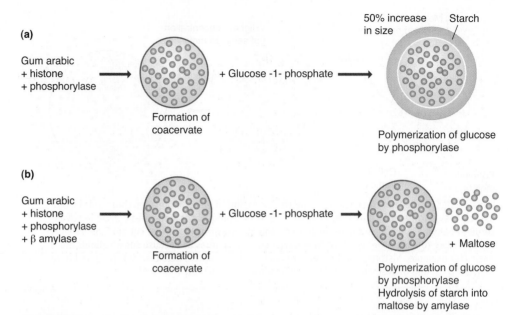

FIGURE 4.5 Synthesis and hydrolysis of starch in coacervate systems in which enzymes have been included in the droplets. In (**a**) the phosphorylase enzyme acts to polymerize phosphorylated glucose into starch, while in (**b**) the starch formed this way is hydrolyzed into maltose by the enzyme amylase.

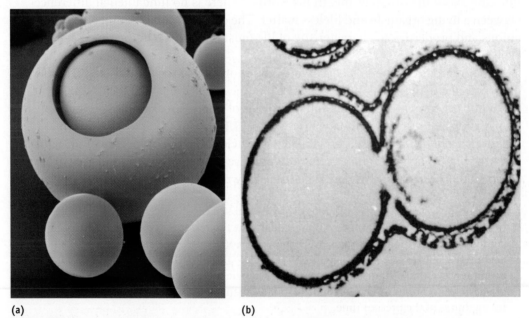

(a) (b)

FIGURE 4.6 Microspheres produced in the laboratory. (**a**) Polymer microspheres for drug delivery. (**b**) Protenoid microspheres form a double-membrane and undergo budding, as visualized with transmission electron microscopy.

Selection

An overview of how selection relates to the essential units of life is used to introduce the notion of **individuals**, **populations**, variation and selection in microspheres and protocells.

As introduced in Chapter 1, the two essential organizations of life above the level of the cell are individuals and populations. Individuals develop, grow, may divide, but ultimately die. Individuals do not evolve, in that an individual exists only for one generation (see Chapter 1). Individuals within each generation do, however, respond to natural selection.

Central to biological evolution are inherited changes in populations of organisms over time leading to differences among them; a population is the basic ecological and reproductive unit of a species (see Chapter 1). Variation provides the raw material of evolution (see Chapter 12). Differential survival and reproduction of individuals in a population from one generation to the next are the essential bases of selection (see Chapter 15).

Genes within individuals (**genotypes**) in a population, which are passed down from generation to generation, and the features (**phenotypes**) of individuals in successive generations of organisms, evolve (see Chapter 1). The accumulation of heritable responses to selection of the phenotype, generation after generation, leads to evolution — Darwin's descent with modification — as outlined in Chapter 1 and discussed in greater depth in Chapter 6.

- Selection characterizes living systems because of the expansion of populations through reproduction, resource limitation, variation among individuals in a population and environmental change over time.

- Selection is the sum of the survival and fertility mechanisms that affect the reproductive success of genotypes (see Chapter 15).

The Origin of Selection

From the experiments discussed in the previous section, we see that the microspheres produced from one large molecule such as a polymer differ from the millions of microspheres produced from another polymer. The protocells produced by combining two specific chemical mixtures differ from the protocells produced by combining two other chemicals. Each set of microspheres or protocells is a population composed of individual microspheres or protocells that differ from one another (Figure 4.6). Furthermore, variation upon which selection acts would have existed in these populations.

When variation and selection occur in the laboratory or as a consequence of artificial breeding of animals or plants or exposure to drugs or insecticides, we refer to the process as **artificial selection**. When they occur in nature, we refer to the process as **natural selection**. Fundamentally, the processes are the same.

With respect to the evolution of selection

- Selection would have originated as soon as the first non-homogeneous population of molecules appeared on Earth. We could call that molecular selection.
- Selection among protocells would have originated as soon as the first non-homogeneous population of protocells appeared (non-homogeneous because of differences in size, and/or in efficiency of obtaining/utilizing energy). We could call that protocellular selection.
- If raw materials and/or energy were limited, we would have the classic conditions for Malthusian selection — struggle for life as applied to the natural world by Charles Darwin (see Chapter 6).

Selection and Inheritance

You will have noticed that selection at these molecular and protocellular levels does not involve a system of genetic inheritance; differential survival is independent of inheritance.

As inheritance is essential to life as we know it, we can ask whether these molecules/protocells and the presence of differential survivals qualifies these early stages to bear the title "first life." Some would argue no; change over time in the composition of non-reproductive molecules or protocells represents chemical, not biological evolution. The forms involved — molecules and protocells — are not living; they are not biological. Biological evolution, they would maintain, is change in inherited differences among organisms that reproduce.

Whether the presence of selection is a sufficient criterion to move organized chemical activity within membrane-bound vesicles from the category "non-living" to the category "living" is subjective, in large part because we define life on the basis of molecules, cells, energy transfer, and/or inheritance. Whether these protocells were living or not is immaterial (no pun intended). Selection is an essential process bridging molecules and organisms, as recognized by Oparin in the 1920s. The finding that a specific ribozyme can be selected from among trillions of random RNA molecules with only few selective steps in the laboratory (see Chapter 3) demonstrates that selection operates on molecules in the absence of any higher levels of organization.

Selection and the Cellular Basis of Life

Selection is not merely a passive agent sifting the good from the bad, the adaptive from the nonadaptive. Because of its historical continuity, selection enables a succession of adaptations that allows new levels of organization to arise and be maintained.

Some would say that the **cellular level of organization** is what separates living from non-living, an approach that leaves viruses in a rather ambiguous living/non-living "no organisms" land (see below). Others would argue for the requirement for a genetic code and stop there. Others would say a genetic code enables the results of selection to be passed to future generations, provided a process of cell division exists to effect the transfer from generation to generation. The essential point is to recognize that the emergence of life was a gradual and incremental process in which new levels of activity and organization appeared at each stage.

Chromosomes and Cell Division

One of the most basic qualities of life is the ability to perform those reactions necessary for survival and replication. The localization of genes on **chromosomes** (nuclear structures that replicate with each cell division) provides a physical mechanism that ties inheritance to cell division — the transmission of genetic information as sets of nucleotides (genes) arrayed along chromosomes accompanies cell division (see Chapter 7).

The organization of genes on chromosomes differs between prokaryotes — which have a single, often circular chromosome — and eukaryotes, which, with only one known exception, have at least two (one pair) and as many as 1,260 (630 pairs) chromosomes. The processes of cell division also differ between prokaryotes and eukaryotes.

In **prokaryotic cells** such as those that comprise bacteria, division is by a process of **binary fission**. Duplication of the DNA and replication of the single circular chromosome (which is attached to the cell membrane) produces two chromosomes, each with its own attachment to the cell membrane. The cell then divides giving rise to two daughter cells, each with a single chromosome. Because prokaryotic bacterial cells can undergo

■ The only eukaryote known to have a single pair of chromosomes is the jack jumper bulldog ant *Myrmecia pilosula*, a basal species found in Australia. Being haploid, the males have only a single chromosome in each cell.

this process of binary fission every 20 to 30 minutes, bacteria can multiply enormously. A single bacterial cell dividing every 20 minutes under ideal conditions can produce over a million bacteria in seven hours, and more than 553 million bacteria in 10 hours.

One type of **eukaryotic** cell division, seen in the three major groups of eukaryotes — fungi, plants (see Chapter 10) and animals (see Chapter 11) is **mitosis**, the division of cells other than germ cells (FIGURE 4.7). One phase of division of the chromosomes during mitosis is shown in cells in the root tip of onion plants in FIGURE 4.8 .

A second type of division, **meiosis**, is found in those cell lineages that produce germ cells (Figure 4.7): eggs and sperm in animals and fungi, ovules and pollen in plants. Meiosis consists of (i) a number of sequential divisions in which cells first divide by mitosis to produce many gamete precursor cells; (ii) two meiotic divisions during which the chromosome number is halved so that each daughter cell contains only one of the two pairs of chromosomes; and (iii) further waves of mitosis, cell growth and differentiation to produce mature male or female germ cells (see Chapter 7).

A third type of cell division — **cleavage** — characterizes the earliest stages of the development of animal embryos (Figure 4.7). Chromosome number remains the same following each mitotic or cleavage division, but is halved during meiosis.

■ The largest number of chromosomes found so far is 1260 in the adder-tongue fern *Ophioglossum reticulatum.* The horsetail, *Equisetum arvense,* a plant with an ancient evolutionary history back to the Mesozoic Era (see Figure 10.9) has 216 chromosomes. Among animals, the crayfish, *Cambarus clarkii,* has 216, and the Brazilian ant, *Dinoponera lucida* 106 chromosomes (the largest number for any ant).

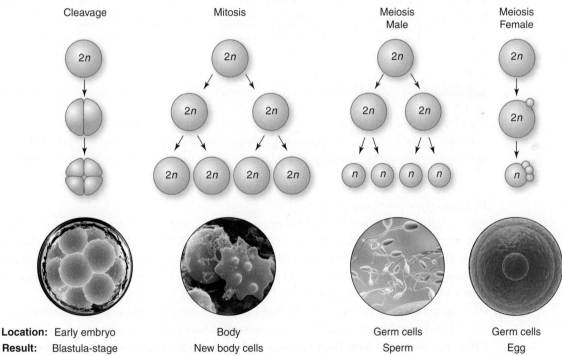

Location:	Early embryo	Body	Germ cells	Germ cells
Result:	Blastula-stage	New body cells	Sperm	Egg

FIGURE 4.7 The essential differences between the three types of cell division (cleavage, mitosis and meiosis) in animals, with emphasis on cell size and chromosome number. In cleavage of the fertilized egg, cell size is reduced at each division, most simply by equal division of cytoplasm of cells that remain attached to form the multicellular embryo. All cells are diploid (shown as $2n$). In mitosis of body cells, cell size remains the same after each division because of the synthesis of new cytoplasm (growth). All cells are diploid ($2n$). Although the mechanism of chromosome reduction during meiosis is the same in germ cells of males and females, the fate of the cells differs. Equal distribution of cytoplasm in the male germ line results, in the example shown, in the production of four sperm-producing cells of equal size. In contrast, unequal distribution of cytoplasm in the female germ line results, in the example shown, in the production of one large and three small cells. Only the large cells go on to become an egg. In both male and female germ lines, the reduction (meiotic) division reduces chromosome number from diploid ($2n$) to haploid (n).

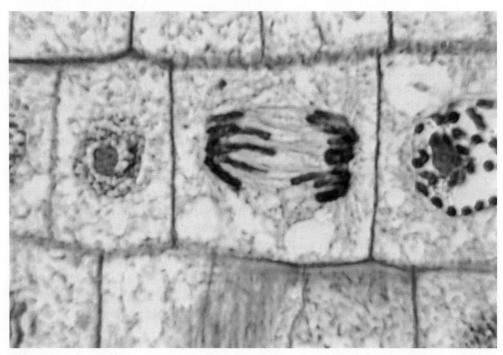

FIGURE 4.8 Separation of chromosomes across the mitotic spindle (middle image) during the division of cells at the tip of the root in an onion plant.

Life Without Cells?

This background on the importance of chromosomes, cellular organization and cell division as the basis of biological organisms and as providing the means to transfer genetic information from generation to generation places us in a position to examine whether two non-cellular entities are living. The two are **prions** and **viruses**.

Prions

Prions (and viruses; see below) function as living entities when associated with live (cellular) organisms, although prions contain neither DNA nor RNA. A prion, believed to be the smallest agent of infection, is a particle derived from the cell membrane protease-sensitive prion protein PrPC. Prions are only functional when incorporated into a living cell where they convert cellular proteins into new prions.

The prion protein PrP exists in both a soluble enzyme (protease)-sensitive form (PrPc) and a highly insoluble protease-resistant form, PrPSc. Conversion of PrPc → PrPSc is associated with the deposition of aggregates of amyloid fibers in the brain and acquisition by PrPSc of infectious properties. Although prion protein sequences are genetically determined, the interactions that change their conformation can be transmitted and reproduced between cells and individuals without further genetic information.

The prion protein, normally harmless, can undergo a pathogenic change in its 3-D shape, which converts other such proteins to a similar form. Through such "domino effect," a **prion disease** progressively develops. Prions act remarkably like a non-nucleic acid infectious agent, as prion proteins increasingly change to the pathogenic form. By entering brain cells and taking over protein synthesis, prions are thought to

be the agent responsible for neurological degeneration in humans (Creutzfeldt-Jakob disease) and for "mad cow" disease (bovine spongiform encephalopathy) in cattle.

Prions also are thought to be the agents for kuru or laughing death in the Fore tribe of New Guinea, whose custom was to eat the brains of dead relatives. Kuru, which is a form of transmissible spongiform encephalopathy, is characterized by loss of coordination, dementia, and, paradoxically, outbursts of laughter as the disease progresses (hence the name "laughing death").

■ Prions containing the same sequence of the prion protein (PrPSc) can give rise to different disease states, an aspect of prion function that is not understood.

Viruses

Named from the Latin *virus* for a toxin or poison, viruses are usually referred to as agents associated with organisms rather than as organisms per se. Typical definitions would be "a virus is a sub-microscopic infectious agent that can neither grow nor replicate (reproduce) outside a host cell," or "a virus is an acellular entity that can only replicate inside the cells of a host organism." The **tobacco mosaic virus (TMV)** was the first virus to be discovered (1899). Since then, over 5000 viruses have been described from all groups of organisms. All viruses consist of a protein coat and genes that are comprised of DNA or RNA. Hence, we speak of **DNA-** and **RNA-viruses.**

■ This is why software programs that "infect" computers where they reproduce and do great harm, even destroying the host program, are known as *viruses.*

No organism is immune to viral infection. Infection of tobacco and related plants by tobacco mosaic virus leads to a mottled discoloration of the leaves. The protein and nucleic acid components of tobacco mosaic virus spontaneously aggregate into the exact configuration needed to produce an active virus.

■ The mutual interactions and co-evolution of a myxoma virus and rabbits in Australia is discussed in Chapter 14.

A new type of virus — a **mimivirus** (<u>mi</u>micking <u>mi</u>crobe <u>virus</u>) — infecting amoebae discovered in 2003 and named *Acanthamoeba polyphaga* mimivirus (APMV) is the largest virus known and has the largest viral genome known (900 protein-coding genes). A second small virus with 21 genes is associated with the mimivirus in the amoebae, but it infects the mimivirus, not the amoeba. Although APMV is the only such virus known (yet), a virus infecting a virus opens up a new window on the link between life and non-life. Viruses therefore have much to tell us about the level of 3-D sophistication that can be generated with seemingly small amounts of information.

The simplest form of cellular life that has been produced artificially had around 300 genes; see Box 1.3 for definitions of a gene. Using readily available DNA, a "virus" containing 5,386 base pairs and capable of infecting and killing bacterial cells has been produced in the laboratory. Of course, it was produced using the DNA-protein system that characterizes modern life. Even this simple life form must be much more complex that the first life form on which selection would have acted.

Life

Can a prion or a virus be said to be alive when not within a live cell? Not if we follow the definition that the units of life are cells. Prions are misshapen proteins and the products of cells, but cannot function unless incorporated into a live cell. Such attributes of prions suggest to some that prions (and viruses) lie on the edge of life.

Discussion of viruses and prions naturally leads us to amplify the attributes of life outlined at the end of Chapter 3 by adding the following:

1. Cells are the units of life and life is based on membranous structures. The first protocells represent the origin of life and prions and viruses are not alive.

2. Life involves acquiring energy from outside to drive internal chemical processes.

3. Living things can perform the reactions necessary for survival and replication.

4. The two essential organizations of life above the level of the cell are individuals and populations.

These four points are combined in [BOX 4.1] with the points on the attributes of life listed at the end of Chapter 3 as a summary of approaches to answering the two questions of "what is life?" and "when did life arise?"

It is instructive to compare the conclusions here and in Chapter 3 with the criteria for life (living agents) set out in 2005 by the philosopher Robert A. Wilson, who concluded that living agents

- have parts that are heterogeneous and specialized;
- include a variety of internal mechanisms;
- contain diverse organic molecules, including nucleic acids and proteins;

BOX 4.1 Life
What Is Life?

Attributes that pertain to a **definition of life** and that allow us to **distinguish living from nonliving** are summarized below. The evidence for these conclusions is laid out in Chapters 3 and 4.

1. Life is associated with three classes of molecules — DNA, RNA and proteins — but viruses contain either DNA and protein or RNA and protein; proteins can be synthesized *in vitro*; and amino acids are found in meteorites and so were present very early in the life of the universe.

2. If life is defined as the transfer of hereditable information from generation to generation then nucleic acids (DNA and RNA) are the essential molecules of life.

3. If life is defined as the organization of structures, function and metabolism based on carbon, then carbon is the element of life and living things.

4. Life can be defined as the ability to transform external sources of material and energy into such processes as metabolism, growth and development.

5. Cells are the units of life and life is based on membranous structures; viruses and prions are not alive.

6. Living things perform the reactions necessary for survival and replication.

7. Living things exist as individuals and collections of individuals (populations).

When Did Life Arise?

The answers listed below **depend on how life is defined.** Examination of the evidence for these conclusions may be found in Chapters 3 and 4.

1. Life can be considered to have arisen with the first molecules around 4.5 Bya, with the first cells arising 3.5 to 3.8 Bya, or with the first organisms.

2. Because molecules produced by organisms (biotic organic molecules) can be formed abiotically in the absence of life, and because they arose before the origin of organismal life, we recognize a continuum between the abiotic production of molecules and first life.

3. Because life requires cellular organization, life arose with the first protocells.

an Austrian Augustinian monk, Gregor Mendel (1822–1884) provided the essential basis for understanding the material basis of biological inheritance in his genetic experiments using pea plants (see Chapter 7). As discussed in Chapter 7, Mendel's experiments went unnoticed for four decades. Furthermore, a physical (cellular) basis for transmitting hereditary information from generation to generation was unknown. Chromosomes were not discovered until 1888, their constant number in the cells of an individual species not until 1910, and their function in heredity not until even later.

BOX 5.1
Fossils

An essential basis for understanding evolutionary relationships between organisms of the past, and for appreciating their lengthy history, was (and is) the study of their fossil remains.

During the sixteenth and seventeenth centuries **fossils** (stones as they were called) were regarded by many as images of God's creation, placed on Earth for man's admiration and use but formed naturally by God (FIGURE B5.1). Some wondered whether fossils indicated errors in the plan of nature, causing some species to become extinct? For them fossils were *lusi naturae* ("jokes of nature").

When Thomas Jefferson (1743–1826), agriculturalist, botanist, fossil hunter, and third President of the United States, discovered the extinct clawed giant sloth, *Megalonix jeffersoni* (FIGURE B5.2), he mistakenly thought it was a giant lion that perhaps still existed in the unexplored areas of North America. Like many others, Jefferson proposed that these species were not really extinct, only rare. "Such is the economy of nature, that no instance can be produced of her having permitted any one race of her animals to become extinct; of her having formed any link in her great works so weak to be broken."

Other hypotheses sought to explain fossils as caused by the Noachian flood described in Genesis or having purposely been implanted into Earth at the time of creation in order to test humanity's faith in religion. Contrary arguments proposed the reality of fossils and led to more naturalistic attempts to understand fossil origins. As discussed in Chapter 2, Cuvier saw that fossils from younger rocks were more like living organisms than were fossils from older rocks.

In Darwin's day, the fossil record was sporadic, the result of serendipitous collecting and random finds. It was known that fossils predominantly occurred in sedimentary rocks originally laid down by a succession of deposits in seas, lakes, riverbeds, deserts, and so on. Fossils were known to occur in some areas but not in others (FIGURE B5.3). Even in appropriate sedimentary environments, many dead organisms decompose before they fossilize or are destroyed by the erosion of sedimentary rocks even when they have fossilized.

Also, because the isolation of populations encourages and sustains their differences, we rarely find intermediate forms in the same place as the original forms; this was Jefferson's dilemma, discussed above. A complete evolutionary progression of fossils from most ancient to most recent has never been found in a single locality. Nevertheless, fossils provide the hard evidence for evolution, evidence used by Charles Darwin who first offered an acceptable explanation for historical changes among organisms and thereby helped tie all organisms together by a community of descent: evolution. *(continued)*

BOX 5.1
Fossils (*continued*)

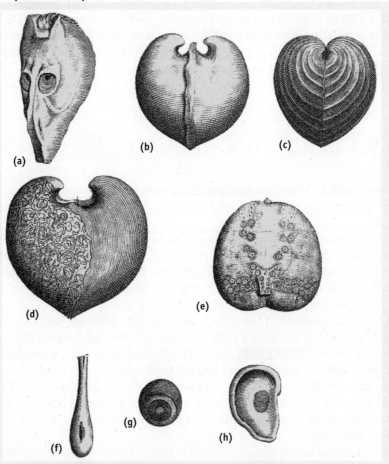

FIGURE B5.1 A portion of Plate 7 from *The Natural History of Oxfordshire* by Robert Plot (1676) illustrating naturally formed stones interpreted as representing parts of the human body, including: the heart (**a–d**), the holes in **a** being interpreted as the major artery taking blood away from the ventricles; the brain (**e**, showing the cerebellum and medulla oblongata); nerves (**f**, the olfactory nerve); the eye obscured by a cataract (**g**); and the external ear or pinna (**h**). (Modified from Plot, R., *The Natural History of Oxfordshire*, Plate 7, 1676.)

BOX 5.1
Fossils (*continued*)

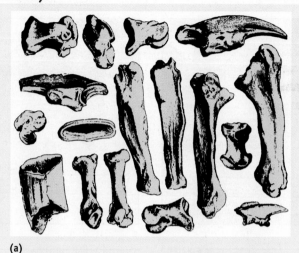

(a)

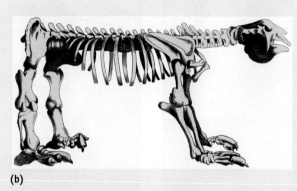

(b)

FIGURE B5.2 (a) Some of the bones of the extinct giant sloth, *Megalonix jeffersoni,* discovered in western Virginia in 1796. (b) Reconstruction by Cuvier of the skeleton of a similar extinct South American giant sloth, *Megatherium*. Both sloths were edentates, clawed mammals without cutting teeth. (Reproduced from Greene, J. C. *The Death of Adam.* Iowa State University Press, 1959. With permission from Wiley-Blackwell.)

(*continued*)

BOX 5.1

Fossils (*continued*)

FIGURE B5.3 The process of fossilization in which an organism (in this case, an animal) dies in a watery environment that protects it from scavengers. Reduced oxygen levels in deeper water further resist deterioration. The remains are gradually silted over (**3**) and eventually covered by successive layers of soil that compact into sedimentary rock (**5**). In time, because of erosion, the fossil surface may become exposed (**6**). (Reproduced from Kardong, K. V. *Vertebrates: Comparative Anatomy, Function, Evolution, Fourth edition*. McGraw-Hill, 2006. Used with permission of The McGraw-Hill Companies.)

Defining Species

The problem of how to recognize a species has a history that is thousands of years old. Despite devoting several decades of thought to the problem, Darwin never did resolve the problem of the nature of species. Why is that?

For almost two millennia the prevailing philosophy in Western Europe was Plato's **idealistic concept** that all natural phenomena are imperfect representations of the true essence of an ideal unseen world (BOX 5.2). Following Platonic ideas, Aristotle proclaimed that species were immutable and that there was a hierarchical order of species from most imperfect to most perfect, a concept refined over the centuries as the "Great Chain of Being" (BOX 5.3).

Such philosophies maintained that the world of essences is perfect and all change is illusory. This unchanging order remained unquestioned until inexplicable gaps in the chain of nature prompted philosophers such as Gottfried Leibniz (1646–1716) to propose that the universe was not perfect. To Leibniz, the evolution of species was part of the perfection toward which the universe continually progressed. His philosophy represented a major shift from a perfect universe to one in the process of becoming perfect.

Progress toward the perfection of species was expressed by natural historians such as Charles Bonnet (see Chapter 1) who maintained that the development of any organism from its seed was the unfolding of a preconceived plan inherent in the seeds of previous generations. The notion of progress therefore fitted into a teleological framework of necessity and direction toward some particular end (Box 5.3).

Species Concepts

Before Darwin, concepts of species reflected the philosophies outlined above and in Box 5.2. Today, we have three major species concepts. These are introduced below and discussed in more detail in Chapter 14.

- A morphological species (sometimes morphospecies) is based on uniting individuals who share more characters (features) with one another than they do with any other organism (see Chapter 16).
- A biological species is based on the inability of individuals in populations to interbreed with individuals in other populations and produce viable offspring. Because of its nature, the biological species concept cannot be used for **fossils** or for many unicellular organisms. Because of hybrid viability, the concept can only rarely be applied to plants or fungi.
- Evolutionary species are defined on the basis of their evolutionary isolation from each other, using morphological, genetic, molecular, behavioral and ecological features.

■ As noted in Chapter 1, the terms *feature* and *character* are used interchangeably for any distinguishable attribute of an organism, be it morphological (blue eyes), behavioral (burrows), functional (breathes oxygen) or molecular (has the gene for hemoglobin).

We could consider a fourth category, paleontological species, as a subset of morphological species. Because reproductive isolation cannot be determined for fossils, paleontological species are not biological species. In situations in which a set of paleontological species can be identified as representing a series of ancestral and descendant populations, paleontological species also may be recognized as evolutionary species (see Chapter 16).

BOX 5.2
Plato, Idealism, and the Concept of Species

Attempts to understand the world in a rational way, through thought and logic, began about the fifth century BC in Greece.

Plato (428–348 BC), the philosopher who along with Aristotle (384–322 BC) had the greatest impact on Western thought, suggested that our experience of the world is no more than a shadowy reflection of underlying "ideals" that are true and eternal for all time. Most things, according to Plato, were originally in the form of such eternal ideals, a philosophy we now know as **idealism.**

Plato and his successors assumed that only ideal generalizations are real; all else is merely a shadowy illusion. In Plato's famous parable in his dialogue, *The Republic,* humans deprived of philosophy are depicted as prisoners confined to a cave and facing a wall upon which are displayed their own shifting, distorted shadows as well as shadows of objects they cannot see directly. Their chains prevent them from turning their heads toward the light. As a result, the prisoners interpret the observed deformed, shadowy aberrations as reality but the actual unchanging humans and objects are the true "essences" or "forms."

The Platonic goal for human society was to use experience to understand and strive for ideal perfection. Plato's writing had a goal that was founded in his belief in the centrality of beauty, truth and justice, and the need to shape a society in which all could attain those goals. The concepts of perfect circles to explain the motions of the heavenly bodies, perfect numbers such as 6 (1 + 2 + 3) and 10 (1 + 2 + 3 + 4), and the four "elements" — earth, water, fire, and air — to which all matter could be reduced, were among aspects of this search for perfection.

Variations on this theme were common. To the four elements, Empedoceles (c. 490–430 BC) added two active principles: love, which binds elements together, and hate, which separates them. With respect to mystical numbers, Lorenz Oken (1779–1851), a German Natural Philosopher, proposed that the highest mathematical idea is zero, and God, or the "primal idea," is therefore zero.

To a large extent, idealism originates from the practice of abstracting concepts from experience; to think of "cat" in a way that includes all cats, not just one particular animal. Abstraction allows us to generalize our experience, to differentiate between cat and tiger, to pet the cat and run from the tiger, and to communicate these general concepts or universals to others through symbolic language.

Despite such abstract reasoning, however, generalizations are not always reliable. Experiences can modify generalizations — not all cats or tigers are the same. The dilemma for **natural scientists** has always been to recognize the reality of differences among members of a group and yet to recognize the reality of the group itself. Idealism offered practically no means of reconciling these two aspects of reality. No sooner do we conceive of some new generality than we discover further instances that may force us to modify our original concept.

To Plato, the form of a structure could be understood from its function because function dictated form; the form of the universe derives from its function of goodness and harmony imposed by an external creator. Aristotle, whom many regard as the founder of biology (among other sciences), modified this notion to accommodate the embryonic development of organisms, pointing out that the last stage of development — the adult form — explains the changes that occur in the immature forms. This type of explanation is **teleological** (goal-oriented), where the adult represents the "*telos,*" or final goal, of the embryo.

■ The terms *natural scientist* and *natural historian* are used when referring to individuals working before the twentieth century, the term *scientist* having been introduced in 1830.

■ *Embryology* was the term used for the study of embryonic development until the mid-twentieth century. In the early twentieth century, embryology was divided into *descriptive* and *experimental* and often taught as such in separate classes. In the mid 1950s, embryology was renamed *developmental biology.*

Charles Darwin

Many biographies have been written of Charles Darwin (FIGURE 5.1), a man who defied his social and religious background by espousing a radical concept and by becoming the instrument that brought it to the world. As described by the Darwin biographers Adrian Desmond and James Moore (1991), the enigma was

> How could an ambitious thirty-year-old gentleman open a secret notebook [in 1837] and, with a devil-may-care sweep, suggest that headless hermaphrodite mollusks were the ancestors of mankind? A squire's son, moreover, Cambridge-trained and once destined for the cloth. A man whose whole family hated the "fierce & licentious" radical hooligans.

■ This is an interesting comment; some scholars consider that Erasmus Darwin (Charles' grandfather) was a bit of a fierce and licentious radical.

Darwin's life is the history of a curious, driven and intellectually creative man who was courageous yet fearful, living in a society undergoing considerable change.

Darwin was born to an affluent, upper-middle-class British family whose fortunes derived largely from his father, Robert Darwin (1766–1848), and paternal grandfather, Erasmus Darwin (1731–1802), both of whom were prosperous physicians. Erasmus was an early thinker on and writer about evolution. Although his contemporaries judged him wildly speculative in science, a contemporary of Charles Darwin's, Herbert Spencer, who coined the phrase "struggle for existence" (see Chapter 6) was much influenced by the views of Erasmus Darwin. When searching for evolutionary explanations, Charles was aware of his grandfather's interest in evolution, although some historians believe that earlier evolutionary ideas, no matter what the source, had little effect on Charles' early development.

At age 16, Charles left grammar school in Shrewsbury for Edinburgh University to study medicine. Surgical procedures at that time were dreadfully brutal. Having witnessed two operations — one of them on a child — Darwin realized that he could never become a surgeon. He transferred to Cambridge University with the intention of becoming a minister of the Church of England. However, his interests were neither in academic nor clerical pursuits but in hunting, collecting, natural history, botany and geology. Darwin despised formal classical education and was no more than a mediocre student. His father, who believed Charles had betrayed the family-trust of industrious professionalism, castigated him: "You care for nothing but shooting, dogs, and rat-catching, and you will be a disgrace to yourself and all your family."

In 1831, through the recommendation of John Henslow (1796–1861), Regius Professor of Botany at Cambridge University, and the intercession of an uncle, Josiah Wedgwood (1769–1843), Darwin was able to put off further study for the ministry and set off on his now famous voyage around the world on *H.M.S. Beagle* (FIGURE 5.2). *H.M.S. Beagle* was a 10-gun brigantine, 27-meters long, weighing 240 tons. In 1831 (the year Darwin began his voyage), it had been refitted for circumnavigation in order to fix world longitudinal markings and to chart the coast of South America.

■ A brigantine or brig is a sailing ship with two masts and square sails (square-rigged) (Figure 5.2).

Darwin's post was *not* that of naturalist, a special unpaid position created by the British Admiralty on naval ships making broad geographical surveys. Darwin's primary role on the voyage was to serve as a dining companion to the Captain, Robert FitzRoy, to whose service his duties as naturalist were secondary. For this reason, it was of no importance to the British Admiralty that Darwin had not distinguished himself academically or even finished his studies at Cambridge. His qualifications as a gentleman, good shot, and sportsman were quite sufficient. There was a Navy-appointed naturalist on board, the ship's surgeon Robert McCormick (1800–1890).

■ Robert FitzRoy (1805–1865), a career naval officer, rose to become Vice-Admiral. An extraordinarily able navigator, Fitzroy pioneered meteorological forecasting and the development of barometers to measure atmospheric pressure. He was Governor of New Zealand for two years.

BOX 5.3
The Great Chain of Being

The idealistic concept of a species became entrenched once species were used to explain the divine origin and design of nature. Plato defined the species as representing the initial mold for all later replicates of that species, "The Deity wishing to make this world like the fairest and most perfect of intelligible beings, framed one visible living being containing within itself all other living beings of like nature." Aristotle expanded this view to a chain-like series of forms — the **Scale of Nature** — each form representing a link in the progression from least perfect to most perfect (see Figure 6.1). This concept continued long into the history of European thought, merging with other ideas into the **Ladder of Nature** and the **Great Chain of Being**, forerunners of our search for the **Tree of Life** or **The Universal Tree of Life** (UToL; see Chapter 8).

Philosophically satisfying as it was, the concept of the Great Chain of Being did not necessarily put humans on the highest rung of the Ladder of Nature. Many who contemplated the innumerable steps between humans and perfection (God) felt the despair of occupying a relatively lowly position and only consoled themselves with the thought that there were even more lowly organisms. Nevertheless, the Great Chain of Being was accepted well into the 18th century.

According to Johann von Goethe (1749–1832), the polymath who coined the term morphology, the origination of each level of organisms was based on a fundamental plan — an archetype or *Bauplan* (pl. *Baupläne*). Goethe conceived the morphology of plants, for example, as founded on an *Urpflanze* (ancestral plant) that had only one main organ, the leaf, from which the stem, root, and flowers derived as variations (FIGURE B5.4a). Similarly, the bones of the skull were modifications of the vertebrae of an Urskeleton, an animal archetype, composed of vertebrae (Figure B5.4b).

To its exponents, the Ladder of Nature stressed a precisely ordered regularity of relationships among organisms, relationships that could be used to support and justify prevailing social and political orders. Among the relatively few who disputed this concept was François-Marie Arouet (1694–1778) better known by his pen name, Voltaire, who saw that the many gaps between species were not in accord with the expected innumerable steps in the continuous progression from imperfect to perfect. Although there were no living species to fill these gaps, Voltaire proposed that such gaps were real, perhaps caused by the extinction of species.

The Great Chain of Being had important effects on plant and animal classification, in part from the search for the multitude of organisms that many assumed occupied all the rungs of the Ladder of Nature. Despite its idealistic nature, the Great Chain of Being led almost directly to the idea that the perfection of organisms may demand multiple intermediate stages. By the early 19th century the basic concept of evolution — the transformation of one species into another — required only the philosophical acceptance of actual change between the innumerable steps in the Great Chain of Being.

■ Underappreciated by many is that classification and an appreciation of the fundamental branching nature of evolution underpinned Darwin's theory of evolution (see Winsor, 2009).

The *Beagle* voyage lasted almost five years (27 December, 1831 to 2 October, 1836), during which time Darwin transformed himself from a casual amateur to a dedicated geologist and naturalist. His letters to John Henslow recounting many of the observations made during the voyage, along with his collections of plants, animals, fossils, and minerals, excited considerable scientific interest, even before the *Beagle* returned to England. Darwin's account of the voyage was published as a *Journal of Researches into the Natural History and Geology of the Countries Visited During the Voyage of H.M.S. Beagle Round the World, Under the Command of Capt. FitzRoy.* An abridged

BOX 5.3
The Great Chain of Being (*continued*)

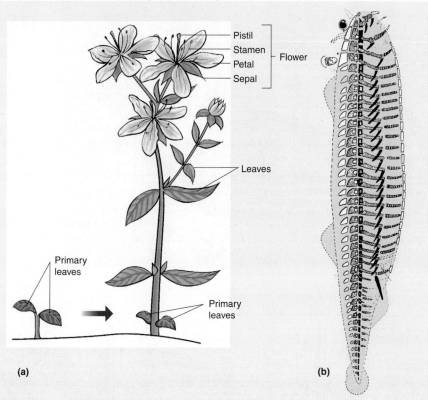

FIGURE B5.4 Archetypes of plants and vertebrates. The idealized plant (**a**) shows Goethe's concept of the derivation of all plant parts from the leaf. The segments in the vertebrate skeleton pictured by Owen (**b**) are alike from head to tail. (**a** Adapted from Wardlaw, C. W. *Organization and Evolution in Plants.* Longmans, 1965. **b** Modified from Owen, R. *On the Archetype and Homologies of the Vertebrate Skeleton.* Voorst, 1848.)

(a)

(b)

(c)

FIGURE 5.1 Portraits of Charles Darwin at 8 (**a**), 40 (**b**), and in later life (**c**).

(a)

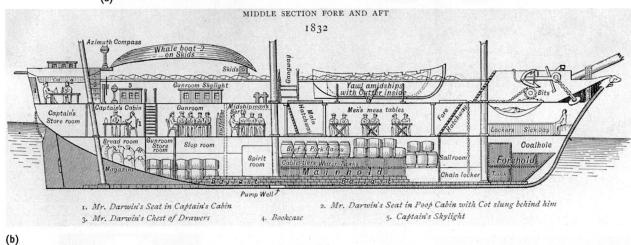

MIDDLE SECTION FORE AND AFT

1832

1. *Mr. Darwin's Seat in Captain's Cabin* 2. *Mr. Darwin's Seat in Poop Cabin with Cot slung behind him*
3. *Mr. Darwin's Chest of Drawers* 4. *Bookcase* 5. *Captain's Skylight*

(b)

FIGURE 5.2 **(a)** *H.M.S.* (His Majesty's Ship) *Beagle* in the Strait of Magellan at the southern tip of South America. **(b)** Side elevation of the *Beagle,* based on a drawing by one of Darwin's shipmates showing the general plan of the ship and the cramped quarters occupied by the crew of about 70.

popular version, *Voyage of the Beagle,* remains one of the most perceptive chronicles of exploration published in the 19th century.

This voyage of the *Beagle* was no cruise and certainly no picnic. In an account of the voyage published in 1995, Keith Thomson noted, "To say that the *Beagle* was extremely cramped, even given the expectations of the time, would be a supreme understatement. The ship was, after all, no longer than the distance between two bases on a baseball field," that distance being 27.4 m. Darwin slept in the poop cabin at the stern of the ship, which he shared with two officers. This cabin also held a 3 × 1.8 m chart table and various chart lockers, as well as drawers for his equipment and specimens. Darwin wrote, "I have just room to turn around and that is all."

On his return to England, a substantial income and inheritance enabled Darwin to forgo financial pursuits and dedicate himself entirely to biology. He married his

cousin, Emma Wedgwood (1808–1896), and in 1842 settled near the town of Downe in Kent, 23 km from London, where he and Emma began to raise a large family.

For 40 years to the time of his death in 1882, Darwin lived at home, mostly as a semi-invalid subject to heart palpitations, rashes and gastric discomfort. The cause(s) of his disability (or disabilities) is not known. Hypotheses range from parasitic infection — trypanosomes that cause trypanosomiasis (Chagas disease), which affects heart and intestines — to heavy metal (arsenic) poisoning from some of the so-called cures of his time, to psychosomatic illness involving severe symptoms of panic disorder. Despite his physical discomforts, Darwin displayed a warm personality to his family and was much comforted by the sympathetic concern of his wife. It has often been said that Darwin was the perfect patient and Emma the perfect nurse.

Whatever the cause, his illness/es isolated him — or was/were used by Darwin to isolate himself — from most of the world about him. He maintained contacts through letters and publications, through which he was extraordinarily well connected socially and academically, corresponding extensively with naturalists, scholars and breeders. This exchange of ideas was enormously important to his work. As Janet Browne (2006) argues, the relatively advanced development of the British postal system, which was founded in 1660, was a critical background element to Darwin's success.

■ Darwin spent £20 (£1,000 [US$1,890] today) on stationery, stamps and newspapers in 1851, and £53 (£2,650 [US$5,000]) in 1877, when a postage stamp cost a penny.

■ Voyage of the *Beagle*

The five-year voyage on the *Beagle* (FIGURE 5.3) enabled Darwin to observe and think about a wide range of organisms and geological formations. He collected birds, insects, spiders and plants in the Brazilian tropical forests. At Punta Alta on the coast of Argentina, Darwin unearthed fossil bones of a six-meter-high giant sloth, *Megatherium* (see Figure B5.2), discovered the hippopotamus-like ungulate, *Toxodon*; a giant armadillo, *Glyptodon* (see Figure B18.2), and other animals resembling extant species, yet recognizably different. The primitiveness and wildness of the Tierra del Fuego Indians at the southern tip of South America impressed Darwin, as did the severity of their struggle for subsistence in a meager and unrelenting environment.

During the voyage, Darwin carried with him the first — and at the time — the only published volume of Charles Lyell's *Principles of Geology*, a gift from the *Beagle*'s captain Robert FitzRoy. Darwin assiduously noted the geological features of many terrains he covered. To explain some of the geological uplifting processes that shaped the South American landscape, Darwin gathered evidence showing the distribution of marine shells at various locations above sea level (recall the frontispiece to Lyell's book reproduced in Figure 2.6) and the terracing of land by erosion as the land was elevated. While Darwin was ashore near the Bay of Concepción on the coast of Chile, a severe earthquake raised the level of the land in some places from 1 m to as much as 3 m above sea level. This experience had a deep effect on Darwin, although it did not shake his belief in uniformitarianism or gradualism.

■ Robert FitzRoy arranged, at his own expense, for three Tierra del Fuegian Indians to be taken back to England.

■ Although initially known as geology and an ancient science most university departments of geology have been renamed departments of earth sciences to reflect the multi- and interdisciplinary approaches brought to bear on the study of Earth.

> A bad earthquake at once destroys our associations: the earth, the emblem of solidity, has moved beneath our feet like a thin crust over a fluid; one second of time has created in the mind a strange idea of insecurity, which hours of reflection would not have produced.

An experience that years later had great impact on Darwin's thinking about evolution was the month he spent in the bleak, lava-covered Galapagos Islands off the coast of Ecuador. Here, 800 km from the mainland was an exotic collection of

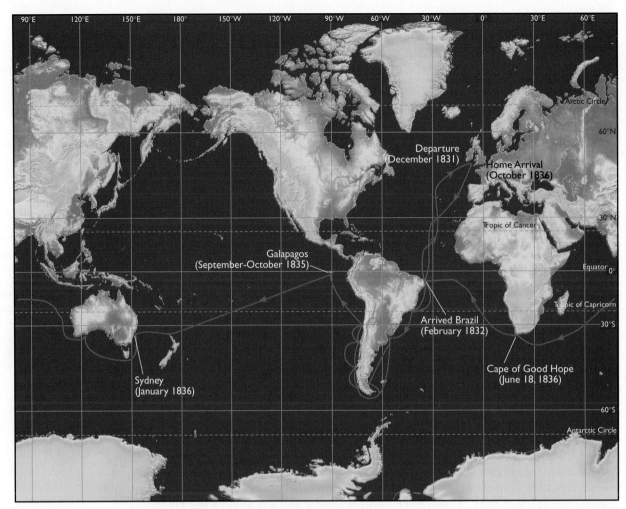

FIGURE 5.3 Route of the five-year voyage of the *Beagle,* beginning at Plymouth, England, in December 1831, and ending in Falmouth, England, in October 1836. Almost four years were spent in South America, including one month (September–October 1835) among the Galapagos Islands.

organisms: giant tortoises, meter-long marine and land iguanas—large tropical lizards with a characteristic spiny crest along the back (**FIGURE 5.4** and chapter opening photo)—as well as many unusual plants, insects and seashells.

As he had noted on the mainland, different geographical localities on the islands with environmentally similar habitats were not always occupied by similar species. Darwin was particularly struck by the situation in the Galapagos where different species of tortoises were found on each island. Insect-eating warblers and woodpeckers were absent; various species of finches, usually seed eating, assuming the insect-eating mode of the missing species (**FIGURE 5.5**). Darwin's observation that each island appeared to have its own constellation of species raised the important question: What could account for this distribution of organisms? In Darwin's words,

> It is the circumstance that several of the islands possess their own species of the tortoise, mocking-thrush, finches, and numerous plants, these species having the same general habits, occupying analogous situations, and obviously filling the same place in the natural economy of this archipelago, that strikes me with wonder.

FIGURE 5.4 A marine iguana from the Galapagos Islands.

Did separate and different creations make one species in one place slightly different from another species in another place? Why, and more importantly, how?

Darwin's account of his voyage on the *Beagle* was published in 1838 and revised some years later. The ornithologist David Lack and various Darwin historians have pointed out that although Darwin's *Journal of Researches* expresses evolutionary forethoughts, the significance of his observations on the Galapagos Islands and elsewhere did not become apparent to him until after his return to England. This was especially true for the various Galapagos finches, which were first classified in England by John Gould (1804–1881), a British ornithologist who pointed out to Darwin that these were related species of finches and species from different islands had different morphologies (BOX 5.4). Darwin had not kept notes as to which island was home to which species, and so did not immediately see the significance of geographical isolation.

The *Beagle* voyage did stir Darwin. In 1837 he began his first notebook on the transmutation of species. Darwin came to the view that only changes among species could explain the observation that present species resemble past species and that different species can share similar structures. "The only cause of similarity in individuals we know of is relationship." The differences between the flora and fauna of different geographical areas, he thought, must have arisen because not all plants or animals are universally distributed. For the Galapagos Islands, for example, Darwin raised the question

Why on these small points of land, which within a late geological period must have been covered with ocean, which are formed of basaltic lava, and therefore differ in geological character from the American continent, and which are placed under a peculiar climate — why were their aboriginal inhabitants . . . created on American types of organization?

Darwin realized that islands such as the Galapagos contained only those organisms able to reach them, and that evolution could transform only those species available. "Seeing this gradation and diversity of structure in one small, intimately related group

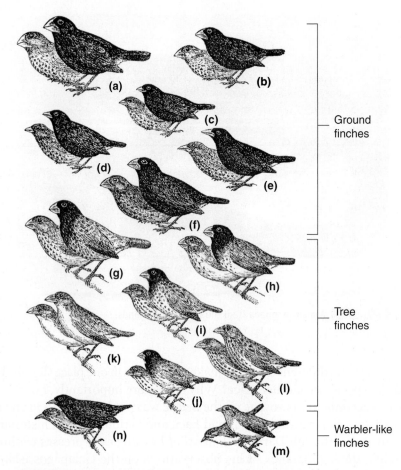

Ground
finches

Tree
finches

Warbler-like
finches

FIGURE 5.5 Species of finches (male on left, female on right; about 20 percent
of actual size) observed by Darwin on the Galapagos Islands. (**a–f**) Species in the
genus *Geospiza*. (**g–l**) Species in the genus *Camarhynchus*. (**m**) *Certhidea olivacea,*
the warbler finch. (**n**) *Pinaroloxias inornata,* the Cocos finch. Evolutionary relation-
ships among these finches are shown in Figure B6.3. (Reproduced from Lack, D.
Darwin's Finches: An Essay on the General Biological Theory of Evolution. Cambridge
University Press, 1947. Illustration by Lt. Col. William Percival Cosnahan Tenison.)

of birds, one might really fancy that from an original paucity of birds in this archipelago
one species had been taken and modified for different ends." Uncovering the mechanism
for the transformation of species, however, was nowhere near as obvious as was the
realization that such transformation had occurred.

Why do species change? In seeking an answer over the next 20 years, Darwin
explored a variety of theories. One of the most persistent, a theory that later had many
adherents in France and the United States, was most fully developed by Jean-Baptiste
Lamarck (see Chapter 2). We know it as *Lamarckian inheritance.*

Lamarckian Inheritance

As discussed above and in Chapter 2, Charles Darwin's grandfather, Erasmus, and
Lamarck were among the first to actively advocate evolution. Where others had looked
at fossils and saw species extinction, Lamarck made the intellectual leap of proposing
the continuity of species by gradual modification through time.

BOX 5.4
Darwin's Finches

The young Charles Darwin was profoundly impressed with the diversity of wildlife he saw during his brief visit to the Galapagos Islands. Whether his observations on these remote islands provided seeds for his later insights into the mechanisms of evolution or not, Darwin drew heavily on them for examples throughout *The Origin of Species*. Prominent among these were the birds he brought back from the *Beagle* expedition, analysis of which was actually performed by John Gould, a prominent English ornithologist.

Darwin assigned the birds he collected to different species. Darwin had not recorded the island of provenance for the birds he collected on the Galapagos. Darwin's manservant and shooter, Syms Covington (c. 1816–1861), however, collected his own specimens, carefully recording the Island from which each specimen had been obtained. Gould pointed out to Darwin that the diverse assemblage of birds Darwin had mistakenly assigned to different species was in fact a remarkably varied group of finches. By examining the birds collected by Covington, Gould showed that these varied finch species, which varied dramatically in body size and beak shape (FIGURE B5.5), came from different islands.

Once Darwin was made aware of this important observation, he began to realize that a homogeneous population of finches must have colonized the islands after their volcanic origin from the ocean floor and then evolved into different forms on different islands. In this way, the Galapagos finches and other varied groups that he encountered on these islands influenced Darwin's thinking on how geographic isolation relates to the formation of new species. (Modern research into Darwin's finches, research initiated by Peter and Rosemary Grant 40 years ago, is discussed in Box 14.2.)

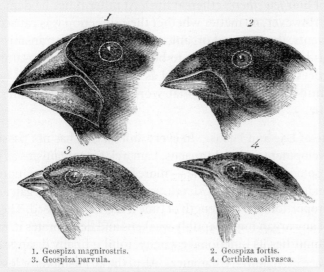

1. Geospiza magnirostris.
2. Geospiza fortis.
3. Geospiza parvula.
4. Certhidea olivasea.

FIGURE B5.5 Drawings by Charles Darwin of four species of Darwin's finches showing variation in size and beak morphology.

By the early 19th century, most naturalists accepted the inheritance of acquired characters, the utility (adaptedness) of features, and the concept of some internal force toward change. To explain how modifications occurred and the exquisite relationships (adaptations) through which organisms exploited their environments, Lamarck is credited with proposing that

- variations among organisms originate because of the organisms' response to the needs of the environment; and
- this ability to respond in a particular direction accounts for the adaptation of new features and their passage to subsequent generations

For example, Lamarck suggested that the long legs of water birds such as heron and egret arose through the following mechanism.

> We find . . . that the bird of the waterside which does not like swimming and yet is in need of going to the water's edge to secure its prey, is continually liable to sink in the mud. Now this bird tries to act in such a way that its body should not be immersed in the liquid, and hence makes its best efforts to stretch and lengthen its legs. The long-established habit acquired by this bird and all its race of continually stretching and lengthening its legs, results in the individuals of this race becoming raised as though on stilts, and gradually obtaining long bare legs, denuded of feathers up to the thighs and often higher still.

A central feature of Lamarck's hypothesis was that, when confronted with new environments, organisms moving toward improvement learn new habits and can change appropriately by exercising an unknown, inner perfecting principle. This special power could sense the needs of the environment and respond by developing new traits in appropriate adaptive directions, mostly from simple to complex. The source for such directional orientation was not clear to Lamarck, who had no mechanism comparable to natural selection that could lead to progressive improvement.

At times Lamarck ascribed his belief in evolutionary "progress" to an inner vitalistic property of life (*feu éthéré,* ethereal fire). At other times he denied such supernatural causes. However, no matter whether their direction was caused by natural or supernatural events, the origin of organic changes and their transmission to further generations was believed by Lamarck to be aided by two universal mechanisms in two natural processes, which he codified as two **Laws of Nature**. Both laws were incorporated into *Zoonomia,* a popular work by Charles Darwin's grandfather, Erasmus Darwin.

1. Principle of Use and Disuse. In every animal that has not passed the limit of its development — like Linnaeus, Lamarck saw organisms as progressing through some internal principle — more frequent and continuous use of any organ gradually strengthens, develops and enlarges that organ and gives it a power proportional to the length of time it has been so used. The permanent disuse of any organ imperceptibly weakens and deteriorates it, and progressively diminishes its functional capacity, until it finally disappears.

2. Inheritance of Acquired Characters. All of the acquisitions or losses wrought on individuals through the influence of the environment, and hence through the influence of the predominant use or permanent disuse of any organ, are passed on by reproduction to the new individuals that arise, provided that the acquired modifications are common to both sexes, or at least to the individuals producing the young.

Some would list the **drive toward perfection** as a third process central to Lamarck's thinking.

According to Lamarck, what is called a species is merely a continuum between organisms at different stages in the process of change. Fossil species, he maintained,

■ Although abandoned by practically all biologists in modern countries, a form of Lamarckism was adopted by the Soviet Union as official policy between 1948 and 1963 as a result of experiments by the Russian agronomist T. D. Lysenko and his supporters, who claimed to have shown inheritance of environmental adaptations in crops.

are not truly extinct but had been modified in time and thereby become later, more complex organisms.

Georges Cuvier (see Chapter 2) marshaled what at the time seemed to be the most telling argument against Lamarck's evolutionary proposals: no intermediate forms bridging the gaps between different species were known, either alive or as fossils. When a species hybrid such as the mule occasionally occurred, it was always sterile. The Lamarckian concept that organisms strive for perfection seemed ludicrous: How could new habits of swimming or flying produce organs enabling such habits without these organs already being present? Furthermore, Cuvier argued, despite 4,000 years of recorded history, no new species had evolved. Cats and wheat found in the earliest Egyptian tombs resembled present-day cats and present-day wheat. Why assume they can transform? Like begets like.

Although later proved incorrect, the processes Lamarck espoused — the effects of use and disuse and the inheritance of acquired characters — had the important advantage that they were uniformitarian in principle and did not rely on supernatural or catastrophic events. By proposing a **materialistic explanation for evolution** (inheritance of acquired characters), an explanation that reflected the **close interactions known to occur between organisms and their environment,** Lamarck fostered a climate of opinion in which evolution could be understood in the same fashion as any other natural event.

The organisms and natural events he had seen on the voyage of the *Beagle* set Darwin thinking about transformation. The result, his theory of evolution by natural selection, is discussed in the next chapter, as is a very similar theory proposed independently by another naturalist on the basis of his travels in the tropics.

Recommended Reading of Books by and on Darwin

Appleman, P. (ed.), 2001. *A Norton Critical Edition. Darwin: Texts, Commentary,* 3d ed. W. W. Norton, New York.

Browne, E. J., 1995. *Charles Darwin: Voyaging.* Alfred A. Knopf, New York.

Browne, E. J., 2002. *Charles Darwin: The Power of Place. Volume II of a Biography.* Alfred A. Knopf, New York.

Darwin, C., 1845. *The Voyage of the Beagle* (originally published as *Journal of Researches,* it has now appeared in numerous editions).

Darwin, C., 1859. *On the Origin of Species by Means of Natural Selection or the Preservation of Favoured Races in the Struggle for Life.* Murray, London.

Desmond, A., and J. Moore, 1991. *Darwin.* Warner Books, New York.

Fortey, R., 2002. Fossils: The Key to the Past. Natural History Museum, London.

Rudwick, M. J. S., 2005. *Bursting the Limits of Time. The Reconstruction of Geohistory in the Age of Revolution.* The University of Chicago Press, Chicago.

Stewart, P. D., 2007. *Galápagos; The Islands That Changed the World.* Yale University Press, New Haven, CN.

Thomson, K. S., 1995. *HMS Beagle. The Story of Darwin's Ship.* W. W. Norton, New York.

Winsor, M. P., 2009. Taxonomy was the foundation of Darwin's evolution. *Taxon* **58,** 43–49.

6

Darwin and Wallace's Evolution by Natural Selection

KEY CONCEPTS

- Charles Darwin proposed a theory of evolution by natural selection.
- Alfred Russel Wallace proposed a very similar theory. The two theories were communicated to a meeting of the Linnaean Society on 1 July 1858.
- Ideas of artificial selection were in the air centuries before Darwin's time.
- Darwin made enormous use of artificial selection in his theory.
- Darwin was the first to see the significance of small, continuous variations in Nature.
- Darwin grappled with how selection could get started, invoking the concept of preadaptation.
- The writings of the Rev. Thomas Malthus showed Darwin and Wallace (independently) how limited resources could lead to competition for these resources.
- Competition would result in some individuals leaving more offspring than other individuals.
- Slowly, natural selection would increase the numbers of individuals best adapted to the resources and to the environment: survival of the fittest.

Above: The common red rock crab, *Grapsus grapsus*, from the Galapagos Islands.

- Darwin saw that such slow change could lead to the origin of new species.
- Darwin placed his emphasis on mechanisms other than geographical isolation (season, habitat, behavior, sexual selection) as necessary for speciation.
- Darwin lacked a mechanism by which changes could be inherited.
- Publication of *On the Origin of Species* in 1859 revolutionized understanding of the relationships between organisms and how organisms arise and change.
- *On the Origin of Species* initiated widespread public discussion concerning science and the place and role of science in society.

Overview

By extrapolating from the artificial selection carried out by animal breeders, Darwin proposed a theory of evolution by natural selection, a theory he held along with Lamarck's mechanism of use and disuse. Darwin saw that an excess of offspring would set up competition among those offspring for limited resources; competition leading to differential survival and reproduction provided Darwin with a process for changing the composition of a population.

Organisms with traits better suiting them to their environment tend to leave more offspring than organisms less suited to the environment. Consequently, the traits of the former are passed on in higher proportions to future generations than are the traits of the latter. In this way populations adjust to changing environment. Populations unable to meet changing environments decline and may become extinct. Reproductive success is the ultimate judge of survival. Darwin recognized that humans, in seeking to perpetuate desirable traits in domestic plants and animals, could select (over a relatively short period of evolutionary time) radical alterations, albeit artificially rather than naturally. The enormous amount of data and examples Darwin analyzed provided the evidence to support his theory for evolution.

By coincidence, and in no small part because of exposure to the diversity of organisms in the tropics, the naturalist Alfred Russel Wallace independently proposed a similar evolutionary mechanism. Two communications were read (in their absence) to the Linnaean Society in London on July 1, 1858, and Darwin's *On the Origin of Species by Means of Natural Selection* was published a year later. Darwin had not previously published a single word on evolution. In 1855, on the other hand, Wallace had published a paper on the origin of species in relation to their geographical distribution, a paper of which Darwin was aware. Darwin reacted to the possibility that he might be scooped by Wallace by beginning to write what was to be his big, multivolume treatise on evolution. Following the receipt of Wallace's paper in June 1858, Darwin worked feverously to produce the single volume, *On the Origin of Species*.

Natural Selection

If we accept the basic idea that organisms can change through time and that use and disuse affects organisms during their lifetimes but discard the Lamarckian explanations set out in Chapter 5 for how they change, the essential question becomes, "How do organisms

change?" Darwin elaborated an evolutionary mechanism — **natural selection** — but was not the first to think about selection in relation to living organisms.

Empedocles (c. 490–430 BC) suggested that life initially appeared in the form of parts and organs floating freely and combining together to form whole organisms. Those organisms that adapted to "some purpose" survived, and those that did not, "perish and still perish." Aristotle disagreed with Empedocles' concept of randomness and selection with an argument often used since. According to Aristotle, the **Scale of Nature**, like any other goal-directed process, arises through a fixed progression of steps from lowest to highest stages (FIGURE 6.1). There cannot, therefore, be anything arbitrary or random in this progression that would require selection.

In the 18th century, Buffon saw natural selection — and artificial selection by humans of domesticated animals and plants — as the agent responsible for the extinction of species. He did not, however, see natural selection as responsible for the generation of new species. Buffon believed that new species arose by spontaneous generation (BOX 6.1) and that differences in the conditions under which spontaneous generation occurred caused the differences between species.

Lamarck was a materialist, perhaps one of the last 18th century naturalists to seek systematic, comprehensive and materialistic explanations for natural phenomena, although he never completely abandoned the notion of a mystical drive, vital force or *feu éthéré*. In the Lamarckian view, environmental effects initiated variations that occurred only in an adaptive direction (see Chapter 5). There could be no extinction of "imperfect" or "defective"

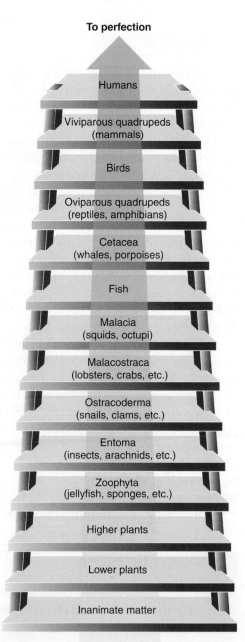

To perfection

Humans

Viviparous quadrupeds (mammals)

Birds

Oviparous quadrupeds (reptiles, amphibians)

Cetacea (whales, porpoises)

Fish

Malacia (squids, octupi)

Malacostraca (lobsters, crabs, etc.)

Ostracoderma (snails, clams, etc.)

Entoma (insects, arachnids, etc.)

Zoophyta (jellyfish, sponges, etc.)

Higher plants

Lower plants

Inanimate matter

FIGURE 6.1 Aristotle (above) and his Scale of Nature (adapted from descriptions in Guyénot, E., 1941. *Les Sciences de la Vie: L'Idee d'Evolution*. Albin Michel, Paris.)

BOX 6.1

Spontaneous Generation

An ancient concept is that life can arise spontaneously, for example, insects from sweat, crocodiles from mud, or insect larvae from inside sealed jars. Until perhaps the middle of the nineteenth century, the common belief was that although most large organisms reproduce by sexual means, smaller organisms could arise spontaneously from mud or organic matter. Some folklore suggested that larger dead organisms decomposed into smaller ones. There were even legends that magical transitions could change an individual of one species into another, a human into a werewolf, for example.

About 300 years ago the physician and chemist Johann van Helmont (1577–1644) offered a classic expression of spontaneous generation.

> If you press a piece of underwear soiled with sweat together with some wheat in an open mouth jar, after about 21 days the odor changes and the ferment, coming out of the underwear and penetrating through the husks of wheat, changes the wheat into mice. But what is more remarkable is that mice of both sexes emerge, and these mice successfully reproduce with mice born naturally from parents. . . . But what is even more remarkable is that the mice which come out of the wheat and underwear are not small mice, not even miniature adults or aborted mice, but adult mice emerge!

Two serious and somewhat contradictory obstacles to the development of evolutionary concepts therefore prevailed almost simultaneously. The Linnaean concept of species constancy (see Chapter 5) helped raise the question of the origin of species, but by insisting on species fixity, prevented consideration of any evolutionary transformations. Belief in spontaneous generation, in contrast, seemed contrary to species fixity, but at the same time cast doubt on any permanent continuity between organisms. If species could arise *de novo* at any time or be changed into another species, could there ever be a rational mechanism to explain their origin or the sequence of their appearance?

By the late seventeenth century a number of experimentalists showed that, at least for insects, spontaneous generation did not occur.

In 1668 it was demonstrated that maggots (larvae) arise only from eggs laid by flies, and flies arise only from maggots. If meat is protected so that adult flies cannot lay their eggs, maggots and flies are not produced. A year later, it was shown that the insect larvae found in plant galls arise from eggs laid by adult insects. Within a century, further experiments demonstrated that even the appearance of the microscopic "beasties" in decaying or fermenting solutions and broth observed by Anton van Leeuwenhoek (1632–1723) could be explained as originating from previously existing particles. The Abbé Spallanzani (1729–1799) heated various types of broth in sealed containers and observed no appearance of tiny organisms. Only when the containers were open to airborne particles did organisms appear.

The French chemist and microbiologist Louis Pasteur (1822–1895), after whom pasteurization is named, and others, put spontaneous generation as a mechanism for the origin of life to rest in the 1860s. Pasteur's evidence came incidentally from experiments he was undertaking to understand the fermentation process used in making beer and wine. Nothing grew when Pasteur sealed broths of beer in airtight glass vessels.

species because organisms could always adapt themselves to changing environments by inheriting acquired characteristics. To Lamarck, variations were not separate from evolution, and therefore could not be random. Thus, selection was not needed for adaptive traits to appear.

In the early 19th century, a number of natural historians, including the physiologist William Wells (1757–1817) and the Scottish horticulturalist and self-proclaimed

■ Matthew, who wrote on the properties of timber used in shipbuilding, claimed he had introduced the concept of natural selection in 1831.

evolutionary theorist, Patrick Matthew (1790–1874), separated the origin of variations from the forces responsible for preserving them and used the principle of natural selection to explain changes within species. Because insufficient evidence was provided, their ideas seemed highly speculative when published. Furthermore, their papers were published in obscure publications that did not come to the attention of later workers.

Artificial and Natural Selection

To support his theory of evolutionary change, Darwin discussed a number of examples of evolution by selection, although the selection involved human choice rather than natural events.

In artificial selection, the breeder — whether of dogs, cats, pigeons, cattle, horses, peas or wheat — selects the parents deemed desirable for each generation and culls (destroys or fails to breed) the undesirable types. Because the selected parents may produce a variety of offspring, the breeder can usually continue to select in a particular direction until the traits in which he or she is interested are consistently present (FIGURE 6.2). For example, for thousands of years humans have selected dogs, all of which are variants within the single species *Canis lupus*. Dogs now range in size from the St. Bernard to the Chihuahua, and in body forms from greyhound to bulldog (FIGURE 6.3). (Dogs may respond rapidly to selection because individual features are controlled by few genes; body size in small breeds has been shown by Sutter and colleagues to be controlled by a single gene, while coat color in 80 breeds has been shown by Cadieu and colleagues to be controlled by three genes.) Fanciers have long bred pigeons, which now show a wide variety of beak and body shapes, and feather patterns (Figure 6.2). The same is true for sheep, cattle, and all the many agricultural species of plants and animals, including the hybrid corn and hybrid wheat many of us consume for breakfast.

One of the main reasons for the effectiveness of Darwin's book is that it is written in a style and uses language and concepts that were familiar to its readers. Darwin argued by analogy, especially to natural selection from his own reading on artificial selection, research that provided much of his primary data. Darwin knew that in a few generations a pigeon breeder could produce a pigeon with a head so big it couldn't fly. Such "sports" and variations often were so different from their stock a few generations before that they would have been considered different species if found in the wild. And remember, Darwin had done the calculations for the denudation of the Weald (see Box 2.1), which gave more than enough time for evolution to occur.

Artificial selection demonstrated to Darwin and others that continued selection was powerful enough to cause observable changes in almost any species. The claim that natural selection for particular environments could accomplish even greater changes than artificial selection and lead to speciation, therefore, seemed reasonable, given the much longer periods of time in evolutionary history and the "unrelenting vigilance" of natural selection. Darwin saw selection and survival rather than extinction as important, and collected and analyzed an enormous amount of evidence for artificial selection. From these data and his inquiries on breeding domesticated species, Darwin obtained clear evidence that selection (in this case, artificial selection) could have marked hereditary effects. Indeed, Darwin was amazed at the ease with which domesticated plants and animals could be changed and the small number

FIGURE 6.2 A sample of the range of varieties of pigeons known to Charles Darwin, as depicted in the *London Illustrated News*.

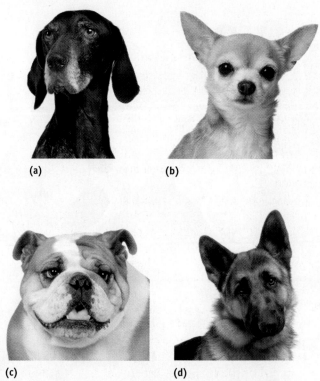

FIGURE 6.3 Heads of four breeds of dogs produced by artificial breeding: (**a**) German Shorthair Pointer, (**b**) Chihuahua, (**c**) English bulldog, (**d**) German Shepherd. Molecular evidence from mitochondrial DNA sequencing shows that the domestic dog is more closely related to the gray wolf than to any other species of the canid family and that domestication started independently on several occasions 10,000 to 15,000 years ago.

of generations of selection it took to do so. He continually referred to artificial and natural selection throughout his writing. Through the combination of

- the ease of effecting change via artificial selection,
- the vastness of geological time,
- the application of the principle of uniformitarianism by which present day processes could be extrapolated back in time, and
- identification of a natural selector — the pressure of continuously limited resources — and with brilliant intuition,

Darwin saw that selection acted by choosing for reproduction those individuals or types with increased chances of survival, thereby changing the composition of the population.

The lack of randomness that we now know selection brings to artificial breeding or to life itself is outlined using a simple example in BOX 6.2 . Complex organs such as the eye and the brain can arise by naturally selecting successively improved adaptations for preservation. The variation on which the choice is exercised is random (the letters in Box 6.2, or mutations, are of different kinds, adaptive and nonadaptive), but the structure that is built over many generations of selection is historically molded and so is not at all random. In the words of the population geneticist R. A. Fisher in 1930, "natural selection is a mechanism for generating an exceedingly high degree of improbability." Or to paraphrase the molecular biologist Jacques Monod, mutation provides the random noise from which selection draws out the nonrandom music.

BOX 6.2

Selection, Randomness and Chance

Although chance events arise, evolution is a historical process; what evolves depends on what has evolved before, is still present and so can be modified by further selection. Furthermore, and as outlined in Box 6.3 for the eye, no complex structure arises all at once by a lucky combination of events. Evolution builds new structures from old ones. For example, if one had a large bowl full of ten different letters (A, C, E, I, L, N, O, T, U, V) with each letter present in equal frequency, nine letters randomly drawn from this bowl can be arranged in many millions of ways. The chance of getting the exact word EVOLUTION from a random draw of nine letters is small indeed: 1 in 10^9 = 0.000000001.

However, if we assume that a selection mechanism exists that will perpetuate certain combinations, the "evolutionary" attainment of the word EVOLUTION is far greater than by chance. Assume that E is the only letter that can survive by itself. The chance of drawing an E from the bowl is, on average, 1 in 10. Assume next that a V in combination with E has additional survival value. The chance of achieving this particular EV combination is again 1 in 10.

Similarly, if we assume that the next adaptive combination consists of adding an O to EV, and that further successive combinations are selected because of their enhanced likelihood of surviving (their fitness), the entire word, EVOLUTION, can eventually be selected with relatively high probability without the intervention of any agent other than the strictly opportunistic one of what is adaptive at each separate stage. Because there may be other combinations in this bowl of letters, similar selective mechanisms can also lead to words such as EVOCATION and ELEVATION. Like EVOLUTION, the chances are extremely small for such words to arise by choosing nine letters at random in a single selective event, yet they can easily be produced by successive selection of adaptive combinations.

Evolution by natural selection marks events whose occurrence depends on previous events, so that a succession of such events can lead to organized structures and increased complexity. The non-organismal (abiotic) synthesis of amino acids and other basic organic molecules (see Chapter 3) indicates the ease with which such molecules can arise and that natural selection can operate to increase their complexity and organization. Selection is a sequential process that ties individual chance events into a sequence because particular steps in the sequence persist. Selection for sight leads to improved visual apparatus (Box 6.3), selection for hearing leads to improved auditory apparatus, and so forth.

■ Once E has been drawn and survives, the chance of getting the EV combination is the 1 in 10 chance of drawing V and not the 1 in 100 chance of drawing EV together.

Variation and Initiating Change

"Darwin revolutionized the study of nature by taking the actual variation among actual things as central to the reality, not as an annoying and irrelevant disturbance to be wished away" (Lewontin, 2000).

Small Continuous Variations

In a notebook started in 1836 (the red notebook), Darwin recorded that he had convinced himself that, in some instances, differences between species were so great that they could only be achieved per saltum — by huge leaps. By 1859 in *The Origin of Species*, Darwin explicitly confined the origination of species by natural selection to small, continuous variations; in the earlier editions Darwin also excluded larger variations as not being useful, adopting the dictum that "nature makes no leaps."

■ The red notebook (our name, not Darwin's) was the first of a series of notebooks in which Darwin noted his ideas on evolution. Begun while on the *Beagle* in 1836, Darwin had filled his first notebook by early 1837.

Darwin's concept of variation in relation to evolution focused on individuals rather than populations. Today, however, we recognize variation as a property of populations rather than of individuals. The current view of variation — populations contains reservoirs of individual-level variation on which selection can act — finally emerged with the modern synthesis in the 1930s and 1940s.

We now know that many traits stem from small heritable changes ascribed to many different genes (*polygenes*), each with small effect (see Chapter 12).

Except for monstrosities (which are highly abnormal or sterile), most observed variations involve only small changes and do not depart from the species pattern. How then can new species arise? Darwin replied that no limits apply to variation because each stage in the evolution of a species entails further variation upon which selection acts; a succession of changes through time, rather than a single simultaneous set of changes, leads to species differences.

A problem with Darwin's theory is the difficulty of determining how selection recognizes each small modification. With many features, such as size, a small modification might hardly be enough to confer significant advantage to an organism, whereas a large modification might well be selected. Although Darwin could not successfully reply to this argument, he doggedly held to his concept of the gradual accretion of small modifications. The rediscovery in 1900 of Mendel's breeding experiments and the subsequent development of the concept of the gene (see Box 1.2 and Chapters 7 and 13) added considerable support for gradual change, although not for small modifications; Mendel's experiments involved larger scale characters (see Chapter 7).

Getting Selection Started

A further aspect of this same problem was how to determine the initial adaptive level that a trait or organ would have to reach to be selected. If the trait already existed before selection acted on it, perhaps some quantitative expression of the trait would suffice for further evolution; a larger eye might function better than a smaller eye. But if the trait did not exist, or only barely existed, how could selection act on it? Many critics thought that the earliest incipient stages of complex organs such as the eye (BOX 6.3; see Figure B6.1) or brain would have no function at all and could hardly be selected. How could one possibly imagine an appropriate adaptive function in only one cell of an eye, brain, or leg, or in "half a wing?"

The phrase "half a wing" appeared in a challenge to Darwin's theory of evolution posed in 1871 by the English biologist, St. George Mivart (1827–1900). Mivart questioned the adaptive value of intermediate forms when he asked, "What use is half a wing?" The half a wing problem arises when considering the origin of flight in birds, a transformation that involved turning a reptile limb adapted for locomotion on land into a wing capable of sustaining locomotion in the air — flight (see Chapter 18). In recent experiments young birds have been shown to use a flapping motion to maintain traction when running up an oblique or vertical slope, activity that has been used to help explain the adaptive value of half a wing. If flapping by extant young birds is a basal character, it could provide an evolutionary explanation for the origin of "half-wings."

Evolving from Something

A more general evolutionary answer to the question of the origin of new traits came from **preadaptation**, a principle of which Darwin was aware. According to preadaptation, a new organ need not arise *de novo* but may be present in an organism but being used for a purpose other than that for which it is later selected. The reptilian limb preadapted for flight is one example (see Chapter 20). Darwin cited others: In his monograph on barnacles, Darwin suggested that the cementing mechanism by which present-day barnacles attach to their substrate is related to the cementing mechanism by which the barnacle oviduct coats its eggs to attach them to the substrate. Only after its earlier evolution in oviducts was this mechanism available to be adapted for attaching the barnacle itself. Similarly, Darwin pointed out that the evolution of lungs from

Darwin spent eight years (1846–1854) before publication of *The Origin* working on the systematics and classification of barnacles.

BOX 6.3
Eye Evolution

The various types of mollusks in Figure B6.1 show a wide range of light-gathering organs ("eyes") from the simplest (a) to the most complex (f); see Chapters 9 and 18 for discussions of complexity in evolution. Six types are outlined in FIGURE B6.1 .

(a) Perhaps the simplest form of light-gathering organ in the animal kingdom is a pigment spot with neural connections stimulated by light. The spot could comprise only a small number of cells, theoretically a single cell.

(b) Folding of pigment cells concentrates their light-gathering activity, providing improved light detection.

(c) A partly closed, water-filled cavity surrounded by pigment cells allows images to form on the pigmented layer as in a pinhole camera.

(d) A transparent fluid, secreted by the cells — and so a fluid extracellular matrix — rather than water forms a barrier that protects the pigmented layer (the retina) from injury.

(e) A thin film or transparent "skin" covers the entire eye apparatus, adding further protection. Some of the fluid extracellular matrix within the eye hardens into a convex lens that improves the focusing of light on the retina.

(f) A complex eye, as seen in species of squid, which has an adjustable iris diaphragm and a focusing lens.

■ Now we know that this is a function of a four-member family of lens proteins, the crystallins.

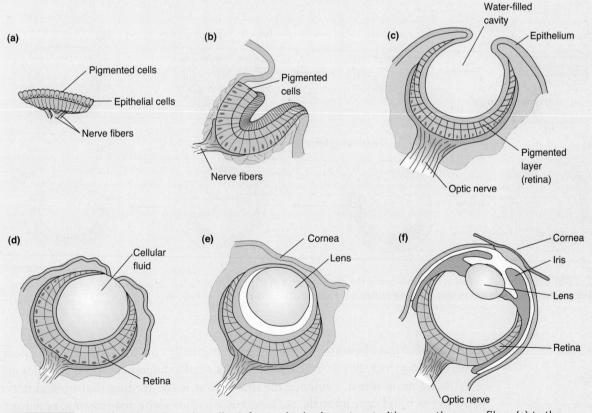

FIGURE B6.1 Light-gathering organs in molluscs from a simple pigment spot with connecting nerve fibers (a) to the complex eye of the squid (f).

(continued)

BOX 6.3
Eye Evolution (*continued*)

Interestingly, genetic studies indicate that a similar inherited factor (the *Pax-6* gene) regulates the development of anterior sense organ patterns in invertebrates and vertebrates. Nevertheless, despite some common regulatory features, specific cellular pathways in embryonic eye development differ substantially between squid and vertebrates. Squid photoreceptor cells derive from the epidermis; vertebrate retinae derive from the central nervous system.

As explained by the process of convergent evolution (see Chapter 16), the structural similarity of squid and vertebrate eyes does not come from an ancestral visual structure in a recent common ancestor of mollusks and vertebrates, but rather from convergent evolution as similar selective pressures led to similar organs that enhance visual acuity. Such morphological convergences may have arisen independently in numerous animal lineages subject to similar selective visual pressures. What is shared is deep in metazoan ancestry and is the ability to form light-gathering cells or organs.

The series of images in FIGURE B6.2 shows stages in eye evolution as depicted by a computerized model in which random changes in eye structure are followed by selection for visual acuity. Beginning with a light-sensitive middle layer of skin backed by pigment (a), successive selective steps for improved optical properties lead to a concave buckling that enhances light gathering (b–e), a focusing lens (f, g), and an eye with a flattened iris in which the focal length of the lens equals the distance between lens and retina (h).

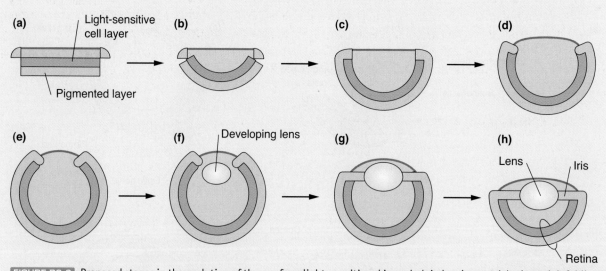

FIGURE B6.2 Proposed stages in the evolution of the eye from light-sensitive skin underlain by pigment (**a**), through infolding (**b–e**), evolution of a focusing lens (**f–h**). (Adapted from Nilsson, D. E. and S. Pelger, *Proc. Roy. Soc. Lond.* **256**, 1994; 53–58.)

fish swim bladders illustrated, "that an organ originally constructed for one purpose, namely flotation, may be converted into one for a wholly different purpose, respiration."

Among more recent evidence for this notion is the finding that optical nerve pathways of blind cave animals, no longer needed for sight, may assume new functions associated with smell and touch. Positive selection for enhancement of these other sensory structures results in loss or reduction of the eyes to rudiments in cave-dwelling fish (FIGURE 6.4).

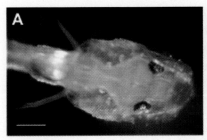

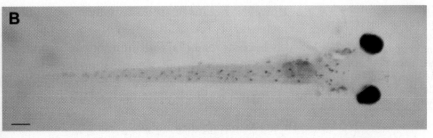

FIGURE 6.4 Mexican tetras (blind cave fish), *Astyanax mexicanus*. (**A**) A blind cave-dwelling form, nine days after fertilization showing the reduced eyes. Compare with (**B**) a sighted surface form, six days after fertilization. The scale bar is 100 μm.

Given continued environmental pressure and selection, evolutionary intermediates are a commonly expected feature. A highly specialized organ like the vertebrate eye did not arise all at once but represents a succession of changes of a previous light-gathering organ and its ancillary tissues, which may have originally involved only a few cells. A turn-of-the-twentieth century illustration of a likely set of stages is shown in Figure B6.2. As outlined in Box 6.3, a flat patch of light-sensitive cells could change into a complex squid-like eye with a focused refractive lens in less than 500,000 years.

Darwin's search for small modifications led him to place less emphasis on the many traits involving distinct steps and differences: different colors, presence and absence of structures, and different numbers of structures. These large variations may be important for selection, as various biologists, including Thomas Henry Huxley, suggested. Interestingly, large observable differences in plant traits were used by Mendel to develop the basic laws that explain inheritance (see Chapter 7). Not until the 20th century did Darwinists resolve the problem of where, how, and to what extent variations originated, issues that were the most often attacked element in Darwin's theory.

> The term *biologist* introduced in the nineteenth century by the English scientist, philosopher and historian of science, William Whewell (1794–1866), did not come into general usage until the twentieth century.

Malthus and Darwin

Many more individuals of each species are born than can possibly survive. Consequently, there is a recurring struggle for existence.

Any individual that varies however slightly in any way that is advantageous will have a better chance of surviving, of being selected, and of leaving offspring. Behind this simple explanation is a complex set of causative events, which Darwin spent most of his life investigating. Darwin ascribed the origin of his idea that species tend to produce more members than resources can sustain — the primary population pressure that leads to competition and selection — not to biological literature but to the sociology of his time.

Largely as a consequence of the rapid increase in the number of poor people resulting from the Industrial Revolution, a variety of social and economic problems became apparent in Victorian England. This tide of poverty began with the impoverishment of small handicrafts establishments and was continually fed by small farmers who were pushed from their lands (the "commons") by the Enclosure Acts. One consequence, famously expressed by the Rev. Thomas Malthus (1766–1834) was that the fate of the poor is inescapable; their reproductive powers will always exhaust their means of subsistence.

Food supplies, Malthus pointed out, at best increases arithmetically ($1 \rightarrow 2 \rightarrow 3 \rightarrow 4 \rightarrow 5 \ldots$) by the gradual accretion of land and improvement of agriculture.

■ Malthus was not the first to see the link between numbers of individuals and resources. On the first page of his treatment Malthus paraphrases an identical proposal made by Benjamin Franklin.

However, because parents usually produce more than two children, the population increases geometrically ($2 \rightarrow 4 \rightarrow 8 \rightarrow 16$). Famine, war and disease, thus, inevitably become major factors limiting population growth. In 1798 Malthus circulated an essay on the principle of population increase. What will amaze you when you read it is that Malthus applies his reasoning not only to human populations but to animals and plants, as the excerpt below demonstrates.

> Assuming then my postulata as granted . . . First, That food is necessary to the existence of man. Secondly, that the passion between the sexes is necessary, and will remain nearly in its present state . . . , I say, that the power of population is indefinitely greater than the power in the earth to produce subsistence for man.
>
> Population, when unchecked, increases in a geometrical ratio. Subsistence increases only in an arithmetical ratio. A slight acquaintance with numbers will shew the immensity of the first power in comparison of the second.
>
> By that law of our nature which makes food necessary to the life of man, the effects of these two unequal powers must be kept equal. This implies a strong and constantly operating check on population from the difficulty of subsistence. This difficulty must fall somewhere and must necessarily be severely felt by a large portion of mankind.
>
> Through the animal and vegetable kingdoms, nature has scattered the seeds of life abroad with the most profuse and liberal hand. She has been comparatively sparing in the room and the nourishment necessary to rear them. The germs of existence contained in this spot of earth, with ample food, and ample room to expand in, would fill millions of worlds in the course of a few thousand years. Necessity, that imperious all pervading law of nature, restrains them within the prescribed bounds. The race of plants and the race of animals shrink under this great restrictive law. And the race of man cannot, by any efforts of reason, escape from it. Among plants and animals its effects are waste of seed, sickness, and premature death. Among mankind, misery and vice (Malthus, *An Essay on the Principle of Population*, 10th ed., 1803).

The only hope that Malthus held out for the poor was self-restraint: delay marriage and refrain from sexual activity. All other solutions — the Poor Laws (welfare), redistribution of wealth, improvements to living conditions — were, in his view, inadequate. Such measures would only stimulate further increase in the number of poor people and begin again the cycle of famine, war, and disease.

As were many others of the time, Darwin was deeply impressed by the Malthusian argument, although Malthus did not espouse evolution. Malthus believed that limiting population growth would *prevent* evolutionary change — individuals who departed from the population norm would be more susceptible to extinction — an idea that had been spelled out in the 1830s and with which many biologists would have been quite comfortable. To Darwin, however, the importance of Malthus' theory lay in revealing the conflict between a population's limited natural resources and its continued reproductive pressure.

In contrast to Malthus' proposals for alleviating the impact of population increase, Darwin pointed out that plants and animals had no such alternatives. "There can be no artificial increase in food, and no prudential restraint from marriage." Malthus therefore played an important role in Darwin thinking out natural selection as a mechanism of evolutionary change. In his autobiography Darwin wrote

> I soon perceived that selection was the keystone of man's success in making useful races of animals and plants. But how selection could be applied to organisms living in a state of nature remained for some time a mystery to me.

In October 1838, that is, fifteen months after I had begun my systematic enquiry, I happened to read for amusement "Malthus on Population" and being well prepared to appreciate the struggle for existence which everywhere goes on from long-continued observation of the habits of animals and plants, it at once struck me that under these circumstances favourable variations would tend to be preserved, and unfavourable ones to be destroyed. The result of this would be the formation of new species. Here then I had at last got a theory by which to work; but I was so anxious to avoid prejudice, that I determined not for some time to write even the briefest sketch of it. In June 1842 I first allowed myself the satisfaction of writing a very brief abstract of my theory in pencil in 35 pages; and this was enlarged during the summer of 1844 into one of 230 pages, which I had fairly copied out and still possess.

Vestiges of the Natural History of Creation and Darwin's Delay

Despite having read Malthus in 1838, having written out a 35-page abstract of his theory in 1842, and a 230-page extended abstract in 1844, Darwin delayed publishing his theory. Why did Darwin not publish his 1842 sketch or his 1844 manuscript? In part his delay reflected Darwin's desire to amass a wealth of evidence that would overcome any opposition to the generality of the theory. But only in part.

A major factor that made Darwin cautious of potential public reception to his theory was the publication in 1844 of the book, *Vestiges of the Natural History of Creation*. Published anonymously, *Vestiges* was shown after the author's death to have been written by Robert Chambers (1802–1871), a member of a prominent Scottish publishing house. Vestiges elaborated the idea that all matter, inorganic and organic, evolved out of inorganic dust by the accumulation of accidental mutations caused somehow by changes in nutrition or environment.

Although enormously popular for a time — there was much public discussion of the 14 editions in Britain, especially in drawing rooms and salons — considerable religious and scientific denunciation focused on this work. These denunciations made Darwin fearful of exposing his own ideas to potential ridicule. Although Darwin had written his lengthy essay setting out the theory, he set the essay aside and never published it. Reading *Vestiges* and the responses it elicited convinced Darwin that he needed to amass as much evidence as possible before exposing his theory to public scrutiny.

Darwin devoted the next decade (1844–1855) to documenting variation in nature, differences between varieties and species, the geographical distribution of organisms, and accumulating and analyzing evidence for artificial selection. During this time, Darwin did not publish a single word on his theory of evolution. Some historians believe that the delay was because Darwin's theory that natural selection was the mechanism of speciation was incomplete until 1852 or perhaps later. Only in the early 1850s did Darwin devise the **principle of divergence**, which he represented in the only figure in *The Origin of Species* (see Figure 1.4). By 1855, however, evolution was in the air as Darwin was to learn to his dismay, even mortification. A competitor appeared on the scene. Another British naturalist, Alfred Russel Wallace (1823–1913) was also seeking to understand species transformation.

The story of how Charles Darwin's and Alfred Wallace's theories on the origin of species were presented as a joint communication to the Linnaean Society of London on 1 July 1858 is a complex one.

- For a start, neither man was present at the meeting. Indeed, Wallace was unaware that his work was reported at the Linnaean Society meeting or

■ It took Darwin some time to accept and use Herbert Spencer's term and metaphor *"survival of the fittest"* (see Chapter 15). In several letters, Wallace tried to convince Darwin to use the phrase, which he (Wallace) had adopted. Only after he had completed several editions of *Origin of Species* did Darwin adopt the term for the processes that lead to the differential survival of organisms from one generation to the next. Darwin wrote to Thomas Huxley that it might have been preferable had he used *"struggle for reproduction"* rather than struggle for existence.

that his manuscript would be published in the society journal along with a summary of Darwin's theory taken from extracts from Darwin's 25-year-old essay and letters to an American botanist.

- Each arrived at a theory of the transformation of species by natural selection independently, in both cases as a result of extensive analysis of variation in nature.

- Both were influenced by Malthus in seeing the importance of competition for resources.

- For Darwin, artificial selection was the key to understanding natural selection. Wallace thought artificial selection an inadequate way to understand natural selection.

- Darwin's principle of divergence asserts that as long as competition between subpopulations exists, subpopulations can begin to specialize and diverge to the point of speciation. With this principle, Darwin maintained that competition was the most powerful selective force. Geographical isolation featured much less prominently in Darwin's thinking than it did for Wallace. So although both derived a theory of evolution by natural selection, the theories were not identical.

- A final complexity is the way Darwin behaved when he received a manuscript on species transformation from Wallace in June 1858.

Alfred Russel Wallace

The lives and situations of these two British naturalists could not be more different.

Darwin was a member of a wealthy, upper-middle-class family. Educated at Cambridge, with one trip abroad as a naturalist, Darwin married money and lived the life of a financially secure but continuously ill independent scholar.

Wallace was born in Wales to a respectable but poor middle-class family. Forced to leave Grammar School at age 13 because of his father's financial difficulties, he was apprenticed to one of his older brothers as a surveyor. Self-educated by attending lectures at the London Mechanic's Institute, Wallace worked as a surveyor until age 20. A close friend of the naturalist, future explorer and discoverer of Batesian mimicry (see Chapter 15), Henry Walter Bates (1825–1892), Wallace was so inspired by reading Darwins' *Journal of Researches* and other journals of naturalists/explorers/collectors, that he and Bates set off for Brazil in 1848 to collect insects to sell to museums and collectors back in England and to seek evidence for species transformation.

Wallace's entire collection, all his notes and diaries were lost when the ship transporting them back to England caught fire and sank on 12 July 1852. Darwin, on the other hand, returned to England with collections that he distributed to the experts of the day and with his *Journal of Researches* to establish his reputation as a naturalist to be reckoned with. Wallace spent a further six years (1854–1862) exploring and collecting in what is now Malaysia and Indonesia, this time collecting 125,000 specimens — 1,000 of which were species new to science — and laying the foundation for his ground-breaking studies in the distribution of animals (**zoogeography**). The Wallace line separates species with links to Australian from species with Asian connections in the Indonesian region.

Wallace had already read Malthus' essay before leaving for Brazil. Both he and Bates had discussed and were groping toward a theory of the transformation of species. In 1855 Wallace published a paper on the distribution of species in relation to their geographical distribution, a paper of which Darwin was aware. In his paper Wallace concluded "every species has come into existence coincident both in space and time with a closely allied species."

In 1858, Wallace was collecting birds, insects, and mammals in the islands of Southeast Asia and continuing to seek a mechanism for evolution. Remarkably, Malthus performed the same function for Wallace as he had for Darwin 20 years earlier.

> At that time [February 1858] I was suffering from a rather severe attack of intermittent fever at Ternate in the Moluccas . . . and something led me to think of the positive checks described by Malthus in his "Essay on Population," a work I had read several years before, and which had made a deep and permanent impression on my mind. These checks — war, disease, famine, and the like — must, it occurred to me, act on animals as well as on man. Then I thought of the enormously rapid multiplication of animals, causing these checks to be much more effective in them than in the case of man; and, while pondering vaguely on this fact there suddenly flashed upon me the idea of the survival of the fittest — that the individuals removed by these checks must be on the whole inferior to those that survived. In the two hours that elapsed before my ague fit was over I had thought out almost the whole of the theory, and the same evening I sketched the draft of my paper, and in the two succeeding evenings wrote it out in full, and sent it by the next post to Mr. Darwin (Introductory note to Chapter II of *Natural Selection and Tropical Nature*, revised edition, 1891).

Wallace sent Darwin a manuscript prepared for publication in which he described the theory of natural selection in the essential form Darwin had envisaged. In the accompanying letter Wallace asked Darwin to transmit the manuscript to the Linnaean Society for publication if he felt it had merit.

History should have taken a different turn at this stage. Had Wallace sent the manuscript directly to the Linnaean Society or to any other society or journal, the manuscript would have been reviewed and a decision made as to whether to publish or not. Clearly, had Darwin been a reviewer in such a circumstance, he would have had no choice but to recommend publication of the manuscript. After all, it laid out the theory Darwin had been groping toward for 20 years, a problem that Wallace but not Darwin already had published on.

■ Darwin did not publish a single word on the problem of the transformation of species until the letter and notes read to the Linnaean Society were published.

However — and this is one of the most significant "howevers" in the history of science — Wallace sent the manuscript to Darwin, who did not pass it along to the Society but sent it to his good friend Charles Lyell, asking for advice on what to do. To prevent Darwin losing his priority in applying natural selection to evolution, his friends Charles Lyell and Joseph Hooker, without contacting Wallace, arranged for a joint communication on the topic by Darwin and Wallace to be presented to the Linnaean Society in London (in the absence of both authors) and to be published in 1858 in *The Journal of the Linnaean Society*. Today, such an action would be regarded as highly unethical. Somehow, the reputations of Darwin, Lyell and Hooker have not been sullied. When he finally heard about "the arrangement," Wallace was grateful to have his paper associated with Darwin and the body of evidence Darwin had accumulated. Later in life, Wallace came to a more nuanced opinion.

■ So, with reference to the previous note, it could be argued that Darwin still had not written a word intended for publication about the transformation of species.

■ Although crammed with evidence, including much data gleaned from studies by others, Darwin provided almost no documentation of the sources of the evidence he amassed, something over which he was taken to task.

The communication presented to the Linnaean Society on 1 July 1858 is strange indeed, consisting as it does of Wallace's paper along with portions of a document Darwin had sent to Hooker in 1847 (asking Hooker not to divulge its contents) and excerpts from a letter Darwin had sent in 1857 to an American botanist, Asa Gray.

Surprisingly, the joint publication evoked little response from either the scientific or the nonscientific communities, indicating perhaps that the theory itself, without supporting evidence and without enlisting large-scale evolutionary phenomena, neither invited serious interest nor threatened established views. Only with the publication in November 1859 of Darwin's expanded work with its huge weight of supporting evidence, *On the Origin of Species by Means of Natural Selection or the Preservation of Favoured Races in the Struggle for Life,* did the world take notice. The conventional story is that the first printing of 1,500 sold out within hours of its release. In fact, 389 copies were given away by the publisher or by Darwin to promote the book. The other 1,111 copies were bought by booksellers several days before the official publication date.

The evolutionary principle expressed by Darwin and Wallace is outlined in FIGURE 6.5 . We can summarize the process proposed as twofold.

1. Excess reproduction + limited resources → competition, which, because of natural variation and natural selection allows those best adapted to pass their existing characters to the next generation.
2. Changing environments + hereditary variation + natural selection results in the modification of existing characters or the origin of new characters that become established, and spread throughout a population/species (FIGURE 6.6).

Note that the evolutionary process is a continual one: Enhanced reproductive ability of some individuals is followed by further competition for the limited resources and further natural selection in subsequent generations. Of course chance and random change also come into play; an individual killed by a falling tree is prevented from reproducing whether well adjusted to the environment or not. Because each evolutionary stage builds on the one before, the process spirals in the direction of improved adaptation for any particular environment.

You should not be left with the impression that Wallace and Darwin came up with identical theories. They did not. They differed over the use of Spencer's term "struggle for existence," but more fundamentally, Wallace was unconvinced of the utility of artificial selection as a means to understand natural selection, a coupling that was essential for Darwin. Wallace also placed more importance on geographical isolation than did Darwin.

Aspects of the Natural World Explained by the Theory of Evolution by Natural Selection

Many recognized evolution as the cause of the diversity of species on Earth, although only some accepted, and to various degrees, that natural selection was the primary *mechanism* for evolution. Acceptance of evolution as a fact came because the theory conceived by Wallace and Darwin explained much that had puzzled naturalists about the natural world.

- Intermediate forms observed by taxonomists could be explained as transitions between species.

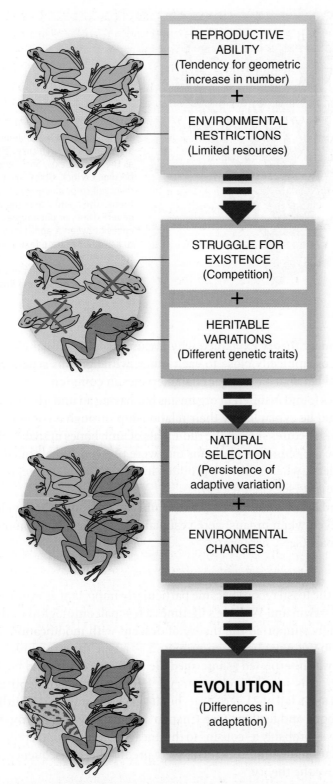

FIGURE 6.5 Schematic of the main conceptual arguments for evolution by natural selection given by Charles Darwin and Alfred Russel Wallace. (Adapted from Wallace, A. R. *Darwinism: An Exposition of the Theory of Natural Selection with Some of Its Applications.* Macmillan, 1889.)

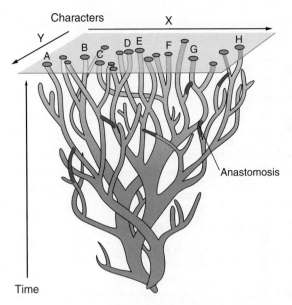

FIGURE 6.6 A phylogenetic tree shown as continuous branches undergoing evolutionary changes through time. Time is shown as the vertical axis. Some populations have become extinct. Others have merged or diverged to produce new and different forms. The X and Y axes represent different genetic traits or phenotypes. The differences between A and H may be sufficient to warrant separate taxonomic designations. (Adapted from Levin, D. A. (ed.). *Hybridization: An Evolutionary Perspective.* Dowden, Hutchinson and Ross, 1979.)

- Evolutionary theory explained why organisms with similar characteristics occur geographically close to each other, while groups separated by geographical barriers have fewer characteristics in common.
- Structures found in different organisms but having an underlying similarity of plan could be explained by their relationship through a common ancestor. Vestigial structures such as the rudiments of limb bones in snakes and whales would be the remnants of organs that were present, fully formed, in ancestors. Indeed, whale embryos produce hind limb buds that regress early in development (see Chapter 16).
- The similarities among vertebrate embryos during early developmental stages, even when organs were vestigial, also suggested a common evolutionary past (see Chapter 16).

Fossils

Evidence from the fossil record became particularly important as evolutionary thinking spread. In Darwin and Wallace's lifetimes a few paleontological findings came to light that strongly supported the theory of descent with modification. One was the discovery in 1861 of what had been proposed as a true "missing link," in this case, an animal that was interpreted as intermediate between reptiles and birds. As shown in Figure 18.6, this fossil, *Archaeopteryx*, had a number of reptilian features, including teeth and a tail of 21 vertebrae, but it also had a number of birdlike features, such as a wishbone and feathers (see Chapter 18). Huxley argued convincingly that *Archaeopteryx* was probably a "cousin" to the lineage running from reptiles (dinosaurs) to birds, and that such "primitive" forms were predictable consequences of evolution that helped prove the theory.

Horses provided the most complete and continual fossil record of the evolution of any group of animals. Extant horses evolved from a small, four-toed browsing animal to a large, single-toed creature with continuously growing teeth, adapted to chewing

tough grasses, through a long evolutionary process featuring many speciation and extinction events (see Figures B16.1 and B16.2 and Box 16.1).

Geographical Distribution and Classification

After 1859 it seemed clear that the gradation of different organisms observed through attempts at classification, whether from simple to complex or from one type to another, could most reasonably be explained by evolutionary relationships (FIGURE 6.7 ; BOX 6.4). Every extant or extinct organism has a phylogenetic (evolutionary history) and only one phylogenetic history. **Phylogeny** happened. **Classification** did not happen but is a way(s) to arrange or order the results of evolution (Box 6.4; see Chapter 16). Classification arises and exists only in our minds.

Many became aware that groups of organisms that are evolutionarily related are usually connected geographically. Large geographical barriers such as oceans and mountain ranges isolate groups from one another and are associated with considerable differences among the separated groups. Colonizers that transcend such barriers — usually by some mode of isolation — often become the ancestors of new lineages or groups. This shows up in the wide evolutionary radiation of species descended from a small number of individuals of a single finch species that colonized the Galapagos Islands. Beginning with what was probably an ordinary mainland finch, new kinds of finches evolved in the Galapagos (Figure B6.3, and see Boxes 5.4 and 14.2 and Figure B14.1).

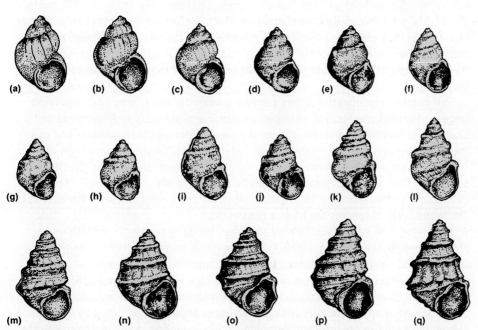

FIGURE 6.7 A 19th century illustration of evolutionary relationships among fossil species of the mollusk genus *Paludina,* ranging from the oldest form *P. neumayri* (**a**), to the youngest *P. hoenesi* (**q**). To Darwin and many others, differences of this kind "blend into each other in an insensible series; and a series impresses the mind with the idea of an actual passage." (Adapted from Romanes, F. J. *Darwin, and After Darwin.* Open Court, 1910.)

BOX 6.4
Classification

Ideally, the most informative phylogenetic picture of a particular population of organisms would be a portion of a multi-limbed tree that has branched connections to all extant and ancestral populations, and that indicates, through these connections, its degree of relationship to all other populations (FIGURE B6.3). However, ten to thirty million species of organisms may exist today. Some are identified and named as species. Many are not. Hundreds of millions of species may have existed in the past. Creating a Universal Tree of Life (see Chapter 8) that includes even the named species appears an impossible task to contemplate, yet alone attempt. But then, sequencing the genome of an organism seemed an impossible task two decades ago.

Taxonomists call each unit of classification, whether it be a particular species, genus, order, or whatever, a **taxon** (plural, **taxa**) and give it a distinctive name. Taxa are arranged in nested categories so that a taxon in a "higher" category includes one or more taxa in "lower" categories. Some of the categories and taxa used in classifying modern humans (*Homo sapiens*) and fruit flies (*Drosophila melanogaster*) are shown in FIGURE B6.4 . The arrangement in the figure is not a phylogenetic tree but shows how classification produces nested categories.

The approximately two million named species of animals are arranged into some 37 phyla, examples of most of which are discussed in Chapters 16 to 18. Some have enormous numbers of species. The phylum *Arthropoda* (arthropods) has more than a million species, *Nematoda* (roundworms) half a million. At the other extreme, the phylum *Placozoa* (from the Latin for flat animal) consists of a single marine species, *Trichoplax adhaerens,* 300 µm "long" (perhaps we should say short or across as it has no obvious symmetry) with four cell types, 12 chromosomes, a low DNA content but 11,514 protein-coding genes, many of which are shared with other animals (see Chapter 18). The phylum *Cycliophora* (from the Latin for small wheel-bearing) also consists of a single microscopic marine species, *Symbion pandora,* which spends its life attached to the mouth parts and appendages of the Norway lobster *Nephrops norvegicus.*

This mode of classification offers a simple scheme for identifying and cataloging large numbers of species. For example, we can use phyla such as Arthropoda and Chordata to identify two major groups of animals from all other animals, and use mammalian orders such as Primates and Rodentia to distinguish two groups of mammals from all other mammals. To a significant extent, these classification schemes reflect an underlying phylogenetic pattern in which each taxon "seems" to have originated from the one in which it is included. Why "seems?" Because taxa above the level of species are human constructs.

Furthermore, as each traditionally recognized group in such a classification is not always **monophyletic** (derived from a single common ancestor), these classification schemes need not reflect evolutionary history. Mammals, a long-recognized group of animals, may have had a polyphyletic origin (see Figure 8.2 and Chapter 18). Some invertebrate biologists have concluded that the arthropod grade of organization, characterized by an exoskeleton with jointed appendages, was achieved by several segmented wormlike organisms undergoing parallel evolution, some giving rise to the extinct trilobites, others to crustaceans or insects (see Chapter 18). Major plant taxa — bryophytes, tracheophytes, gymnosperms, and angiosperms — also may have had multiple origins, that is, be **polyphyletic** (see Chapter 10), weakening any correlation between traditional classification and phylogeny.

■ A **phylum** (division, in plants), the taxonomic category immediately below the kingdom in traditional classification schemes, is a group of animals (plants) that share a common plan and have a closer evolutionary relationship to one another than they do to animals (plants) in another phylum (division).

■ A better way to express this type of classification is to speak of taxa nested within other taxa.

BOX 6.4
Classification (*continued*)

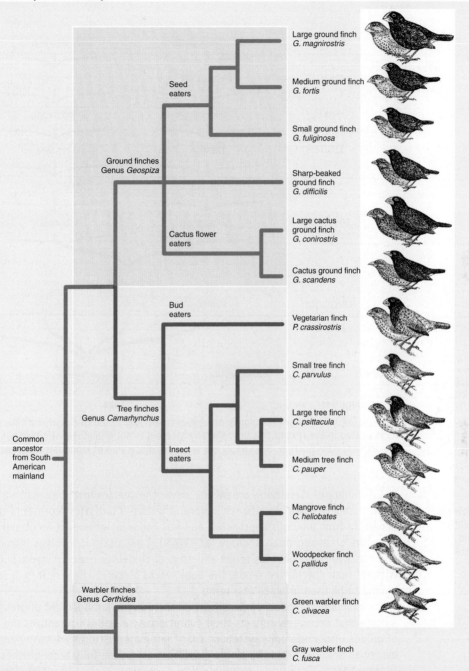

FIGURE B6.3 Evolutionary tree of Darwin's finches showing adaptations of the beaks of different species. A molecular study published in 1999 supports some aspects of this phylogeny. Using comparisons among mitochondrial DNA sequences, the molecular study distinguishes between tree finches and ground finches, but shows that distinctions among members within each group have not yet been firmly established. This molecular study also indicates that these finches are all descended from a single species, now identified as a warbler-type, "dull-colored grassquit" (genus *Tiaris*). (Adapted from Lack, D. *Darwin's Finches: An Essay on the General Biological Theory of Evolution*. Cambridge University Press, 1947. Illustration by Lt. Col. William Percival Cosnahan Tenison.)

(*continued*)

BOX 6.4
Classification (*continued*)

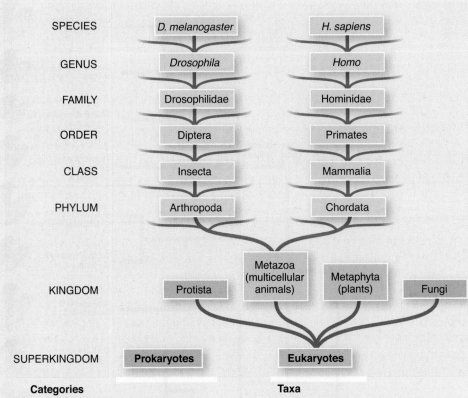

Categories

Taxa

FIGURE B6.4 The nested taxonomic categories traditionally used to classify fruit flies and humans. As indicated in the text of this box, the arrangement is not a phylogenetic tree but a nested hierarchical classification. See Chapters 8 and 9 for discussion of kingdoms and super kingdoms.

■ Cladistics also is known as phylogenetic systematics to distinguish it from traditional systematics (taxonomy).

The difficulty of classifying organisms above the species level has led to a variety of proposals to make classification more objective. **Cladistics** represents the latest departure from previous approaches and has had a major impact on classification and therefore on evolutionary biology (BOX 6.5). I am afraid it involves introducing new, and not intuitively obvious terms and ways of thinking. I see no way around this. Without such specialized terms, the book would be twice the size it is, a sentence having to be used in place of a term.

In summary, the goals of classification include the arrangement of groups into a pattern that accurately reflects their evolutionary relationships and/or the placement of groups into a reference system so that their major features are easily and efficiently described and identified. No single classification system fully accomplishes either of these purposes. Traditional morphological classification simplifies the placement of organisms into a classification scheme but runs the risk of ignoring their evolutionary relationships. Cladistic classification offers hypotheses of relationships that can be tested as new data arise.

BOX 6.5
Cladistics

Cladistics is an approach to classification that relies on branching. The original proponent of this method was the German entomologist and systematist, Willi Hennig (1913–1976), whose pivotal role in founding cladistics is recognized in the name of a professional association of cladisticians, the Willi Hennig Society.

Cladistics stresses the separation of ancestral (**plesiomorphic**) characters from newly derived (**apomorphic**) characters, and emphasizes the latter to establish phylogenies. The sharing of derived characters (**synapomorphy**) dictates the phylogeny. For example, how species A, B and C are related can be determined by noting which share a newly derived character (FIGURE B6.5). Thus, specific character X shared by species A and C indicates a closer relationship than to species B; that is, A and C branched off together from a common ancestor, whereas species B branched off separately (see Figure 16.6a). Emphasis on shared-derived characters contains no suggestion as to which species is ancestral (cf. Figure B6.5a and b). Indeed, because evolutionary patterns can be networks — more like a shrub with many branches from a single node (see Figure 6.6) than like a tree with only two branches from each node — dichotomous branching may oversimplify the evolutionary process, especially for groups with many species or rapid evolution. It is important to bear in mind that each phylogenetic tree is a hypothesis about relationships, which, like any hypothesis, requires constant testing.

Every significant evolutionary step marks a dichotomous branching event that produced two genetically separated **sister taxa** equal to each other in rank. For example, as birds and crocodiles derive from a common reptilian stem ancestor, cladistics consider them to be sister groups of equal rank within a group (archosaurs, "first lizards,") which in turn ranks lower than the group Reptilia, within which it nests (see Chapter 18).

For a long time cetaceans — whales, dolphins, porpoises — were classified as a separate mammalian group. More recent analyses of fossil, morphological and molecular data (12S and 16S ribosomal RNA sequences) supports the hypothesis that cetaceans are closely related to the extant artiodactyls (even-toed ungulates), a group that includes camels, pigs, hippos and ruminants. Early whale fossils morphologically similar to artiodactyls support this phylogeny. Further, evidence from mitochondrial, nuclear and chromosomal DNA shows that cetaceans are not only related to artiodactyls but arose deep within the artiodactyls.

So far, so good. As noted above, many of the traditional groups of organisms are not monophyletic lineages and so do not accurately reflect evolutionary lineage relationships. Arthropods are one such group (see Chapter 18).

■ The Willi Hennig Society (http://www.cladistics.org/) publishes the journal *Cladistics*, which keeps alive Hennig's name and pivotal role in founding cladistics.

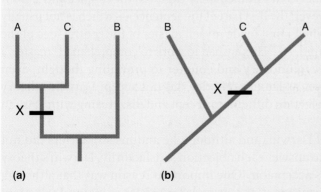

(a) **(b)**

FIGURE B6.5 Relationships between three species, two of which share a newly derived character (X), as depicted using a cladistic approach (**a**), which does not make any statement of ancestry, and as depicted in a more traditional tree (**b**) in which species B is ancestral to both species A and species C. Note that in both trees species A and C are more closely related to one another than either is to species B.

(continued)

BOX 6.5
Cladistics (*continued*)

Perhaps the most difficult change to get our head around is the term *birds*. We all know what birds are; we know their identifying features (feathers, flight, wings) and can see how they must be related. Cladistics, however, does not follow the traditional classification of ranking birds (Aves) separate from reptiles (Reptilia). In cladistic analysis, groups that do not include all the descendants of a common ancestor — for example, the class Reptilia does not include their mammalian or avian offshoots — are **paraphyletic,** and so are not valid taxa. Similarly, cladistic schemes exclude polyphyletic groups such as arthropods and others that consist of convergent or parallel lineages. In cladistic analysis, a group of taxa (a clade) can only achieve taxonomic status if it is monophyletic (Figure B6.5). Furthermore, the group must include its common ancestor and all of the common ancestor's descendants. Therefore, birds, which are related to dinosaurs and descended from dinosaurs, are dinosaurs (and reptiles). To distinguish birds as a monophyletic group we have to speak of avian-reptiles and non-avian reptiles, meaning those reptiles that gave rise to birds and those that did not.

The name given to this process, **adaptive radiation**, signifies speciation following expansion of one or a few forms into a new geographical area. The radiation of marsupial mammals in Australia (FIGURE 6.8) shows how protection from competition with placental mammals by the isolation of the continent facilitated an array of species with widely divergent functions, ranging from herbivores to carnivores, and paralleling placental mammals on other continents.

The term **microevolution** is sometimes used for adaptive radiation within a species, **macroevolution** for evolution above the species level (see Chapter 17).

Wider Impacts of *On the Origin of Species*

Darwin's detailed and elegant exposition of his theory, supported by 20 years of thought and documentation, was impossible to overlook. Its contents initiated the first major public discussions concerning science and the place and role of science in society.

Thomas Huxley (1825–1895), who later became Darwin's main public defender, wrote, ". . . How extremely stupid of me not to have thought of that!" This oft-quoted phrase gives the impression that Huxley immediately captured the essence and significance of Darwin's theory. However, the first part of the sentence — which is not usually quoted — is, "My reflection when I first made myself master of the central idea of the *Origin* was, how incredibly stupid. . . ." Huxley had to work to comprehend ("master") Darwin's ideas, which were revolutionary and counter to prevailing thought, even to one as versed in the sciences as Huxley. Huxley did not accept Darwin's theory uncritically, finding natural selection difficult to accept and disagreeing with Darwin on the reality of species.

Although critics opposed Darwin, and although he and his supporters did not always have the information to answer each objection satisfactorily, Darwin's theory made the evolution of species acceptable. One important reason was that although Darwin presented many mechanisms — the struggle for existence, natural selection, divergence between species — each relied on the existence of variation and natural

Sugar glider
(*Petaurus*)

Tasmanian wolf
(*Thylacinus*)

Tiger cat
(*Dasyurus*)

Marsupial mole
(*Notoryctes*)

Banded anteater
(*Myrmecobius*)

**Marsupial
radiation**

AUSTRALIA

Koala
(*Phascolarctos*)

Marsupial rat
(*Dasyuroides*)

Bandicoot
(*Macrotis*)

Wombat
(*Phascolomys*)

Kangaroo (*Petrogale*)

FIGURE 6.8 Diversity of Australian marsupials. Many divergent forms evolved independently of but often in parallel to changes occurring among placental mammals on other continents because selection for survival in similar habitats can lead to similar adaptations in parallel or convergent evolution.

processes and could readily be supported by observations. These natural processes were in strong contrast to previous, more speculative, evolutionary theories such as Lamarck's (see Chapter 5), which were tied to nonmaterial agents impossible to observe.

Another attractive feature of the theory of evolution by natural selection was expansion of the role of biology to include the study of relationships among all living creatures (Box 6.4), including humans, formerly thought to be divine, and therefore separate from the animals. From anthropology to botany to paleontology to zoology,

Darwinism paved the way for new lines of thought and new areas of investigation. Individuals now could ask, for example

- What are the relationships among different kinds of cells, different parts of cells, different flowers?
- How did the remarkable range of pollination mechanisms in orchids evolve?
- What are the steps in the evolution of organs such as the circulatory and nervous systems?
- Why do some species mimic others?
- How did sterile insect castes such as worker bees evolve?
- What accounts for the geographical distribution of specific organisms?

Although each topic would demand separate techniques and study, and although some are only now being understood, all could be approached against a background of descent with modification. By offering an overall view of adaptation and evolution, Darwin and Wallace provided a fundamental theoretical framework to bind all of biology together. Consequently, many lines of direct and indirect evidence began to accumulate in support of an evolutionary view, many of which are discussed in the remainder of the present book.

Arguments Raised Against the Theory of Evolution by Natural Selection

Some of the major reasons for resisting accepting evolution and or evolution by natural selection were

- The prevalence of the idea that an individual's traits were a blend of those of the parents' (see Chapter 7) made it difficult to see how traits could be selected for.
- Given that geologists had yet to establish unambiguously Earth's great antiquity, there did not seem to be sufficient time for evolution to occur.
- Darwin's insistence that evolutionary forces act on small and continuous variations led to criticism that selection could not recognize such slight variations and therefore could not lead to the formation of new species. While selection might account for the evolution of one species into another, it remained difficult to understand how many species might arise from a single one.

An additional problem was the appearance of new traits, which Darwin thought might form by preadaptation or the conversion of an existing structure to a new use.

A branching pattern of evolution (Figure B6.4) comes about through the isolation of groups from one another, something that Darwin did not emphasize, but which others such as Wallace, Gulick and Wagner did. Because geographical isolation plays such a large role in theories of speciation (see Chapter 17), this concern is explored in more detail.

Geographical Isolation

Critics singled out Darwin's strong emphasis on the transformation of a single species into another single species, a process often called **phyletic evolution** (see Chapter 16).

Although Darwin's approach accounted for the evolution of a particular species in time, critics argued that it did not easily account for the multiplication of species in

BOX 6.6
The Real Father of Geographical Isolation

Although almost unknown today, during the late 19th and early 20th centuries Rev. John Gulick (1832–1923) was one of the world's most well-known and influential evolutionary biologists.

In the early 1850s, while still in his early twenties, Gulick collected over 200 species of land snails of the genus *Achatinella* from Oahu, Hawaii (FIGURE B6.6), where his father was a missionary. John, too, spent his life as a missionary. Gulick's analysis of the restricted geographical distributions of each species — often to a single valley or region within a valley — and publication of his results beginning in 1872, provided what many considered the missing mechanism in Darwin's theory of evolution, namely *geographical isolation*. One or more individuals became isolated from the rest of the population and their descendants subsequently diverge from that population to the extent that they become separate species.

A century later the evolutionary biologist Ernst Mayr (1904–2005) called this process the **founder effect** (see Chapter 17), acknowledging Gulick's origination of the concept and recognizing Gulick as, "the first author to develop a theory of evolution based on random variation" (Mayr, 1988, *Towards a New Philosophy of Biology*, p. 139).

■ As these snails are hermaphrodite, a single individual could be sufficient to establish a new population.

FIGURE B6.6 Tree snails (*Achatinella mustelina*) on a leaf of a *Nestegis sandwicensis* (a native Hawaiian plant in the olive family) in Makaha Valley, island of Oahu. The zigzag markings are where the snails have grazed for food.

geographical space. What explains the origin of many new species rather than the transformation of one old species? The German naturalist and geographer, Moritz Wagner (1813–1887), among others — especially John Gulick (see below) — maintained that Darwin did not even fully explain the evolution of a single species into a single new species.

A major problem was how a new species could evolve in the same locality as its parents,

> Free crossing of a new variety with the old unaltered stock will always cause it to revert to the original type. . . . Free crossing, as the artificial selection of animals and plants uncontestably teaches, not only renders the formation of new races impossible, but invariably destroys newly formed individual varieties.

Such criticisms were not unfounded. Missing from Darwin's 1859 argument is a strong emphasis on the barriers that prevent exchange of hereditary material between different groups, enabling each isolated group to follow its own evolutionary path. Barriers such as season, habitat, behavior, sexual selection, mating incompatibility, and species sterility are discussed in Chapter 18.

Because Darwin did not emphasize isolation among groups as a primary cause for evolution, he also dismissed the notion that sterility among separately evolved groups might be beneficial. As the co-discoverer of natural selection, Alfred Wallace showed, it would be advantageous for isolated populations (although not necessarily for individuals in the population), each with its special adaptations, to produce sterile hybrids if they mated; sterility would allow each group to maintain its specific adaptations without dilution. Darwin insisted that sterility was primarily accidental, an insistence that blocked him from explaining the almost universal prevalence of sterility among hybrids and from using this important isolating barrier to account for the divergent evolution of closely related species.

Although Darwin knew isolation could be important in helping a population evolve, he thought that it was more essential to establish that speciation could occur through time without isolation. It therefore remained the task of later evolutionary biologists to explore the role of isolation in species formation (see Chapter 17). Furthermore, Darwin was not the first to develop a theory of evolution based on random variation and geographical isolation. He was scooped by John Gulick (1832–1933), a missionary and evolutionary biologist much influenced by Darwin's writings. Gulick's research is outlined in BOX 6.6 .

Nevertheless, Darwin's theory had the advantages of relying on explicable mechanisms, being supported by a massive amount of data, and successfully explaining many features of natural systems.

Recommended Reading

Appleman, O. (ed.), 2001. *A Norton Critical Edition. Darwin: Texts, Commentary,* 3rd Edition. W. W. Norton, New York.

Cadieu, E., M. W. Neff, P. Quignon, et al., 2009. Coat variation in the domestic dog is governed by variants in three genes. *Science* **326**, 150–153.

Darwin, C. R., 1859. *On the Origin of Species by Means of Natural Selection or the Preservation of Favoured Races in the Struggle for Life.* John Murray, London.

Lewontin, R., 2000. *The Triple Helix: Gene, Organism, and Environment.* Harvard University Press, Cambridge, MA.

Malthus, T., 2008. *An Essay on the Principle of Population,* edited by G. Gilbert. Oxford University Press, Oxford, England.

Mayr, E., 1988. *Toward a New Philosophy of Biology. Observations of an Evolutionist.* The Belknap Press of Harvard University Press, Cambridge, MA.

Quammen, D., 2006. *The Reluctant Mr. Darwin.* W. W. Norton, New York, 304 pp.

Raby, P., 2001. *Alfred Russel Wallace, A Life.* Princeton University Press, Princeton, NJ.

Shermer, M., 2002. *In Darwin's Shadow: The Life and Science of Alfred Russel Wallace.* Oxford University Press, New York.

Slotten, R. A., 2004. *The Heretic in Darwin's Court: The Life of Alfred Russel Wallace.* Columbia University Press, New York.

Sutter, N. B., C. D. Bustamante, K. Chase, et al., 2007. A single IGF1 allele is a major determinant of small size in dogs. *Science* **316**, 112–115.

Wallace, A. R., 1889. *Darwinism: An Exposition of the Theory of Natural Selection, with Some of Its Applications.* Macmillan, London.

Darwin, Mendel and Theories of Inheritance

- Neither Darwin nor Wallace knew how variation could be produced in Nature.
- The accepted theories of inheritance were the blending of parental traits in each generation and the inheritance of acquired characters.
- Darwin continued to hold to Lamarckian inheritance.
- In seeking an alternate theory of inheritance, Darwin revived the theory of pangenesis in which gemmules traveled from all the organs to accumulate in the germ cells.
- Discovery of the separation of germ plasm and body plasm disproved both Lamarckian inheritance and pangenesis.
- Gregor Mendel's breeding experiments with pea plants provided a mechanism of inheritance in particular factors, subsequently shown to be genes.
- Mendel proposed two laws of what would later be known as genetics; the principles of segregation and of independent assortment of hereditary units, now known to be genes.

Above: Like begets like is fundamental to the theory of inheritance.

- The presence of two sexes in many species and the discovery of varied mechanisms of sex determination demonstrated the role of genetics, the environment, and environment-gene interactions.

Overview

Although natural selection explained how species survive, and could explain changes within a species, how species transform into other species remained unresolved. Darwin and Wallace's theory of natural selection depended on heritable variations on which selection could act. Neither Darwin nor Wallace knew how such variation arose. Not until the science of genetics developed in the early 20th century was the source of inherited variation shown to reside in genes, but genes did not resolve all the issues surrounding the transformation and origination of species.

Darwin was aware of the problem of the prevailing view of blending inheritance by which offspring were thought to represent blended copies of their parents' features. Most 19th century biologists advocated blending inheritance. If blending inheritance operated, however, natural selection would be incapable of maintaining a trait for more than a few generations. The idea of blending inheritance therefore confronted evolutionary theory with a serious enigma. If characters are blended out when individuals mate with other members of the population, how can beneficial variations be preserved by natural selection? Darwin spent many years seeking an alternative to blending inheritance, settling on the theory of pangenesis, a theory disproved by the discovery of the separation of germ cells from body cells.

In the mid 1860s, Gregor Mendel broached two principles of inheritance that contradicted blending inheritance: (1) the *principle of segregation* — what we now know to be alleles of a single gene, discrete entities that segregate from each other into gametes — and (2) the *principle of independent assortment* — the independent segregation of genes on different chromosomes within the gametes that form from germ cells. Although the phenotypic effects of some genes may blend with those of other genes, genes do not blend with each other.

The first two to three decades of the 20th century witnessed an explosion of knowledge of genetics and modes of inheritance. I say "modes" because some inheritance was discovered to be extranuclear and some was linked to genes that determine sex. Surprisingly, in some groups, sex is determined not by genes on sex chromosomes but by a diverse array of environmental signals (environmental sex determination). Gene-environment interactions therefore emerged as an important aspect of biological organization.

Seeking a Mechanism of Heredity

Darwin's most fundamental insight was that small differences among individuals within populations are the raw material for evolution (see Chapter 6). Before and during his time, most scholars, even Darwin's most staunch supporter, Thomas Huxley held the view that **saltations** (jumps; large discontinuous changes in organisms or parts of organisms) were the changes most relevant to evolution. Resolving the important issue of small versus large differences was not possible because the mechanisms of inheritance were not understood.

Lamarckian Inheritance

At the base of the evolutionary process is the continual introduction of new heritable variations on which selection can act. Darwin did not know the biological basis of heredity or of variation. Consequently, his arguments were weakest in these areas.

At times, Darwin proposed that environmental change, a large increase in numbers, or some disturbance of the reproductive organs might enhance variation among a population of individuals. At other times, he adopted a version of the Lamarckian view of the inheritance of use and disuse outlined in Chapter 6. We can gain an appreciation of how Darwin and Lamarck's approaches to variation and heredity differed by their positions on several key issues.

- Both Darwin and Lamarck held that the diversity of species descended from a common ancestor; that is, both believed in evolution.
- As to why some features persisted while others disappeared, Lamarck invoked use and disuse and the inheritance of acquired characters. Darwin invoked **natural selection.**
- As to how new variations arose Lamarck invoked use and disuse. So did Darwin although he came to place more emphasis on environmental change rather than use and disuse. Darwin saw no directionality imposed by environmental changes but proposed natural selection as the guiding principle in evolution.

Blending Inheritance

The fundamental problem for Darwin, as for many of his contemporaries, was that offspring were thought to represent perfect or *blended copies* of their parents' parts. Because blended parts will tend to produce features that average out those of the parents — just as mixing black and white produces gray — **blending inheritance** sets variation and inheritance in opposition.

Because the prevailing concept of inheritance in the 1850s was that maternal and paternal contributions blend in their offspring, a number of critics objected that new adaptations would be successively diluted with each generation of interbreeding. According to this argument, natural selection would be incapable of maintaining a trait for more than a few generations. Darwin made various replies/responses/rebuttals among which the five outlined below are perhaps the most interesting.

A beneficial character could maintain itself if the individuals with the character became isolated from the rest of the population. Darwin pointed to the familiar practice among animal breeders of isolating newly appearing "sports" and their offspring. This mechanism was commonly used to develop new stocks of domesticated animals or plants. Darwin placed major emphasis on artificial selection as engineered by man as evidence supporting his conception of natural selection (see Chapter 6).

Some characters are prepotent (we would now say *dominant*) and appear undiluted in later generations.

A new trait does not appear only once in a population. Rather, such traits must arise fairly often; witness the large amount of variation in most populations. Because variation is common, went the argument, it would not dilute out as easily as it would if it was rare. Moreover, Darwin believed, some forms that carry a particular variation pass on to future generations the tendency for the same variation to arise again.

Natural selection both enhances the reproductive success of favorable variants and diminishes the reproductive success of unfavorable ones. The frequency of favorable variations, thus, increases when unfavorable ones die out, reducing the likelihood of diluting out favorable variations.

As discussed below, Darwin developed the concept of pangenesis by gemmules to help explain the inheritance of traits, which he believed were affected by use and disuse, and to provide constancy for the determining agents of inheritance (see Chapter 6). There were presumably many gemmules for each particular trait, and their numbers could vary during passage from one generation to another providing a source of variation. Gemmules could be lost but were not changed by "blending."

Pangenesis

Some ten years after *The Origin* was published and as an alternative to blending inheritance, Darwin revived an old theory, **pangenesis**.

According to pangenesis, small particulates known as gemmules or pangenes were produced by all the tissues of a parent and incorporated into the developing eggs or sperm. When fertilization occurred and parental gametes united, gemmules disperse to form the tissues of the offspring (FIGURE 7.1a). By suggesting that changes can arise in the frequencies of particular gemmules while their structure remained constant, pangenesis helped account for the presumed effects of use and disuse and for the observation that not all traits become blended.

There was no convincing evidence for pangenesis, however. Indeed, The German biologist and evolutionary theorist, August Weismann (1834–1914) believed he disproved pangenesis when he cut off the tails of 22 generations of mice and showed that tail length was not affected by the presumed loss of tail gemmules in each generation. Although proof against the inheritance of acquired characters, this is indirect

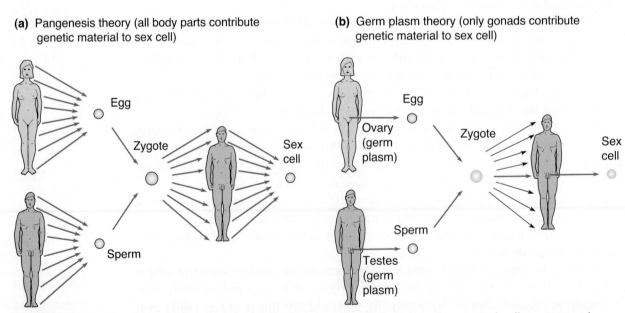

(a) Pangenesis theory (all body parts contribute genetic material to sex cell)

Egg

Zygote

Sex cell

Sperm

(b) Germ plasm theory (only gonads contribute genetic material to sex cell)

Egg

Ovary (germ plasm)

Zygote

Sperm

Testes (germ plasm)

Sex cell

FIGURE 7.1 Comparison between **(a)** pangenesis and **(b)** germ plasm theories in humans. In pangenesis, all structures and organs throughout the body contribute particles (pangenes, gemmules) to a sex cell. In the germ plasm theory, only the sex cells contribute hereditary units (genes) to the next generation. (From *Genetics*, third edition, by Monroe W. Strickberger. © 1985 by Monroe W. Strickberger. Reprinted by permission of Prentice Hall, Inc., Upper Saddle River, NJ.)

proof, at best, against pangenesis. A better test was undertaken by Francis Galton (1822–1911), a cousin of Charles Darwin, who transfused blood between different breeds of rabbits without altering the nature of their offspring. Darwin's response was that he had not proposed that gemmules circulated in the blood. He did, however, investigate pangenesis using rabbits.

Nineteenth-century Lamarckians understood "acquired characters" to be somatic traits that could be reproduced in each generation without instruction from germ-line tissue and that could be transmitted through somatic or through germinal cells. Weismann devised the **germ plasm theory** of inheritance, in which only the reproductive tissues (testes and ovaries) transmit the heredity factors of the entire organism; changes that occur in nonreproductive somatic tissues are not transmitted (Figure 7.1b). Thus, changes in heredity cannot simply be explained by inheritance of acquired characters or by use and disuse.

■ Weismann's theory of the continuity of the germ plasm is applied far too uncritically as if it applies to all animals. Separation of germ plasm and soma is true only for a minority of phyla within the animal kingdom and is not true in plants, fungi or protists, in which a new individual could arise from any body cell.

With respect to heredity and inheritance, the rediscovery in 1900 of Mendel's work produced a generation of Mendelians, many of whom were opposed to Darwin, especially to his mechanisms of natural selection. Evolution meant something very different to them. It would be the 1930s and 1940s before many of the difficulties Darwin faced were resolved.

Constancy and Variation

Organismal evolution relies on two fundamental aspects of genetics or biological inheritance: **constancy** and **variation**. Constancy resides in the observation that **like produces like**. Constancy has the evolutionary significance that all life processes depend on the transmission of information from previous generations. In contrast, variation resides in the observation that **like can produce unlike**.

Replication of biological information is not always constant or exact. This is because of changes (mutations) in the DNA of the genes, which, if they occur in gamete-producing cells are passed on to the next generation (see Chapter 12). The evolutionary significance of genetic variation is its potential to fuel evolution.

Constancy and variation are indelibly intertwined in evolution through hereditary information (the genotype) that primarily determines organismal features (the phenotype). However, there is no one-to-one correspondence between genotype and phenotype. As developmental and environmental signals influence (perhaps even determine) which genes, gene pathways and gene networks are expressed at which times and stages of the life cycle, multiple phenotypes can arise from a single genotype. Examples are tadpole and frog, caterpillar and butterfly, and queen, soldier and worker ants (see Chapter 11).

Mendel's Experiments

As discussed above, the pre-Mendelian view prevailing throughout the 19th century was that heredity followed blending inheritance in which offspring inherited a dilution, or blend of parental characteristics derived primarily from their appearance (phenotype). Because the molecular basis of genetics long remained unknown, fundamental genetic principles were derived primarily from observations on the transmission of more obvious biological features. In the early 1860s, Gregor Mendel in Brno, Czechoslovakia developed the genetic laws derived from such observations.

Biologists were unaware of Mendel's work until it was discovered independently by three individuals in 1900. Soon after 1900, discovery of the association between Mendelian factors and the distribution of chromosomes during meiosis formed the foundation for the new science of genetics. Mendel's exceptional contribution was to demonstrate that organisms have a distinct hereditary system (now known as the genotype), which transmits biological characteristics through discrete units (now known as genes) that remain undiluted in the presence of other genes. The time line of discoveries leading to this knowledge is outlined in TABLE 7.1 .

TABLE

7.1 Major Discoveries Leading to Our Current Concepts of the Nature of the Gene

Year(s)	Discovery
1856	Gregor Mendel's first crossbreeding experiments with the garden pea
1865	Presentation by Mendel of his results at the monthly meetings, 8 February and 8 March, of the *Naturforschenden Vereins* [The Natural Science Society] in Brünn, Austria-Hungary (now Brno, Czech Republic)
1866	Publication of Mendel's paper "Versuch über Pflanzenhybriden" ("Experiments on Plant Hybridization") in the journal of the *Natural Science Society;* his work attracts little attention
1869	Isolation by Friedrich Miescher of nuclein (DNA) from nuclei, marking the first research with what we now know to be nucleic acids
1900	Mendel's 1866 experiments rediscovered independently by three researchers and publicized by William Bateson
1902–03	Chromosomal theory of inheritance proposed independently by William Sutton and Theodor Boveri
1909	Demonstration by Archibald Garrod that some human diseases are inborn errors of metabolism, inherited as Mendelian recessive characters
1910–11	X-chromosomes, sex linkage, mutant gene for eye color in *Drosophila* discovered by Thomas Hunt Morgan, who proposes exchange of chromosome segments by crossing over
1913	Chromosomes shown to contain genes in a linear array
1927	Discovery by Hermann Muller that X rays induce mutations in *Drosophila* genes
1941	Discovery that genes code for enzymes in the mold *Neurospora;* proposal by George Beadle and Edward Tatum of the "one gene–one enzyme" hypothesis
1943–44	Evidence that bacteria contain genes; isolation of DNA and demonstration that DNA is genetic material
1952	Proteins eliminated as basis of genes
1953	Proposal by James Watson and Francis Crick of the three-dimensional double helical molecular structure of DNA, based in part on unpublished x-ray crystallographic data obtained by Rosalind Franklin and by Maurice Wilkins
1957–58	Proposal by Francis Crick of the Central Dogma of molecular biology: DNA → RNA → protein
1950s	François Jacob and Jacques Monod establish control functions located on chromosomes that turn the expression of genes on or off

TABLE 7.1	Major Discoveries Leading to Our Current Concepts of the Nature of the Gene (continued)

Year(s)	Discovery
1958	Isolation of the first DNA replicating enzyme
1960	Discovery of messenger RNA in bacterial cells
1961	Determination by Jacob, Crick, Sydney Brenner and others of the triplet nature of the genetic code, transcription of information in DNA into messenger RNA, and translation of mRNA into protein
1962	Use of synthetic RNA to unravel the genetic code (genetic information is carried in three-nucleotide sets of 64 codons, each coding for one of the 20 amino acids); discovery of repressor and transcriptional control of genes
1964	Crick's Central Dogma shown not to hold for viruses
1966	Complete genetic code translating codons into amino acids established
1975	DNA copied from messenger RNA (cDNA)
1983	Discovery of homeobox as basic element of homeotic genes; direct detection of the nucleotide sequence of a specific mutant allele (sickle cell)
1989	First human gene sequenced, and a defect in the gene product shown to "cause" cystic fibrosis
1999	Sequencing of a human chromosome (see Chapter 20), and of complete fruit fly (*Drosophila melanogaster*) genome
2001	First draft sequences of human genome completed independently by the Human Genome Project and Celera Genomics
2003	Sequencing of 99 percent of the human genome completed to 99.9 percent accuracy
2006	Sequencing of the human genome completed; publication of a global map of p53 transcription-factor binding sites in the human genome, and of the first map of DNA methylation of a genome, the water cress *Arabidopsis thaliana*.

■ Methylation of cytosine is an important means of silencing and regulating gene function. Methylation is found within transcribed regions in over one third of expressed genes in *Arabidopsis,* which has undergone repeated whole-genome duplications during its evolution.

Mendel experimented with a number of characters in the pea plant, *Pisum sativum*, in which each character possesses two alternative appearances (traits): smooth or wrinkled seeds, yellow or green seeds, tall or short plants. Mendel bred pea plants for many generations, observing the appearance and counting the numbers of each different trait in every individual of every generation in a large number of plants.

Mendel was a very careful experimentalist; some have made the comparison between Mendel as the first great modern experimentalist in biology and Darwin as the last great descriptive naturalist (BOX 7.1). Mendel spent from 1856 to 1858 ensuring that his plants bred true; he was using the scientific method outlined in Chapter 1. This was important, both for the characters he chose and because he was working with 37 varieties of peas from this one species. In case you want to repeat some of Mendel's experiments, the characters he found most distinctive and easiest to distinguish were

- seed shape — round(ish) or angular and wrinkled
- seed color — pale yellow/bright yellow/orange-colored, or green

BOX 7.1

Styles in Science; Mendel and Darwin

The structure of Mendel's paper is worthy of a short comment. Unlike the discursive literary "natural-history" style employed by Darwin with a minimum of analyses of data beyond the descriptive, Mendel's paper is in the plant hybridization tradition (Mendel was not the first to hybridize plants) with data, analyses, ratios, citations of past work, and a more direct style. The notation S for the dominant character, s for the recessive, and Ss for the hybrid are used throughout Mendel's paper. Darwin also did experiments in plant hybridization, reported some traits in 3:1 ratios (as did Mendel), and used statistical methods in his analyses. Whether Darwin was one of the last of the "old-style" natural historians in biology and Mendel one of the first of the "new" quantitative biologists would make an interesting topic for a term essay. See the Mendel Web site (http://www.mendelweb.org/) for further information.

- seed coat color — white or grey/grey brown/ leather brown; presence or absence of purple spots
- mature seed pods — inflated or deeply constricted and less wrinkled
- upright pod color — light to dark green or vividly yellow
- flower position — along the main stem, or at the tip of the stem
- stem length — 0.3 to 2.1 m

From the results obtained breeding peas, recording characters and calculating ratios between 1856 and 1863, Mendel developed two fundamental principles of heredity: the **principle of segregation**, and the **principle of independent assortment**.

Principle of Segregation

When two plants with different expression of a character are crossed, the first filial (F_1) generation carries the genes for each of these traits. One of the characters used by Mendel was the appearance and "feel" of the seed's coat (its outer covering), either smooth or wrinkled.

We now know that each pea plant has a pair of genes for seed coat texture as well as pairs of genes that govern other characters such as seed color and plant size. Pea plants are diploid, containing a copy of the gene from the female parent and one from the male parent. In current terminology, the two members of a particular gene pair (the gene for smooth or wrinkled seeds, for example) are known as alleles of the gene. Each gene has at least two alleles but may have many alleles, greatly increasing the source of variation over that which a pair of alleles could provide.

Geneticists use symbols to distinguish alleles, a system introduced by Mendel himself (Box 7.1). For simplicity, we will label the smooth and wrinkled alleles as S and s, respectively, the system used by Mendel. These alleles segregate in the gametes of the F_1 generation, which, if crossed, unite to form a second filial generation (F_2) in predictable proportions of $1\ SS : 2\ Ss : 1\ ss$. In this experiment Mendel obtained an average of 2.96 smooth seeds for every one wrinkled seed. In a second experiment involving flower color — red or white — he obtained an average of 2.89 red flowers for every one white flower. Experiments involving these two characters along with five other different characters gave an average ratio of 2.98:1, or rounding off, 3:1. Why a 3:1 ratio?

In breeding experiments the parents (parent generation) is designated the parental (P) generation. The generation obtained by crossing the parents is the first filial (F_1) generation, where filial refers to offspring. The generation obtained by crossing two individuals from the F_1 generation is the second filial (F_2) generation, and so on.

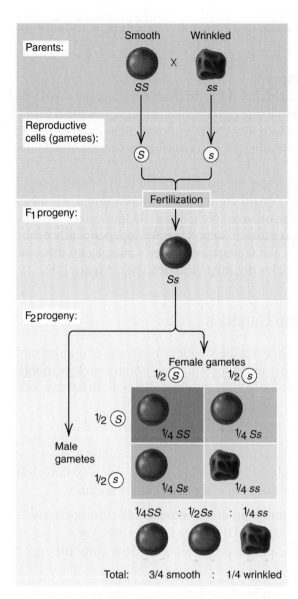

FIGURE 7.2 Explanation of Mendel's results for the inheritance of seed shape — smooth (S) or wrinkled (s) — in pea plants (*Pisum sativum*) beginning with a parental cross of homozygous smooth (SS) and homozygous wrinkled (ss) plants, where S alleles are dominant and s alleles recessive. Self-fertilization of Ss gametes from the first filial (F₁) generation results in a 3:1 ratio of smooth to wrinkled seeds in the second filial (F₂) generation.

As shown in FIGURE 7.2, these F_2 proportions arise because the allele for one trait S is **dominant** over the allele s (known as the **recessive** allele), when both are present in Ss individuals. That is, SS and Ss seeds both are smooth. The only wrinkled seeds are those with the ss genotype. Hence, the 3:1 ratio. In general, the alleles found in nature in the majority of individuals (and therefore known as **wild type alleles**) are dominant, presumably because they produce advantageous products in the presence of other alleles whose products may not be as advantageous or may be deleterious.

Mendel's experiments showed that the factors (genes) responsible for heredity are neither changed nor blended in the heterozygote, but segregate from each other to be transmitted as discrete and constant particles between generations. Within a few years of the rediscovery in 1900, many other studies in a variety of plants and animals supported Mendelian ratios and their variations, and supported the

Individual organisms that carry two different alleles for any particular character (Ss) are **heterozygotes**. Those that carry two identical alleles (SS, ss) are **homozygotes**.

principle of segregation and Mendel's second principle, the principle of independent assortment.

Principle of Independent Assortment

When Mendel cross-pollinated plants with two different characters, he discovered that the observed results could be predicted if one character had no effect on the segregation of the other. The characters behaved as expected of particulate, non-blending inheritance. We now say that genes for different characters segregate independently of one another.

The cellular explanation for such independent assortment is the localization of the genes for each of the two characters on different pairs of chromosomes, as shown diagrammatically in FIGURE 7.3 . During meiosis (see Chapter 4), the two halves of a pair of homologous chromosomes move toward opposite poles independently of any other pair of chromosome. As a result, the segregation of genes on one chromosome is independent of the segregation of genes on other chromosomes (Figure 7.3).

Deviations from Mendelian Genetics

In the decades that followed the rediscovery of Mendel's research, two important deviations from Mendelian genetics became apparent. One is **extranuclear inheritance**. The other is an **unpredicted pattern of segregation of alleles**. Both have genetic and evolutionary consequences.

Extranuclear Inheritance

The first unusual finding was that some traits do not follow a nuclear pattern of inheritance but rather transmit through the cytoplasm of the egg. Such extranuclear inheritance (also called *cytoplasmic* or *maternal inheritance*) arises because

- cytoplasmic organelles such as mitochondria in animals and chloroplasts in plants have their own genetic material (DNA; see Chapter 10), and
- the female parent deposits gene products (proteins, mRNAs) into the egg as it is being formed.

During animal development, practically all the cytoplasm of the zygote, including organelles such as mitochondria and ribosomes, is maternally derived (see Figure 4-7). The sperm contributes nuclear genetic material and, at most, a few mitochondria. Mitochondria, the nucleus and nuclear membrane, and the microstructure of the zygote are all present pre-formed in the cytoplasm of the egg at the start of every generation. Such inherited structures provide evidence that we should not abandon entirely the notion of preformation (BOX 7.2; see Chapter 1).

Sex-Linked Genes and Sexual Reproduction

The second unusual departure from Mendelian genetics — actually an elaboration rather than a departure — is that genes do not necessarily assort independently of each other if they are linked together on the same chromosome. Gene linkage can be appreciated in several contexts, perhaps the most relevant being the **evolution and possession of two sexes** (male and female) in a population, and the **sexual reproduction** that accompanies the evolution of male and female sexes. In fact, lack of independent assortment was discovered when it was found that some genes — **sex-linked**

genes — are localized to the X or Y (sex) chromosomes, which are not partitioned equally to females and males.

Sexual reproduction provides two important sources of new genetic variation, both of which provide advantages when environments are changing. First, because of **recombination or genetic exchange**, sections of chromosomes can exchange material (known as crossing over) thereby forming different linear arrays of nucleotides. As nucleotide sequences constitute genes (see Box 1.2 and Chapter 9), recombination can produce different combinations of genes along a chromosome. In humans with a normal chromosome number of 23 pairs of chromosomes, many millions of different kinds of gametes can form by the mechanism of recombination alone.

Despite close to two dozen different proposals, we do not fully understand whether the origination of sex and the evolution of genetic recombination were connected nor why they persist. We expect natural selection to preserve associations between alleles that benefit the fitness of an individual. Sexual reproduction, however, creates new association and genetic recombination breaks up those alleles.

With respect to evolutionary origins, one proposal is that sex (here considered as the origin of meiosis; see Chapter 4) originated as a way of overcoming DNA damage by using recombinational DNA repair mechanisms. Any genetic variation that resulted is regarded as an accidental by-product. However, even if recombination arose as a DNA repair mechanism, the production of genetic variation is the primary reason for the persistence of sex.

Of the many hypotheses to explain the persistence of sexual reproduction (and as a consequence, genetic recombination), **removing deleterious mutations** is perhaps the most reasonable. In the water flea, *Daphnia pulex* — which can modulate its mode of reproduction from sexual to asexual — more mutations accumulate with asexual than with sexual reproduction. It also has been proposed that sex is advantageous to organisms that exist in variable environments, and that parasitic elements, such as transposons and plasmids (see Chapter 12), initiated or promoted sexual fusion as a mechanism to infect other cells.

Various population geneticists point to populations rather than individuals in seeking a cause for sexual reproduction. Among the most popular proposals, sexual crossing in a population allows single individuals to incorporate different beneficial mutations from other members through mating and genetic recombination. Without sex, combinations of such beneficial mutations are more difficult to achieve. Furthermore, mating and recombination in sexual populations enable relatively rapid combinations of existing mutations. In both cases, the advantages achieved include additional genetic variation allowing a population to persist in a changing environment. For example, parasites usually evolve rapidly to counter resistance in their hosts and hosts usually evolve rapidly to counter infectivity of their parasites. This "**arms race**" places a premium on rapid evolution through sexual recombination.

Sex Determination

Sexual reproduction involves differentiation into two different sexual forms in which reproduction occurs only as a result of union between gametes of different individuals to form a zygote (see Figure 4.7). This simple statement fails to reveal the amazing range of mechanisms for sex determination, a few of which are outlined in BOX 7.3. The

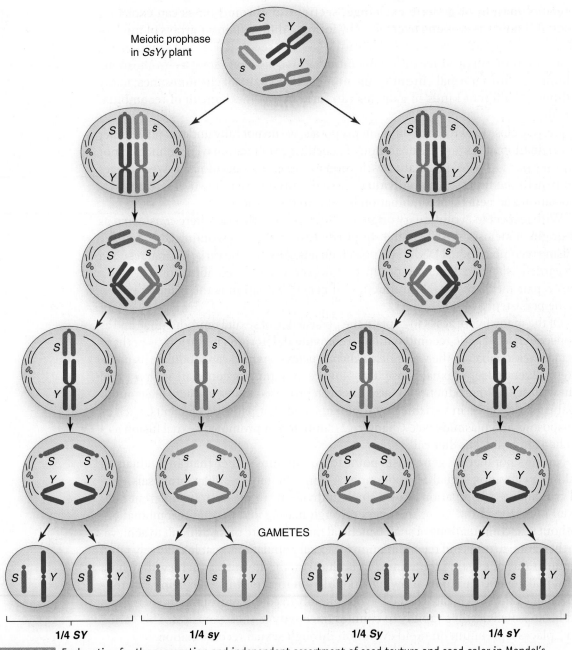

Meiotic prophase in *SsYy* plant

GAMETES

1/4 SY **1/4 sy** **1/4 Sy** **1/4 sY**

FIGURE 7.3 Explanation for the segregation and independent assortment of seed texture and seed color in Mendel's experiments in terms of genes (*S*, *s*, and *Y*, *y*) on different chromosomes. Because of independent assortment in the four stages of meiosis shown, all four combinations of chromosomes in the gametes correspond to all four possible combinations of the genes *S* and *s* (smooth and wrinkled seeds) and *Y* and *y* (yellow or green seeds). Because of dominance-recessive relationships, three quarters of the gametes produce smooth yellow seeds and one quarter produce wrinkled green seeds — the classic 3:1 ratio obtained by Mendel. (Adapted from Strickberger, M. W. *Genetics, Third edition*. Macmillan, 1985.)

BOX 7.2
Preformation

Preformation was introduced in Chapter 1 in two contexts. One was as the original meaning and usage of the word evolution for the gradual unfolding of a preformed body plan, as seen when a butterfly emerges from a chrysalis. The second was the theory of encapsulation (*emboîtment*) proposed by Charles Bonnet for all members of all future generations as preformed within the egg, seed or larva. A third context for preformation (preformationism) is as an alternative to spontaneous generation.

Although the theory of spontaneous generation was not abandoned until the crucial experiments of the chemist Louis Pasteur and the physician John Tyndall in the 19th century (see Box 6.1), serious attempts to replace spontaneous generation with a theory of preformationism began much earlier.

Under preformation, at conception each embryonic organism was regarded as preformed as a perfect replica of the adult structure, which gradually enlarged through the nourishment provided by the egg and the environment. Some preformationists, now known as ovists, proposed that the miniature adult was contained within the egg. Others (now known as spermists or animalculists) imagined that the adult in miniature was contained within the paternal seminal fluid or sperm (FIGURE B7.1). In its most extreme form, preformationism led to the *emboîtment* (encasement) theory, espoused by Bonnet and others (see Chapter 1), in which the initial member of a species encapsulates within it the preformed "germs" of all future generations; Eve's ovaries contained the entire preformed human species nested like an infinite set of Russian dolls. Although preformation had the satisfying quality of explaining the many different plans of organismal growth and discounting the idea of spontaneous generation, it led to belief in the fixity of species and left the origin of species in the hands of a creator.

FIGURE B7.1 A fully formed human (homunculus) encased within the head of a sperm. (Adapted from Nicholas Hartsoeker, 1694.)

genetic and phenotypic variation produced by sexual reproduction appears essential for the long-term survival of those groups facing changing environments.

Sex Chromosomes

Early in the twentieth century, geneticists discovered that particular genes were located on those chromosome(s) associated with sex determination, the **sex chromosomes**.

BOX 7.3

Modes of Sex Determination

The range of mechanisms of sex determination among animals is illustrated with five examples.

Penis Fencing in Flatworms

Even though they are **hermaphrodite** — both male and females organs are found within each individual — marine flatworms of the species *Pseudobiceros hancockanus* hunt and fight for mates. Having found another individual with which to mate, they begin penis fencing. Indeed, some individuals have two penises, giving them a decided advantage.

Why bother, when you can fertilize yourself? Aside from the obvious disadvantage of inbreeding, a disadvantage of which worms may be unaware, it turns out that your sex depends on whether you are the first to pierce your opponent (mate) with your penis. The first to do so functions as a male, delivering a package of two-tailed sperm to the other individual, which, perforce, becomes the female and then invests resources in egg production.

Temperature-Dependent Mechanisms

The sex of alligators and of many other reptiles is set during embryonic development by the temperature at which the eggs develop. In some species, a higher temperature determines maleness while in other species maleness is set at lower temperatures and femaleness at higher temperatures.

Fish That Change Their Sex as Adults

You probably know about the many chemicals that when released into the environment, lead to the appearance of female characteristics (feminization) in male animals. Not all may know that the majority of tropical reef fish change sex during their lifetimes, often in a matter of hours, some switching from male to female, others from female to male, and some switching without inhibiting the other sex, becoming hermaphrodites.

We assume the signals are hormonal but they await investigation; hormonal changes are required for sex reversal in birds, although the change requires experimental manipulation and does not occur in nature.

Compensatory Mechanisms in Chickens

Male chickens have right and left testes of similar sizes, which is the normal vertebrate condition. Females have left and right gonad rudiments of similar sizes, but only the left becomes a functioning ovary, the right remaining rudimentary. If the functional ovary is removed, the rudiment develops, sometimes into an ovary, sometimes into a testis and sometimes into an "ovotestis" with features of both sexes. Hormones provide the trigger here; treating males with estrogen leads to the development of an "ovotestis."

Fertilized Versus Unfertilized Eggs

As the final example, male honeybees — indeed, all hymenopteran insects — develop from unfertilized eggs and so are haploid. Females develop from fertilized eggs and so are diploid, a mode of sex determination known as **haplodiploidy**, and which facilitates the evolution of sociality in which only one female becomes a queen and lays eggs. All other females help raise the queen's eggs and so contribute, albeit indirectly, to her reproductive success, a phenomenon named kin selection and discussed in Box 15.1. The queen is not merely a passive recipient of the contributions of these sterile workers; she releases a hormone that suppresses fertility in the workers (see Chapter 15).

Although the haplodiploidy hymenopteran system may produce different sex ratios based on various environmental factors, in the parasitoid wasp, *Lariophagus distinguendus,* which parasitizes the larvae of common granary weevils, *Sitophilus granaries,* the

BOX 7.3

Modes of Sex Determination (*continued*)

female lays differently sexed eggs in response to the size of the wheat grain containing the larval host. If the wheat grain is relatively large, the wasp inserts a single fertilized female egg into the weevil larva; if the grain is relatively small, an unfertilized male egg is injected. This difference can be traced to the difference in resources needed for the fertility of male and female offspring: a larval host in a larger grain enables female offspring to be more fecund because the larger grain supplies more resources, whereas a larval host in a smaller grain still supplies enough resources to enable a male offspring to produce a large number of viable sperm.

For many organisms, especially mammals, sex determination is associated with chromosomal differences between the two sexes, typically XX females and XY males, with the Y chromosome often smaller and mostly inactive, except for male determining and male-fertility genes. An independently evolved XX:XY system of sex chromosomes also exists in *Drosophila* (see the following section). XX:XY chromosomes are not the only system of chromosomal sex determination, however. In the nematode *Caenorhabditis elegans* females are XX and males XO. Snakes and birds independently evolved a ZZ:ZW system in which females are ZW and males ZZ.

Monotremes diverged from placental mammals between 170 and 210 Mya (see Chapter 18). The platypus, *Ornithorhynchus anatinus*, has five X and five Y chromosomes, a system thought to be partly a modification of the mammalian X chromosome system and partly a carryover from a more ancient reptilian (and avian) sex-determining system; one of the X chromosomes carries the master sex-determining gene *doublesex and mab-3 related transcription factor 1* (*Dmrt1*), which is located on the Z chromosome in male birds; indeed this entire platypus X chromosome is equivalent to the Z chromosome.

Among the many questions these various findings raise are how sex determination became tied to sex chromosomes; how the sex with only a single X chromosome (XY or XO) compensates for the other sex having two copies of the X chromosome; and whether there is a selective advantage to the sex ratio being 1:1 in most species.

Interestingly, genes not associated with sex determination are linked to sex chromosomes in many animals. Furthermore, many of the sex-linked genes identified in modern humans are sex-linked in other mammals. Three examples are muscular dystrophy in humans, mice and dogs; hemophilia in humans, dogs, cats and horses; and testicular feminizing syndrome in humans, cattle, dogs, mice, rats and chimpanzee.

A prominent example is the "bleeder" disease, hemophilia, which is localized to the X chromosome of humans, dogs, cats and horses in which males are XY and females are XX. Males produce sperm that contain either X or Y chromosomes, females produce only X-bearing eggs. Because males carry only a single X chromosome, and because the Y chromosome is mostly inactive, alleles present on the X chromosome express their effects in males, although such alleles may be recessive in females. Thus, a single hemophilia-producing allele on the X chromosome in an XY male causes the classic hemophilia disease, whereas in XX females, two such alleles are necessary to cause hemophilia, a situation that rarely happens.

Human Y chromosomes have been studied especially intensively. While human X chromosomes are some 165 Mb in size with around 1,000 genes, the Y chromo-

■ Hemophilia is a little more complex than indicated because there are two forms: Hemophilia A, a deficiency in the blood clotting factor VIII affecting humans, dogs, cats and horses, and Hemophilia B, a deficiency in clotting factor IV affecting humans, dogs, cats and mice.

some is only 60 Mb and has far fewer genes. It is becoming apparent that genes on the Y chromosome have been and continue to be subject to specific forces of selection and higher levels of mutation, deletion and insertion than any other gene in any mammalian genome. Interestingly, and perhaps telling us something quite important, in flowering plants the mutation rate in pollen grains is also higher than in ovules.

It seems paradoxical to find, therefore, that mammalian Y chromosomes are undergoing systematic and ongoing degeneration to the point that the human Y chromosome has been regarded as close to extinction. Other mammals — indeed other vertebrates — have gone even further down the road to Y-chromosome extinction; males of two species of mole voles (*Ellobius*), and the Japanese spinous country ray, *Tokudaia,* have lost their Y chromosomes, the sex-determining genes having translocated to non-sex chromosomes (autosomes).

Autosomes and Sex Determination

The almost universal presence of sex chromosomes among sexual species does not necessarily mean these are the only chromosomes affecting sexual development. Sex is a complex developmental character affected by numerous non–sex-chromosome (**autosomal**) genes. In *Drosophila*, for example, the sex of an individual is determined by the ratio of X chromosomes to sets of autosomes (A). Females have an X/A ratio of 1 (2X/2A) and males have an X/A ratio of 0.5 (1X/2A). The function of sex chromosomes or the X/A ratio is to act as part of the "switch" mechanism that directs development into male or female. Ratios that differ from 1 or 0.5 are associated with sexual abnormalities; for example, triploids with an X/A ratio of 0.66 (2X/3A) are intersexes.

In another variant found in many species, the potential for becoming male or female exists at the time of fertilization, no matter what the sex chromosome complement; the gene that causes testicular feminization syndrome in humans turns XY zygotes into females, despite the presence of the Y chromosome. An evolutionary change considered at the end of the chapter is the replacement of chromosomal sex determination by environmentally induced sex determination in many taxa/lineages.

Environmentally Induced Sex Determination

In contrast to genetic mechanisms of sex determination, **environmentally dependent sex-determining mechanisms** occur in many animals and in most plants in which the sexes are separate. The influence on aspects of sexual reproduction of interactions between organisms and the environment or of interactions between individuals was outlined at the beginning of the section on sex determination. Environmental sex determination demonstrates the importance of **gene-environment interactions.** Five examples in BOX 7.4 provide a flavor of the extraordinary wide range of environmentally dependent sex-determining mechanisms.

If defined as the formation of separate male and female organisms — whether genetically or environmentally determined — sex determination does not apply to most plants; some 90 percent of seed plants produce both male and female gamete, pollen and ovules, respectively. Furthermore, only a minority of the five percent of plant species with separate male and female plants has sex chromosomes; interestingly, one of the most primitive of living plants, *Ginkgo biloba,* has sex chromosomes. Even in the absence of sex chromosomes, sex determination may be genetic or environmental; we just do not know for the vast majority of species.

Interestingly, chromosomal sex determination can transform into environmental sex determination when the sex of one or both of the XY and XX individuals reverses

BOX 7.4

Environmental Sex Determination

The range of modes of environmental sex determination among animals is illustrated with five examples.

Bonellia

In the green spoon worm, *Bonellia viridis,* larvae that are free-swimming and settle on the sea bottom develop into females with a 10- to 20-cm long body and a meter-long proboscis (FIGURE B7.2). Larvae that land on the proboscis of a female metamorphose into tiny, 1-mm-long males that lack digestive organs, existing as a parasite embedded within the genital ducts of the female, and producing virtually nothing but sperm.

FIGURE B7.2 At 10–20 cm long, the female green spoon worm *Bonellia viridis* (shown) dwarfs the 1-mm-long males (not shown).

Osedax

Populations of some 15 to 20 species of tubeworm in the genus *Osedax* consist of populations of females feeding on decaying whale bones at depths of 3,000 m or more (FIGURE B7.3a,b). The females lack a digestive system but contain endosymbiotic bacteria thought to produce the enzymes responsible for digesting the bone. Females are surrounded by a gelatinous capsule, which is colonized by hundreds of dwarf (<1 mm long), paedomorphic, non-feeding males (Figure B7.3c). It appears that larvae that settle on bone develop into females while larvae that settle on females become dwarf males with development arrested at a larval stage.

Painted Turtles

As introduced earlier, in many reptiles, high egg-incubation temperatures result in the production of males. In other taxa, for example, in the painted turtle *Chrysemys picta,* high temperatures result in the production of females.

Social Fish

In certain fish, social behavior can influence sex. Loss of socially dominant males from a group is followed by conversion of the dominant female in the group into a male. In some coral reef fish, sex changes seem to occur as a result of visual stimulation: females become males when the surrounding fish in the group are relatively small.

(continued)

BOX 7.4
Environmental Sex Determination (*continued*)

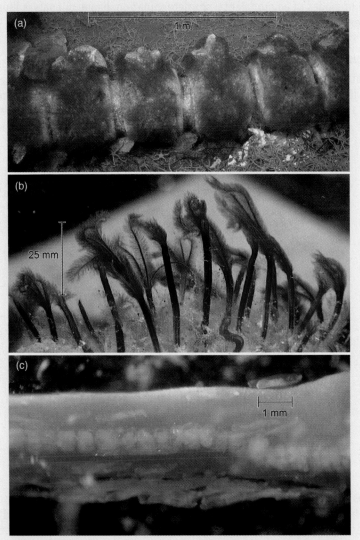

FIGURE B7.3 Tubeworms in the genus *Osedax* consist of large (50 mm long) females that obtain nutrition through digestion of whale bone at depths of 3,000 m or more (**a, b**). Tiny 1-mm–long dwarf males reside within a gelatinous capsule that surrounds the female (**c**).

Parasitoid Wasps

Some wasps deposit their eggs inside insect larvae. Female parasitoid wasps deposit eggs that will develop into females into larger host larvae and eggs that will develop into males into smaller host larvae. This mechanism enabled females to grow to larger sizes in the larger hosts and therefore to produce more eggs. Large versus small host larvae seem to be recognized by the wasp as a relative property of the larvae available; as long as there is a size difference, wasps will differentiate between them. An example, the parasitoid wasp, *Lariophagus distinguendus,* which parasitizes the larvae of common granary weevils, *Sitophilus granaries,* is discussed in Box 7.3.

because of sensitivity to agents such as temperature or hormones. An example of environmental impact on sex ratio was described in a 1997 study on Seychelles warblers, *Acrocephalus sechellensi,* birds that commonly use their daughters as "helpers" in raising additional offspring. When food is plentiful, helper daughters increase their parents' reproductive success, producing broods with a female:male ratio of about 6:1. When food is scarce, such daughters hinder their parents' reproductive success by competing for the limited supply, and the female: male offspring ratio drops to about 1:3.

Recommended Reading

Fisher, R. A., 1958. *The Genetical Theory of Natural Selection,* 2nd ed. Dover, New York.

Grützner, F., W. Rens, E. Tsend-Ayush, N. El-Mogharbel, et al., 2004. In the platypus a meiotic chain of ten sex chromosomes shares genes with the bird Z and mammal X chromosomes. *Nature,* **432,** 913–917.

Hallgrímsson, B., and B. K. Hall (eds.), 2003. *Variation: A Central Concept in Biology.* Elsevier/Academic Press, Burlington, MA.

Hartl, D. L., and E. W. Jones, 1998. *Genetics: Principles and Analysis.* Jones and Bartlett, Sudbury, MA.

Mendel, G., 1866. *Versuch über Pflanzenhybriden. Verh. Natur. Brünn,* 4, 3–47. (This is Mendel's classic paper, originally published in the *Proceedings of the Brünn Natural History Society.* It was translated into English by William Bateson in 1901 under the title *Experiments in Plant Hybridization.*) Available at http://www.mendelweb.org/Mendel.html.

IV

The Universal Tree of Life

From Single-Celled Organisms to Kingdoms

8

KEY CONCEPTS

- The first organisms may have been single-celled methane-producing bacteria that existed more than 3.5 Bya.
- The earliest known fossils are single-celled cyanobacteria that deposited stromatolite reefs 3.5 Bya.
- Both methanogens and cyanobacteria modified Earth's atmosphere, resulting in rising levels of O_2 and declining levels of CO_2.
- Organisms, speciation, and ecosystems are ancient, having arisen at least 3.5 Bya.
- Cells with nuclear membranes and organelles (eukaryotic cells) arose 2.5 Bya.
- The ability to form organelles with specific functions was a major factor in the origin and diversification of eukaryotes.
- Until about 50 years ago organisms were classified into two kingdoms, plants and animals.
- Life was classified into prokaryotic and eukaryote domains about 50 years ago using cellular characters.

Above: Darwin's grand analogy: evolution as a branching tree.

- About 30 years ago life was reclassified using molecular evidence into three kingdoms: Eubacteria, Archaea and Eukarya.
- Although we can classify life into kingdoms, all organisms are connected by having shared a common universal ancestor and as branches of a Universal Tree of Life (UToL).
- Shared fundamental cellular organization, metabolic pathways, and genes are three lines of evidence for the UToL.
- Constructing a tree of life for cells with prokaryotic organization may be impossible because of horizontal gene transfer throughout the history of life.

Overview

On the basis of the fossil record we can determine that prokaryotic cells existed 3.5 to 3.8 Bya, and that about two billion years later (2.5–2.8 Bya) some had diversified into eukaryotic cells (see Chapter 9). Extant prokaryotes include eubacteria and archaebacteria (Archaea). All are small (about 1 μm across), contain no nuclear membrane, cytoskeleton or complex organelles, and divide by binary fission. Fossilized prokaryotic cells (cyanobacteria) that formed complex structures known as stromatolites are known from rocks that are at least 3.5 By old.

Prokaryotic and eukaryotic cellular organization and the three kingdoms of organisms — Eubacteria, Archaea and Eukaryotes — are hypothesized to have originated from a single common ancestor. While similar in organismal complexity to eubacteria, Archaea share several core similarities with eukaryotic cells. Horizontal transfer of genes between prokaryotic lineages is now known to have been common, a finding that complicates establishing a **Universal Tree of Life (UToL)**. Recovering a tree of life for prokaryotes may be impossible.

All organisms, no matter how we name, classify or arrange them on The Tree of Life, are bound together by four essential facts.

1. They share a common inheritance.
2. Their past has been long enough for inherited changes to accumulate.
3. The discoverable relationships among organisms are the result of evolution.
4. Discoverable biological processes explain how organisms arose and how they were modified through time by the process of evolution.

Although each of these aspects has been studied and discussed at various times in human history, only after Charles Darwin developed and published his theory in the mid-nineteenth century did biological evolution become an acceptable scientific alternative to earlier explanations. The acceptance that organisms could change over time brought about an enormous change in the way we view the world and explain natural phenomena.

Kingdoms of Organisms

Organisms have long been arranged (classified) into major groups using various schemes and types of evidence: (A) unicellular and multicellular; (B) bacteria (microbes), plants and animals; (C) prokaryotes and eukaryotes (FIGURE 8.1). Such large groups

(a) Prokaryotic cell

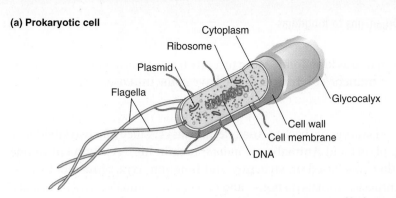

Cytoplasm
Ribosome
Plasmid
Flagella
Glycocalyx
Cell wall
Cell membrane
DNA

(b) Eukaryotic (animal) **cell**

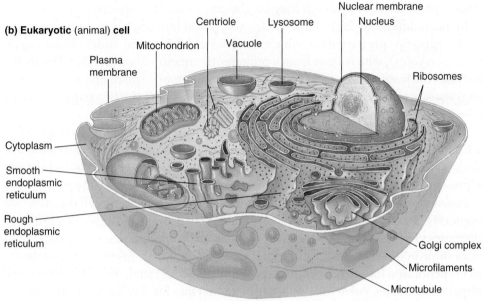

Centriole
Lysosome
Nuclear membrane
Nucleus
Mitochondrion
Vacuole
Plasma membrane
Ribosomes
Cytoplasm
Smooth endoplasmic reticulum
Rough endoplasmic reticulum
Golgi complex
Microfilaments
Microtubule

(c) Eukaryote (plant) **cell**

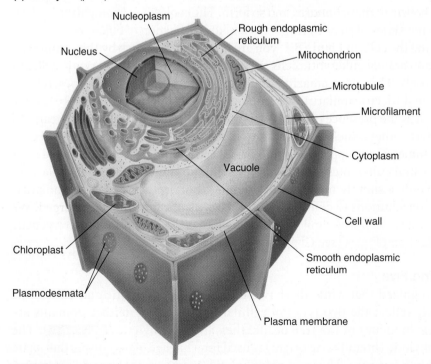

Nucleoplasm
Rough endoplasmic reticulum
Nucleus
Mitochondrion
Microtubule
Microfilament
Cytoplasm
Vacuole
Cell wall
Chloroplast
Smooth endoplasmic reticulum
Plasmodesmata
Plasma membrane

FIGURE 8.1 Diagrammatic representations of generalized prokaryotic (**a**) and eukaryotic animal (**b**) cells and eukaryotic plant (**c**) cell showing cross sections through various important cellular organelles.

have long been referred to as **kingdoms**. More recently terms such as super-kingdoms, domains, lineages or branches of the tree of life have come into use.

Two

For over two thousand years going back to Aristotle, life was classified into two kingdoms, **Plantae** (*L. planta,* plant) and **Animalia** (*L. anima,* breath, life). Assignment to one kingdom or the other was based on structure and function, type of metabolism — plants use photosynthesis, animals do not — and movement; animals move from place to place, plants do not, although they do disperse as seeds or spores.

In the middle of the 20th century, life was divided into two broad domains: single celled organisms (**prokaryotes**), which arose 3.5 to 3.8 Bya, and multicellular organisms (eukaryotes), which arose from a prokaryotic ancestor about 1.5 Bya. Figure 8.1 illustrates the features of typical prokaryote and eukaryotic (plant and animal) cells. Fungi also are eukaryotes. Prokaryotes and eukaryotes were the two major divisions of the tree of life.

The most obvious differences between prokaryotic and eukaryotic cells (Figure 8.1; see Chapter 4) are the absence of any internal membranous networks such as a nuclear membrane, organelles or cytoskeleton in the small prokaryotic cells (0.5–10 μm) and the presence of all three in the generally larger eukaryotic cells (10–100 μm).

Prokaryotes reproduce by binary fission, eukaryotes by mitotic cell division (see Chapter 4). Prokaryotes lack the membranous structures and organelles found in eukaryotes. Those organelles include endoplasmic reticulum associated with ribosomes in protein synthesis (see Chapter 3), a Golgi apparatus as secretory body, and mitochondria for energy production (Figure 8.1b, c). Such membrane-enclosed **compartments** allow eukaryotic cells to isolate enzymes for specific reactions, confine transcription of DNA to the nucleus, translation of DNA into RNA to the cytoplasm, aerobic metabolism to mitochondria, and so forth. The evolution of compartmentalization was a major series of steps in the origin and diversification of eukaryotes.

Comparing the cells in Figure 8.1, you are immediately struck by the complexity of the plant and animal eukaryotic cells in comparison with the prokaryotic cell. But look more closely. There is an amazing similarity between the three cell types, reflecting fundamental (deep) similarities (homologies; see below and Chapter 16), some of which, such as the presence of a cell membrane and genes, are billions of years old (see Chapter 4). Others, such as the organelles seen in plant, fungal and animal cells (nuclei, ribosomes, Golgi apparatus, endoplasmic reticulum, cytoskeleton) had their origin in ancestral eukaryotes. Yet others (mitochondria in animals, chloroplasts in plants) have only a slightly more recent origin(s), but, in their essential similarity, reflect common solutions to common problems as plants and animals emerged. We can conclude that a deep structural homology underlies life as much as do deep genetic and molecular homologies (see Chapters 3 and 4).

Two Became Five

It is now recognized that while these two domains of prokaryotic and eukaryotic life accurately reflect the two types of cellular organization, neither domains are monophyletic branches of the tree of life; they are polyphyletic (FIGURE 8.2). The term **monophyletic** (literally one origin) is used for a lineage of organisms that share a single common ancestor. The term **polyphyletic** (many origins) is used for groups that have more than one (often multiple) independent origins.

The use of light as an energy source ushered in the mode of metabolism — *photosynthesis* — that characterized virtually all plants (see Chapter 10).

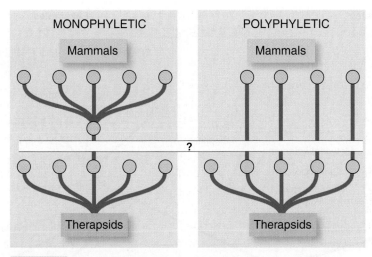

FIGURE 8.2 Monophyletic and polyphyletic schemes used to explain the evolution of mammals from therapsid reptiles. In monophyletic evolution, only a single reptilian group served as ancestor to the mammalian radiation. Two or more groups gave rise to mammals in the polyphyletic scheme.

Because they are not monophyletic, not all unicellular organisms are closely related to one another. Indeed, some unicellular organisms are more closely related to multicellular organisms than they are to other unicellular organisms. Prokaryotic and eukaryotic domains do not reflect these evolutionary realities.

Classification began to change in 1959 when the American plant ecologist, Robert Whittaker (1920–1980) presented evidence that the two domains, prokaryotes and eukaryotes (Figure 8.1) were more naturally represented as **five kingdoms**, one for prokaryotes and four (protists, fungi, plants and animals) to cover what were previously known as eukaryotes (see Box 9.1). This system lasted for about 25 years when molecular studies transformed our understanding of relationships yet again.

Five Became Three

In the early 1980s the American microbiologist, Carl Woese found that rRNA sequences from an **archaebacterium** (which is a prokaryote) were sufficiently different from those of other prokaryotes that archaebacteria should be placed into a separate domain or division of life termed *Archaea*. With further molecular analyses, especially sequencing of entire organismal genomes, both the prokaryote and eukaryote kingdoms were found to be artificial constructs. Consequently, a classification of life into three kingdoms was proposed. The three — **Eubacteria, Archaea and Eukarya** — cut across the prokaryotic and eukaryotic domains (**FIGURE 8.3**); organisms with prokaryotic cellular organization reside in more than one kingdom.

- Eubacteria (Bacteria) includes the major forms of bacteria and the cyano-bacteria, the latter being the earliest organisms known as fossils (see Chapter 4 and below).

- Archaea (Archaebacteria) are unicells with cell walls made of different molecules than those found in Eubacteria. Archaea often live under more rigorous environmental conditions, as in hot sulfur springs or extreme salt concentrations (**FIGURE 8.4**).

■ Some 10 percent of the unicells in the photic zone of the open ocean are Archaea, but because they are difficult to culture they are little understood.

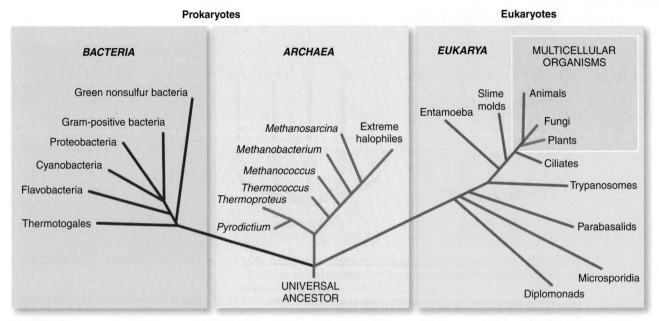

Prokaryotes **Eukaryotes**

FIGURE 8.3 A simplified version of the three domains of life, Eubacteria, Archaea and Eukarya, showing some representative groups within the Eukarya.

- Eukarya (Eukaryota), the kingdom that includes some unicellular organisms (slime molds, ciliates, trypanosomes, and others) and the three groups of multicellular organisms: fungi, plants and animals.

◼ A word of warning lest you become too complacent with having mastered the three kingdoms of life: recent schemes organize eukaryotic life into five or six kingdoms (see Chapter 9).

We can use one of the first organisms whose complete genome was sequenced to illustrate why prokaryotes were subdivided. The organism is the methane-producing marine "bacterium," *Methanococcus jannaschii,* which is found at ocean depths of 2,600 m where the pressure is over 200 atmospheres. The kingdom to which *M. jannaschii* was assigned, the Eubacteria, was erected partly for organisms that were united by lack of histone proteins. Subsequently, *M. jannaschii* was found to contain genes coding for histones. Furthermore, 56 percent of the genome of *M. jannaschii* is not found in other groups of eubacteria. For such reasons organisms allied to *M. jannaschii* were placed into a separate domain of life, the Archaea, which are more closely related to Eukaryotes than they are to Eubacteria (Figure 8.3).

◼ Some algae and some protists can use both photosynthetic and non-photosynthetic modes of metabolism.

The Eukarya (Figure 8.3) includes photosynthetic and non-photosynthetic protozoans and algae, multicellular plant, animals (also known as metazoans), and fungi (Figure 8.1 and FIGURE 8.5). The tree in Figure 8.5 shows a number of interesting features, including a close relationship between animals and fungi and the widely separate positions of some eukaryotic lineages (flagellates, ciliates and diplomonads, for example), lineages traditionally classified in a single kingdom.

◼ *Giardia lamblia* is a parasite of the small intestine of many birds and mammals, including humans, in whom it causes giardiasis (beaver fever). Because it can form cysts, *Giardia* can survive for long periods in quite cold water. Ingesting cysts from contaminated water is the most common route of infection.

The diplomonads are represented by *Giardia,* a protozoan whose position as the earliest eukaryotic offshoot is supported by nucleotide and amino acid sequence analyses but not by other analyses. Although *Giardia* was thought to lack mitochondria, mitochondrial organelles have now been found in *Giardia* and in *Entamoeba histolytica,* another parasitic protozoan estimated to infect 50 million people worldwide.

Eukarya share mechanisms of cell division and protein synthesis. Almost all are aerobic, although some can function anaerobically by adapting to aerobic conditions: A mutation in the yeast *Saccharomyces* is found in individuals at the surface of the

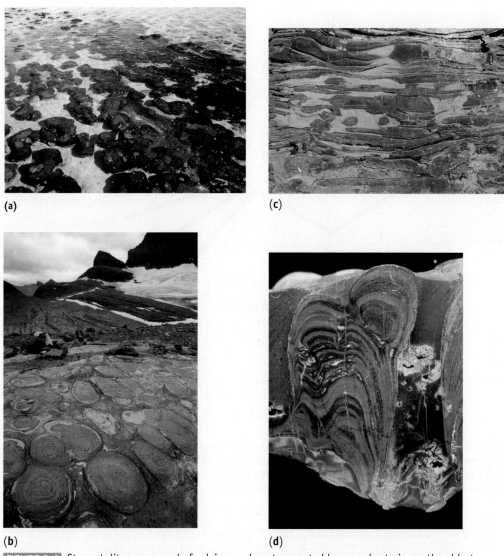

(a)

(c)

(b)

(d)

FIGURE 8.4 Stromatolites composed of calcium carbonate secreted by cyanobacteria are the oldest organisms known (3.5 By old) and the longest lasting. Compare the living stromatolites in Namibia (**a**) with the 2-By-old fossils from the Helena Formation in Glacier National Park, Montana (**b**). Cross-sections of the Namibian (**c**) and fossil (**d**) stromatolites show the similar organization of the internal layers of calcium carbonate.

culture medium where they utilize oxygen. Individuals without the mutation occupy the more oxygen-poor medium beneath the surface and metabolize anaerobically.

With this background in classification and broad relationships between uni- and multicellular organisms, we can explore the first signs of unicellular and multicellular organisms on Earth.

Early Fossilized Cells/Unicellular Organisms

The earliest cells preserved as fossils are of unicellular organisms with features of prokaryotes, now classified as Eubacteria. As discussed above, prokaryotes are not a single monophyletic group (Figure 8.2). The term is best used for the level of organization of

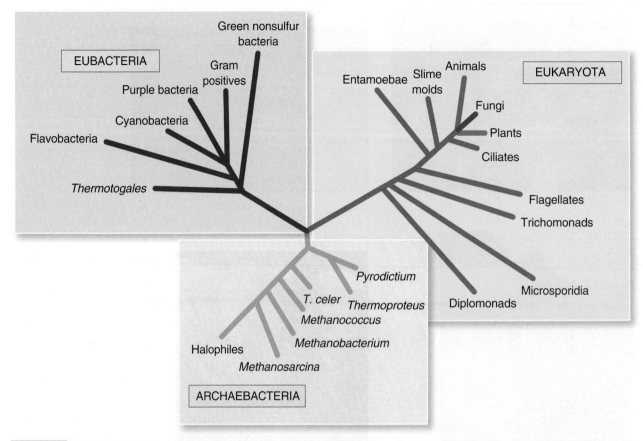

FIGURE 8.5 One possible phylogenetic tree for all cellular life ("the universal tree") based on nucleotide sequence similarities among some ribosomal components (small-subunit RNA, 16S rRNA). In this illustration, the tree is "unrooted," meaning that no source ("root") is indicated for the origin of the common ancestor. See text for details. (Adapted from Sogin, M. L., *Current Opinion Genet. Devel.* **1**, 1991: 457–463 and Wheelis, M. L., et. al., *Proc. Natl. Acad. Sci. USA* **89**, 1992: 2930–2934.)

■ **Cherts** are fine-grained sedimentary rocks that were deposited as sediments on ocean floors as a crystalline aggregate of silica or quartz either as a chemical precipitate or derived from the skeletons of microscopic marine organisms; that is, chert may be biotic or abiotic in origin, an important point when seeking fossils in rocks that are billions of years old. Chert was used by early hominins to make stone tools (see Chapter 19).

■ DNA can be extracted from ancient organic remains and used to identify individual genes, and most recently, to reconstruct genomes (see Box 9.2).

those single celled organisms that lack a nuclear membrane separating the DNA and surrounding cytoplasm (nucleoplasm) from the rest of the cytoplasm.

Cyanobacteria and Stromatolites

A high content of carbon in rocks along with eubacterial cells in the same rocks were discovered in 3.5-By-old cherts in a Precambrian geological eon known as the **Archean**, which existed from 3.8 to 2.5 Bya (FIGURE 8.6). One of the oldest Archean formations is the Warrawoona group in Western Australia, in which we find layered biotic organic deposits known as **stromatolites** (Figure 8.4 and FIGURE 8.7).

How do we know the age of these organic deposits and cells?

Usually we would use isotopes of carbon to estimate the age of organic remains. Carbon-14 (C-14), which decays into nitrogen-14 (N-14) with a half-life of 5,730 years (see Box 2.2), is especially useful for dating organic material. Produced continuously in the upper atmosphere, C-14 eventually finds its way via photosynthesis into all living plants, and thence into animals, in a fixed ratio to non-radioactive carbon. After an organism dies the amount of C-14 slowly decreases, allowing the geological age when death occurred to be calculated. In this way, we've been able to date Egyptian mummies, mammoths, ancient trees, and more. However, at 5,730 years, the half-life of C-14 is far too short to age Archean rocks.

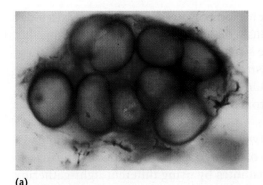

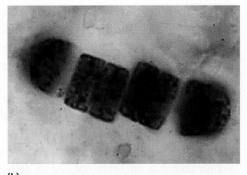

(a) (b)

FIGURE 8.6 Ancient fossil bacteria. Two kinds cyanobacteria from the Bitter Springs chert of central Australia, a site dating to the Late Proterozoic, about 850 million years old. (a) is a colonial chroococcalean form, and (b) is the filamentous *Palaeolyngbya*.

Fortuitously, because two other isotopes of carbon (C-13 and C-12) differ in their participation in cellular metabolism, the ratio of C-13 to C-12 can be used to tell whether stromatolites had a biological origin. Almost all the stromatolites dating from 3.5 Bya and later have carbon isotope ratios similar to rocks from the much later Carboniferous Period and other strata in which living forms are found.

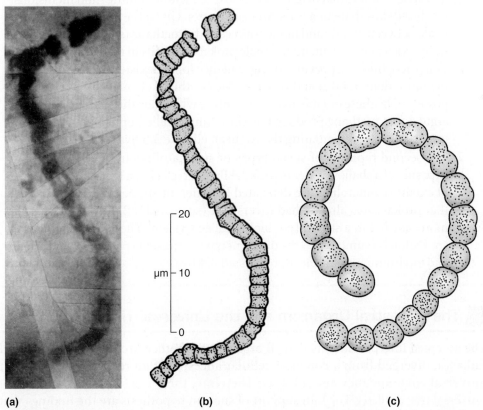

(a) (b) (c)

FIGURE 8.7 (a) Thin section of a filamentous unicellular fossil found in stromatolite chert in the 3.5 By-old Warrawoona formation in Western Australia. (b) Reconstruction of this fossil. (c) Diagram of a phase-contrast microphotograph of a filament of cells of an extant cyanobacterium. Among other extant prokaryotes that appear similar to the microfossil in (a) are colorless sulfur-gliding and green sulfur bacteria.

Stromatolites did not disappear as other life forms evolved. Stromatolites exist today as mats of microorganisms trapping aqueous sediments and cementing them together to form characteristic laminated knob-like structures. As shown in Figure 8.4, these living structures — usually referred to as stromatolite reefs or reef complexes — are remarkably similar to fossil stromatolites and fossilized reef complexes. Furthermore, many of the fossil organisms found in stromatolite deposits are remarkably similar in organization to extant prokaryotic cells.

Living stromatolites are deposited by two types of cyanobacteria (see Chapter 4), by other forms of bacteria, and by a variety of eukaryotic algae. Aerobic cyanobacteria and nonaerobic bacteria coexist in stromatolites by using different light-gathering pigments that enable them to occupy different levels in the water column. Oxygen-producing cyanobacteria reside in the uppermost layers, non-oxygen-producing forms beneath. Extant stromatolites are limited in their distribution to extreme and what appear to us to be inhospitable environments — marine waters with salinities twice that of normal seawater and temperatures greater than 65°C — conditions that limit the presence of animals that could (do) remove microorganisms in more hospitable waters. Ancient stromatolites were distributed more widely. We assume this is because of the absence of competitors or predators in these ancient seas.

Two fascinating finds, both from Western Australia and both reported in 2006, indicate that the hunt for Early Archean life is in full swing and yielding important findings.

1. **Methane of microbial origin** was discovered within minute fluid inclusions in silica dykes that cut across Archean cherts. Given the careful distinction made between abiotic and biotic sources of the methane (see Chapter 4), and the coexistence of filamentous single prokaryotic cells in the same geological unit, these findings provide strong evidence for the existence of prokaryotic life more than 3.5 Bya and provide clues to the likely metabolic pathways possessed by these organisms. The scientists who made this discovery further argued that methane produced by these single-celled organisms played an important role in regulating the Archean climate 3.5 Bya (see Chapter 4).
2. The second report is of **seven types of stromatolites** from a 10-km–long exposure of a shallow water marine 3.4 By-old reef. Given that different forms of extant stromatolites are deposited by different species we can conclude that prokaryotes already had diversified (speciated?) 3.4 Bya, and what's more, existed in a structured, biological ecosystem. Therefore, organisms, speciation (assuming that the different types represent species), ecosystems, and environment modification, existed 3.4 Bya.

The Ancestral Organism and the Universal Tree of Life

The accepted interpretation is that all organisms, whether Archaea, Eubacteria and Eukarya, diverged from a common cellular ancestor often referred to as the last universal common ancestor (LUCA). Therefore, there is a single tree of life: the universal tree of life (UToL). In support of such an hypothesis are the findings that all three kingdoms of organisms share common fundamental attributes of life: a common genetic code, a similar mode of DNA replication, a similar protein-synthesizing system, many similar metabolic pathways, a similar phospholipid bilayered cell membrane, and a similar mechanism of molecular transfer across membranes (known

as *active transport*). The likelihood that all these attributes arose independently in separate lineages is infinitesimally small.

In addition to these shared **deep homologies** (see Chapter 16) relationships have been detected and phylogenetic trees constructed using the enormous amount of information now available from DNA, RNA and protein sequences (BOX 8.1), and increasingly also from classes of molecules that regulate gene activity (see Chapter 13).

In 1990, Carl Woese and coworkers catalogued the presence and frequencies of various sequences in the 16S rRNA component of ribosomes in representatives of the three kingdoms of organisms (Figure 8.5) and showed that they have distinctive differences (Figure 8.4). Although accumulating such molecular and other data show that Archaea occupy an intermediate evolutionary position between Eubacteria and Eukarya, these data are not sufficient to determine the source of their common ancestor. The hypothesis is that the tree of life has a single trunk. Or does it?

■ This is how pathogenic bacteria undergo rapid adaptive evolution as they adapt their spectrum of resistance to newly encountered antibiotic drugs, often adapting in a year or two, which, as illustrated in Chapter 1, can be as many as 10,000 to 15,000 generations/year for some bacteria.

Horizontal Gene Transfer and the Tree of Life

A major complication in reconstructing the tree of life has emerged in recent years. The complication is the discovery of widespread **horizontal (lateral) gene transfer (HGT)**, which is the transfer and incorporation of one organism's or species DNA

BOX 8.1
Using Genes and Gene Products to Construct Phylogenetic Trees

For centuries we established relationships between organisms and constructed phylogenetic trees using **morphological information**. In my very first zoology laboratory we collected animals from the hill behind the zoology building and then classified them using a simple dichotomous key — has a head/does not have a head, has legs/does not have legs. It is amazing how quickly you learn the critical features of an organism using such a simple key.

We can also order organisms and construct trees by comparing amino acid sequences of polypeptides and proteins, nucleotide sequences from nuclear and mitochondrial DNA and from the various types of RNA, and sequences of entire organismal genomes. The rationale behind using **molecular data** is that organisms are bound together through the genetics of evolution by innumerable threads of inherited molecular sequences and their variations (see Figure B11.3 for an example). Molecular approaches also have the advantage that we can compare organisms, such as many prokaryotes, in which morphological or physiological differences between "species" are minimal or nonexistent, and species themselves difficult to identify.

DNA sequencing, and increasingly, sequences of entire genomes from individual species reveal evolution at extraordinary deep scales, allowing us to explore evolution in individual lineages and to attempt to reconstruct the Universal Tree of Life.

From an evolutionary view, advances in sequencing technology and the availability of entire organismal genomes have produced a flood of molecular information, offering opportunities for a wide range of comparative genetic research. Important evolutionary features obtained from comparisons of sequenced prokaryotic genomes include the discovery of extensive HGT between genomes (see text); the greater similarity of archaebacterial protein sequences to eubacterial than to eukaryotic proteins (this chapter); and gene duplication (see Chapter 11), which accounts for as much as 25 percent of the genome of the bacterium *Bacillus subtilis*.

■ 16S rRNA molecules are essential for ribosomal protein synthesis. Like 5S rRNA, portions of their sequences are conserved across a wide range of organisms. The 16S rRNA molecule or prokaryotes is homologous to the 18S rRNA molecule of eukaryotes, both now being referred to as SSUrRNA (small subunit ribosomal RNA).

into the DNA of a different organism or species. Horizontal gene transfer has occurred between prokaryote lineages and from prokaryotes to eukaryotes. Once thought impossible, then possible but insignificant, HGT is now known to have taken place during the early evolution of unicellular organisms and, what's more, continues today. Horizontal transfer of genes between individuals contrasts dramatically with the vertical transmission of genes from generation to generation. Evidence for such transfer is found in the mosaic nature of eukaryotic nuclear genomes, which contain large numbers of genes that are similar to genes from Archaea and from Eubacteria (see Chapter 9). Such genes are unlikely to have been present before the separation of the three lineages from the last universal common ancestor.

HGT is effected by a virus or a small, circular DNA particle known as a **plasmid** that contains a foreign gene that can be transferred. Antibiotic resistance genes are transmitted between bacterial strains by such a mechanism. So too are **transposons**, which are nucleotide sequences that promote their own movement (transposition) among different genetic loci (see Chapter 12). A second mode of HGT in eukaryotes (see below) is via intracellular symbionts (which live in harmony with their host) or parasites (which harm their host).

Between Unicellular Organisms

Successful gene transfer is most common between closely related unicellular organisms. Many bacterial clones and species groups (*Escherichia, Salmonella,* and *Shigella*) have similar genes and gene sequences indicative of HGT among distantly related groups. Once established in their new hosts, mutation of horizontally transferred genes can provide a further source of new genetic information and a potential source for new characters. As an important example of rapid adaptive evolution following HGT, many pathogenic bacteria can widen their spectrum of resistance to antibiotic drugs in a decade or less, providing a dramatic demonstration of the speed at which evolution can occur.

Estimates of the amount and rate of HGT have been obtained. *E. coli* diverged from the *Salmonella* lineage 100 Mya. More than 200 horizontal gene transfers occurred during the subsequent 100 My so that today, almost 20 percent of the *E. coli* genome can be traced to HGT. HGT obscures phylogenetic relationships when otherwise distantly related organisms share a gene or sequence obtained through HGT rather than through common ancestry. Estimates such as these are obtained by the analysis of enormously large data sets, increasingly of completely sequenced genomes. Here are two further examples:

In a major study published in 2005, over 220,000 proteins from the completely sequenced genomes of 144 species were compared. The patterns of gene sharing revealed were so prevalent that the researchers referred to them as *"highways of gene sharing."* As part of the analyses, closely related taxa were compared with one another and with distantly related taxa that inhabited similar environments, the latter to check for parallel evolution. Genes that code for 16S rRNA, cell wall, cell division and ribosomal proteins were found to have rarely been subject to horizontal transfer. By contrast, other genes were found to have been more subject to transfer. Overall, as much as one third of the genome of some prokaryotic organisms has been acquired through HGT.

To or Between Multicellular Organisms

Among eukaryotes investigated to date, HGT appears most prevalent among bdelloid rotifers, a group of asexually reproducing fresh-water invertebrates. The rotifer species

■ Data are available from 5,287 species ranging from viruses to eukaryotes. They are accessible from GenBank (http://www.ncbi.nlm.nih.gov/Genbank/). Complete genome sequences, now known for over 860 organisms (83 eukaryotes, 55 Archaea and 723 bacteria) are accessible in GenBank.

Adineta vaga has inherited genes from bacteria, fungi and plants. Despite lacking sexual reproduction, HGT enhances the genetic diversity in this species. A recent survey of some 8,000 protein domain families between and across kingdoms estimated that greater than 50 percent of members of the Archaea, 30 to 50 percent of bacteria but less than 10 percent of eukaryotes acquired one or more protein families by HGT. The marine green alga *Ostreococcus tauri* acquired an entire chromosome by HGT. Even more dramatic, the fruit fly *Drosophila ananassae* has acquired the entire genome of a bacterial parasite (*Wolbachia*) by HGT.

Determining relationships between Archaea and bacteria on the basis of molecular data and then generating a tree based on those relationships may be an impossible task if much of the DNA and so many of the genes in these organisms are present because of horizontal transfer rather than having been inherited from a common ancestor. Even depicting the tree of life as "unrooted" (Figure 8.5), a depiction that makes no claims about which lineage was ancestral may not be possible.

Because HGT is not a major factor in eukaryote evolution, recreating a eukaryote tree of life is a more realistic project, but still daunting. Reconstructing a Tree of Life or Universal Tree of Life (UToL) is severely compromised by HGT among prokaryotes. Indeed, because so much genetic exchange occurred among prokaryotes — perhaps as much as a third of their genomes (above and see Chapter 9) — the very notion of a prokaryote Tree of Life may be an entirely inappropriate way to represent prokaryote evolution.

Recommended Reading

Knoll, A. H., 2003. *Life on a Young Planet: The First Three Billion Years of Evolution on Earth.* Princeton University Press, Princeton, NJ.

Maynard Smith, J., and E. Szathmáry, 1995. *The Major Transitions in Evolution.* Freeman, Oxford, England.

Morowitz, H. J., 2002. *The Emergence of Everything: How the World Became Complex.* Oxford University Press, New York.

Wilson, R. A. 2005. *Genes and the Agents of Life. The Individual in the Fragile Sciences.* Cambridge University Press, Cambridge, England.

Woese, C. R., O. Kandler, and M. L. Whellis, 1990. Toward a natural system of organisms: Proposal for the domains Archaea, Bacteria and Eucarya. *Proc. Natl. Acad. Sci. USA.*, 87, 4576–4579.

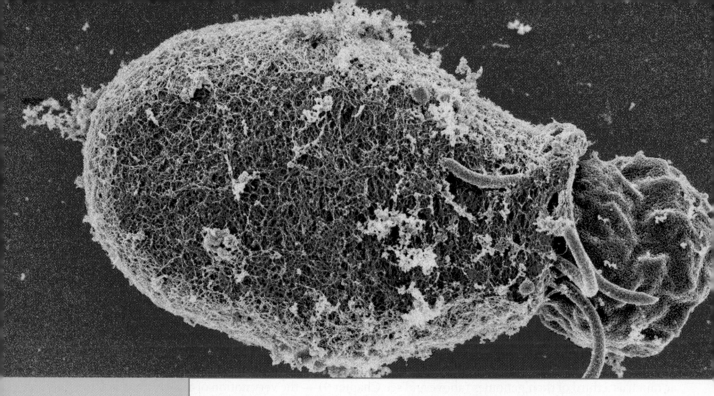

9

Eukaryotic Cells and Multicellular Organisms

KEY CONCEPTS

- Most eukaryotic cells are multicellular and all contain a nuclear membrane and organelles.
- Single-celled eukaryotes (protistans) arose 1.6 to 1.8 Bya and link ancestral prokaryotic cells with all multicellular eukaryotes.
- Five supergroups of eukaryotes are now recognized, replacing old familiar groups such as animals and plants.
- The five supergroups are Archaeplastida (Plantae), Excavata, Chromalveolata, Rhizaria and Unikonta.
- Eukaryotic cells usually have many chromosomes and genes containing introns.
- Organelles, including the nuclear membrane arose when prokaryotes were engulfed (endosymbiosis) and transformed into organelles.
- Endosymbiosis occurred in several waves: the formation of mitochondria and chloroplasts (primary endosymbiosis); and secondary and tertiary endosymbiosis in which eukaryotic algae were engulfed by other eukaryotes.

Above: *Giardia* (here pseudo-colored), a member of the supergroup Excavata, causes the intestinal illness giardiasis (traveler's tummy).

- Organelles such as mitochondria in animals and chloroplasts in plants retained their DNA and so provided eukaryotic cells with additional genomes.
- Some genes were transferred from the organelles to the nucleus, others from the nucleus to the organelles.

Overview

In the previous chapter we explored the origin of the first unicellular organisms based on prokaryotic cell organization. In this chapter we explore the origin of some 2.5 Bya of cells with **eukaryotic** organization, an organization that facilitated the **evolution of multicellularity**. Five supergroups of eukaryotes are now recognized.

Eukaryotes are larger than prokaryotes and are usually multicellular, generally aerobic and more complex. They contain many organelles, internal membranes, a cytoskeleton, and a microtubular apparatus for mitotic cell division. The evolution of a nuclear membrane and intracellular organelles that allowed functions to be allocated to specific regions within the cell — energy generation to mitochondria and chloroplasts, protein synthesis to ribosomes, protein processing to Golgi bodies — are fundamental features of eukaryotic cells. As with the origination of prokaryotic cells discussed in Chapter 8, global events such as changes in Earth's atmosphere facilitated the evolution of eukaryotic cellular organization.

The large amount of DNA in eukaryotic cells is dispersed among several linear chromosomes, in contrast to the single (usually circular) chromosome of prokaryotic cells. Eukaryotic genes almost always contain nucleotide sequences known as introns that do not translate into peptide sequences. Prokaryotic genes do not contain introns. Because meiosis is almost universal in eukaryotes, meiosis also must have arisen early.

Paradoxically, we know more about the origin of the mitochondrial and chloroplast organelles of eukaryotic cells than we do about the origins of the organisms that contain them. Ancient anaerobic eukaryotic cells engulfed prokaryotic organisms in a process known as endosymbiosis, established a symbiotic relationship with the prokaryote. Subsequently, the prokaryotes were retained as cellular organelles — mitochondria and chloroplasts — providing eukaryotes with additional source of DNA. The shared features of mitochondria, chloroplasts and cyanobacteria provide evidence that cyanobacteria were the primary source of these organelles, although some eukaryotes obtained their chloroplasts from other eukaryotes (algae) later in eukaryote evolution.

Evolution of Eukaryotic Organization

As early as 1.5 Bya or even earlier, eukaryotic cells appear as fossils (FIGURE 9.1). The approximately 2 By interval between the age of these fossils and the earliest prokaryotes 3.5 to 3.8 Bya may indicate the length of time needed to incorporate and coordinate the many profound changes in cell structure and function associated with the origin of eukaryotic cells.

As discussed in Chapter 8 genes and gene products are used extensively to reveal relationships between organisms and to uncover phylogenetic history. In 1979

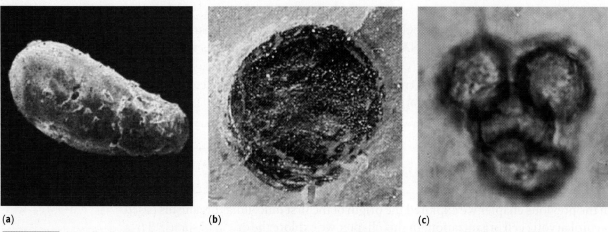

(a) (b) (c)

FIGURE 9.1 Microfossils of probable eukaryotic cells that date back to the Proterozoic era. The cells in (a) and (b) are many times larger than any known prokaryotic cells and are considerably more complex. The group of cells in (c) is in a characteristic tetrahedral arrangement, suggesting they formed through either mitotic or meiotic eukaryotic cell division mechanisms.

a **Tree of Life** was established using nucleotide sequences from 5S rRNA of over 30 species of prokaryotes and eukaryotes. Based on these data, prokaryotes and eukaryotes diverged around 2 Bya and major groups of organisms form discrete clusters: animals, plants, fungi and bacteria (FIGURE 9.2). The tree in Figure 9.2 shows a divergence between early prokaryotes and early eukaryotes at a time perhaps 50 percent earlier than the divergence between fungi (for example, yeast) and plants and animals. Because biologists estimate the latter divergence to have occurred about 1.2 Bya, the earlier prokaryote–eukaryote separation can be estimated to have occurred 1.8 Bya, a finding supported by the existence of fossil eukaryotic-type cells of Proterozic age (Figure 9.1, and see Chapter 4).

Some small multicellular eukaryotes may have evolved as long as one billion or more years ago, although larger visible multicellular eukaryotic fossils with differentiation into several cell types or tissues are only apparent in the late Precambrian. About 545 Mya in the Early Cambrian, numerous forms of multicellular invertebrates made a sudden marked appearance as forms that could fossilize (see Chapter 18). In an interval of 10 My or less, an explosive radiation of eukaryotes occurred in which a large number of animal lineages (phyla) appeared (see Chapter 11).

Single-Celled Eukaryotes: Protistans

Early eukaryotes were single-celled organisms or simple filaments (Figure 9.1). Today, most eukaryotes are **multicellular**. The changes in classification discussed in the previous chapter reflect this reality that life cannot be divided simply into single-celled and multicelled organisms. The five-kingdom scheme introduced in Chapter 8 classifies all unicellular eukaryotes into the kingdom **Protista**, which includes photosynthetic algae, and some saprophytic fungi that ingest their food directly. Protists form a large assemblage of more than 100,000 species; their identification with one of five (or six) **superkingdoms** in the latest Tree of Life is discussed in BOX 9.1. Evolutionarily, protistan ancestry is old, dating back 1.6 to 1.8 Bya. An ancestor-descendant relationship between protists and multicellular eukaryotes is well established. Aside from their diverse forms and the many ways in which they affect other forms of life, protistans are

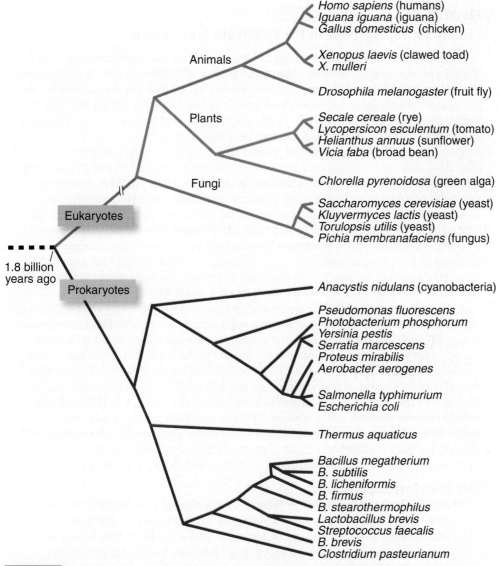

FIGURE 9.2 A phylogenetic tree based on early data gathered from comparing 5S rRNA nucleotide sequences among over 30 species. The separation between eukaryotes and prokaryotes appears to be close to about 2 Bya. Other studies comparing amino acid sequences from 57 prokaryotic and eukaryotic enzymes also indicated that the two groups shared a common ancestor about 2 Bya. (*Source*: Hori, H. and S. Osawa, *Proc. Natl. Acad. Sci. USA* **76**, 1979: 381–385.)

the essential link between early unicellular organisms and all multicellular eukaryotes. (The origins of the three groups of multicellular organisms — plants, animals and fungi — are discussed in Chapters 10 and 11.)

Endosymbiotic events (see the following section) provided protistans with mito-chondria, chloroplasts, and perhaps other constituents (see Chapter 10). In addition to these organelles, protistans share many features with other eukaryotes, including a nuclear membrane, cilia and flagella.

The near-universal presence of cytoskeletal microtubules in eukaryotes reflects the nuclear chromosomal division (mitosis) that replaced the prokaryotic method of

BOX 9.1
Five or Six Supergroups in the Eukaryote Tree of Life

In the text of Chapter 8 under "Kingdoms of Organisms," we saw how increasing understanding of relationships between organisms has been reflected over time in changes in the way we **classify living things**. Classification has moved from the division of life into two broad divisions (prokaryotes and eukaryotes), to two kingdoms (plants and animals), five kingdoms (prokaryotes, protists, fungi, plants and animals) and now three kingdoms or domains (Bacteria, Archaea and Eukarya; see Figure 8.3).

Four eukaryotic branches of the tree of life — protists, fungi, plants and animals — have been recognized for some time (see Chapters 8, 10 and 11). Over the last two decades, however, as molecular and evolutionary biologists have discovered new types of data and more and more organisms to analyze, increasing concern has been voiced that the four eukaryotic kingdoms do not accurately reflect evolutionary relationships among eukaryotes. For example, the Kingdom Protista includes organisms that are closely related to animals (choanoflagellates), organisms closely related to plants (e.g., red algae) and other major lineages that are extremely distantly related (e.g., ciliates and slime molds, both classified as protists, are more distantly related to each other than animals are to fungi; see Figure 8.5 and FIGURES B9.1 and B9.2).

Analyses of the molecular data, genes and gene networks, morphology and development have led to the replacement of the four eukaryotic kingdoms with **five supergroups** (Figure B9.1). Relationships between the supergroups are not fully resolved. As a consequence, a large group of experts has divided one of the five supergroups into two to give a **six-superkingdom** scheme (Figure B9.2). The fact that we can present two alternatives indicates that this area of biology is in flux and will continue to change. Although it is unclear how the five (or six) supergroups branched off the Tree of Life (Figure B9.2), affinity between members within a group is greater than their affinity to any other supergroup. Both schemes reflect accumulating information on phylogenetic relationships, and the dynamic state of classification and of nomenclature, which, after all, should change as our knowledge of organismal relationships, origins and evolution changes.

Five Eukaryote Supergroups
- **Archaeplastida** (also known as **Plantae**). A supergroup that includes red and green algae land plants and Charophyta (the latter the ancestors of plants [see Chapter 10]), characterized by chloroplastids that arose by primary endosymbiosis and whose ancestors are thought to have been the first photosynthetic organisms (see Chapter 4).
- **Excavata**. Organisms that previously were members of the **Protista**. Commonly known excavates include *Giardia,* which causes the intestinal illness giardiasis (beaver fever, traveler's tummy), *Trypanosoma brucei,* which causes sleeping sickness, and *Trichomonas,* which causes trichomoniasis. Excavates are not united by any single morphological or molecular feature. Rather, ultrastructural or molecular features unite overlapping subsets of the 10 groups within this supergroup (Figure B9.1). Consequently, there is considerable controversy about this supergroup and relationships are in flux, although a recent analysis by Hampl and colleagues supports Excavata as monophyletic.
- **Chromalveolata**. A supergroup of some 23 previous groups, including various types of algae (kelp, dinoflagellates, diatoms) that possess chloroplasts acquired by secondary endosymbioses (see text) as well as some important non-photosynthetic groups, notably ciliates and parasites such as *Plasmodium falciparum,* which causes one form of malaria (see Figure 15.10). One subgroup, the **alveolates** (ciliates, dinoflagellates and others in Figure B9.1) is especially well supported through phylogenies of nuclear genes, but the monophyly of Chromalveolates as a whole is controversial.

■ I am afraid that the names of these supergroups will be unfamiliar. I don't see any alternative but to learn them. On the other hand, when a friend calls a rose a plant you can correct him/her by telling them, "it is not a plant, it is an archaeplastid."

BOX 9.1

Five or Six Supergroups in the Eukaryote Tree of Life (*continued*)

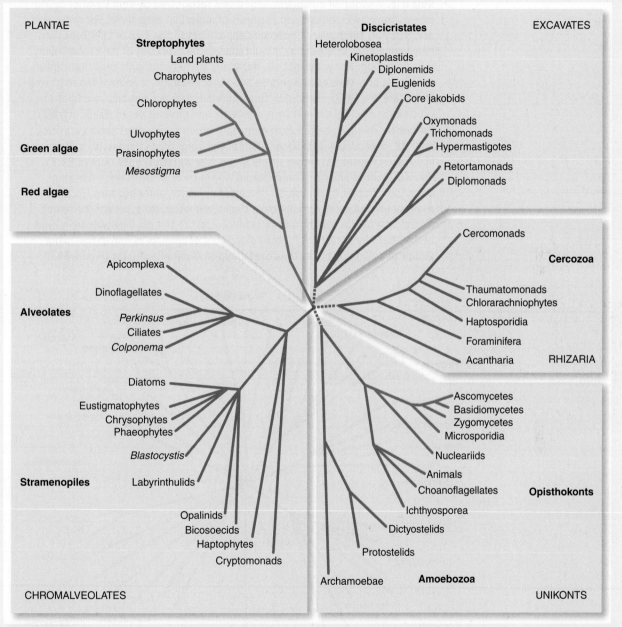

FIGURE B9.1 The eukaryotic tree of life as five supergroups. Relationships between the five supergroups are unresolved. (Modified from Keeling, P. J., *Trends Ecol. Evol.,* **20**, 2005: 670–676.)

(*continued*)

BOX 9.1
Five or Six Supergroups in the Eukaryote Tree of Life (*continued*)

- **Rhizaria.** A group of eukaryotic organisms recognized by Tom Cavalier-Smith (one of the world's experts) on the basis of molecular data alone. For some, this demonstrates the power of molecular approaches to the Tree of Life. For others for whom groups of organisms should share some aspect(s) of their morphology, it is disturbing. Many, but not all, rhizarians are heterotrophic cells that capture and digest prey such as prokaryotes and other eukaryotes. Perhaps the most well known are the 4,000+ species of foraminiferans and radiolarians, examples of which Ernst Haeckel drew in such wonderful and glorious detail (FIGURE B9.3).

- **Unikonta.** This may be the most surprising group for it unites some parasitic protists, choanoflagellates (see Chapter 8), fungi (see Chapter 10) and animals (see Chapter 11) into a single group, the **opisthokonts** (Figure B9.2). Previously these organisms were placed within three of Whittaker's five kingdoms (see Chapter 8). The second group of unikonts, **amoebozoans**, includes slime molds and many types of more typical amoebae. No aspect of the search for the Tree of Life stands in more stark contrast to the phylogenetic trees drawn in the 1880s by Ernst Haeckel than placement of animals as one of five major groups of unikonts, and as the sister group to choanoflagellates (Figure B9.2).

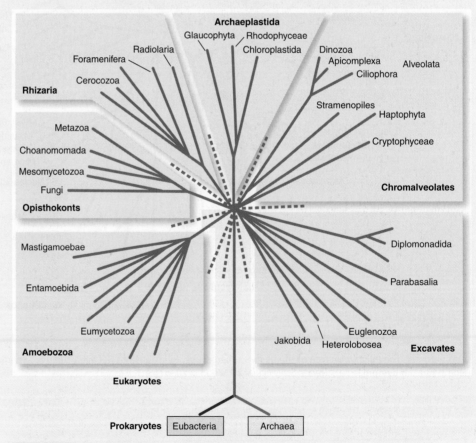

FIGURE B9.2 The eukaryotic tree of life as six supergroups. Relationships are unresolved. (Adapted from Adl, S. M., Simpson, A. G. B., et al. *J. Eukaryot. Microbiol.*, **52**, 2005: 399–451.)

BOX 9.1

Five or Six Supergroups in the Eukaryote Tree of Life (*continued*)

Six Eukaryote Supergroups

In the five supergroup classification, opisthokonts and amoebozoans are grouped within the Unikonta. After extensive consultation with specialists in fungi, algae, parasites, and many more, a subgroup within the International Society of Protistologists identified

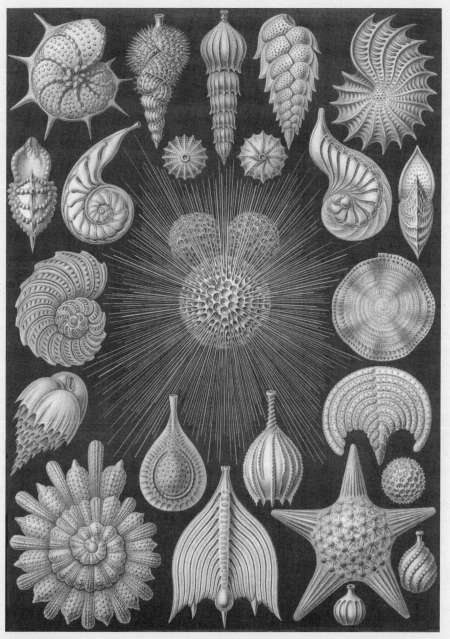

FIGURE B9.3 Part of the diversity of forms of foraminiferans drawn by Ernst Haeckel. (Reproduced from E. Haeckel. *Art Forms in Nature*. New York: Dover Publications, Inc., 1974.)

(*continued*)

BOX 9.1

Five or Six Supergroups in the Eukaryote Tree of Life (*continued*)

sufficient differences between opisthokonts and amoebozoans to place each into a separate supergroup, creating six supergroups (Figure B9.2). Opisthokonta and Amoebozoa therefore replace Unikonta to give

- Archaeplastida, Excavata, Chromalveolata and Rhizaria (as above), and
- **Opisthokonta** (animals, fungi, sponges, choanoflagellates, Mesomycetozoa (parasitic protists), Mesozoa, *Trichoplax*) and
- Amoebozoa (most amoebae, slime molds, some amoeboflagellates and several species lacking mitochondria).

In both schemes, plants (Plantae) have been expanded and renamed Archaeplastida to include several monophyletic protist lineages as close allies of plants and of red and green algae (see Chapter 10). Both five- and six-supergroup classifications recognize the origin of the **three multicellular groups — animals, plants, fungi** — from monophyletic lineages of protists: animals and fungi within the Opisthokonta, and plants from within the Charophyta, a multicellular group within the Archaeplastida. Both systems leave relationships between the five or six supergroups unresolved; Figure B9.2 shows only broad ancestral relationships to Eubacteria and Archaea. In part this is a consequence of uncertainty associated with (a) the phylogenetic methods employed, (b) the statistical approaches used to determine significant results, and (c) horizontal gene transfer, which clouds ancestral relationships.

Recall from Chapter 7 that in prokaryotic cells, dividing chromosomes are individually attached to a lengthening cell membrane, from which they separate after they divide.

chromosomal division. It seems likely for two reasons that once eukaryotic mitosis evolved, sex cell division (meiosis) would have quickly followed: (1) Meiosis and sexuality almost universally appear among the major eukaryotic taxonomic groups. (2) All asexual multicellular eukaryotes are derived from sexual forms with meiosis (see Chapter 4).

All eukaryotic cells share a fundamental gene structure known as **split genes** with **exons** (*expressed sequences*) and **introns** (*intervening sequences*). In a split gene, amino-acid coding nucleotide sequences (exons) are separated by hundreds of bases of noncoding DNA known as introns. In order to translate mRNA into a single polypeptide product, exons and introns are combined in mRNA. Split genes transcribe their nucleotide sequences from DNA to RNA, but the RNA is processed so that the introns are removed and the exons spliced together. Imagine making a chain of paper clips that are silver and red in color. Once the chain is assembled the red paper clips are removed and the silver chains splice together.

Origin and Evolution of Mitochondria and Chloroplasts

Paradoxically, we know more about the origin of the mitochondrial and chloroplast organelles of eukaryotic cells than we do about the origins of the organisms that contain them.

A proposal that had little support initially but now is strongly supported by several lines of evidence is of ancient anaerobic eukaryotic cells evolving the ability to engulf (**endocytose**) molecules and supramolecular assemblies. With this newly

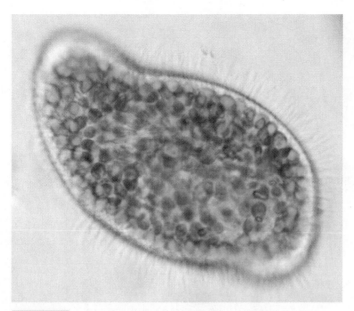

FIGURE 9.3 Symbiotic relationships between a eukaryote and its photosynthetic organelles. The protozoan ciliate *Paramecium bursaria* harbors hundreds of symbiotic algae (green) that may be released from the cell and cultured independently.

evolved ability eukaryotic cells engulfed prokaryotic organisms, both aerobes capable of oxidative metabolism and cyanobacteria capable of photosynthesis. This is not as outlandish as it sounds; some organisms have this ability today — some paramecia can engulf and retain up to hundreds of single-celled algae (FIGURE 9.3).

In 1982, Barnabas and coworkers produced a comprehensive Tree of Life that took into account nucleotide sequences from 5S rRNA and amino acid sequences of two enzymes, ferredoxin and the c-type cytochromes. This phylogeny provided strong support for the **symbiotic theory of organelle function**. Such endosymbiotic events occurred more than once, with some eukaryotic algae receiving their chloroplasts through secondary transfer from other eukaryotes.

According to what has come to be known as **primary endosymbiosis**, the first symbiotic relationship was established with mitochondrion-like aerobic bacteria (FIGURE 9.4). Later, one or more lines of these new aerobic eukaryotes established a similar symbiotic relationship with photosynthetic cyanobacteria, eventually evolving into the various eukaryotic algae and plants (Figure 9.4, and see Chapter 10). Sequence analysis of a gene (*tufA*) coding for an elongation factor used in protein synthesis is consistent with a single symbiotic event, although multiple events cannot be ruled out.

Reinforcing the hypothesis of primary and early endosymbiosis and reflecting the selective advantages that accrue to such organisms, was the discovery of **secondary** and **tertiary endosymbiosis**, later evolutionary events in which eukaryotic algae were engulfed by other eukaryotes (Figure 9.4). Indeed, some eukaryotic algae are thought to have acquired chloroplasts through **secondary endosymbiosis from a eukaryote** rather than from a prokaryote, expanding even further the role of endosymbiosis in evolution.

■ It is estimated that some 18 percent of the genome of one of the most well studied flowering plants, the water cress *Arabidopsis thaliana*, was derived from this primary cyanobacterial endosymbiotic event.

Mitochondrial DNA

Mitochondria and chloroplasts both contain their own DNA/genes. Such organelle DNA is separate from the nuclear DNA spatially, organizationally, functionally and in

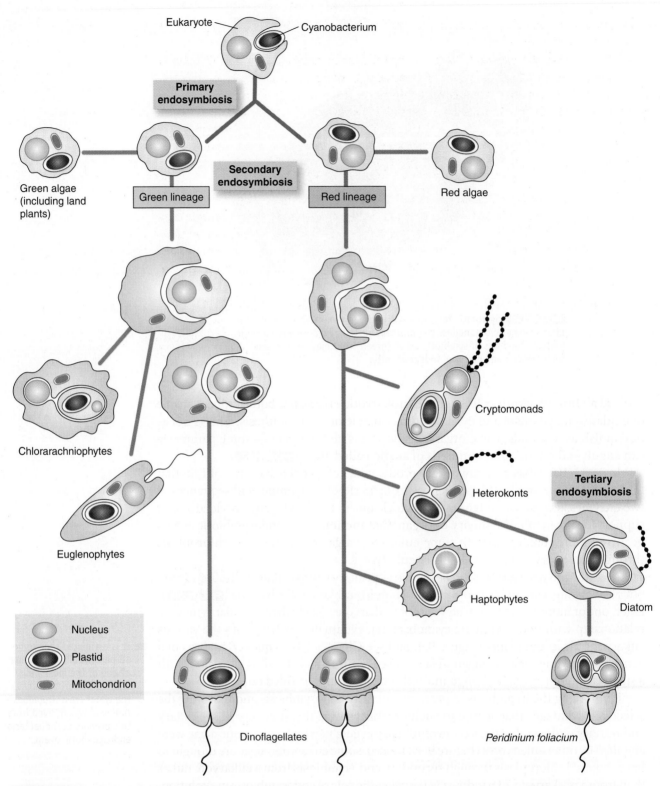

FIGURE 9.4 Primary, secondary and tertiary endosymbiosis and the origin of organelles in various lineages of eukaryotes including algae, land plants, dinoflagellates and cryptomonada, all of which are chimeras. As discussed in the text, horizontal gene transfer between lineages complicates reconstruction of a phylogenetic tree of organelle evolution. (Adapted from Cracraft, J. and M. J. Donoghue [Eds.], *Assembling the Tree of Life*. Oxford University Press, Oxford, 2004.)

mode of inheritance, all of which provide evidence for the separate origins of organelle and nuclear DNA and genes.

One of the major features distinguishing **mitochondrial DNA** (commonly abbreviated as **mtDNA**) from nuclear DNA is that mtDNA is a single circular molecule, typically of some 16,000 base pairs; human mtDNA is 16,569 base pairs long (FIGURE 9.5). A second is the **mode of inheritance of organelle DNA.** Because sperm or pollen contain little cytoplasm and therefore few mitochondria or chloroplasts, these organelles are contributed by the female not the male parent. Therefore, mitochondria and chloroplasts are transmitted from generation to generation by a process known as **uniparental inheritance.**

Two copies of nuclear DNA are inherited, one from each parent. A major consequence of uniparental inheritance of mitochondria and chloroplasts is that only **a single copy of the DNA in each organelle is inherited.** Furthermore (a) because mtDNA does not recombine with nuclear DNA or with the DNA in other mitochondria, and (b) because mitochondrial number increases by the division of preexisting mitochondria, the thousands of copies of mtDNA in each cell are the same throughout the body.

Changes in nuclear DNA result from genetic recombination at meiosis and from random mutations. Because of uniparental inheritance, only mutations create differences among copies of mtDNA. Because of the lack of genetic recombination in mtDNA, more changes accumulate over time in mtDNA than in nuclear DNA. Another

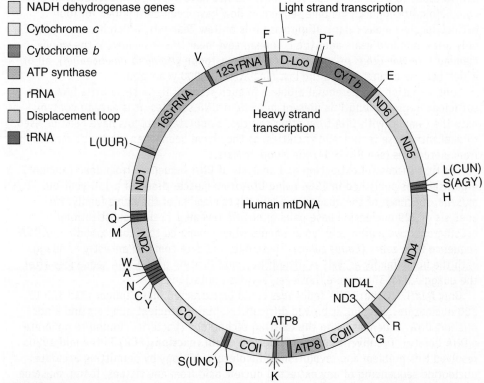

FIGURE 9.5 Genetic organization of the human mitochondrial genome showing the location of the genes for seven products as colored areas in the circular DNA. Most genes are transcribed clockwise along the light strand, but some are transcribed counterclockwise along the heavy strand (shown by the red arrows at the top). (Reprinted with permission from the *Annual Review of Genetics*, Volume 29 © 1995 by Annual Reviews [www.AnnualReviews.org]. Illustration courtesy of David A. Clayton, Howard Hughes Medical Institute.)

■ The 2009 Nobel prize in medicine was awarded to three scientists for their work on telomeres and telomerases, work that demonstrated how chromosomes can be "copied in a complete way . . . and how they are protected against degradation."

■ An analysis in which mtDNA was used to reveal relationships between East African fishes is outlined in Box 17.2.

difference between nuclear and mtDNA that contributes to the differential accumulation of changes in the DNA is that nuclei contain DNA repair enzymes but mitochondria do not. Helicases repair DNA and a second class of enzymes, telomerases, maintain the sequences at the end of DNA molecules, sequences whose disruption is associated with premature aging. DNA repair can compensate for some mutations in nucleotide sequences by eliminating the mutated sequence. As a consequence, mutations in mtDNA accumulate as much as 10 times faster than mutations in nuclear DNA. Therefore, mtDNA changes (evolves) faster than nuclear DNA. This is why mtDNA has proven to be such a powerful tool for revealing evolutionary change and relationships between organisms, mtDNA serving as a molecular clock (see Box 11.2 and Chapter 19). Analysis of mtDNA on museum specimens, extant organisms, and even fossils ("ancient DNA") has proved to be an excellent tool for reconstructing patterns of evolution (BOX 9.2).

BOX 9.2

Ancient DNA

DNA (so-called **"ancient DNA"**) can be extracted from dead or fossil organisms. After the death of an organism, DNA degrades rapidly, initially because of enzymes present within cells. DNA becomes chopped into smaller and smaller fragments with increasing time post-mortem. Because it is much more abundant than nuclear DNA and so tends to be better preserved, most studies have analyzed **mitochondrial DNA**. Complete mitochondrial genome sequences now have been determined for four extinct species: moas (giant, flightless birds of New Zealand), which became extinct only a few hundred years ago; the European cave bear (*Ursus spelaeus*); the wooly mammoth (*Mammuthus primigenius*) and the mastodon (*Mammut americanum*), all of which became extinct during the last glaciation 15,000 years ago.

Contamination is a perennial problem in such studies. Because so little DNA with sufficient fragment length is present in ancient DNA samples, it is easy to contaminate the samples with DNA from other sources. Elaborate protocols have been devised to deal with these technical difficulties, as they have been when retrieving molecules from meteorites (see Box 3.1), the Moon or Mars.

The first successful extraction and analysis of DNA sequences from dead (ancient) organisms was published in 1984 using DNA from muscle tissue of a 140-year-old museum specimen of the quagga, a now extinct member of the horse family. An analysis of 229 nucleotide base pairs of mtDNA revealed 12 base substitutions, causing only two amino acid replacements when compared to a corresponding mtDNA sequence from zebra (*Equus zebra*). These data indicate common ancestry of quagga with the horse family as well as little if any modification of the DNA sequences after the quagga died. There were, however, serious limitations.

Only relatively short DNA sequences could be extracted, none longer than 100 to 200 nucleotides. Second, it proved difficult to obtain sufficient numbers and kinds of short DNA sequences using the existing technology (bacterial cloning to generate a DNA library). The invention of **polymerase chain reaction (PCR)** in the mid 1980s resolved this problem and revolutionized molecular biology by permitting accurate nucleotide sequencing of any extracted nucleic acid from any tissues, living, museum or fossil. How PCR works is outlined in FIGURE B9.4 .

By the 1990s, PCR had been used to investigate many fossil animals, plants, seeds and fungal spores; recently extinct animals such as the marsupial (Tasmanian) wolf; insects embedded in amber; human remains from the Arctic and from peat bogs in Florida and

BOX 9.2
Ancient DNA (*continued*)

Ireland; resolved archaeological controversies and resolve problematic phylogenetic relationships, to name but few. Ancient DNA from human bones on Easter Island indicates that its settlers were Polynesians from other Pacific islands, and not the South American Indians as the Danish explorer Thor Heyerdahl suggested. Isolation of DNA from Neanderthal skeletons is discussed in Chapter 19. A study of flightless New Zealand birds shows that kiwis and extinct moas were much more distantly related than previously proposed, and that kiwis are more closely related to the Australian emus and cassowaries.

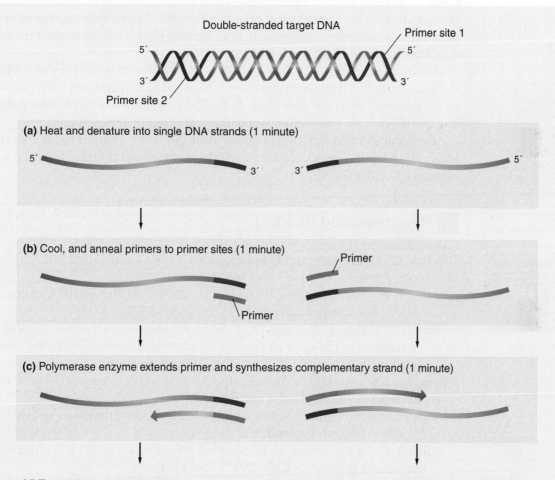

FIGURE B9.4 Simplified diagram of the polymerase chain reaction (PCR) showing the basic steps in replicating a DNA target sequence. (**a–c**) represents one PCR cycle. Short primer sites at each end of the target are identified (**a**), and nucleotide sequences synthesized (usually about 20 base pairs long) that can pair with the primer sites when "melting" the DNA into single strands (**b**). A heat-resistant DNA polymerase enzyme extends nucleotide synthesis from the primers along each complementary strand (arrows in **c**), forming two double-stranded replicates of the original target DNA sequence (**d**). By alternately heating and cooling the mixture, each cycle (**a–c**) exponentially replicates the DNA target. An original sequence can potentially be amplified more than one million times in 25 cycles.

Transfer of Genes between Organelles and Nucleus

Once incorporated in eukaryotic cytoplasm, symbiotic organelles retained some of their genes but many were transferred to the eukaryotic nucleus. Conversely, some nuclear genes were transferred to organelle genomes.

Especially susceptible to transfer were genes in organelles acquired by endosymbiosis that already were present in the nucleus. Two examples are genes for anaerobic glycolysis and genes for amino acid synthesis. Such transfers to the nucleus and deletions from mitochondria had the advantage of maintaining and replicating only two copies of a symbiotic gene in a diploid host nucleus instead of sustaining a separate gene copy within each of many cellular organelles. Because some cells carry enormous numbers of organelle genomes — more than 8,000 copies of the mitochondrial genome in some human cells and even more copies of chloroplast genomes in some plant cells — reductions in organelle gene number by deletion or nuclear incorporation must have been highly selected for.

Chloroplasts synthesize only a small portion of the proteins they use. Many of the genes introduced by the original cyanobacterial endosymbionts were transferred to the nuclei of what are now plant cells. The protein products of these genes are transported into the chloroplast where they carry out their function(s).

Gene transfer also occurred in the opposite direction. Transfer to the appropriate organelle of nuclear genes coding for symbiotic organelle proteins has been widely identified in eukaryotes.

Recommended Reading

Burger, G., M. W. Gray, and B. F. Lang, 2004. Mitochondrial genomes: anything goes. *Trends Genet.*, **19**, 709–716.

Green, R. E., A. W. Briggs, J. Krause, et al., 2009. The Neandertal genome and ancient DNA authenticity. *EMBO J.* **28**, 2494–2502.

Hampl, V., L. Hug, J. W. Leigh, et al., 2009. Phylogenomic analyses support the monophyly of Excavata and resolve relationships among eukaryotic "supergroups." *Proc. Natl. Acad. Sci. U.S.A.*, **106**, 3859–3864.

Lane, N., 2005. *Power, Sex, Suicide. Mitochondria and the Meaning of Life*. Oxford University Press, Oxford, England.

Maynard Smith, J., and E. Szathmáry, 1995. *The Major Transitions in Evolution*. Freeman, Oxford, England.

Venter, J. C., M. D. Adams, E. W. Myers, P. W. Li, et al., 2001. The sequence of the human genome. *Science*, **291**, 1304–1351.

Plants and Fungi as Branches of the Tree of Life

10

- Plants, fungi and animals are the three kingdoms of multicellular organisms.
- Multicellularity evolved multiple times, either by cells failing to separate after division or by aggregation of two or more cells.
- Land plants evolved around 460 Mya from ancestors with some features found in green algae alive today.
- Horizontal gene transfer, endosymbiosis, and the separation of somatic and germinal tissues were important processes in land plant evolution.
- Many of the adaptations of land plants reflect the transition from water to land.
- Leaves evolved multiple times independently in different groups.
- The first large leaves appeared in tree ferns during the Carboniferous Period.
- Plants and the insects that pollinate them co-evolved.
- Fungi are not simple plants but a sister group of animals.

Above: Co-evolution of insect and flower: the exquisite process of pollination.

165

Overview

Land plants are known to have evolved from organisms similar to some living green algae. The more than 500,000 species of land plants and green algae store starch, form a cell plate during cell division, and use the pigments chlorophyll a and chlorophyll b to obtain energy through photosynthesis. Other land plant characteristics that may have been present in the algal-like ancestor(s) are alternation of diploid and haploid generations and the production of spores. Ranging from simple mosses and liverworts (bryophytes) to more complex flowering plants (angiosperms), land plants are characterized by reproductive structures consisting of one or more multicellular layers that help protect the developing gametes. These layers also provide embryonic nutrients for the egg.

The earliest land plants with living representatives but obscure origins are bryophytes, which live in moist areas and have simple water- and food-distributing tissues. Bryophytes exhibit a distinctive life cycle in which the diploid generation (the sporophyte) parasitizes the dominant stage, which is the haploid plant or gametophyte; the sporophyte is the dominant life history phase in other land plants.

The earliest vascular plants — the ancestors of club mosses, horsetails, and ferns — were leafless and rootless. Evolutionary trends in these plants include the development of two types of spores, the formation of large leaves, and the differentiation of increasingly elaborate vascular tissues, xylem to transport water from the roots and phloem to transport nutrients from the leaves.

Plants developed techniques to minimize water loss when they became terrestrial. When plant embryos gained complete independence from water and became enclosed in seeds, the gametophyte was greatly reduced in gymnosperms (land plants with naked seeds and cones) such as conifers and spruce, and in angiosperms, which are flowering land plants with covered seeds. Elaborate flower structures attract insect and bird pollinators.

No longer regarded as simple plants without chlorophyll, fungi are the third major lineage of multicellular organisms (although some are unicellular), a sister group of animals. The lack of chlorophyll is reflected in fungal life styles; fungi obtain nutrition either from parasitizing live hosts or by digesting dead organic matter.

An Abominable Mystery

Although fossil pollen has been discovered in 136-My-old Early Cretaceous deposits, comparatively little is known about the origins of flowering plants (**angiosperms**). Because of the success of their reproductive features, especially mechanisms of pollination and the formation of seed coats to protect their seeds, angiosperms became dominant in many environments. A phylogenetic tree of 24 families of land plants produced by the United States Botanic Garden is shown in FIGURE 10.1 . You will find it useful to refer to this tree as you work through the chapter.

Interest in the origin of land plants has a long history. In a letter written in 1879, Charles Darwin called plant origin an abominable mystery. "The rapid development as far as we can judge of all the higher plants within recent geological times is an abominable mystery." The mystery of land plant origins remains incompletely resolved to this day, although origination from green algae, a long-standing hypothesis, has received support recently.

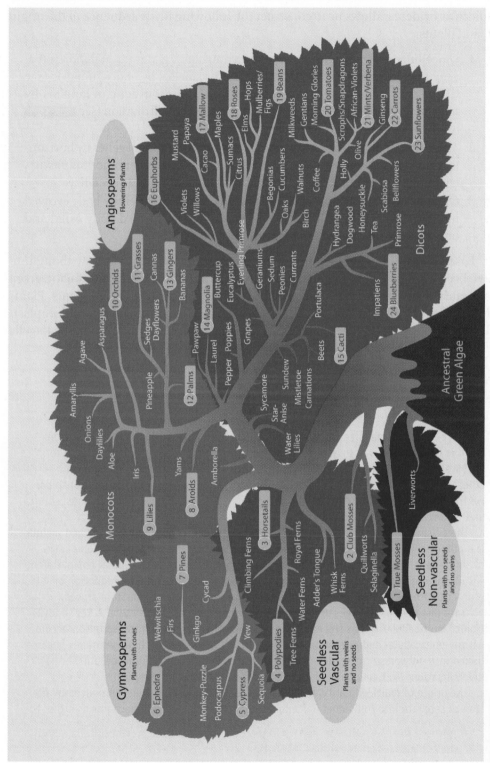

FIGURE 10.1 A phylogenetic tree of 24 families of plants, including vascular and nonvascular plants, gymnosperms and flowering plants (angiosperms). (Courtesy of U.S. Botanic Garden.)

An Algal Ancestry

Molecular evidence allows us to construct the following likely sequence of the origin of land plants.

1. **Unicellular green algae** that had incorporated new genes through horizontal gene transfer (see Chapter 8) and gained organelles by endosymbiosis (see Chapter 9), transformed into **multicellular photosynthetic organisms** by remaining attached rather than separating after each cell division (see **BOX 10.1** for theories of how multicellularity arose).
2. These first small multicellular "plants" captured additional genes by horizontal gene transfer, increased in size and in organelle and genome complexity, enabling their component cells to compartmentalize and specialize.
3. Organisms in which most or all cells could reproduce the entire body evolved into organisms in which most cells were somatic (formed body cells) and only a few were reproductive, forming gametes.

Based on ultrastructural features, the ancestor of land plants is now thought to have been a **single-celled flagellated green alga** not unlike the living aquatic green flagellated alga, *Mesostigma viride*, which is the nearest relative of land plants. Plants originated when these eukaryotic algal cells were invaded by prokaryotic chloroplast-like symbionts (step 1 above, and see Chapter 9). With further evolution the flagella

BOX 10.1
Multicellularity

Increase in organismal size can have and usually does have a considerable effect on relative fitness. One way to achieve increase in size is to become multicellular, a term that we use to refer to individual organisms consisting of more than a single cell. Whether we examine genomic and cellular organization or phylogenetic relationships, we find that **multicellularity** has evolved numerous times. Plants, animals, most fungi and brown algae are multicellular.

The advantages of multicellularity over the unicellular condition are many and include the potential for increase in size beyond the limits set by the surface-to-volume ratio of single cells, specialization into distinct cell types, each (or each cellular compartment within the organism) with its own function such as food gathering, reproduction or protection from predators, and enhanced dispersal, especially of immature stages.

Once distinct cell lineages arose, selection would have acted on them. A key innovation of multicellularity was the ability to regulate where and when transcription occurs within multicellular embryos or organisms. Within distinctive cell lineages genes became linked into networks or cascades, a different cascade for each cell lineage, furthering diversification. Multicellularity therefore facilitated increasing complexity.

The origin of multicellularity would have required single cells that reproduced by fission to separate into individual unicellular organisms and develop mechanisms either to

- prevent the two cells from separating, producing an organisms whose cells would have had the same genetic constitution, *or*
- facilitate aggregation with a cell(s) from another individual, potentially producing an organism whose cells would have different genetic constitutions.

of the aquatic alga were lost, producing sessile organisms and initiating the land plant lineage.

This transition occurred more than once. Evidence from molecular phylogenies established using 18S rRNA, the chloroplast gene for the large subunit of the enzyme ribulose bisphosphate carboxylase (the gene for this enzyme is abbreviated *rbcL*), and nuclear and mitochondrial genomes' sequences, demonstrates that numerous lineages of aquatic algae invaded the land. Transformation of one of these lineages gave rise to two major lineages, the **chlorophyte green algae** and the **charophyte algae and the land plant**. Green algae (Chlorophyta in older classifications) have been reclassified into two major lineages.

Five species in the genus *Coleochaete,* which are terrestrial green algae that live as epiphytes (air plants) on aquatic plants, appear to be the closest extant algal relatives of stoneworts (which includes some of the largest green algae) and **vascular plants**. "Vascular" refers to the presence of conductive tissue (**xylem**) that enables water to reach the erect parts of the plant and a second tissue (**phloem**) that enables nutrients to be distributed from the leaves and stems.

Green algae and land plants share various similarities reflecting their common origins.

- Green algae store their carbohydrate reserves as starch.
- Many species of algae have rigid, cellulose-reinforced cell walls, as do all land plants.
- Green algae and vascular plants use similar types of pigments in metabolic pathways, both green chlorophyll (a and b) and yellow-orange carotenoids (α and β).

Other similarities appear at first sight to be the result of *parallelism* (see Chapter 16). Both sea lettuce (*Ulva* sp.) and razor algae (*Caulerpa* sp.) have membranous forms that simulate the morphology of some vascular plants (FIGURE 10.2) but show their separate evolutionary ancestries by passing through an alga-like, filamentous stage.

The transformation of marine algae to terrestrial forms may have occurred as early as the Ordovician 460 Mya, when falling sea levels exposed aquatic plants in shoreline communities to selection for resistance to desiccation. Increase in atmospheric

■ For the relationships between these lineages of algae and land plants see the U.S. National Science Foundation-funded program, "Assembling the Tree of Life" (http://atol.sdsc.edu/).

■ Epiphytes (air plants) grow on other plants but obtain water and nutrients, not from the host plant, but from the atmosphere and rain.

FIGURE 10.2 *Caulerpa,* a green alga with leaf-like form.

O_2 levels and the formation of the ozone layer (see Chapter 8) also were important environmental factors driving the move to land.

Bryophytes

The "simplest" land plants, limited in distribution to moist environments, are known collectively as **bryophytes.**

The name bryophytes does not signify a monophyletic group but is a collective name for some 17,000 species in several lineages: liverworts (7,000 species), mosses (10,000), club mosses, and hornworts (100 species) (Figure 10.1 and FIGURE 10.3). Bryophytes and "more complex" vascular land plants both have multicellular reproductive structures, a cuticle protecting their aerial parts, and epidermal pores known as *stomata* to permit the transfer of CO_2, water vapor, and O_2 between their tissues and the atmosphere. Bryophytes lack the transporting phloem and xylem tissues found in vascular plants.

The earliest unequivocal appearance of bryophytes in the fossil record is 420 Mya in the Devonian for liverworts and 350 Mya in the Carboniferous for mosses. Fossils of

(a)

(c)

(b)

FIGURE 10.3 A typical liverwort, *Marchantia* (**a**), a moss showing the characteristic growth habit (**b**) and a hornwort (**c**).

vascular land plants appear in 440 Mya in the Early Silurian. A recent comprehensive phylogenetic analysis based on morphological characters provides strong support for liverworts as the sister group to all other land plants (Figure 10.1) and hornworts as the sister taxon to vascular plants. An important implication of these data is that bryophytes were important in two major transitions, the transition from water to land, and the transition from a haploid gametophyte-dominated life cycle to a diploid sporophyte-dominated life cycle (FIGURE 10.4). These two life history phases, which characterize all plants, are discussed in BOX 10.2.

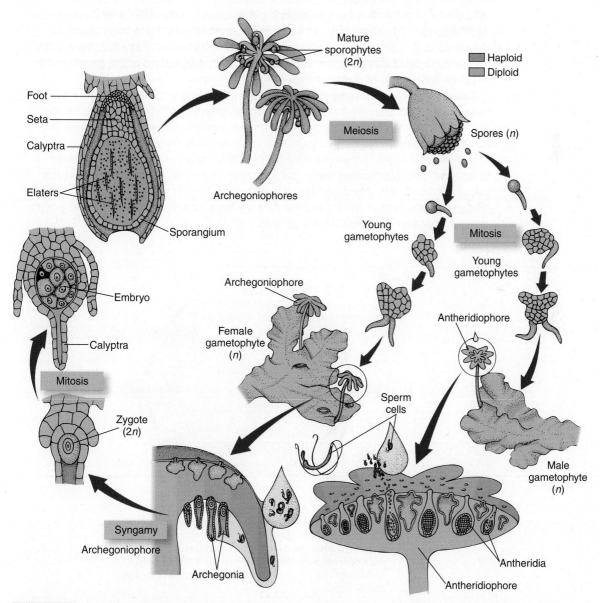

FIGURE 10.4 Life cycle of a liverwort in the genus *Marchantia*. The gametophyte is the haploid (*n*) stage shown at the bottom of the cycle. Fertilization of the "egg" by a "sperm" produces the diploid (2*n*) zygote, embryo and sporophyte (left-hand side). The sporophyte produces sporocytes, each of which divides by meiosis to form haploid (*n*) spores. Germination of the spore (lower right) leads to the gametophyte, and the cycle is reinitiated. Also see Figure B10.1 for alternation between gametophyte and sporophyte.

BOX 10.2
Alternation of Generations

All plants share a life style characterized by **alternation of generations** between a free-living haploid **gametophyte** generation and a diploid and parasitic **sporophyte** generation (see Figure 10.4). The distinctive features of gametophyte and sporophyte life history stages are outlined in Table B10.1.

The sporophyte–gametophyte alternation of generations in plants (FIGURE B10.1) has long been puzzling. An obvious advantage of the sporophyte is that, as the diploid stage of the life cycle, its retention allows genetic recombination to influence future generations. The sporophyte produces dispersible spores that can resist desiccation; meiotically produced spores develop into haploid gametophytes, which later produce gametes (TABLE B10.1). The vulnerability of gametes to terrestrial conditions may account for the persistence of the gametophyte stage in plants. Adaptations of sporophytes enable resistance to desiccation and so facilitate life on land. Adaptations of the gametophyte facilitate the transfer of plant gametes in water.

The sporophyte has become independent of a gametophyte in some lineages. This independence has been regarded as a trade off or compromise between the advantages of diploidy and the needs of spore production. For example, many plants have lost the ability to reproduce sexually. Sexual reproduction has been replaced with asexual reproduction such as vegetative propagation from roots or stems, or reproduction through unfertilized eggs (*parthenogenesis*), also seen in some animals. Although some asexual plants are found over wide geographical ranges, their success is often restricted to specific environments or to conditions that severely limit cross-fertilization; asexual groups rarely survive over long evolutionary periods.

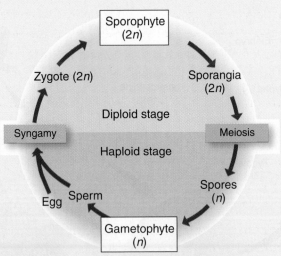

FIGURE B10.1 Alternation of gametophyte and sporophyte generations in the plant life cycle (see also Figure 10.4). In some plants, the gametophyte is unisexual, either male or female. In others, the gametophyte is bisexual or hermaphroditic (as shown here), producing male and female tissues.

BOX 10.2
Alternation of Generations (*continued*)

TABLE

B10.1 **Major Features and Definitions of Gametophyte and Sporophyte Life History Stages in Plants**

Gametophyte	Sporophyte
Haploid (n) chromosome number	Diploid (2n) chromosome number
The haploid stage of the life cycle	The diploid stage of the life cycle
The gamete-producing stage of the life cycle	The spore-producing stage of the life cycle
The stage of the life cycle (generation) in which gametes are produced	The stage in which spores are produced
The stage[a] containing the gamete-producing organs	The stage formed by the union of gametes
The stage with the mitotic phase of gamete formation	The stage with the meiotic phase of gamete formation
The stage that alternates with the sporophyte	The stage that alternates with the gametophyte

The spore-forming and gamete-forming tissues in algae and the similarity between the filamentous growth pattern of some green algae and the branching filamentous stage of many mosses are consistent with an algal origin of bryophytes, although different groups of bryophytes may have arisen independently from different algal ancestors. Also, given that the sporophyte generation of bryophytes depends for nutrients and support on the gametophyte while the sporophyte of vascular plants is independent, bryophytes and vascular plants may have arisen from different algal lineages.

Early Vascular Plants

The earliest of the Silurian plant fossils includes a number of simple vascular plants with leafless stems classified in the genus *Cooksonia*. A diminutive plant about 25 mm or so high, *Cooksonia caledonica* had no distinctive leaves or roots (FIGURE 10.5a). Similar fossils were long known from Devonian rocks in Rhynie, Scotland, and have been classified together as Rhyniophyta, or Rhynia-type plants. The first leafless land plants also appeared in the Devonian. *Zosterophyllum* is an example (FIGURE 10.6a).

Whatever the origin of vascular plants, the fossil record shows their rapid evolutionary radiation after their appearance more than 400 Mya. By the Mid to Late Devonian the evolution of the **cambium** permitted plants to increase in size by an order of magnitude over those present earlier in the Devonian. The

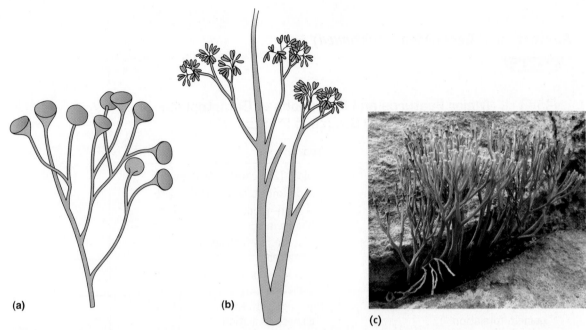

(a)

(b)

(c)

FIGURE 10.5 The sporophytes of two early fossil plants and of an extant member of the whisk ferns. (**a**) As shown in this illustration, this diminutive Upper Silurian plant, *Cooksonia caledonica* had naked, dichotomously branched axes with terminal sporangia. (**b**) *Psilophyton princeps*, a spiny, leafless fossil plant that first appeared less than 10 My after *Cooksonia* had a main stem with lateral branches terminating in sporangia. (**c**) The extant epiphytic whisk fern, *Psilotum nudum*. The simple stems, nondiscernible leaves and absence of a root system resemble the fossil forms.

■ Growth of the tips of stems and roots of vascular plants is from terminal apical meristems of dividing cells. Growth of side branches is from lateral meristems.

cambium is the tissue from which new cells are produced to increase the diameter of plant stems.

The structure of plant stems became more complex serving as conduction, support, and storage structures that facilitated the vertical growth of plants and allowed large trees and shrubs to evolve. By the end of the Devonian forests containing a great variety of woody trees were established. By 350 Mya (the beginning of the Carboniferous), lush and extensive forests had developed in vast tropical swamps along the east coast of North America and similar coastal regions of Europe and North Africa. The death, submergence and decay of these forests produced the enormous Carboniferous coal seams. These successful land plants were vascular with xylem to transport water and phloem to distribute nutrients. Both xylem and phloem arise from the cambium, the evolution of which therefore was a key to land plant diversification. Early plant evolution also is characterized by **evolution in divergence in spore size** (FIGURE 10.7).

The Devonian also saw the origin of sphenopsids (horsetails; Figure 10.1), plants with segmented stems and whorled leaves and branches (FIGURE 10.8a). Horsetails were common until the Mesozoic, contributing huge trees to Carboniferous coal forests. Only the genus *Equisetum* (Figure 10.8b) remains, with about 25 herbaceous species.

Ferns

Ferns (Pterophyta) numbering about 12,000 species are vascular plants with true leaves. Ferns alternate generations between a free-living gametophyte and a sporophyte (FIGURE 10.9). Ferns lack the flowers and seeds that evolved in other lineages of land

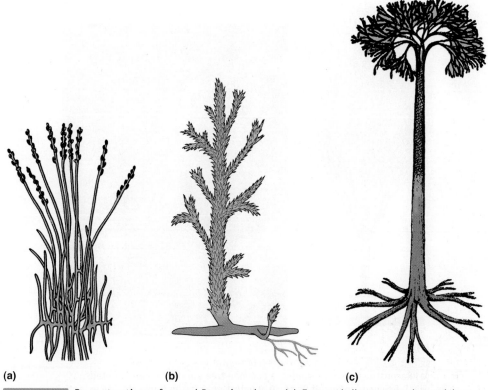

(a) **(b)** **(c)**

FIGURE 10.6 Reconstructions of several Devonian plants. (**a**) *Zosterophyllum myretonianum* (about 10.5 cm). (**b**) *Asteroxylon mackiei* (about 0.6 m tall). (**c**) Lepidodendron species (about 46-m tall). The scars along the upper stem of *Lepidodendron* are leaf cushions, where the long, filamentous leaves attached during earlier growth. (**a** and **b** adapted from Foster, A. S., and E. M. Gifford, Jr. *Comparative Morphology of Vascular Plants, Second edition*. Freeman, 1974; **c** adapted from Stewart, W. N., and G. Rothwell. *Paleobotany and the Evolution of Plants, Second edition*. Cambridge University Press, 1993.)

FIGURE 10.7 Proposed stages in the evolution of megaspores that develop into egg-bearing gametophytes as shown in a reconstruction of a seed from a Devonian gymnosperm, *Archaeosperma arnoldii* (right). The selective advantage of this trend was the development of large eggs providing increased embryonic nutrition. (Adapted from Niklas, K. J. *The Evolutionary Biology of Plants*. University of Chicago Press, 1997.)

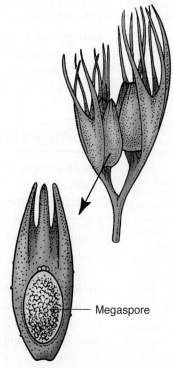

Megaspore

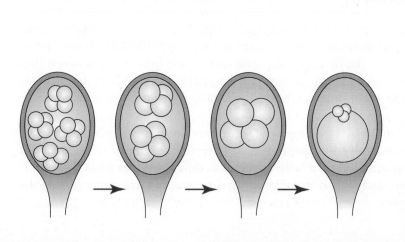

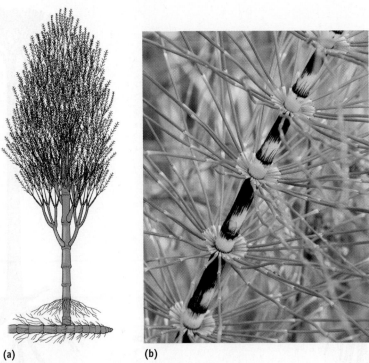

(a) **(b)**

FIGURE 10.8 Ancient and extant representatives of horsetails.
(**a**) Reconstruction of *Calamites,* a common tree during the Carboniferous,
which reached heights of 27 to 30 m with trunks 0.6-m thick. (**b**) The extant
horsetail, *Equisetum arvense,* showing vegetative and fertile shoots. (**a** Adapted
from Niklas, K. J. *The Evolutionary Biology of Plants.* University of Chicago
Press, Chicago, 1997.)

plants. From their origin 350 Mya in the Early Carboniferous, ferns included both
small forms and large tree ferns (FIGURE 10.10). Mitochondrial gene and morphologi-
cal evolution in trees ferns has been so slow over the past 200 My that tree ferns can
be regarded as *molecular living fossils.*

Ferns radiated enormously about 70 Mya in the Late Cretaceous. The four major
fern lineages and many of the extant families arose in this "age of ferns." Ferns were
the first plants to evolve large, prominent leaves.

Origins of Leaves

The two major hypotheses for the evolutionary origin of leaves are grounded in our
understanding of leaf development in extant plants (FIGURE 10.11). According to
the **telome hypothesis**, the first leaf-like structures arose from webs between thin
flattened branches (telomes). According to the **enation** (extension) **hypothesis**
(Figure 10.11b), the earliest leaves arose from small flaps or extensions of tissue
along the stem.

On the basis of molecular phylogenetic reconstructions of land plant evolution,
we know that **leaves evolved multiple times.** Therefore, either of the two modes
proposed for leaf origination may have operated at different times and/or in differ-
ent lineages. At least six independent evolutionary origins of leaves emerge from the
phylogenetic analyses.

■ Although the evolutionary
origin(s) of leaf types remain
unresolved, the hormonal
and genetic basis for leaf
shape is being revealed.
Transcription factors bind to
DNA to activate or repress
genes. The transcription
factor, Knox (the knotted-
like homeobox family of
genes) establishes a gradi-
ent of the plant hormone
auxin, which regulates leaf
shape. Knox family members
are active in apical meri-
stems where they regulate
leaf and stem development.

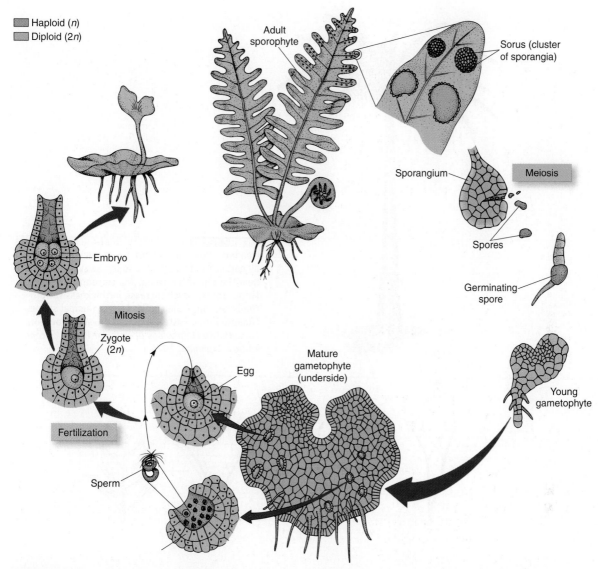

FIGURE 10.9 Life cycle of a fern, the common polypod, *Polypodium vulgare,* showing alternation between haploid gametophyte and diploid sporophyte. Compare with liverworts (Figure 10.4) and vascular plants (Figure B10.1).

- Three times in the gametophyte generation: twice in liverworts and once in the line that gave rise to the mosses.
- Three times in the sporophyte generation: once in the lycopods (club mosses), once in ferns and related forms, and once in seed plants.

Moving onto Land

Successful as they were, Carboniferous plants were limited to a moisture-laden environment; motile pollen must swim through water to the female.

Three evolutionary changes — reduction in the size of the gametophyte, the evolution of easily dispersible pollen, and encasement of spores in seeds — allowed plants to avoid desiccation and so move away from water. Evolution of these structures

FIGURE 10.10 Reconstruction of the Carboniferous tree fern, *Psaronius,* about 7.6-m tall. Leaf scars left by earlier fronds that have fallen away are visible near the top of the trunk. The surrounding adventitious roots, which increase in thickness toward the base, cause the trunk's long pyramidal shape. (Adapted from Foster, A. S., and E. M. Gifford, Jr. *Comparative Morphology of Vascular Plants, Second edition.* Freeman, 1974.)

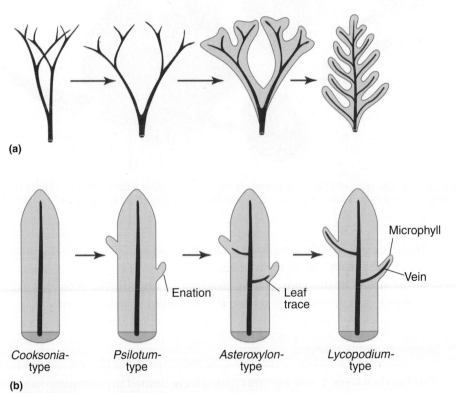

(a)

Cooksonia-type

Psilotum-type Enation

Asteroxylon-type Leaf trace

Lycopodium-type Microphyll Vein

(b)

FIGURE 10.11 Diagrammatic representations of two hypotheses for the origin of leaves. (a) Flattening (planation) of a branch system according to the telome hypothesis, followed by webbing between the branches to form a flat, veined large leaf. (b) Evolution of microphylls according to the enation hypothesis. (Adapted from Bold, H. C., C. J. Alexopoulos, and T. Delevoryas. *Morphology of Plants and Fungi, Fourth edition.* Harper & Row, 1980.)

provided a basis for interbreeding, and wind dispersal of seeds. Many land plants also have fruity tissues, adhesive burrs, feathery parachutes, or other devices that allow animals or air and water currents to disperse them.

We don't yet fully understand how early vascular plants evolved into the two major lineages of pollen-producing, seed-bearing land plants, which are **gymnosperms** (FIGURE 10.12) and **angiosperms** (TABLE 10.1). **Convergent evolution** was common as selection under similar environmental conditions produced similar plant phenotypes in different lineages residing in different geographical localities. Prominent examples are some New World cacti and African euphorbs, both occupying desert environments and both highly similar in appearance, with sharp spines or thorns to dissipate heat and to guard their succulent, water-laden stems (FIGURE 10.13). Such evolutionary convergences, like those of animals derive some similarities by modifying different genetic pathways: the cactus spine is a modified leaf and the euphorb thorn a modified branch.

Evolutionary changes contributing to the rapid evolution of angiosperms were (1) the evolution and elaboration of **flowers** as reproductive organs, and (2) the evolution of flower structures that enable insects or birds to pollinate them and to disperse their seeds are prominent evolutionary changes (FIGURE 10.14). Plants and insects co-evolved adaptations that facilitated pollination of flowering plants, a topic of much

■ The angiosperm phylogeny group maintains a web site that contains the latest version (AGP II, 2003) of angiosperm classification (http://www.f-lohmueller. de/botany/apg/apg_ii.htm).

(a) (b)

FIGURE 10.12 (**a**) Reconstruction of the progymnosperm *Archaeopteris,* about 23-m high. (**b**) Reconstruction of a cycad gymnosperm, *Williamsonia sewardiana,* from Jurassic rocks in India. (**b** Adapted from Andrews, H. N., Jr. *Studies in Paleobotany.* John Wiley, 1961.)

TABLE

10.1 Primary Features Distinguishing the Two Major Lineages of Land Plants: Gymnosperms from Angiosperms

Gymnosperms	Angiosperms
Naked-seed plants	
Non-flowering plants[a]	Flowering plants
Seeds hidden within cones	
Seeds not enclosed in an ovary	Seeds enclosed in an ovary
Seeds not enclosed by a protective fruit	Seeds enclosed by a protective fruit
Examples: pine trees, conifers	Examples: maple trees, roses
Double fertilization	

[a]This criterion applies to other plants such as bryophytes and ferns.

interest to Charles Darwin, especially in relation to orchids and moths (BOX 10.3). With the evolution of leaves into the petals and sepals of flowers, competition for pollinators arose. Large-scale analyses of insect pollinator-plant evolution support origination and/or radiation of insect pollinators soon after origination of the plant group pollinated by those insects. Pierid butterflies, members of a family that includes the cabbage white butterfly *Pieris rapae*, lay their eggs on plants such as mustard and cabbages. The plants evolved chemicals that act as insecticides when

(a) (b) (c) (d)

FIGURE 10.13 Convergent evolution between representative desert species of American Cactaceae, illustrated by the American saguaro (**a**, **b**) and African Euphorbiaceae (**c**, **d**), illustrated by two species of African euphorbs.

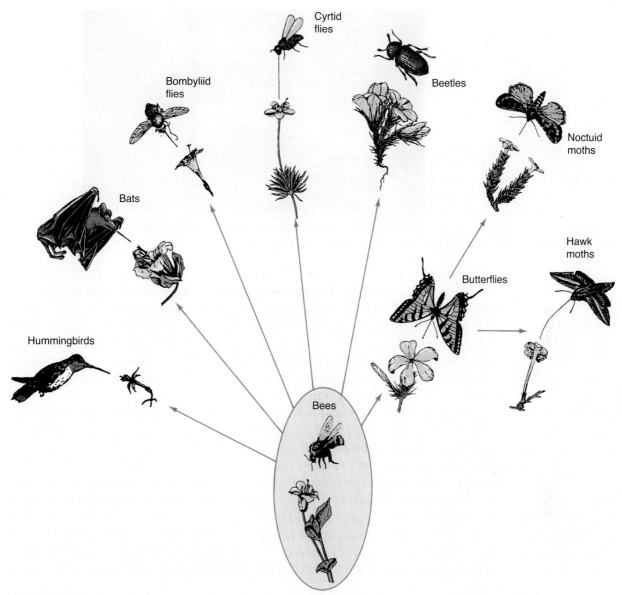

FIGURE 10.15 The range of flower types and their animal pollinators in phlox plants (Polemoniaceae). Only relatively few mutations are required for the transition from a species pollinated by bees to a new species pollinated by hummingbirds. In more exotic cases, different plant species deposit pollen on different parts of the pollinator's body, allowing the same animal species to pollinate different plant species. (Adapted from Grant, V., and K. A. Grant. *Flower Pollination in the Phlox Family*. Columbia University Press, 1965.)

Fungi

The advantages of multicellularity as outlined in Box 10.1 are evident in **fungi**, a group that contains both unicellular and multicellular organisms. Fungi first appeared along with the first vascular plants in the Silurian, 440 Mya, although according to some interpretations of the fossil record, fungi can be traced back to the Precambrian. By the Paleozoic (554 Mya) fungi had evolved into most of the lineages represented by extant forms (FIGURE 10.17).

FIGURE 10.16 *Amborella trichopoda,* a small shrub found in New Caledonia.

■ The term *arms race* origi-
nated for the buildup of
military weapons by the
Soviet Union and the
United States after the
Soviet Union tested an
atomic bomb on August
29, 1949. The term has
expanded (evolved) to any
competition in which the
goal is to stay ahead of
the competition. Winning
may never be an option.
The 1983 movie *War Games,*
which starred Matthew
Broderick and Ally Sheedy,
is a must see for the futility
of an arms race.

The presence of cell walls and the production of spores led early biologists to include fungi within the plant kingdom, albeit as simple plants lacking chlorophyll. The lack of chlorophyll is reflected in fungal lifestyles; fungi obtain nutrition either from live hosts by parasitism or from the digestion of dead organic matter. Not all fungi are multicellular. Some of the 120,000 fungal species such as yeasts are unicellular. Other fungi have vegetative stages of branched multicellular or multinuclear filaments called *hyphae* that aggregate into a mass called the *mycelium.*

Adopting a parasitic existence on live hosts would have offered the opportunity for early aquatic fungi (perhaps related to extant forms, the water molds) to resist desiccation in the host tissues of land plants and evolve subsequently into forms that use aerially dispersed spores. A recent comprehensive analysis of six gene regions in 200 species of fungi showed that the spore flagellum used in swimming may have been lost four times independently during fungal evolution, each of these losses being accompanied by the evolution of new mechanisms of spore dispersal. Parasitic fungi are still actively evolving on the gene level, in what has been called an "**arms race**" between host and parasite. Each new genetic variant of a host that confers resistance against a fungal parasite is often overcome by

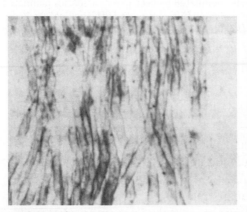

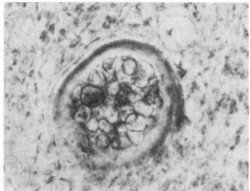

FIGURE 10.17 Fossils of different parts of a Devonian fungus, *Palaeomyces gordonii.* (left) Hyphae. (right) Spore.

selection for increased frequency of a fungal genetic variant that increases host susceptibility.

With enhanced knowledge of morphology, molecules and genes, fungi have been reclassified as the third major group of multicellular organisms. The weight of evidence is that fungi: (1) share a choanoflagellate ancestor with animals, (2) are more closely related to animals than they are to plants and (3) that fungi are not simple plants but a sister group to animals (see Box 9.1 and Figure 9.4).

The evolution of animals discussed in the next chapter further reinforces the evolutionary advantages of multicellularity. Increase in surface area and the development of food-gathering surfaces provides a stable food supply and allows multicellular organisms to digest larger particles of food than can be ingested by single cells. Multicellularity facilitated the developments of signaling systems (see Chapter 13) that coordinated movement. Regions of multicellular organisms can specialize into different tissues (*division of labor*) with different functions. Elaboration of regulatory pathways provided morphological and functional innovations that broadened the scope of protection, dispersion, food gathering, reproduction, excretion, and other functions.

> ■ Recall from Chapter 8 that the ability of prokaryotic cells to ingest other unicells gave rise to eukaryotic organelles.

■ Recommended Reading

Alcock, J., 2006. *An Enthusiasm for Orchids: Sex and Deception in Plant Evolution.* Oxford University Press, Oxford, England.

Darwin, C., 1862. *On the Various Contrivances by Which British and Foreign Orchids Are Fertilized by Insects, and on the Good Effects of Intercrossing.* Murray, London, England.

Donoghue, M. J., 2008. A phylogenetic perspective on the distribution of plant diversity. *Proc. Natl. Acad. Sci. U.S.A.* **105**, 11549–11555.

Frohlich, M. W., and M. W. Chase, 2007. After a dozen years of progress the origin of angiosperms is still a great mystery. *Nature* **450**, 1183–1189.

Gensel, P. G., and D. Edwards (eds.), 2001. *Plants Invade the Land: Evolutionary and Environmental Perspectives.* Columbia University Press, New York.

Kenrick, P., and P. Davis, 2004. *Fossil Plants.* Natural History Museum, London, England.

Margulis, L., and K. V. Schwartz, 1998. *Five Kingdoms. An Illustrated Guide to the Phyla of Life on Earth,* 3rd ed. Freeman, New York.

Niklas, K. J., 1997. *The Evolutionary Biology of Plants.* University of Chicago Press, Chicago.

Thompson, J. N., 2005. *The Geographic Mosaic of Coevolution.* The University of Chicago Press, Chicago.

Wang, H., M. J. Moore, P. S. Soltis, et al., 2009. Rosid radiation and the rapid rise of angiosperm-dominated forests. *Proc. Natl. Acad. Sci. U.S.A.* **106**, 3853–3858.

Waser, N. M., and Ollerton, J. (eds.), 2006. *Plant–Pollinator Interactions: From Specialization to Generalization.* The University of Chicago Press, Chicago.

Willis, K. J., and McElwain, J. C., 2002. *The Evolution of Plants.* Oxford University Press, Oxford, England.

11

Animals as a Branch of the Tree of Life

KEY CONCEPTS

- The entire Earth was encased in a sheet of ice 1 km thick (snowball Earth) two to four times between 725 and 635 Mya.
- Communities of enigmatic organisms — the Ediacaran Biota — existed in the Upper Precambrian more than 565 Bya, before animals appear in the fossil record.
- Based on molecular evidence we can conclude that animals originated in the Precambrian 700 to 750 Mya.
- Animals of all major phyla first appear as fossils 545 Mya in the Early Cambrian Burgess Shale and equivalent faunas worldwide.
- Stem taxa of animal phyla therefore must be Precambrian in age.
- Reasons for the origination of animals include the elaboration of genes and embryonic development, environmental and climate change and changes in the level of atmospheric O_2.
- Homeobox genes that pattern animal bodies originated before animals, plants and fungi arose.

Above: Radiolarians: one of the many branches of animal evolution.

Overview

A fascinating and enigmatic assemblage of organisms known as the Ediacaran Biota appears in the Upper Precambrian-Lower Cambrian some 565 Mya. Assigned to various domains, these organisms formed complex ecosystems but disappeared early in the Cambrian. Any relationship to the multicellular organisms we know as animals (metazoans) is unresolved and may be nonexistent.

About 545 Mya in the Early Cambrian, an explosive radiation of multicellular organisms with soft and hard tissues appears in the fossil record. Multicellularity facilitated cellular specialization within organisms and structural differences between organisms, the evolution of different modes of food gathering, and enormous increases in body size. The Burgess Shale fauna in Canada and the Chengjiang fauna in China are the major, but not the only sites that record this Cambrian Explosion. These, the first animals, were widespread in distribution. The dramatic appearance of animals a billion years or more after prokaryotic life arose (see Chapter 4) raises important questions concerning the origin of animals.

The Cambrian fauna is composed of derived (crown) taxa. Therefore, animals originated from stem taxa in the Precambrian. Global climate change known as "snowball earth" may have limited evolution in the Precambrian. The appearance of so many "fully formed" body plans in what may have been as little as 10 million years raises the question of how animals radiated and diversified so rapidly in the Early Cambrian. Explanations offered include a rise in atmospheric oxygen, changing geological features, recovery from snowball earth, origination or elaboration of embryonic development, diversification of mechanisms of gene regulation and the evolution of predator-prey interactions associated with the opening up of new ecological zones.

■ The terms *metazoans* and *animals* are used interchangeably throughout this chapter.

Animal Origins

We saw in the previous chapter that the multicellular eukaryotes we know as fungi and land plants originated around 440 Mya. What of those multicellular eukaryotes we know as animals?

As discussed in Chapters 8 to 10, we can trace unicellular eukaryotic fossils back 1.6 to 1.8 By (see Figure 9.1). Less generally accepted are claims of eukaryotic algae dated 2.1 Bya or older, although diverse and multicellular forms of algae are present in the fossil record about 1 Bya. The evolution of multicellularity from unicellular organisms occurred more than once, giving rise to different lineages of organisms — colonial ciliated protozoans (protists); various groups of algae, plants and fungi (see Chapters 9 and 10), and animals — which are the organisms discussed in this chapter.

Protistan ancestry is old, dating back to before 2 Bya (see Chapter 9). An ancestor-descendant relationship between protists and multicellular eukaryotes (including animals) is well established from analysis of 16S rRNA, alpha-tubulin and heat-shock protein-90 gene sequences. This data set indicates that single-celled choanoflagellates either were animal ancestors or choanoflagellates and animals descended from a common protistan ancestor. The discovery of members of a family of cell adhesion molecules (cadherins) in the extant choanoflagellate *Monosiga brevicollis,* but in neither plants nor fungi is consistent with a choanoflagellate ancestry of animals.

Before discussing the origin of animals another group(s) of multicellular organisms that remain enigmatic will be introduced. They are known as the **Ediacaran Fauna**

from the Ediacara Hills in South Australia where they were discovered. In hindsight, fauna is not the most appropriate term to describe this and similar associations of organisms. Fauna refers to the animal life found in a region or at a particular time. Flora is a similar term for the plants of a region or period. As the Ediacaran organisms are not animals (with perhaps one exception) the term *fauna* is inappropriate. *Biota* is a more appropriate term for the organisms found in a region or at a geological period. So we will speak of the **Ediacaran Biota**.

The Ediacaran Biota

A diverse and enigmatic assemblage of organisms, the Ediacaran Biota (FIGURE 11.1) existed from early in the Precambrian into the Early Cambrian.

The soft-bodied fossils found in the Precambrian Ediacaran strata of South Australia — and those found more recently in deposits at other locations around the globe — are strange indeed. The fossils at Mistaken Point, Newfoundland, are preserved beneath a thick layer of volcanic ash that has been dated at 565 ± 3 My. Amongst the oldest Ediacaran organisms, these fossil communities provide amazing insights into a *Precambrian ecosystem* (see below).

Any relationship between Ediacaran and other multicellular organisms has been hard to uncover, perhaps because the Ediacaran Biota is an independent evolutionary experiment. I say "experiment" rather than lineage because these biotas (assemblages) appear to contain multiple lineages of organisms whose relationships to one another are as obscure as are their relationships to other multicellular organisms. We cannot even

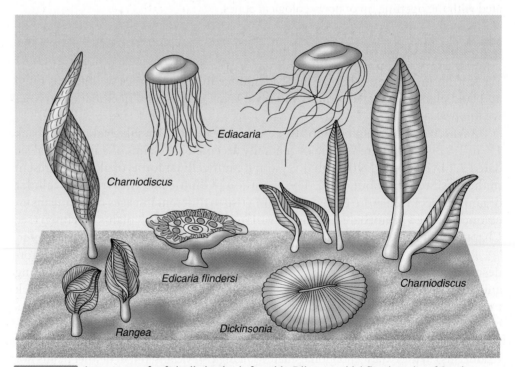

FIGURE 11.1 A panorama of soft-bodied animals found in Ediacaran tidal flat deposits of South Australia and 10 or so other places throughout the globe, most occurring in the Precambrian, 545 to 650 Mya. *Ediacaria* resembles extant jellyfish. Others, like *Rangea* (see also Figure 11.2) and *Charniodiscus,* are frond-like and bear no resemblance to extant animals or plants.

be sure whether they are eukaryotic in their cellular organization. These organisms may be as much an independent experiment in cellular evolution as in organismal design.

As depicted in Figure 11.1, some Ediacaran fossils superficially resemble jellyfish. Others bear likenesses to segmented worms and, more distantly, to mollusks and echinoderms. Whether these resemblances mean that Ediacaran organisms (or some Ediacaran lineages) were animals is another matter entirely. Because paleobiologists have not detected any basic animal anatomical features in any Ediacaran fossil — no eyes, mouths, anuses, intestinal tracts, or locomotory appendages — their relationship to Cambrian animals remains obscure.

Numerous interpretations of the Ediacaran Biota have been proposed. Any interpretation must distinguish between **stem** and **crown taxa**. A stem taxon is the earliest representative of a lineage. There can only be one stem taxon for a lineage. Crown taxa are the terminal branches of a lineage arising from a stem taxon. A single lineage may have more than one crown taxon. By definition and reflecting their evolution, stem taxa do not show all the features of the crown groups into which they evolve. The existence of divergent lineages of organisms among Ediacaran fossils is illustrated by the identification and reconstruction of *Kimberella* from the Ediacaran Biota of the White Sea in Russia and by what Cohen and colleagues have identified as resting stages of animals. *Kimberella* has been interpreted as having features consistent with it being a common ancestor (stem taxon) of mollusks and allied invertebrates. A German paleontologist, Adolf (Dolf) Seilacher placed Ediacaran organisms into a distinctive group, the **Vendozoa**, which he and others regard as unrelated to any animals; a separate evolutionary experiment. The absence of features usually associated with prey capture, and the absence of digestive tracts, led to proposals that many Ediacaran organisms may have depended on photosynthetic or other types of symbiosis (as lichens do today, with an alga and a fungus cooperating to form a lichen).

The Canadian paleontologist, Guy Narbonne, invoked repeated structures (**modularity**; see Box 18.1) based on fractal repeats of a frond structure in his analyses of early Ediacaran organisms to identify a major group which he named **rangeomorphs** (FIGURE 11.2). Rangeomorphs existed in complex ecological communities with more than one trophic level. One species, *Charnia wardi,* found in the Drook Formation in Newfoundland, grew to heights of two meters, and is the oldest representative of this assemblage, dating to 570 to 575 Mya. Narbonne is cautious when it comes to saying whether rangeomorphs were animals or an alternative (and earlier) evolutionary experiment in multicellularity.

■ A fractal is a shape that can be split into parts each of which has the same shape as the original. A fern leaf is an example of a fractal.

Another species, *Funisia dorothea,* is the most abundant species in the shallow water marine Ediacaran fauna of South Australia. *Funisia* consists of tubes up to 30-cm long and 12-mm wide, consisting of modular serially repeated elements (segments?) anchored to a microbial mat by holdfasts (as you might see in a living kelp). The closest similarity of *Funisia* to animals is as a stem sponge or jellyfish, but that is no more than conjecture at this stage. The contrasting body form and ecology of rangeomorphs and *Funisia* attests to the diversity of species and ecosystems present in the Precambrian.

Whatever their affinities, the morphological diversity seen in the Ediacaran biotas shows that a considerable diversity of body plans and types of organisms existed up to or across the Cambrian boundary, and that these organisms had a considerable Precambrian history. Although some survived into the Cambrian, they became extinct. Why? We don't know. One suggestion is that their lack of armor made them easy prey for the many mobile predators that appeared in the Early Cambrian. Another is that

they did not make it into the Cambrian because they fell victim to "**snowball Earth**" in the late Precambrian (BOX 11.1).

Animals Arise

We do not yet know the evolutionary changes that led to the origination of animals. We do know that by the beginning of the Cambrian 545 Mya many different types of animals were present, recognizable, and can be classified on the basis of their different body forms.

Within a relatively short geological time span, perhaps as little as 10 My, an explosive radiation marked the emergence of a large number of distinctive and different animal body plans representing virtually every known animal phylum (FIGURES 11.3 and 11.4). We can identify these organisms as animals and place them into phyla and lower taxa based on their possession of phylum-specific characters seen in their extant descendants. The major groups of animals originated and were well established before they appear in the fossil record. Indeed, as much as 200 My of evolution in the Precambrian may predate the fossil record of animals.

Burgess Shale Fauna

FIGURE 11.2 A rangeomorph frondlet to show the organization based on repetitive branching from a central stalk. Compare with the frond-like forms in Figure 11.1.

The most well-known formation in which Cambrian organisms are preserved is a limestone reef, 160-m deep and more than 20-km long, in the **Burgess Shale** in Yoho National Park, British Columbia, Canada. The animals of the Burgess Shale were neither an isolated evolutionary experiment nor unrepresentative of the general situation in the Early to Middle Cambrian. At least 12 other sites contain animals with an equivalent range of body plans, including faunas in Pennsylvania, north Greenland, China, Spain, Poland, and the Chengjiang fauna in China, which is some 10 My older than the Burgess Shale fauna.

Although discovered in 1909, not until an expedition in the late 1960s did the Burgess Shale reveal its true story, a story with a cast of 124 genera and 140 species. Just over a third of the genera are arthropods (especially trilobites), but sponges, brachiopods, polychaete worms, echinoderms, cnidarians and mollusks all are represented. So well known is the Burgess Shale fauna that the scientific names of some species are known outside paleontology — *Marrella splendens* and *Canadapsis perfecta* (arthropods) and the velvet worm, *Hallucigenia sparsa* (FIGURE 11.5) — may be familiar to you.

Our ability to assign organisms from the Burgess Shale to clades such as phyla is *prima facie* evidence for three important and interrelated conclusions.

1. These organisms are crown taxa that already had evolved the characters that define the phyla of living animals.
2. Morphological gaps separated these crown taxa.
3. Origination of phyla from stem taxa has to be sought in earlier forms in older deposits.

Assessing Morphology

The only evidence we have to recognize and classify Early Cambrian organisms is morphological. How do we separate individuals into crown taxa and decide whether the range of morphologies in Cambrian animals is similar to the range of morphologies in living members of the same groups? One approach measures features and plots them in three dimensions.

BOX 11.1
Snowball Earth

A relatively recent hypothesis for the lack of evidence of Precambrian animals goes under the name "snowball Earth," a name for the presence of a kilometer-thick sheet of ice over the entire Earth 635 Mya. This ice persisted for 10 My during which the temperature of the entire Earth is calculated to have hovered around −40°C. Subsequent research has revealed evidence for at least two and as many as four episodes of snowball Earth between 725 and 635 Mya.

Until 1964, when glacial deposits dating from 600 Mya were reported from what are now the baking deserts of Namibia (FIGURE B11.1), it was not known that such tropical regions had an ice age(s) in the past. The discovery of similar glacial deposits from every present day continent led to the hypothesis of a long period of intense, worldwide cold. Mathematical models developed in the 1970s showed how, under certain conditions, runaway freezing of the whole planet was almost inevitable.

How did Earth and its evolving organisms recover from the effects of prolonged universal temperatures as low as −50°C? The hypothesis is as follows.

- The intense heat of emerging volcanoes penetrated the ice releasing CO_2, which as we know is a greenhouse gas that raises the temperature of the atmosphere.
- CO_2 levels increased because it was too cold for rain to wash the CO_2 from the atmosphere.

■ Snowball Earth may have been a more regular feature of Earth's history than previously anticipated. An even earlier snowball Earth, dating from about 2.2 Bya and detected on the basis of glacial deposits in South Africa, would have had profound impacts on life.

FIGURE B11.1 Carbonaceous rocks in Namibia (indicated by Rudy Raff of Indiana University) deposited following the melting of ice associated with "snowball Earth."

(continued)

BOX 11.1

Snowball Earth (*continued*)

- Eventually, atmospheric CO_2 reached such high levels — hypothesized to be as much as 10%, which is 260 times current levels — that the greenhouse effect kicked in, elevating temperatures to levels where the ice began to melt and disappeared over as short a time as a few thousand years.

This is also how geologists explain another puzzle in Namibia: the presence, immediately above the glacial deposits, of thick layers of carbonaceous rock that form only in water and that bear all the marks of having been deposited very quickly (Figure B11.1). Again, these carbonaceous rocks are found around the globe. With these findings, the hypothesis of recovery from snowball Earth has been extended.

- As temperatures continued to rise from the coldest conditions ever experienced on Earth (−50°C) to the hottest (+50°C), evaporation from the oceans resulted in torrential rains that may have lasted for centuries, washing CO_2 from the air.
- CO_2 dissolved in water forms a weak acid, carbonic acid. Large amounts of carbonic acid would have bathed the globe, interacting with particles of rock eroded by the torrential rains. The resulting carbonaceous deposits formed the cap on the glacial remains seen in the geological record and illustrated in Figure B11.1. Eventually, conditions would have returned to temperatures and CO_2 levels more typical of Earth over the past 500 to 600 My.
- Although the waters of the deep oceans were anoxic after snowball Earth, recent studies are consistent with atmospheric oxygen and marine sulfate concentrations having been higher after than before snowball Earth, providing a marine environment that could have been colonized by the Precambrian (Ediacaran) biota known to have arisen between 635 and 540 Mya.

When you consider that global climate change today is being driven by quite small increases in CO_2 levels — from less than 0.03% before the Industrial Revolution to 0.038% today — just imagine the impact of 10% CO_2. Runaway global warming would be the only logical outcome.

Imagine plotting the height, weight and waist measurements of all modern humans in three dimensions, using the height, weight and diameter of all mammals as the three axes. If some humans were as big as elephants or as small as mice, the entire 3-D space would be filled. However, because humans represent only a small portion of animal body sizes, they occupy only a small region of the 3-D space. Elephants would cluster at one extreme, mice at another.

In 1966, the American paleontologist and paleobiologist David Raup developed such an approach to plot the morphologies of organisms in three dimensions in what he termed **morphospace**, which he represented as a cube with different features of the organisms along each axis (FIGURE 11.6). Morphospace is essentially the universe of possible forms. If all morphologies were possible, the cube (morphospace) would be filled. Occupancy of only a portion of morphospace by known extinct and extant forms would indicate that only a subset of morphologies was possible. Raup's analysis showed that the known morphologies of shelled invertebrates (bivalves, brachiopods, cephalopods and gastropods) cluster in one region of morphospace (Figure 11.6), a cluster that demarks the limits of the body plan in these organisms. Unoccupied morphospace could represent impossible morphologies, nonadaptive morphologies, and/or constraints on morphology.

A second class of morphological evidence comes from the preservation of **cleavage-stage animal embryos** in Late Precambrian and Early Cambrian rocks. Interpretation

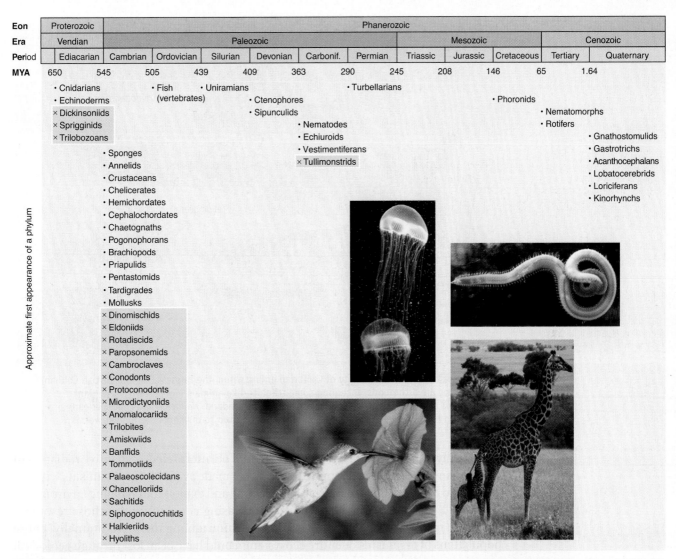

Eon	Proterozoic		Phanerozoic										
Era	Vendian		Paleozoic							Mesozoic			Cenozoic
Period	Ediacarian	Cambrian	Ordovician	Silurian	Devonian	Carbonif.	Permian	Triassic	Jurassic	Cretaceous	Tertiary	Quaternary	
MYA	650 545	505	439	409	363	290	245	208	146	65	1.64		

(Left axis label: Approximate first appearance of a phylum)

- Cnidarians
- Echinoderms
- × Dickinsoniids
- × Sprigginids
- × Trilobozoans

- Fish (vertebrates)
- Uniramians
- Ctenophores
- Sipunculids
- Turbellarians
- Phoronids
- Nematomorphs
- Rotifers
- Nematodes
- Echiuroids
- Vestimentiferans
- × Tullimonstrids
- Gnathostomulids
- Gastrotrichs
- Acanthocephalans
- Lobatocerebrids
- Loriciferans
- Kinorhynchs

- Sponges
- Annelids
- Crustaceans
- Chelicerates
- Hemichordates
- Cephalochordates
- Chaetognaths
- Pogonophorans
- Brachiopods
- Priapulids
- Pentastomids
- Tardigrades
- Mollusks
- × Dinomischids
- × Eldoniids
- × Rotadiscids
- × Paropsonemids
- × Cambroclaves
- × Conodonts
- × Protoconodonts
- × Microdictyoniids
- × Anomalocariids
- × Trilobites
- × Amiskwiids
- × Banffids
- × Tommotiids
- × Palaeoscolecidans
- × Chancelloriids
- × Sachitids
- × Siphogonocuchitids
- × Halkieriids
- × Hyoliths

FIGURE 11.3 Approximate times at which various major metazoan lineages first appear in the fossil record. Unshaded groups marked with filled-in circles have extant descendants, although the original species representing these groups are long extinct. Shaded groups marked with xs represent extinct clades (orders or phyla) that have no known surviving descendants. (Adapted from Conway Morris, *Nature* 361 (1993): 219–225 and from Conway Morris. *The Crucible of Creation: The Burgess Shale and the Rise of Animals*. Oxford University Press, 1998.)

of these embryos as fossils rather than, for example, giant bacteria, is becoming more firmly based as novel technological approaches such as X-ray tomography are applied to the specimens. While not numerous, these specimens have given us a glimpse of jellyfish development 530 Mya, and of the presence of segments in early embryos. The absence of any fossilized **larval stages** strongly suggests that early animals developed directly without a larval stage; that is, they had direct development.

■ Direct development refers to a life cycle without a larval stage. Humans are a good example. Indirect development is the term used for a life cycle with a larval stage. Butterflies are a good example.

A Cambrian "Explosion?"

How did so many body plans arise and why have so few, if any, arisen since the Early Cambrian?

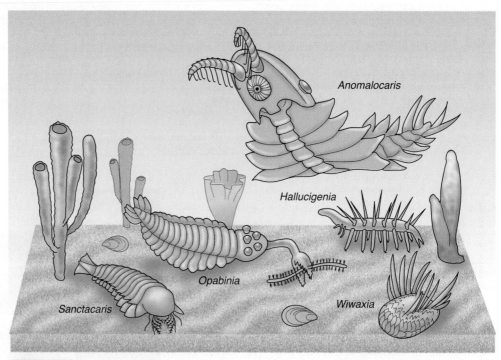

FIGURE 11.4 Part of the diversity of Cambrian animals from the Burgess Shale of British Columbia, Canada. Many are arthropods, including some trilobite-like animals (*Sanctacaris*), the giant *Anomalocaris* and *Opabinia* with its anterior feeding appendage. *Hallucigenia* is another early arthropod in the ancestral arthropod lineage perhaps related to onychophora (velvet worms).

The diversity of Cambrian organisms could be interpreted as adaptive radiation in which many new ecological opportunities were made available for organisms with the capacity to evolve in diverse ways so as to occupy and exploit the changing environment (Figure 11.4). If resources were limited, increasing competition for those resources among so many groups would have led to selection among them. Additionally, entire populations of organisms or entire ecosystems could have been lost through geological events; as at other times, both selective and accidental factors led to extinction or to adaptation, survival and possible diversification.

Although ancestral connections are unclear, many if not all the forms present in the Cambrian must represent new adaptive radiations of Precambrian stem taxa; each form is too specialized upon appearance in the fossil record to have originated in the Cambrian. These early soft-bodied stem taxa are not preserved in the fossil record, with the possible exception of *Kimberella* and early stages of animals in the Ediacaran Biota (see above).

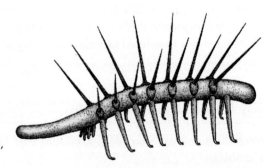

FIGURE 11.5 A reconstruction of *Hallucigenia sparsa,* from the Burgess Shale, thought to be related to the velvet worms (onychophorans) or an early arthropod. (Adapted from Ramsköld, L., and X.-G. Hou. *Nature,* **351**, 1991: 225–228.)

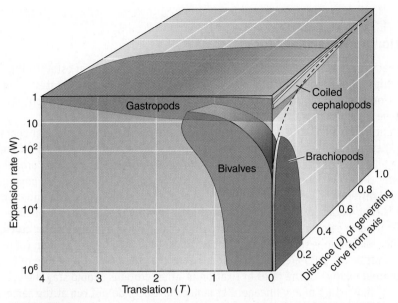

FIGURE 11.6 Only a portion of available morphospace (represented as three-dimensional space) is occupied by known shell morphologies of bivalves, gastropods, brachiopods and coiled cephalopods. The axes represent three measures of shell morphology/growth. (Adapted from Hall, B. K. *Evolutionary Developmental Biology, Second edition*. Kluwer Academic Publishers, 1999; modified from Raup, D. M., *J. Paleontol.*, **40**, 1966: 1178–1190.)

One way to find the time of origin of these (or indeed of any forms) is to use nucleotide sequence data as an **evolutionary molecular clock** (BOX 11.2). Based on an enormous amount of data, we can conclude that divergences among major Cambrian lineages began in the Precambrian about 700 Mya. Modifications during the Cambrian added hard parts and mineralized skeletons that provided leverage for evolving muscles, support for body organs, enclosures for gills and filtering systems, and protective shells and spines.

Causes of the Cambrian Radiation

At least five hypotheses have been produced and lines of evidence amassed to explain the evolutionary burst of multicellular animals in the Cambrian radiation.

1. Geological, environmental and climatic conditions
2. Changing O_2 levels
3. Predator-prey relationships
4. Evolution of embryonic development and body plan specification
5. New sources of genetic variation and changes in gene number and regulation

Evidence for numbers 1 to 4 are discussed below. Sources of genetic variation (number 5) are discussed in Chapter 12.

Geological Conditions

A change in oxygen concentration is one of several factors hypothesized to have facilitated the Cambrian Explosion.

BOX 11.2
Evolutionary Molecular Clocks

Phylogenetic trees based on molecular characters provide a means of establishing relationships independent of morphology. Reconstruction of ancestral states using a combination of morphological and molecular characters provides further hypotheses that can be tested as new data arise.

Inherent in the reconstruction of any phylogenetic tree bases on molecular characters is that (1) evolutionary differences between organisms are related to molecular differences, and (2) generally, the greater the number of molecular differences between organisms, the greater the evolutionary distance between the organisms. In many past phylogenies mutational differences were used to provide evolutionary time scales. The underlying premise is that mutations are incorporated (fixed) at fairly regular rates over time, and so can be used to calculate an evolutionary time scale known as a **molecular clock**.

A basic assumption is that molecular clocks can be calibrated for any protein (or gene) for which nucleotide data are available. This means that the proportional rate of fixation of one protein relative to the rates of fixation in other proteins should stay the same throughout the history of any lineage. But molecular clocks do not run at the same speed for all proteins. Averaging changes among different genes to obtain a single rate has been done, but caution has to be exercised; no single molecular clock applies to every nucleotide sequence. The molecular clock does not tick at the same rate in all taxonomic groups, nor for all genes or all proteins.

Animal fossils may be absent from Precambrian rocks because geological conditions prevented fossilization or destroyed any fossils present; the heat and pressure involved in Precambrian mountain building has been proposed as an important factor limiting fossilization. However, prokaryotic and eukaryotic organisms (and the Ediacaran Biota) were preserved in the Precambrian. Therefore, it is considered unlikely that the Cambrian discontinuity in animals is unreal and merely the consequence of geological metamorphism or imperfect fossilization.

Another physical cause offered to explain the Cambrian Explosion is plate tectonics (see Chapter 2) and the resulting changes in the shape, extent and latitude of shorelines, climate and environment. Sea level changes that accompany glaciation also would have played a role by opening up new environments.

Rising Oxygen Levels

As discussed in Chapter 4, O_2 began to accumulate in the atmosphere a billion years before the Cambrian because of the rise and spread of cyanobacteria. Evidence also exists for major increases in oxygen during the Precambrian. Oxygen forms a protective blanket of ozone (see Chapter 4) that could have facilitated the expansion and radiation of multicellular animals in shallow waters, tide pools and nearby rocky surfaces.

Extrapolating from our knowledge of the descendants of Cambrian organisms we can conclude that aerobic metabolism, which is dependent on oxygen, facilitated the use of new sources of energy, permitting increase in body size. Aerobic metabolism would have required an oxygen atmosphere that had reached perhaps one percent of current atmospheric oxygen. The Early Cambrian atmosphere is hypothesized to have reached such levels (see Chapter 4). It is further hypothesized that animals capable of exploiting Early Cambrian oxygen would have possessed a battery of common genes including those for hemoglobin, which is a conveyor of molecular oxygen.

Predator-Prey Relationships

Another biological factor that could have driven increased organismal diversity is the changing modes of feeding facilitated by rising oxygen levels.

Predators feed on the most abundant **prey** species, reduce the numbers of prey and so allow other species to use resources formerly monopolized by the dominant prey. We see this in modern-day plant communities where removal of a dominant plant species by a predator provides an opportunity for other species previously present in low numbers or excluded from the habitat to expand or migrate into it. Diversification of prey species leads in turn to diversification of predator species. Prey-predator interactions escalate into what has been characterized as a persistent **co-evolving arms race**: successive rounds of selection for predator responses to their prey's protective devices, followed in turn by adaptations by the prey to their predators' devices, promoting diversity.

Shared Embryonic Development and Body Plan Specification

The most obvious and distinctive morphological features of almost all adult animals are

- axes of symmetry, usually **anterior-posterior (A–P)**, **dorso-ventral (D–V)** and **left and right (L–R)**;
- paired appendages; and
- similar tissues and organs.

Such precise and repeatable features arise from precisely ordered embryonic development based on controlled genetic pathways. One of the most informative approaches to understanding animal origins lies in understanding the genetic and cellular bases of embryonic development in different animals. From analyses of patterns and mechanisms of gene expression and of natural or induced mutations we now know that genes that govern developmental processes in all animals share the same nucleotide sequences, similar linkage orders, and even similar developmental targets. For example, and as discussed in the next section, **homeobox genes** provide the primary signals and initiate the pathways that pattern body regions and the formation of particular structures such as wings and legs.

A-P and D-V Patterning. The unequal distribution of maternal gene products deposited into the egg by the female parent establishes localized differences at the very outset of development. Many of these molecules are known as **morphogens** because they activate pathways leading to specific morphologies. An example, using the fruit fly genes *bicoid* and *gurken* that control A-P and D-V patterning, respectively, is shown in (FIGURE 11.7).

■ Patterned distributions of maternal products in the eggs of animals are important early indications of the epigenetic control of gene regulation in development discussed in Chapter 13.

Regional Identity. FIGURE 11.8 depicts further how some cells or tissues obtain information on their position in relation to chemical gradients. Uniformity of the sequence of developmental events in different individuals of the same species lies in the fact that the succession of regulatory events is identical from individual to individual. Change a step — for example, by a mutation — and the outcome changes (unless the mutation is of slight effect and/or can be compensated for). The *Drosophila* mutation *Antennapedia* results in a leg developing from the embryonic region (the imaginal disc) that normally forms an antenna (FIGURE 11.9). A cluster of six *Antennapedia*-linked genes, called the **antennapedia complex**, possess similar regulatory functions. Such mutations are called **homeotic** because they change a particular organ in a particular body region to resemble an organ normally found in a different region along the body axis, a type of change known as **homeosis**.

■ We now also know that Hox genes pattern the developing intestinal tract along an A-P axis and that another class of genes, *distal-less (Dlx)*, patterns vertebrate head development along the D-V axis.

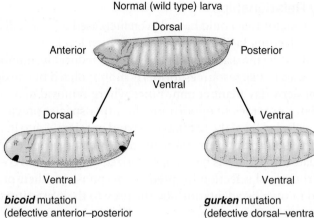

bicoid mutation
(defective anterior–posterior
differentiation)

gurken mutation
(defective dorsal–ventral
differentiation)

FIGURE 11.7 Effects on formation of the body axis of the fruit fly *Drosophila melanogaster* of mutations in the genes *bicoid* and *gurken*. Bicoid mRNA establishes the posterior end of the egg at the high point of a concentration gradient that decreases along the long axis of the egg. Mutations in the *bicoid* gene interfere with A-P patterning, producing a headless individual; the A-P axis becomes a P-P axis. Gurken mRNA acts in a similar manner to establish the D–V axis. Mutations in *gurken* produce a ventralized larva lacking dorsal tissues as the D-V axis is turned into a V-V axis. (Adapted from Ephrussi, A., and R. Lehmann, *Nature* **358**, 1992: 387–392.)

FIGURE 11.8 Representation of how cells or tissues along the major body axes respond to gradients during development. A gradient (**a**) provides cells or tissues A, B, and C (**b**) with information as to their relative anterior–posterior positions. Further growth and differentiation confers dorsal–ventral information producing A1/A2, B1/B2 and C1/C2. Subsequent cell division of these tissues provides α and β subclones with information as to their proximal–distal positions (**d**). As a result of their geographical position and of the activity of selector genes — of which homeotic genes are a major group — cells of tissue A come to possess a specific set of developmental responses and develop differently from tissues B or C, or from other subclones within A. (Adapted from Strickberger, M. W. *Genetics, Third edition*. Macmillan, 1985.)

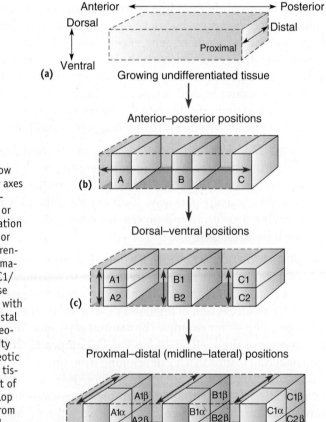

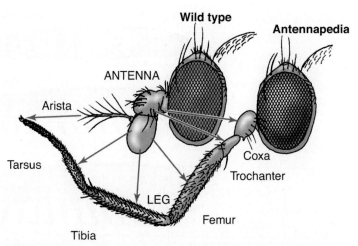

FIGURE 11.9 Positional correspondence between antennae and leg shown in *Drosophila melanogaster* carrying the homeotic *Antennapedia* mutation. The mutational substitution of leg tissues for antenna tissues is shown by the blue arrows; distal cells, for example, interpret their position as tarsal segments, and proximal cells as coxa or trochanter.

As with the *Drosophila* antennapedia complex, a **bithorax** complex of three genes can initiate homeotic changes by regulating many genes that control the fates of various structures from the posterior part of the second thoracic segment to the tip of the abdomen. Should expression of the normal *Ultrabithorax* gene fail in the anterior abdomen, this region develops as though it was thoracic (more anterior) rather than abdominal (FIGURE 11.10).

As shown in FIGURE 11.11 the bodies of embryonic insects are subdivided into compartments (segments) that arise in sequence along the A-P axis. As A-P and D-V axes

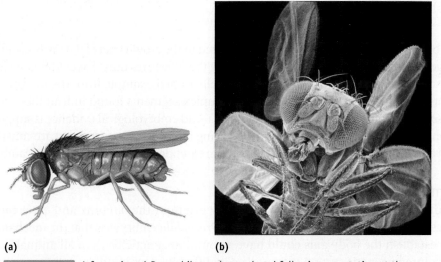

(a) (b)

FIGURE 11.10 A four-winged *Drosophila* can be produced following a mutation at the *bithorax* locus. (**a**) As in all dipteran insects, *Drosophila* has only a single pair of wings, which arise from the second of the three thoracic segments; the second set of paired appendages evolved into balancing organs, *halteres* (blue structure below the wings). (**b**) Certain *bithorax* mutations cause the third thoracic segment to transform into a second thoracic segment complete with wings, a condition similar to ancestral four-winged flies.

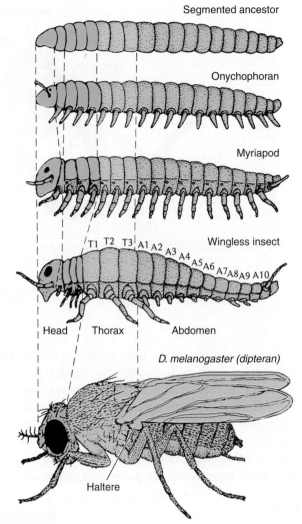

FIGURE 11.11 Proposed evolutionary steps in insect segmental organization as typified by different insect groups. The three thoracic segments are numbered consecutively from the head boundary as T1, T2 and T3. Body segments, at first, relatively uniform, became more complex as different lineages evolved. In dipteran insects such as *Drosophila* (*bottom figure*) the segments bear different complex structures; the second thoracic segment (T2) has wings, the third (T3) has the halteres (balancing organs). A1–A10 are eight abdominal segments numbered consecutively from the thoracic boundary. (Adapted from Strickberger, M. W. *Genetics, Third edition*. Macmillan, 1985.)

and body segments are present in animals found in the crown taxa of the Early Cambrian Burgess Shale, the mechanisms for producing these features must have originated in the Precambrian. Insect evolution provides a well-studied example from the uniform segments in the insect ancestor to the more complex segments found in fruit flies such as *Drosophila* (Figure 11.11). This morphological and embryological evidence is supported by phylogenetic trees of animals produced using molecular data, tress that are consistent with animal phyla having diversified during an extended period during the Precambrian.

Homeobox Genes

An implication of the ancient evolutionary origin of the *bithorax* and *antennapedia* genes that specify the axial organization of *Drosophila* embryos is that the mechanisms that establish the body axis could have a common genetic basis in all animals. If so, the common genetic system is expressed in different ways in different lineages: wings in some, legs in others; segments in some, lack of segments in others. How could such a genetic system be structured?

The discovery of a family of closely related DNA nucleotide sequences in DNA called the **homeobox** answered this question. Homeobox genes (genes that contain

the homeobox) were found in various locations in the *Drosophila* genome, including loci within the *bithorax* and *antennapedia* complexes. Each homeobox gene codes for a polypeptide sequence about 60 amino acids long called a **homeodomain**. The homeodomain, in turn, is part of a transcription factor that binds to DNA, thereby regulating mRNA production.

Fifty or more families of homeobox genes have been identified in animal genomes. Although expressed in different structures in different taxa patterning of the A-P and D-V axes throughout the animal kingdom is controlled by homeobox genes. Homeobox genes in vertebrates, which are known as the **Hox gene family**, are introduced in BOX 11.3. Conserved homeobox genes also are present in flowering plants and fungi (see Chapter 10), both of which produce homeodomain proteins much like those in animals. Homeobox genes, therefore, must have functioned early in eukaryotic history, having arisen before plants, fungi, and animals diverged. It is hypothesized that homeobox genes diversified during the evolution of multicellular organisms by duplication from a common ancestral gene. Early duplications of this gene or gene complex, and subsequent divergent function(s) of each duplicate, permitted different regions of the body to specialize, a topic discussed by Van de Peer and colleagues in a recent review. Duplication is discussed in BOX 11.4 in relation to vertebrate Hox gene evolution.

■ The abbreviation Hox is derived from <u>Ho</u>meobox for the gene family and <u>x</u> for vertebrates.

Surprisingly, the linkage order of these homeobox-containing genes in the *Drosophila antennapedia–bithorax* clusters accords with their expression along the anterior–posterior axis of *Drosophila* embryos (see Figure B11.2). Equivalent homeodomains with equivalent linkage orders of homeobox (Hox) genes occur across the animal kingdom (FIGURES 11.12 and 11.13), indicating that specification of body plans in different lineages has a common and ancient evolutionary origin.

Furthermore, a particular homeobox protein can perform a similar function in different organisms; for example, much or all of the function of the *Drosophila* homeobox gene *Antennapedia* can be assumed by a protein produced by a homologous *Hoxb* gene in mice. Homologous developmental genes (such as *engrailed*, which specifies compartmental distinctions within segments) function throughout the animal kingdom.

Several major conclusions come from these studies of comparative gene structure and function in the animal kingdom.

- These important developmental (regulatory) genes all **share** a common, highly conserved, and evolutionarily **ancient role** as transcription factors.
- A **common genetic evolutionary origin** underlies the conservation of the basic developmental pathways that establish animal body plans.
- What has **changed with evolution is context**; the specific function of these conserved regulators varies from cell lineage to cell lineage, tissue-to-tissue, organ-to-organ, as well as from time to time during development.

Developmental Change and Evolution

Given the deep conservation of developmental genes, and the stability that arises from their conserved functions, we need to understand how the evolution of variation in development is mediated at the genetic level. As discussed in the next chapter, **gene regulation** has emerged as a central mechanism explaining developmental and evolutionary change. In a seminal 1977 paper credited with enticing many researchers into

BOX 11.3
Hox Genes and Vertebrate Origins

Homeobox-containing genes in vertebrates are known as **Hox genes**, Ho from homeo-domain and x for vertebrate. Thus, *Pax-6* is a vertebrate homologue of the *Drosophila paired rule* (*Pa*) gene, which is combined with *x* to make *Pax*. The *6* indicates that 5 other genes in the Pax family were known when *Pax-6* was named. Convention also dictates (although not all follow the convention) that gene names are italicized (*Pax-6*), but the gene protein is not, and that Hox genes in humans are italicized and capitalized (*PAX-6*). The recognition of the importance of Hox genes (see Chapter 13) marks a fascinating episode in the history of the search for vertebrate origins.

Vertebrate homeobox (Hox) genes with sequence homology to such *Drosophila* genes as *Ultrabithorax* and *Antennapedia* (this chapter) are a series of transcription factors organized as homeobox clusters on four chromosomes in vertebrates. As in *Drosophila*, the order of Hox genes within a cluster is paralleled by an anterior-posterior sequence of expression. The patterning role of Hox genes is demonstrated by findings, in mice, that knocking-out or knocking-in a Hox gene to eliminate or enhance its function is followed by transformation of skull, vertebral or other features into a more anterior element in the sequence, a transformation known as homeotic (see Chapter 13). Regenerating amphibian limbs can be duplicated following manipulation of Hox genes. In some species of frogs, an amputated tail can be made to duplicate, homeotically, the posterior portion of the body and to regenerate a limb complete with pelvic girdle and not a tail. Finally, conservation of the roles of vertebrate and *Drosophila* genes has been demonstrated by research showing that, for example, after being inserted into the *Drosophila* genome, the mouse *Hoxb-6* gene elicits leg formation in the place of antennae in *Drosophila*.

The number of Hox clusters varies among vertebrates: four clusters of 39 genes in mice, three clusters in lampreys, and up to seven in teleost fish. Duplication of the genome at the origin of the chordates is the most likely current explanation for the four clusters; duplication sets up the possibility of future structural and functional divergence and *specialization of function among copies of the gene*. Four possibilities, which are not mutually exclusive, could explain this evolutionary change in gene function.

Two involve change in gene number, either in the number of Hox gene clusters or the number of genes per cluster. Duplication of *Hox* clusters before the teleosts arose — perhaps associated with duplication of large portions of chromosomes or entire chromosomes — would take the number from four to eight. This, coupled with subsequent loss of one cluster, would explain the seven clusters found in zebra fish.

The *other two possibilities involve* altered function by modification of the function of individual Hox genes or by increasing the complexity of interaction between gene networks, either of which could come about by alteration in genes that regulate a Hox gene (such genes are said to be upstream) or in genes that are regulated by the Hox gene (such genes are said to be downstream).

molecular biology, François Jacob proposed gene regulation as critically important in development, in evolution, and as a target for natural selection.

Natural selection does not work as an engineer works. It works like a tinkerer — a tinkerer who does not know exactly what [s]he is going to produce.

What makes one [organism] different from another is a change in the time of expression and in the relative amounts of gene products rather than the small differences observed in the structure of these products. It is a matter of regulation rather than of structure.

BOX 11.4

Gene Duplication and Divergence

Observations of the sequence similarities of different hemoglobin chains, strongly suggest that, rather than arising from different genes that converged in sequence and function, hemoglobin and the other **globin genes** such as the gene for myoglobin, arose as **duplications** of an original globin-type gene. A gene terminology has been adopted to reflect two different types of gene duplication (see Chapter 16).

- **paralogous genes (paralogues)** for duplications within a species (for example, α, β and γ hemoglobins in humans), and
- **orthologous genes (orthologues)** for genes shared between species because of shared species ancestry (for example, α hemoglobins in horses and humans).

The time during the evolution of globin genes when the duplications occurred can be calculated from amino acid differences between the molecules; the greater the amino acid differences between any two chains, the further back in time the duplication occurred. FIGURE B11.2 portrays the genetic phylogeny of the five globins in terms of the numbers of nucleotides necessary to account for the differences in amino acid differences. Because we can calculate how long ago mutations occurred, when each of the duplications is inferred to have occurred is also shown in Figure B11.2.

Three data sets enable us to order the stages in the evolution of these five globin genes: (1) The myoglobin chain with distinctive amino acids at more than 100 sites differs most from all other globin genes. (2) The α and β chains of hemoglobin differ from each other at 77 amino acid sites. (3) The β chain of hemoglobin differs from the γ-chain at 39 sites but differs from the δ chain at only 10 sites.

As the most divergent, the gene for myoglobin arose from an early duplication. A later duplication produced two α hemoglobin genes, one of which evolved into the β hemoglobin gene. Because they differ least, the separation between β and γ hemoglobin chains arose from the most recent duplication (Figure B11.2). Subsequent duplications of a different type, known as side-by-side or **in-tandem duplications**, resulted in two copies of the α gene (α1 and α2). Several tandem duplications of the β gene produced seven β genes in what is known as a **gene or multigene family** (FIGURE B11.3).

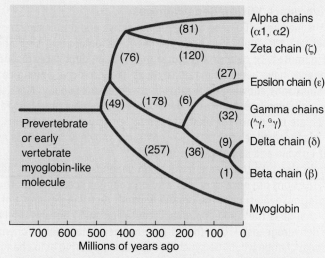

FIGURE B11.2 Phylogenetic relationships between globin-type proteins in humans to show the estimated times when the proteins diverged from each other. The estimated number of nucleotide replacements necessary to cause the observed amino acid changed in each branch of the lineage is given in parentheses. (Adapted from Strickberger, M. W. *Genetics, Third edition*. Macmillan, 1985.)

(continued)

BOX 11.4

Gene Duplication and Divergence (*continued*)

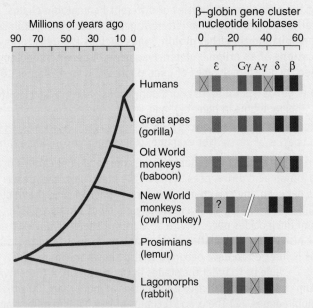

FIGURE B11.3 The clustered organization of the β-globin–type genes in five different primates and in a rabbit, along with a proposed evolutionary tree for the genes. The genes in each of the clusters are linked together on the same chromosome. Each gene, denoted by a small rectangle, is transcribed from left to right. Genes responsible for embryonic and fetal development are on the left (*lighter shading*), genes that produce adult β-globins are on the right (*darker shading*). Genes marked with crosses are nonfunctional pseudogenes. (Adapted from Strickberger, M. W. *Genetics, Third edition*. Macmillan, 1985.)

Functional Divergence Within Gene Families

Sometimes, all the members of a gene family remain active and retain similar or identical functions. Sometimes the duplicated gene become functionless but is retained as a **pseudogene** (Figure B11.3); deleterious mutations in gene duplicates responsible for causing pseudogenes have been estimated to be as likely to occur among early vertebrates as were mutations allowing duplicates to diverge functionally. Most significantly from an evolutionary view are duplicated genes that evolve new functions, a classic example of which are the genes for the transparent lens crystalline proteins. Although gene structure is conserved, each of the 10 crystallins known in animals is associated with a different enzyme and so a different function.

How long duplicates take to diverge in function depends, of course, on how many amino acid substitutions are necessary, but also on changes occurring in other genes and in the adaptation/evolution of the organism itself. A striking example of the speed of evolutionary change is the single amino acid mutation that converts the metabolic enzyme lactate dehydrogenase (LDH) to another enzyme, malate dehydrogenase (MDH). Here the functional change would be synchronous with the amino acid change. Most functional changes are much slower, requiring as they do, changes in many amino acids for the structure and so the function of the gene product to change.

An important issue raised by the existence of gene duplications, especially whole genome duplications, is whether such events facilitated the origin of novel organs, organisms, and/or adaptive radiation. The origins of land plants and bony fish correlate with genome duplications, but the origins of birds and mammals do not, indicating that gene duplication is not a requirement for the origin of new clades.

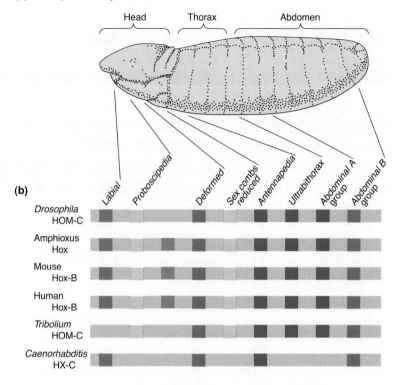

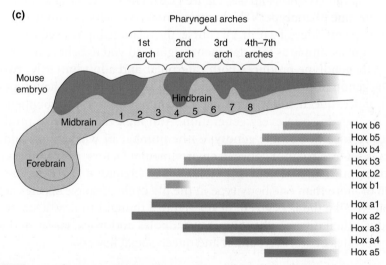

FIGURE 11.12 Homeobox gene relationships in representative animals. Almost all the genes illustrated show a highly conserved relationship between their position along the chromosome (shown in **b**) and developmental function along the anterior–posterior axis (shown in **a** and **c**). (**a**) Regions of action of homeobox genes in *Drosophila melanogaster* along the anterior–posterior axis of the embryo with labial acting most anteriorly and Abdominal B group genes acting most posteriorly. (**b**) Clusters of homeobox (Hox) genes from *Drosophila,* amphioxus (a cephalochordate), mammals (Hox-B from mice and humans), a beetle (*Tribolium castaneum*), and a nematode (*Caenorhabditis elegans*). Homologous (orthologous) genes are aligned vertically. (**c**) Expression of Hoxa and Hoxb genes in the anterior region of a mouse embryo. The left hand end of each bar marks the extent of anterior (rostral) expression of each gene, which determine important boundaries such as the position of pharyngeal arches 1 to 7, divisions 1 to 8 within the hindbrain, and boundaries between populations of migrating neural crest cells (*green*).

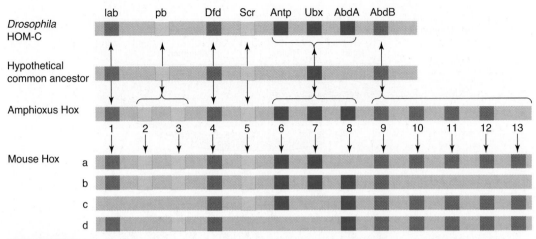

FIGURE 11.13 Relationships between homeotic gene clusters in *Drosophila,* Hox genes in amphioxus and mice, and homeotic genes, and in the hypothetical common ancestor of arthropods and chordates. Expansion of the six gene clusters in the common ancestor gave rise to eight clusters in *Drosophila* and to 13 in mice. (Adapted from Hall, B. K. *Evolutionary Developmental Biology, Second edition.* Kluwer Academic Publishers, 1999; modified from Holland, P. W. H., and J. Garcia-Fernàndez, *Devel. Biol.,* **173**, 1996: 382–395.)

Jacob referred to evolution as **bricolage (tinkering)**, a term that rightly places the emphasis on modification of existing genes, and which has been employed effectively for evolutionary change above the level of the gene.

Genes provide the necessary historical information used by cells as they differentiate. Cells then interact with each other and with their environment to assume their ultimate phenotypic shape and relationship (see Figure 11.8). There is a logical tendency to think **"one genotype–one phenotype."** The human genotype produces a human phenotype; a worm genotype produces a worm. Of course, the relationship between genotype and phenotype is not as simple as this statement might lead you to believe it to be.

A butterfly genotype produces both caterpillar and adult butterfly: two phenotypes. A frog genotype produces tadpole and adult frog, again two phenotypes. An ant genotype produces queen, soldiers and major and minor workers — four phenotypes — and some species do so in every generation. Although the ability to produce four phenotypes resides in a single genotype, the number of workers or soldiers in a population is set by a balance between environmental factors, the numbers of workers or soldiers present, and genetic potential (see Chapter 14). These instances of production of more than one body type in the life cycle of single species are so well known that we take them as given, never giving a thought to how a single genome could produce such diverse phenotypes as tadpoles and frogs, larvae and adults of many invertebrates or worker, soldier and queen social insects.

■ Population structures within social insects vary enormously. The red seed-harvester ant, *Pogonomymex californicus,* for instance, has one queen per colony. When the one queen stops producing workers, the colony dies.

■ **Recommended Reading**

Ayala, F. J., A. Rzhetsky, and F. J. Ayala, 1998. Origin of the metazoan phyla: Molecular clocks confirm paleontological estimates. *Proc. Natl. Acad. Sci. U.S.A.,* **95**, 606–611.

Chen, J-Y., D. J. Bottjer, L. Gang, et al., 2009. Complex embryos displaying bilaterian characters from Precambrian Doushantuo phosphate deposits, Weng'an, Ghizhou, China. *Proc. Natl. Acad. Sci. U.S.A.,* **106**, 19056–19060.

Clapham, M. E., G. M. Narbonne, and J. G. Gehling, 2003. Paleoecology of the oldest-known animal communities: Ediacaran assemblages at Mistaken Point, Newfoundland. *Paleobiology* **29**, 527–544.

Cohen, P. A., A. H. Knoll, and R. B. Kocher, 2009. Large spinose microfossils in Ediacaran rocks as resting stages of early animals. *Proc. Natl. Acad. Sci. USA*, **106**, 6519–6524.

Erwin, D. H., and E. H. Davidson, 2002. The last common bilaterian ancestor. *Development,* **129**, 3021–3032.

Hall, B. K., 1999. *Evolutionary Developmental Biology,* 2nd ed. Kluwer Academic Publishers, Dordrecht, Netherlands.

Hoffman, P. F., A. J. Kaufman, G. P. Halverson, and D. P. Schrag, 1998. A Neoproterozoic snowball Earth. *Science,* **281**, 1342–1346.

Hou, X-G., R. J. Aldridge, J. Bergström, D. J. Siveter, and X-H. Feng, 2004. *The Cambrian Fossils of Chengjiang, China: The Flowering of Early Animal Life.* Blackwell Science, Oxford, England.

Valentine, J. W., 2004. *On the Origin on Phyla.* The University of Chicago Press, Chicago.

Van de Peer, Y., S. Maere, and A. Meyer, 2009. The evolutionary significance of ancient genome duplications. *Nature Reviews Genetics* **10**, 725–732.

V

Principles and Processes of Evolution

Individual Genetic Variation and Gene Regulation

12

KEY CONCEPTS

- Without variation there would be no evolutionary change.
- Evolutionary potential and genetic variation are two sides of the same evolutionary coin.
- How variation at one level (e.g., genetic–genotype) relates to variation at other levels (e.g., tissues and organs–phenotype) is a major unresolved question in evolutionary biology.
- Variation at the genetic level arises from multiple alleles, mutations, change in chromosome numbers, and modification of gene regulation.
- New species of plants have arisen by duplication of chromosome number. (Various species of wheat cultivated for at least 30,000 years are a prime example.)
- Duplication of the entire genome facilitated the origination of some lineages of organisms but not others.
- Gene regulatory changes of all kinds have been among the key agents of organismal evolution.
- Genes are regulated by multiple mechanisms at transcriptional, translational and post-translational levels.

Above: Variation personified and our lives enriched.

211

- Nucleotide sequences (transposons) that promote their own movement (transposition) between genomes are important regulators of gene activity.
- (i) Gene/chromosome duplications, (ii) mutation, (iii) changes in gene regulation, (iv) transposons and (v) horizontal gene transfer are five major means by which **prokaryotic cells** can acquire new genetic information.
- Combine (i) to (v) above with (vi), the origin of eukaryote organelles (and their genes) by endosymbiosis, and we have six major means by which **eukaryotic cells** can acquire new genetic information.

Overview

Natural selection acts on existing phenotypic variation. In most populations up to three-quarters of the gene loci have more than one allele. With this great reservoir of variation, populations can respond to evolutionary pressures without new variants having to arise by mutation. Additional sources of genetic variation can come from alterations in chromosome number — especially in plants — either via changes in entire sets of chromosomes or in individual chromosomes, or through deletions, duplications, inversions, or translocations of nucleotide sequences that affect one or more than one gene. Further variation can arise from gene duplications (which can include duplications of an entire genome) and from modifications in the regulation of gene function, at DNA, RNA or their protein product levels.

Mutations are small, localized changes in nucleotide sequences or regulation and so can lead to changes in a gene or gene product. The term *regulatory mutation* is sometimes used for those changes that affect genes controlling genetic pathways or networks. A classic example is a mutation in the regulatory gene that governs the expression of genes coding for enzymes that digest the sugar lactose in bacteria. Discovery of such mutations laid the foundation of our current understanding of gene regulation. As with any other trait in which genes and their alleles have been selected for their functional (adaptive) value, a multicellular organism's development is the outcome of a historical evolutionary process. Whether new mutations can incorporate successfully depends on how they interact with existing genetic pathways and developmental processes. Such interactions canalize development and both place limits on, and create opportunities for, evolutionary change.

■ Ernst Mayr, one of the most influential evolutionary biologists of the twentieth century, was a staunch defender of the Modern Synthesis, a synthesis that he played a major role in creating. His last book entitled, *What Evolution Is*, published in 2001, is a thorough and thoughtful exploration of evolutionary biology.

Variation: Five Central Questions

Study of the causes of variation is important because variation is the raw material for natural selection. One of the leading 20th century students of evolution, Ernst Mayr saw this early in his career and pursued it over 60 years.

It is amazing to what extent variation in natural populations has been neglected in the study of evolution. Amazing, because natural selection would be meaningless without variation. This conclusion gave me the idea to consider the production of variation as a step in the process of natural selection (Mayr, 2005, p. xviii).[1]

[1]Cited from the Foreword in *Variation: A Central Concept in Biology* (B. Hallgrímsson, and B. K. Hall (eds.), 2005, Elsevier/Academic Press, Burlington, MA).

Genetic variation and evolutionary potential are two sides of the same evolutionary coin. The topics of this and the following chapter — how variation arises in genes and populations — are both central to evolution and are one of the least understood sets of evolutionary mechanisms. I say "sets" because variation can arise in a number of different ways, outlined as four questions below and in more detail in Chapters 13 to 15. Addressing any one of these questions would make a wonderful class project or topic for presentations.

1. What is the relationship between the genetic variation of the **genotype**, which is transmitted from generation to generation, and variation of the **phenotype**, which is the direct object of selection? This question is a major preoccupation of evolutionary developmental biology (evo-devo).

2. What are the mechanisms by which **mutations** and **modifications of gene regulation** serve as sources of variation? What are their respective roles in maintaining the phenotype, modifying the phenotype and/or in the origin of new (novel) phenotypes? This challenging area, which is addressed in this chapter, seeks to understand why some structures/organisms/species change more than others over the course of their evolution.

3. What **other sources of variation are available to populations?** Two sources — gene flow from other populations and random drift of genes within a population — are discussed in Chapter 13.

4. What are the **ecological** and **developmental** determinants of phenotypic variation among species? This includes such questions as how body size, geographic range, home range size, niche width, lifespan, environmental stress, population size and density affect the tendency of populations to exhibit phenotypic variation (see Chapter 14).

Let's begin with variation at the level of entire chromosomes.

Variation in Chromosome Number

Variations in chromosome number are of two major kinds

- changes in the **number of entire sets** of chromosomes, and
- changes in the **numbers of single chromosomes** within a set.

A larger number of chromosomes may allow more genetic recombination and therefore more genetic variation, which would be advantageous in changing environments. Conversely, smaller chromosome numbers may allow genetic combinations to persist, and so be associated with long periods of occupation of specialized environments.

Repetitive doubling can take **chromosome number** well beyond the diploid condition, an outcome known as **polyploidy**, meaning many time the normal ploidy; a triploid is 3n, a tetraploid 4n, and so on. As discussed below, changes in chromosome number led to the evolution of wheat and of tobacco (under very different circumstances). Doubling of chromosome number has led to lineages of vascular plants with greater than 80-fold ploidy. Somewhere between 40 and 70 percent of all plant species have been estimated to be polyploid. The most recent study by Wood and colleagues (2009) concluded that 15% of angiosperm and 31% of fern speciation events are accompanied by polyploidy, a condition that provides evolutionary

advantages with respect to cell and organism size, stability of the genome and tissue/organ-specific gene expression.

Chromosome numbers have been **reduced** in some lineages. For example, in some populations of European wild mice, *Mus musculus,* the normal number of 20 pairs of chromosomes has been reduced to as few as 12 pairs. In contrast, conservation of chromosome number has been a persistent feature in other mammalian lineages. For example, extant species of the 30-My-old lineage that includes Asian (Bactrian) and African (dromedary) camels as well as the South American guanaco, vicuna, llama and alpaca, all have the same number, 74, of morphologically similar chromosomes.

Variation can also be introduced when a single chromosome is added or deleted. In animals, such changes often lead to abnormalities. In plants, however, such changes — especially additions — are common, more tolerated than chromosomal variation in animals, and often associated with the origin of new strains or even new species. Variation in chromosome number between similar animal species may sometimes be associated with the origin of adaptive features. Stressful ecological conditions such as periodic aridity and other unpredictable hardships appear to correlate with chromosome numbers in Israeli and Turkish populations of the mole rat, *Spalax,* increase in chromosome number providing the potential for increased genetic diversity and enhanced potential to respond to ecological variation.

The Evolution of Wheat

The generation of polyploidy as a consequence of hybridization (see Chapter 17) is a common mode of evolution in many plants, including mosses, apples, pears, bananas, tomatoes and corn. Polyploids can arise when different species interbreed naturally or are crossed under field or laboratory conditions. A classic example of both natural and forced interbreeding is the evolution of wheat, *Triticum aestivum,* used to make flour for bread. Evolution of wheat following chromosome duplication is linked to the origin and expansion of human agriculture. Consequently, the evolution of wheat has attracted many researchers from fields as diverse as genetics, plant breeding, agriculture, archaeology and anthropology. The story goes as follows.

At least 30,000 years ago, in the Fertile Crescent of southwest Asia, a natural hybrid formed between two grasses, *Triticum monococcum* (wild einkorn) and a species of *Aegilops* (goat grass). Each had 14 pairs of chromosomes. Hunter-gatherers harvested the seeds of this new 28-chromosome plant (*Triticum dicoccoides,* **wild emmer**) for millennia. Around 10,000 years ago, by which time the ice ages had ended, humans began cultivating wild emmer. By 9,500 years ago, harvesting of the best plants resulted in the selection of desirable qualities and led to a new form of emmer, **cultivated emmer**, now regarded as a new species.

Cultivated emmer was an important crop for 7,000 years, spreading throughout the Near East and into Egypt. When it reached an area southwest of the Caspian Sea around 9,000 years ago, cultivated emmer came into contact and interbred with a second species of goat grass (*Aegilops squarrosa*), which had 14 pairs of chromosome. The new hybrid, known as **spelt** (*Triticum spelta*) had 42 chromosomes (28 + 14; FIGURE 12.1). Neither emmer nor spelt was easy to harvest, but about 500 years later a fortuitous mutation changed the nature of the spike or ear, producing a shell that would allow threshing. An unfortunate side effect was that the plant, a new species (**bread wheat,** *Triticum aestivum*) could no longer be propagated naturally.

Since then, bread wheat has undergone considerable change. For millennia, farmers sowed seed they had collected the previous year, creating in the process a multitude of

The Fertile Crescent is a half-moon (crescent)-shaped region of some 13 million hectares adjacent to the Mediterranean Sea. Extending from the Nile River Valley to the valleys of the Tigris and Euphrates rivers, it is the site of the origin of agriculture.

Many grocery stores carry the grain **triticale** (see Chapter 19). This, too, is a new species we have produced using the principles of polyploidy to cross wheat with rye.

FIGURE 12.1 Spelt (hybrid wheat) to show the habit and details of the seed heads.

"land races" adapted to local conditions (see Figure 13.11). More recently, agronomists have bred improved varieties of wheat, enormously raising yields. About 7,000 years ago, cultivated emmer evolved in a second direction when new mutations resulted in free-threshing grain, facilitating the spread of agriculture. Having evolved directly from spelt, *Triticum durum* (**macaroni** or **durum wheat**) also has 28 chromosomes.

Breeding the Evil Weed in the Laboratory

The first (and perhaps the only) species produced in the laboratory was the product of a cross between two species of tobacco plants; *Nicotiana tabacum* with 48 chromosomes and *N. glutinosa* with 24 chromosomes. The hybrid (FIGURE 12.2) was sterile; the unequal number of chromosomes did not allow normal chromosome pairing during meiosis. But, the hybrid tobacco plant could be, was and continues to be propagated by vegetative cuttings. The next event could neither have been anticipated nor planned. Eventually, and by chance, a chromosome-doubling event produced a fertile plant with 72 chromosomes and normal meiosis (Figure 12.2). The resulting plant, a new species of tobacco, *Nicotiana digluta*, was self-crossed and found to be fertile.

Other Changes in the Phenotype of Chromosomes

A variety of other changes in chromosome structure are shown in FIGURE 12.3 . Because these are changes in the appearance of the chromosome, we can speak of them as the phenotype of the chromosome, just as we speak or the phenotype of the individual when comparing characters.

Deletions and **deficiencies** remove chromosomal material (Figure 12.3a). If functional genes are removed, deletions can be harmful in diploids and haploids but not necessarily in polyploids where such genes may be present on more than one chromosome.

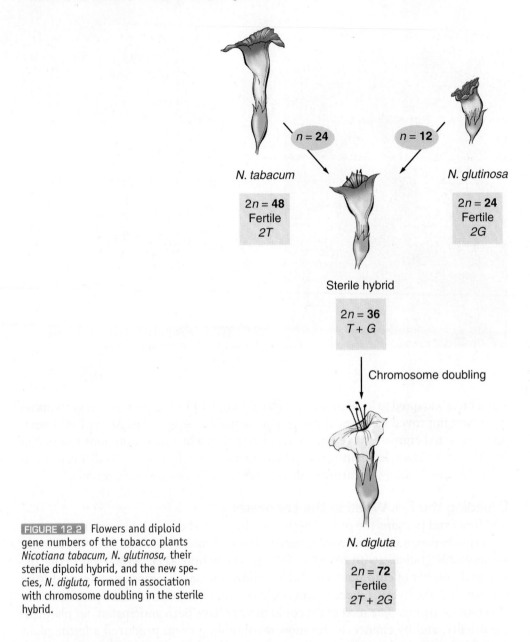

FIGURE 12.2 Flowers and diploid gene numbers of the tobacco plants *Nicotiana tabacum, N. glutinosa,* their sterile diploid hybrid, and the new species, *N. digluta,* formed in association with chromosome doubling in the sterile hybrid.

Duplications are segments of extra chromosome material originating from duplicated sequences in the genome, usually resulting from unequal crossing-over during chromosome pairing (Figure 12.3b and FIGURE 12.4). Duplications have been common during evolution resulting in the formation of numerous **gene families** of similar or identical genes in many species of all multicellular organisms. Indeed, entire genomes have been duplicated in various lineages and at various times during evolution (see Box 11.4).

Inversions are reversals in chromosomal gene order (Figure 12.3c, d). The genes included within an inversion tend to remain together as a nonrecombinant block, called by some a **supergene**.

Translocation moves genes from one chromosome to another (Figure 12.3e). Translocations may change the number and structure of chromosomes and thereby introduce variation. Among Asiatic muntjac deer, the Indian muntjac, *Muntiacus muntjac,* has only three pairs of large chromosomes, the black muntjac, *Muntiacus crinifrons,*

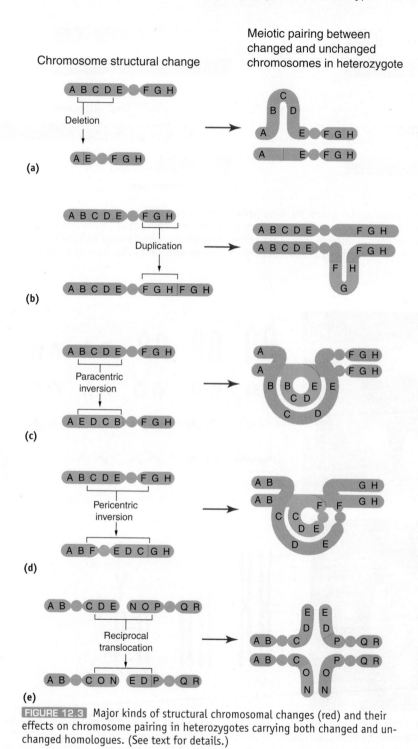

FIGURE 12.3 Major kinds of structural chromosomal changes (red) and their effects on chromosome pairing in heterozygotes carrying both changed and unchanged homologues. (See text for details.)

four pairs and the Chinese muntjac, *Muntiacus reevesi* 23 pairs (**FIGURE 12.5**). The relative amounts of DNA in the two species, however, are about the same, the large muntjac chromosomes arising from successive translocations combining the smaller muntjac chromosomes. (Differences in the amount of DNA in different organisms create what is called "the C-value paradox.")

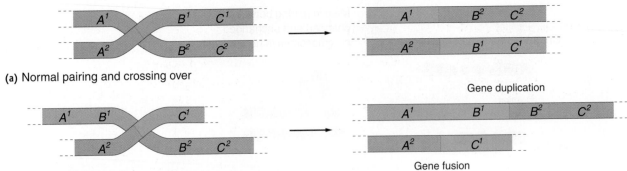

(a) Normal pairing and crossing over

Gene duplication

(b) Unequal pairing and crossing over

FIGURE 12.4 The results of equal and unequal crossing-over for three gene segments on a chromosome. **(a)** When pairing between homologous sections on two chromosomes is equal, the crossover products have the same amounts of chromosomal material. **(b)** When pairing between the two chromosomes is unequal, one of the crossover products carries a gene duplication (the *B* gene segment in this illustration), and the other product shows a fusion between gene segments (*A–C*) that were formerly separated by the intervening *B* segment.

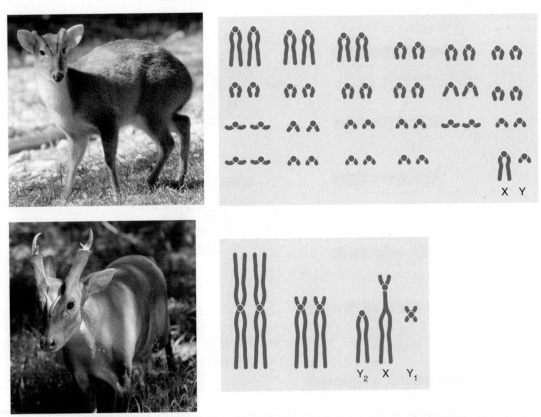

FIGURE 12.5 The Chinese and Indian muntjac deer (*Muntiacus reevesi*, top, and *M. muntjac*, bottom) have 23 and three pairs of chromosomes, respectively. The Indian muntjacs, with two pairs of autosomes and three sex chromosomes, have the lowest known chromosome number of any mammal. (Chromosome spreads are adapted from Austin, C. R., and R. V. Short, 1976. *The Evolution of Reproduction.* Cambridge University Press, Cambridge, England.)

Chromosomal Evolution in *Drosophila* and Primates

In those instances where linear sections of chromosomes can be identified because of their distinctive bandings, chromosomal evolution can be charted in considerable detail. Interest exists at two levels: the evolution of chromosomes themselves as characters of the organism, and how chromosomal evolution influences the evolution of other characters.

In species of *Drosophila* and other similar insects, the chromosomes of salivary gland cells and other tissues have replicated many times, but the replicates remain attached. The resulting giant chromosomes, called **polytene chromosomes**, are enlarged enormously and show highly detailed banding configurations that enable even minor chromosomal changes to be identified. Geneticists have documented practically all the chromosomal changes in the evolution of hundreds of these fly species.

Although polytene chromosomes are absent in many organisms, chromosomal staining techniques enable detailed comparisons, even between relatively small mammalian chromosomes. The G-banding technique illustrated in FIGURE 12.6 was used to compare chromosomes from modern humans, chimpanzee, gorilla and orangutan. These bandings show that some chromosomes — numbers 6, 13, 19, 21, 22 and X — are almost identical in all four species. The difference in chromosome number between humans ($n = 23$) and apes ($n = 24$) derives from a fusion event that combined, in the common ancestor, the two chimpanzee-type chromosomes into the number 2 human chromosome (Figure 12.6). This fusion must have occurred after the human line separated from a human-chimpanzee common ancestor. These banding arrangements provide a source of evidence that modern humans have a closer evolutionary relationship with chimpanzees than with gorillas and a more distant one with orangutans (see Chapter 14).

We turn now from entire chromosomes to genes as sources of variation.

Mutations as a Source of Genetic Variation

Mutations affect the nucleotide structure of a gene. New mutations are usually detected because we observe a harmful effect(s) on the phenotype; even what appear to be small mutational changes may have considerable phenotypic consequences (see below).

The wide range of mutational possibilities has important consequences for organisms. Too many or too few mutations can interfere with the effect that natural selection has on genetic variation (see Chapter 13), which is the basis for adaptation. Too many mutations will generate continued errors in organisms already selected for their environment. Too few mutations will reduce the opportunity for natural selection to initiate changes that could lead to adaptations. Without mutations, we would all be mindless blobs of unicellular protoplasm, or perhaps no more than a few molecules floating in the primordial soup.

Mutations are normally expressed at one of two levels of gene activity,

- changes within a gene product, for example, in the amino acid constitution of a particular protein (this chapter), or
- changes in the regulation of a gene or its product (see Chapter 13).

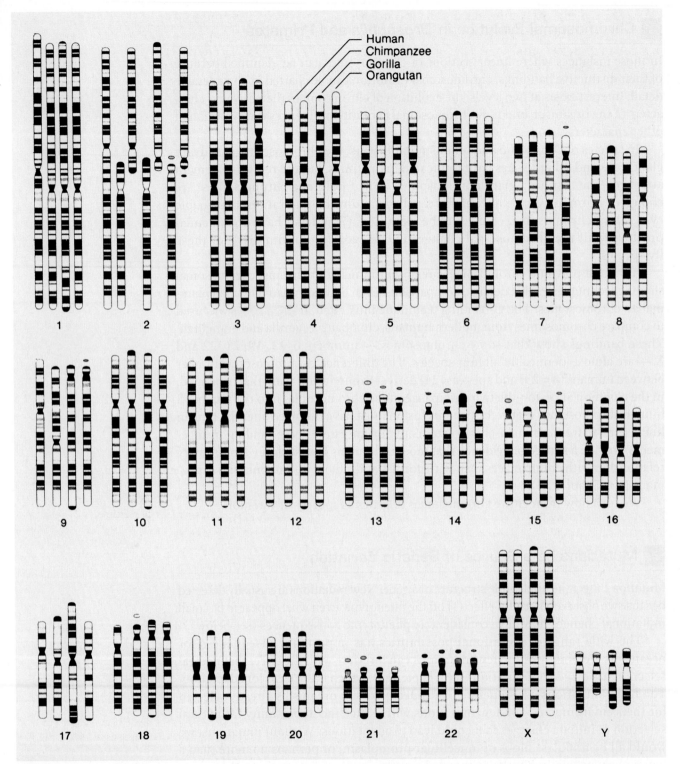

FIGURE 12.6 Banding arrangements of the chromosomes of humans, chimpanzees, gorillas, and orangutans, in respective order from left to right for each chromosome. On the whole, these banding arrangements indicate that humans have a closer evolutionary relationship with chimpanzees than with gorillas and a more distant one with orangutans. (Reproduced from Yunis, J. J., and Prakash, O. *Science* **215**, 1982: 1525–1530. Reprinted with permission from AAAS.)

Mutations may affect the *amount* or *rate* at which a gene product is produced and *not whether* the protein is produced. An example is a mutation in the insulin-dependent growth factor 2 (*Igf2*) gene in pigs. Pigs carrying this mutation have a threefold increase in the production of Igf-2, resulting in a more muscular and leaner phenotype. Breeders take advantage of such mutations to produce strains that fit the current tastes of the marketplace, sometimes even creating strains with the hope of influencing the market place.

Other mutations affect whether the gene product is produced and the rate of production. An example is the *thalassemias,* genetic diseases in which α or β hemoglobin chains are either not produced or are produced in diminished numbers. Like the sickle cell allele discussed below, such mutations often cause death of homozygotes but offer protection against malarial parasites. This special advantage of some thalassemia heterozygotes is presumed to explain the high frequency of these genes in modern human populations.

Sickle Cell Anemia

Perhaps the simplest type of mutation at the molecular level is a change in a nucleotide that substitutes one base for another. Such changes, known as **base substitutions**, may occur spontaneously or from the action of mutagenic agents such as X-rays. Depending on its position, a single nucleotide mutation can have important consequences for a species. A well-studied example is the anemia caused by the **sickle cell** mutation in humans. **Sickle cell anemia** results from a mutated allele of hemoglobin, an allele that is present in highest frequency in Africans. In the United States, the mutant allele is almost entirely confined to Americans with African ancestry (TABLE 12.1).

The hemoglobin molecule in adult human blood cells consists of four polypeptide globin chains, two αs and two βs, each about 140 amino acids long and each with its own specific sequence (see Chapters 8 and 11). In individuals who are homozygous for the sickle cell allele, all β-globin chains differ from normal βs at the number 6 position because of a mutation resulting in a single amino acid substitution. The effects of this single genetic mutation are profound, causing a variety of phenotypic changes (pleiotropy), often resulting in death (FIGURE 12.7).

■ A single gene that results in multiple traits (phenotypes) that are apparently unconnected is called a *pleiotropic gene*.

Sickle cell anemia kills, before the age of 20, more than 10 percent of African-Americans who are **homozygotes** for the allele, and kills even more individuals in Africa, where the frequency of the disease is very high (0.2 vs. 0.04%; Table 12.1) and medical facilities are more limited. So why does the mutation persist? Surely, it should have died out by now. It has not and nor have mutations for other human diseases, some of which are listed in Table 12.1.

The high frequency of the sickle cell mutation in these populations is related to the selective advantage of sickle cell **heterozygotes** in regions in which malaria is endemic. Disadvantages of the allele when homozygous (anemia) are balanced by advantages of the gene when heterozygous (protection against malaria). Heterozygote advantage is not true for all disease-causing mutations; the high incidences of achromatopsia among the Pingelapese of the Caroline Islands (a gene frequency of 0.22) and the rare Ellis-van Creveld syndrome among the Lancaster County Amish, with a gene frequency of 0.07 (Table 12.1) seem to confer no advantage on either their homozygous or heterozygous carriers. Such examples of unique gene frequencies seem best explained by founder or bottleneck effects (see Chapter 17).

■ *Achromatopsia* is a general term for a class of at least five different diseases, the symptoms of which are the inability to tell colors from one another and to see well in bright light. Ellis-van Creveld syndrome is characterized by short-limb dwarfism, extra fingers or toes and heart defects.

TABLE

12.1 Genotype Frequencies for Some Human Diseases Caused by Recessive Alleles

Disease	Population	Gene Frequency	Frequency of Homozygotes	Frequency of Carriers
Achromatopsia	Pingelap (Caroline Islands)	.22	1 in 20	1 in 2.8
Sickle cell anemia	Africa (some areas)	.20	1 in 25	1 in 3
Albinism	Panama (San Blas Indians)	.09	1 in 132	1 in 6
Ellis-van Creveld syndrome	Old Order Amish	.07	1 in 200	1 in 8
Sickle cell anemia	African-Americans	.04	1 in 625	1 in 13
Cystic fibrosis	European-Americans	.032	1 in 1000	1 in 16
Tay-Sachs disease	Ashkenazi Jews	.018	1 in 3000	1 in 28
Albinism	Norway	.010	1 in 10,000	1 in 50
Phenylketonuria	United States	.0063	1 in 25,000	1 in 80
Cystinuria	England	.005	1 in 40,000	1 in 100
Galactosemia	United States	.0032	1 in 100,000	1 in 159
Alkaptonuria	England	.001	1 in 1,000,000	1 in 500

Source: From *Genetics, Third edition* by Monroe W. Strickberger. Copyright © 1985 by Monroe W. Strickberger. Reprinted by permission of Prentice Hall, Inc., Upper Saddle River, NJ.

Gene Regulation

A frequent observation emerging from comparing different organisms on the molecular level is that organisms share the same kinds of proteins. Whether prokaryotic or eukaryotic in cellular organization, organisms share similar enzymes involved in basic biochemical processes such as glycolysis, amino acid synthesis, DNA replication, and protein synthesis. This fundamental conservation raises the question of how genetic control varies between different organisms. When we examine them closely, the distinctive structural features of different organisms within any group are less dependent on differences among the kinds of proteins organisms have than on how they organize and regulate various shared proteins. Which genes are transcribed into messenger RNA, which transcripts are translated into proteins, and which proteins are activated, are the result of interactions that often begin with signals that initiate different genetic regulatory pathways or networks.

Gene Regulation as a Source of Variation

One of the more fascinating developments in evolutionary biology over the past two decades has been the realization that mutations that affect the DNA sequence of genes and the amino acid composition of their protein products may not be as important as mutations that affect when, where and how much of a gene product is expressed

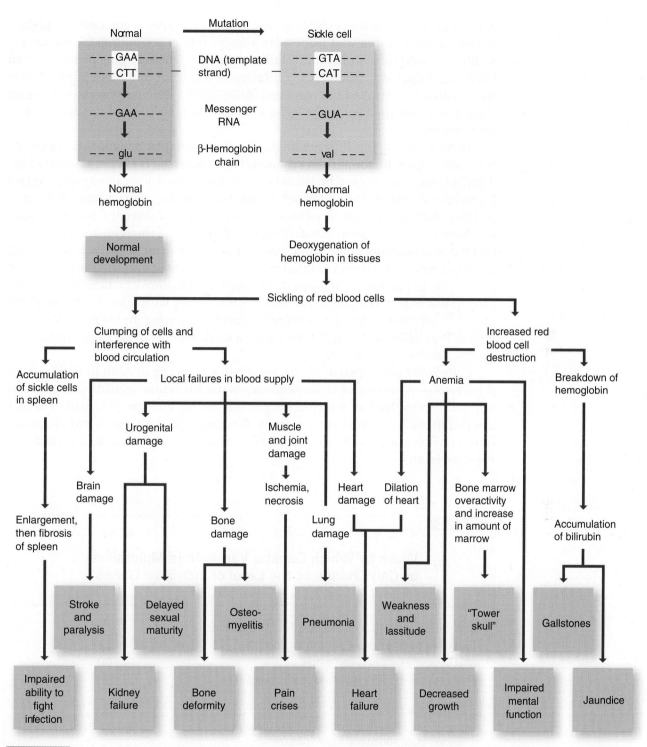

FIGURE 12.7 Varied (pleiotropic) effects of the sickle cell mutation, beginning with the substitution of a thymine by an adenine on the DNA template strand of the β-hemoglobin gene. The resultant GUA triplet coding sequence on messenger RNA translates into a valine (val) amino acid instead of the normal glutamic acid (glu), producing developmental consequences that can seriously affect sickle cell homozygotes. (Adapted from Strickberger, M. W. *Genetics, Third Edition*. Macmillan, 1985.)

in the organism. Regulatory pathways and networks — **gene regulation** — and their specific activity in time and place, play a major role in accounting for the variety of different kinds of cells in multicellular organisms. The emerging picture is of all levels of gene activity, from DNA replication onwards (or upwards), as exquisitely and sensitively controlled and regulated (TABLE 12.2). Selective processes acting on these genes over time guide the form, function, and direction of the evolution of the phenotype.

There are now compelling reasons for the hypothesis that mutations in regulatory genes are of great importance in evolution (this chapter). One is the relatively high degree of gene sequence and therefore gene product conservation among species. This is contrary to what one would expect if structural and functional variation in genes and their protein products accounted for most organismal diversity. Secondly, there are many fewer genes in metazoan genomes than we had thought, and most genes perform multiple functions. Mutations that affect the structure of a gene product are therefore rare; they tend to interfere with several, often unrelated, aspects of the phenotype. In contrast, mutations that influence regulation can affect the role of a gene in one part of the phenotype without interfering with other functions of the gene in other parts of the organisms or at other times during the life cycle. Variation in gene regulation helps explain how such a small number of genes can initiate the endless variety seen in nature.

Embryonic development — the progression from egg to adult in multicellular organisms — is in large part a consequence of the differential regulation of gene and cellular activity, both of which are regulated in time and space. Genetic changes can modify each developmental stage by affecting regulatory agents and processes, from signal reception to transcription and translation, a process named *tinkering* by François Jacob.

TABLE

12.2 Ways by Which Genetic Variation Is Maintained or Can Change at the Level of Individual Genes

Type of Variation	Chapter and/or Box Number
Duplication of genes, chromosome or entire genomes	(12)
Change in function of a duplicated gene	(12)
Shuffling of exons between genes	(9)
The presence of many repeated repeat sequences throughout most genomes, often comprising the bulk of the DNA	(9)
Introduction of novel variation by transposable elements or horizontal gene transfer	(Box 12-2; 8)
Gene regulation whereby more than one product can be produced from a single "gene"	(Box 12-1)
Alternate splicing to produce multiple mRNAs from a single gene	(Box 12-1; 9)
RNAi and post-translational modification to increase the number of products a gene can provide	(Box 12-1)

Although the idea of **regulatory mutations** dates back to the mid 1970s and earlier, a combination of empirical, bioinformatic, and conceptual advances in the past decade has brought regulation into the mainstream. This is why gene regulation is discussed in various contexts throughout the book. Much of evolution is about staying the same, although you don't read much about this in textbooks. The aspects of evolution that excite us are those involving the evolution of new features or new types of organisms, be they feathers, birds, turtles, mammals, humans or whatever. Evolutionary changes in gene regulation are increasingly being appreciated as central to the origin of the new features and/or new lineages.

The following two sections will enable you to appreciate the complexity (and therefore the evolutionary potential) of gene regulation in prokaryotic and eukaryotic cells.

Gene Regulation in the Bacterium *Escherichia coli*

Viruses and prokaryotes have provided a great deal of information on mechanisms of gene regulation. The first study leading to the discovery of gene regulation was by François Jacob and Jacques Monod (1910–1976) in 1961 on the genes governing the production of enzymes involved in lactose sugar metabolism in the bacterium *Escherichia coli* (see Table 7.1 for major discoveries concerning gene function). For this fundamental discovery, they and André Lwoff (1902–1994), the director of the Pasteur Institute where the research was conducted, received the 1965 Nobel Prize in Physiology and Medicine.

Because *E. coli* does not commonly encounter the sugar lactose, a repressor protein prevents the genes used in lac enzyme synthesis from being transcribed into mRNA (FIGURE 12.8a,b). However, when bacteria encounter lactose in the medium, some lactose molecules convert to a form called allolactose, which acts as an inducer and binds with the repressor, allowing the genes necessary to metabolize lactose to be transcribed (Figure 12.8c). The repressor acts as a regulatory protein. Binding of the inducer to the repressor changes the configuration of the DNA binding site, making the repressor inactive.

The ability to adapt to changes in their environment is exploited by bacteria such as *E. coli* when they acquire resistance to drugs. The *lac z* gene in *E. coli* codes for the enzyme β-lactamase. In the presence of the drug ampicillin, *E. coli* can synthesize β-lactamase, inactivate the drug and so acquire resistance to the drug.

Gene Regulation in Eukaryotic Cells

In comparison to prokaryotic gene regulation, genes in eukaryotic cells are regulated at a surprisingly diverse number of levels and by a wide range of mechanisms (Table 12.2). Mechanisms known to play a major role in generating protein and functional diversity in metazoans establish a regulatory code that lies between the DNA transcriptional network and post-transcriptional and translational regulation of RNA.

1. Three major regulatory mechanisms control **transcription of DNA → mRNA**. They are known as *cis-, trans-* and **RNAi-regulation** and are outlined in Box 12.1. A fourth mechanism, insertion of a **transposon** into the genome (see Chapter 8), is discussed at the end of this chapter.

2. Eukaryotic genes also are regulated after mRNA has been produced by **post-transcriptional modification** (also known as translational modification). Differential splicing or post-transcriptional editing of mRNAs can result

(a) Mode of *lac* enzyme synthesis in absence of repressor

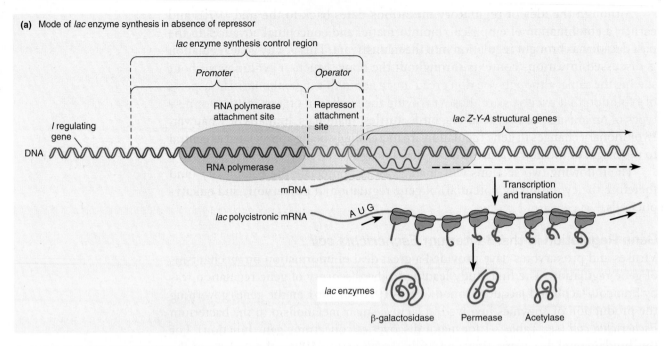

(b) Action of wild-type repressor in absence of inducer

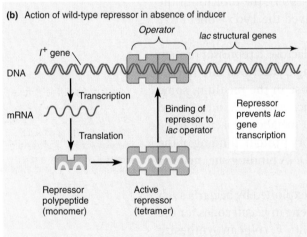

(c) Effect of inducer on repressor: induced *lac* enzyme synthesis

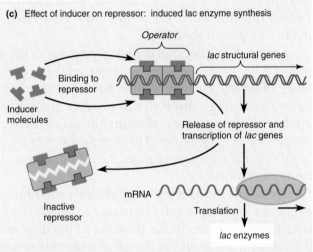

FIGURE 12.8 General scheme of lac enzyme synthesis in *E. coli* and the effects of repressor function or dysfunction on this inducible system. **(a)** The DNA region involved in controlling transcription of the *lac* structural genes *Z, Y,* and *A* consists of two major regulatory sites — an operator and a promoter — each of which binds specific proteins. (A promoter from a eukaryote gene is shown in Figure B12.1.) In the absence of the repressor protein, RNA polymerase begins transcribing at the operator, which is also the site to which the repressor attaches. Transcription of the *lac* genes, followed by translation, leads to synthesis of three *lac* enzymes. **(b)** Transcription and translation of the I+ gene produces a normal repressor protein that binds to the operator site of the *lac* locus, blocking transcription of *lac* genes. This repressed state appears in normal *E. coli* cells not growing on a lactose medium. **(c)** Transferring bacteria to a medium containing lactose leads to the introduction of allolactose inducer molecules, which cause the repressor to dissociate from its DNA binding site on the operator. This allows transcription of the *lac* structural genes, followed by their translation and the synthesis of *lac* enzymes. The repressor acts as a regulatory protein. Binding of the inducer to the repressor changes the configuration of the DNA binding site, making the repressor inactive. (Adapted from Strickberger, M. W. *Genetics, Third edition.* Macmillan, 1985.)

in changes in the protein produced or lead to more than one protein being produced from a single mRNA (BOX 12.1). A newly discovered class of small RNAs, **microRNAs (miRNA)** regulates translation of mRNA into protein in plants and animals (but not in the third multicellular group, the fungi).

The extent to which mutations in *cis-* and *trans-*acting regulatory elements or post-transcriptional regulation produce the phenotypic differences we see among individuals within species and between different species is becoming increasingly apparent. To quote Eric Davidson, one of the major researchers in the field, writing in 2006,

> In my view, *cis-*regulatory information processing, and information processing at the gene regulatory network circuit level, are the real secret of animal development. Probably the appearance of genomic regulatory systems capable of information processing is what made animal evolution possible.

3. **Post-translational modification** of proteins can occur through different splicing patterns that remove amino acid sequences, or by chemically modifying amino acid residues.

Gene Regulation and Evolution

Regulatory changes of all kinds have been among the key agents of organismal evolution.

Like a child's Lego set, with which you can produce differently shaped structures by rearranging the modular blocks, regulatory changes such as changing the signals, pathways, and targets of signal transduction can produce new functions and features. A single regulatory change in a gene that controls other genes can change how a gene network works, with dramatic consequences for the phenotype. An excellent example, the homeobox gene, *Ultrabithorax,* regulates the development of the structure (halteres; FIGURE 12.9) in the abdominal segment in *Drosophila,* by regulating signals and pathways involving at least thirty target genes.

■ The halteres, which are a pair of balancing organs, are found on the abdominal segment in flies, a location where hind wings are located in other insects. Halteres are a homologue of the hind wings.

Other regulatory mutations affect when during embryonic development, and in which tissue or organ, a gene product is produced. Work by Sean Carroll and his colleagues has shown how evolutionary changes in *Drosophila* wing pigmentation and male abdominal pigmentation patterns arise. Changes in *cis-*regulatory elements determine the spatial distribution of expression of the *yellow* gene on the wing. Gain of a homeobox-protein binding site in a *cis-*regulator of the *yellow* gene operates in abdominal cells. Over time, these changes resulted in five independent losses of wing spots within species of *Drosophila melanogaster,* and two independent gains in species of *D. obscura.*

The most dramatic examples of gene regulation involve mutations that alter the development of entire parts of organisms. For example, in *Drosophila melanogaster* the *bithorax* locus specifies a particular segment, the thorax. Mutations of the *bithorax* locus result in the production of an extra set of wings; the abdominal segment is converted into a thoracic segment and abdominal appendages (halteres) are converted into thoracic appendages (wings) (Figure 12.9).

■ Had this mutation leading to a second pair of wings occurred in nature, it may have forced entomologists to erect a special order (a higher level of taxon; see Chapter 11).

Evolution by gene regulation helps explain why evolutionary convergence is more common than expected by chance. Complex developmental pathways can be tinkered with by up- or down-regulating the expression of individual genes or proteins, or by turning on or off cascades of events that determine the fates of individual parts.

BOX 12.1
Major Mechanisms of Gene Regulation in Eukaryotic Cells

Regulation of Transcription

Three major regulatory mechanisms control the transcription of DNA → mRNA in eukaryotic cells; *cis-*, *trans-* and RNAi-regulation. The terminology has to be kept technical, otherwise it would take a sentence each time we wanted to talk about *cis*-regulation or RNAi-mediated regulation. I am afraid you will have to remember these four terms and what they mean. The long-term advantage will be that you will be able to drop *cis*-translation or miRNA into causal conversation with your friends who have not had the advantage of reading this book. More importantly both *cis*- and miRNA regulation are major players in evolutionary change.

cis- and *trans*-Regulation

These two mechanisms of gene regulation are based on DNA sequences either adjacent (*cis*) or apart (*trans*) from the gene they regulate.

cis-regulatory elements are short regions of DNA that lie adjacent to the promoter of the gene they regulate (FIGURE B12.1a). Gene regulation is by interaction between promoter and *cis*-regulatory element(s). Modification of *cis*-regulation is an important genetic mechanism leading to morphological change in evolution (see text).

At a second level are transcription factors that bind to special sites on DNA sequences, called CAAT and TATA boxes (Figure B12.1b). Transcription factors may be transcribed on different chromosomes from the genes they regulate and are therefore referred to as ***trans*-regulatory elements**. Virtually every gene is regulated by one or more transcription factors.

RNA Interference

Regulation of transcription by RNA interference is a much more recently discovered mechanism based on <u>s</u>mall <u>interference</u> <u>RNA</u> (siRNA). RNAi was demonstrated in experiments in which insertion into plants of extra copies of a gene that produces pigment blocked the expression of any pigment formation, a totally unexpected result.

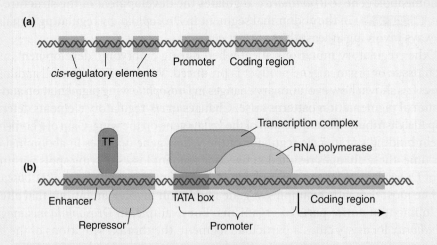

FIGURE B12.1 *cis*- and *trans*-regulatory elements and gene transcription in animals. (a) Location of three *cis*-regulatory elements upstream of the promoter and coding region. (b) Transcription factors (TF) in combination with coactivators (not shown) bind to *cis*-regulatory elements (as do repressors in combination with corepressors). The transcription factor(s) with RNA polymerase II (RNA) forms a transcription complex on the TATA box of the promoter, initiating transcription (arrow). (Adapted from Carroll et al. *From DNA to Diversity, Second edition*, Blackwell Publishing, 2005.)

BOX 12.1

Major Mechanisms of Gene Regulation in Eukaryotic Cells (*continued*)

RNAis are now known to be abundant and highly conserved; several hundred RNAi classes have been identified. *Drosophila* genomes contain more than 100 RNAis and human genomes may encode for more than 800. Importantly, a single siRNA can regulate many transcripts in an RNA interference pathway. Because RNAi regulates two of the most important cellular processes, cell division and differentiation, their regulatory role is enormous. Mutations affecting the activity of RNAis are emerging as important sources of evolutionarily significant variation.

■ The 2006 Nobel Prize in Physiology and Medicine went to U.S. researchers Andrew Fire and Craig Mello for their research on RNAi in animals.

Post-Transcriptional Regulation

Differential Splicing or Degradation of mRNA

Post-transcriptional modification of messenger RNA can occur through **different splicing patterns** that produce different mature mRNAs from the same precursor molecule, or by modification of mRNA nucleotides (**RNA editing**) by transitions, deletions or insertions of individual nucleotides. Because **exon arrangement and intron removal** are flexible, the exons coding for these protein subunits act as **modules** or domains, combining in various ways to form new genes. Single genes can produce different functional proteins by arranging their exons in several different ways through **alternate splicing patterns**. Modified intron splicing during the evolution of domesticated rice (*Oryza sativa*) caused a single base change in the *Waxy* gene, leading to less waxy protein.

Translation of messenger RNA into protein can be regulated by altering the rate of **mRNA degradation**, or by binding proteins or complementary RNA sequences to the mRNA molecule to prevent translation.

As with transcriptional regulation discussed above, post-transcriptional regulation also is regulated by a special class of RNAs known as microRNAs (miRNAs), which degrade mRNA. miRNAs do not produce proteins. Paradoxically, and in apparent contravention of the central dogma of DNA → RNA → protein, miRNAs are encoded by RNA genes that are transcribed from DNA.

miRNAs are 20 to 22 nucleotide sequences of non-coding RNA that regulate the translation of proteins in plants and animals (miRNAs are not found in fungi). miRNAs bind to matching target mRNAs, leading to degradation of the mRNA itself. You can think of miRNA as killer RNAs. Over 6,000 miRNAs have been identified in animals and plants. As each miRNA regulates a number of genes, the role of miRNAs in gene regulation is enormous; one-third or more of human genes are regulated by miRNAs. Recent studies discussed by Wheeler and colleagues (2009) place miRNAs as central players in the origin of animal complexity.

To miRNAs we can add recently discovered **piRNAs (Piwi-interacting RNAs)**, 25- to 30-nucleotide long. piRNAs function to protect the genome of mammalian and *Drosophila* germ cells from transposons inserting themselves.

■ Research into miRNA is at an active stage, with descriptions of 447 new miRNA genes in chimpanzee and human brains, the discovery of a core set of miRNAs in all animals except sponges and jellyfish, and evidence that human miRNAs respond to selection.

Changes in the regulation of conserved gene pathways or networks can explain the repeated occurrence of similar evolutionary changes.

As regulatory pathways extend, change, and interact, organismal integration and complexity increases. This is one of the key reasons proposed for the observations that major changes in embryonic development occurred early in multicellular organismal evolution (see Chapter 11), changes that would not be compatible with continued development in extant organisms.

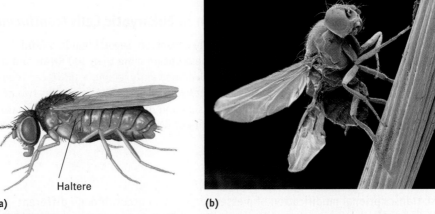

Haltere

(a) (b)

FIGURE 12.9 (**a**) Wild type fruit fly (*Drosophila* sp.) with a pair of wings on the thoracic segment and a pair of halteres (one on each side) on the abdominal segment. (**b**) A four-winged *Drosophila*, the result of a mutation at the *bithorax* locus, which causes the abdominal segment to transform into a thoracic segment with wings instead of halteres.

Similar gene networks can be used by different organs in the same individual, yet respond differentially to selection. An example is a gene network regulated by the gene *Sonic hedgehog* (*Shh*), *Pax-6*, and the homeobox gene *Prox1* (*prospero-related homeobox 1*). This network is used in the development of sensory organs in various organisms. We will discuss its operation in two forms (morphs) of a fish, the Mexican tetra *Astyanax mexicanus*. One morph found in surface pools (the surface morph) is sighted. The other morph, found in caves, is blind. Blind cavefish have reduced and vestigial eyes and no pigment (**FIGURE 12.10**) but have expanded the taste buds and the sensory system found in a line that runs along the body (the lateral line system).

The gene network initiated by *Shh/Pax-6* operates both in the eye and in the taste buds (**FIGURE 12.11**). Over-expressing *Shh* in surface fish results in diminished eye development but increases the numbers of taste buds. Knocking out *Shh* enhances eye development and results in fewer taste buds. As an upstream regulatory gene *Shh* elicits different responses from the two organ systems. Selection could operate

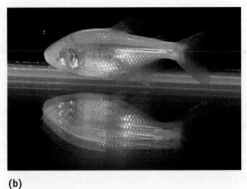

(a) (b)

FIGURE 12.10 Surface (**a**) and cave (**b**) morphs of the Mexican cave fish, *Astyanax mexicanus* showing the failure of eye development and lack of pigmentation in the cave morph.

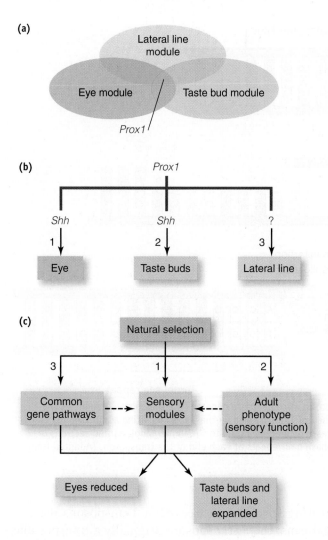

FIGURE 12.11 (a) Three modules (lateral line, taste buds, eyes) in Mexican tetra *Astyanax mexicanus* share an upstream control (*Prox1*) for gene pathways. (b) *Prox1* signals through Sonic hedgehog (*Shh*) in two of the pathways; the equivalent gene for lateral lines modules is not known. (c) The common upstream genetic control shown in (b) operates whether selection is on the adult phenotype, development (sensory modules) or common gene pathway, resulting in expansion of the lateral line and taste bud modules and reduction in the eye module. (Adapted from Franz-Odendaal, T. A., and B. K. Hall, *Evol. & Devel.* **8**, 2006: 94–100.)

on the adult cavefish phenotype, on the sensory modules or on the common gene network (Figure 12.11).

Transposons

Another source of evolutionary change in prokaryotes and eukaryotes are transposons, nucleotide sequences that promote their own transposition among a genome. Transposons were introduced in Chapter 8 as vehicles for horizontal gene transfer between organisms.

A transposon produces special transposase enzymes that allow it to insert copies of itself into various target sites in an organism's nuclear genome. For example, when inserted into the maize genome, the 768-base pair *E. coli* insertion sequence 1 (IS1) makes staggered cuts at each side of a nine-nucleotide base pair sequence. A copy of the IS1 sequence is then inserted into the gap these cuts produce (**FIGURE 12.12**).

Transposons can move from one prokaryotic species to another and from prokaryotic to eukaryotic cells, the latter discussed in Chapter 8 under horizontal gene

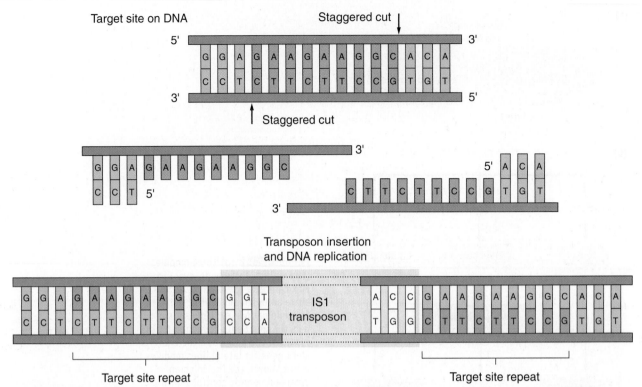

FIGURE 12.12 Mode of insertion of the *E. coli*-derived IS1 transposon into the maize genome. The transposon recognizes the nine-base pair DNA sequence, shown in brown, and cleaves it at the positions marked by the arrows, creating a gap. The IS1 transposon (yellow) inserts into the gap, and DNA sequences are synthesized complementary to the former single-strand sections of the target site. This process produces identical but inverted nine-base pair repeats at each end of the transposon. (Adapted from Strickberger, M. W., *Genetics, Third edition*. Macmillan, 1985.)

transfer and the difficulty of creating a prokaryotic tree of life. Transposons also can move between eukaryotic organisms; the *P* transposon, originally a transposable element common to *Drosophila willistoni*, spread to *Drosophila melanogaster* about 50 years ago and now is found in all wild populations of this species. Evidence that the *P* transposon was acquired horizontally is that the two species differ significantly in their nuclear genes — indicating a separation of about 20 My between the two species — yet their *P* transposons differ by only a single nucleotide. More recent research has shown that *P* elements have repeatedly crossed species barriers in *Drosophila* and spread to related genera, providing new sources of genetic variation. Evidence now supports transposon-mediated horizontal gene transfer between animals, between the mitochondrial genes of plants, and, most recently, between the nuclear genes of millet and rice. The history of the discovery of transposons says much about the difficulty of gaining acceptance for ideas ahead of their time (BOX 12.2).

An important advantage transposons gain by horizontal transfer is circumventing the barriers of reproductive isolation among species, and escaping inevitable extinction in vertical lineages when a species dies out. Therefore, we might expect to find a large number of transposon numbers within a genome, which we do in some cases. In primates, for example, a 300 base-pair sequence with transposon-like features — named an *Alu* sequence because it is recognized by the Alu restriction enzyme — is present in perhaps more than one million copies in each diploid human

■ Restriction enzymes cleave DNA into fragments at specific sites, allowing comparisons of small pieces of DNAs from different species. Different restriction enzymes recognize different sequences and so produce different DNA cleavage products.

BOX 12.2

Barbara McClintock and the Discovery of Transposons

The history of the discovery of transposons is a fascinating one, teaching us much about the openness of the scientific community to new ideas.

Transposons and gene regulation were both discovered in the 1940s and 1950s by Barbara McClintock (1902–1992), an American plant cytogeneticist who worked on maize. If you know the more recent history, especially of gene regulation, this will surprise you for three reasons discussed earlier in this chapter.

1. The discovery of transposons is attributed to research on bacteria in the early 1970s, 30 years after McClintock's research.
2. The knowledge that genes in prokaryotes are regulated is attributed to Jacob and Monod's research on the *lac* operon in *E. coli* published in 1961.
3. The knowledge that eukaryote genes are regulated has an even more recent history, usually attributed to research on animals such as the fruit fly *Drosophila* and the nematode *C. elegans*.

McClintock's name should be known to every molecular biologist, and to an extent it is, though only occasionally do you find citations to her pioneering work on gene regulation. In 1983, she was belatedly awarded the Nobel Prize in Physiology and Medicine for her discovery of genetic transposition.

Amongst her many achievements, McClintock discovered genetic recombination by crossing-over during meiosis and produced the first genetic map for maize, which allowed physical traits to be linked to specific regions of the chromosomes. Her discoveries concerning transposition into the genome were made between 1944 and 1953. They emerged from studies of the unstable inheritance of the mosaic color patterns of maize seeds (FIGURE B12.2). McClintock identified two dominant and interacting gene loci, *Dissociator* (*Ds*) and *Activator* (*Ac*), discovered that *Ds* has an effect on nearby genes only when *Ac* is present, and, perhaps most unexpectedly, that *Ds* and *Ac* can change their positions on the chromosome. Because movement within the genome (transposition) is random, some cells produce pigment but other cells others do not, resulting in the mosaic color patterns so characteristic of maize seeds (Figure B12.2).

By the late 1940s, McClintock proposed these two mobile elements as controlling elements distinct from genes (now known to be transposons). McClintock extended her results to a *theory of gene regulation* applicable to all multicellular organisms. But biology was pre-DNA and pre-molecular and, as discussed in Chapter 3, the nature of the genetic material remained a mystery. McClintock's genome was much too dynamic to fit the then-current view that genes provide preprogrammed instructions to cells.

In 1961, when the discovery of the *lac* operon was published, McClintock responded with an informed comparison of gene regulation in bacteria and maize. But only after the (re)discovery of transposons in the late 1960s did McClintock's prior discoveries on genetic transposition begin to receive the credit they deserved. Her pioneering work on gene regulation remains underappreciated to this day.

Barbara McClintock was decades ahead of the field. Indeed, her results anticipated a molecular genetics field that did not exist. Although her results were published in major journals and although she spoke at meetings and trained and mentored students, her "colleagues" were so skeptical of her findings that in 1953 she stopped publishing on transposition, turning instead to cytogenetics and ethnobotanical studies of races of maize in South America. McClintock articulated her frustration in a letter to a geneticist at the University of Leeds who had invited her to participate in a workshop to be held

(continued)

■ See Table 7.1 for a timetable of the major discoveries leading to our understanding of the nature and function of genes.

■ A biography published in the same year also brought her stellar career to the forefront; see Keller (1993) under recommended reading.

BOX 12.2
Barbara McClintock and the Discovery of Transposons (*continued*)

(a) (b)

FIGURE B12.2 Mosaic color patterns of seeds in cobs of maize (**a**) and heirloom Indian corn (**b**) reflecting differential production of pigment in different seeds.

in September 1973, an invitation she declined. Writing of "my attempts during the 1950s to convince geneticists that the action of genes had to be and was controlled," she continued, "It is now equally painful to recognize the fixity of assumptions that many persons hold on the nature of controlling elements in maize and the manners of their operation. One must await the right time for conceptual change."

cell. Smaller repetitive sequences of the type discussed in Chapter 13 are widely prevalent in various eukaryotes. But this is not always so. The *Drosophila* genome carries only about 30 to 50 copies of the *P* element and a similar number of *copia* family transposons responsible for mutation of the *white* (w^a) locus. Regulatory agents within transposons control their number and so limit their mutagenic effects, a feature that may have been selected to ensure survival of their hosts, and therefore their own survival.

∎ Recommended Reading

Carroll, S. B., 2005. *Endless Forms Most Beautiful*. W. W. Norton, New York.

Carroll, S. B., J. K. Grenier, and S. D. Weatherbee, 2005. *From DNA to Diversity. Molecular Genetics and the Evolution of Animal Design,* 2nd ed. Blackwell Publishing, Malden, MA.

Davidson, E. H., 2006. *The Regulatory Genome: Gene Regulatory Networks in Development and Evolution*. Elsevier/Academic Press, Burlington, MA.

Hallgrímsson, B., and B. K. Hall (eds.), 2005. *Variation: A Central Concept in Biology*. Elsevier/Academic Press, Burlington, MA.

Jacob, F., 1977. Evolution as tinkering. *Science,* **196**, 1161–1166.

Jacob, F., and J. Monod, 1961. Genetic Regulatory Mechanisms in the Synthesis of Proteins. *J. Mol. Biol.,* **3**, 318–356.

Keller, E. F., 1993. *A Feeling for the Organism: The Life and Work of Barbara McClintock,* 10th Anniversary Edition. W. H. Freeman, New York.

Mayr, E., 2001. *What Evolution Is*. With a Foreword by Jared Diamond. Basic Books, New York.

Wheeler, B. M., A. M. Heimberg, V. N. Moy, et al., 2009. The deep evolution of metazoan microRNAs. *Evol. Devel.* **11**, 50–68.

Wood, T. E., N. Takebayashi, M. S. Barker, et al., 2009. The frequency of polyploid speciation in vascular plants. *Proc. Natl. Acad. Sci. U.S.A.* **106**, 13,875–13,879.

13 Genetic Variation in Populations

Above: Variation in the wild: an albino deer.

KEY CONCEPTS

- Without variation there would be no evolutionary change.
- Mutation provides one source of genetic variation.
- Mutation is not entirely random: some parts of the genome are more susceptible to mutation than are others.
- The large amount of polymorphism at gene loci provides a much greater source of genetic variation than do the relatively few new mutations that arise each generation.
- At the population level, allele frequency provides a measure of genetic variation.
- Genetic drift within a population and gene flow between populations provide sources of genetic variation.
- Genetic variation provides the raw material enabling evolutionary change in response to natural selection.
- Large-scale geographical patterns of species distribution can be determined using the genetic history of populations.

Overview

When coupled with the multiple levels of gene regulation discussed in Chapter 12, it is evident that genetic variation can be introduced into populations in multiple ways. Mutation, random drift in gene frequencies (genetic drift) within a population, and gene flow between populations all contribute to enhanced variation at the population level. A further source of variation lies in regions of chromosomes known as *quantitative trait loci* that contain blocks of genes, often influencing characters that show continuous variation, such as height or weight. The randomness of mutation, variation in mutation rates among genes and among species, the frequency of alleles in populations and the ability of DNA to repair itself (and so compensate for otherwise deleterious effects of mutation) all contribute to the maintenance of genetic diversity (genetic polymorphism) within populations. Genetic variation provides the raw material enabling evolutionary change as natural selection operates on populations.

Mutation

Mutations as a source of individual genetic variation were introduced in the previous chapter. Mutation also is an important attribute of populations.

A population that has long been established in a particular environment will have many genes adapted for prevailing conditions. New mutations that arise, if not neutral in effect, will rarely be better, and likely will be worse, than the genes already present — a consequence not much different from the damage we would expect if a random change was introduced into any intricately organized and integrated system, such as a computer or a car engine. Multicellular organisms are constrained by their evolutionary history (see Chapter 15). As a consequence, advantageous mutations are generally confined to few of many intricate developmental processes and functions. For example, although plants and animals are separated by more than a billion years of evolution, there are only two differences between them in the 100 amino acids of histone 4, the protein that binds and folds DNA. For such phylogenetically crucial genes, conserved molecular sequences are commonplace, and viable changes occur only rarely.

For other genes and functions a change in environmental conditions can elicit a genetic response based on the available genetic variation. Alleles formerly in low frequency may now have high relative fitness. We see this in

- rapid genetic changes in many insect populations exposed to pesticides such as dieldrin and DDT (FIGURE 13.1), where resistant alleles appear on all major chromosomes (FIGURE 13.2);
- large increases of black allele frequencies in populations of the peppered moth *Biston betularia* in industrialized regions (see Figure 15.12) and of a receptor involved in pigment cells in light and dark strains of the rock pocket mouse, *Chaetodipus intermedius* in southwestern USA (see Chapter 15);
- increased frequencies of resistant genes in some plant populations exposed to herbicides and metallic toxins; and in
- genes that modify red blood cell physiology offering protection against malaria in the human populations shown in Figure 15.10.

In most populations, many individuals die, leaving those (and it may be those few) individuals with the alleles for resistance, melanism or protection, to pass their

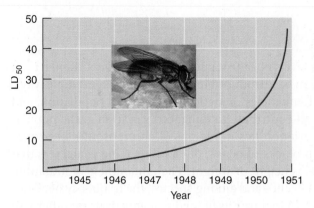

FIGURE 13.1 Resistance to DDT in common houseflies (*Musca domestica*) collected from farms in Illinois between 1945 and 1951 when use of DDT as an insecticide was increasing. The measure is the lethal dose necessary to kill 50 percent of the flies (known as the LD_{50}). (Adapted from Strickberger, M. W. *Genetics, Third edition*. Macmillan, 1985; based on data from Decker, G. E., and W. N. Bruce, *Amer. J. Trop. Med. Hygiene* **1**, 1952: 395–403.)

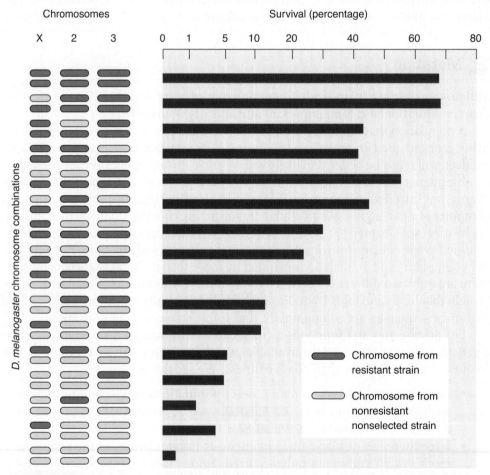

FIGURE 13.2 Percent survival of fruit flies (*Drosophila melanogaster*) with 16 chromosomal patterns of resistance/nonresistance to the insecticide DDT after exposure to a uniform dose of DDT. Each population of flies carries a set of chromosomes X, 2 or 3 derived from DDT-resistant and DDT-nonresistant strains. DDT resistance (shown as % survival in the dark brown bars in the histogram on the right) increases with the increased number of chromosomes in resistant strains. (Adapted from Crow, J. F. *Ann. Rev. Entomol.*, **2**, 1957: 227–246.)

alleles to the next generation. In essence, these individuals become the founders of the next generation (see Chapter 17). Nevertheless, mutations supply an important source of variation upon which selection acts and which selection incorporates into evolutionary change.

Mutation Rates

Although you might have the idea that mutations are bad, optimal mutation rates are advantageous. Interestingly, specific nucleotide sequences may be the site of higher than average mutation rates. Such sequences, known as **hot spots of mutation**, often are sites where the nucleotide change is less readily repaired or compensated for than is a sequence change at another site.

Mutation rates are generally low, of the order of one per 100,000 copies of a gene, although there is much variation. In humans, the mutation rate for achondroplasia, a form of dwarfism, is 0.6 to 1.3 mutations/100,000 gametes. The mutation rate for neurofibromatosis is 5 to 10 mutations/100,000 gametes, although some of these differences may result from lumping, under one name, several syndromes with distinct genetic bases but similar phenotypes. In modern humans, carrying an estimated 25,000 genes per haploid genome, each sperm and egg may well carry less than one new mutation, or an average of 0.4 new mutations in a diploid fertilized zygote.

Mutation rates are not only low they are not constant. As with other essential traits, mutation rates seem mostly selected for optimum values, balancing on the delicate adaptive line between retaining prevailing adaptive features yet facilitating the origin of new features.

A remarkable way in which some organisms accumulate mutations without experiencing their immediate effects is to bind their gene products with **heat shock proteins**, such as heat shock protein-90 (hsp-90). Heat shock proteins are molecular chaperones that help other proteins maintain their normal 3-D conformation and prevent them from degrading. Because of these protective roles, heat shock proteins mask at least some and perhaps many of the effects that mutations in protein-coding genes otherwise would have on the phenotype. Many alleles are unmasked when protective heat shock proteins are disabled, as they are when organisms are exposed to an environmental shock such as a sharp change in temperature or salinity. The new patterns and combinations of proteins expressed can result in significant changes in development, some of which will be adaptive and may open up new evolutionary opportunities.

Neo-Darwinism and Genetic Polymorphism

Evolutionary potential and genetic variation are two sides of the same evolutionary coin.

New mutations with an immediate beneficial effect on an organism seem generally to be rare. As discussed in the previous chapter, some mutations are either neutral in their effect or harmful only when they occur in relatively rare homozygotes. Neutral mutations or deleterious but recessive mutations accumulate in a population and provide a reservoir of potential genetic variation. Such genetic variation, expressed in a population as two or more genetically distinct forms is known as **genetic polymorphism**.

In the fruit fly, *Drosophila pseudoobscura* populations in different localities in the western United States are polymorphic for a wide variety of gene arrangements on the third chromosome (FIGURE 13.3a). The frequencies of such arrangements may

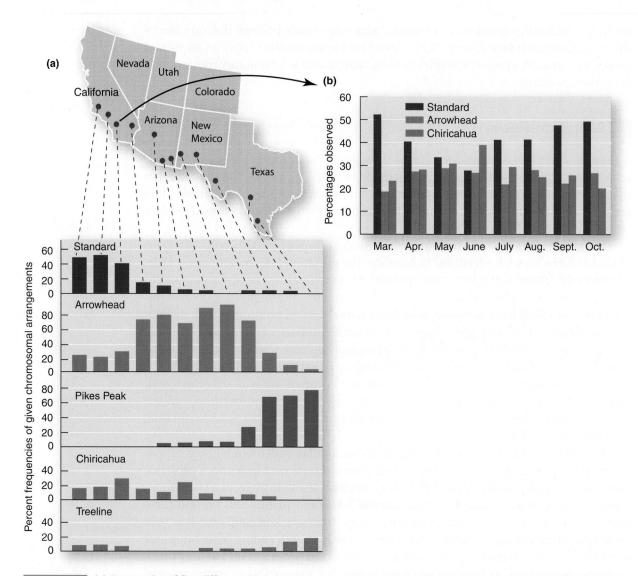

FIGURE 13.3 (**a**) Frequencies of five different third chromosome gene inversions in *Drosophila pseudoobscura* in 12 localities on an east–west transect along the United States–Mexican border. (**b**) Percentages of different third chromosomal inversions found at different months during the year in one locality, Mount San Jacinto, California. Each of these chromosome inversions maintains a specific gene combination that enables adaptive interactions between component alleles (epistasis). (**a** Adapted from Dobzhansky, T., *Carnegie Inst. Wash. Publ.*, **554**, 1944: 47–144. **b** Adapted from Dobzhansky, T., *Heredity* **1**, 1947: 53–64.)

change seasonally (Figure 13.3b), indicating that genetic polymorphism is generally adaptive in this species and that certain polymorphic variations are preferentially adaptive in specific seasons.

About two thirds to three quarters of all protein loci in many species are polymorphic. Of the three billion nucleotides in the haploid human genome, one human may differ from another at an average of about five million sites (see Chapter 14). Therefore, genetic polymorphism provides a much greater source of genetic variation than do the relatively few new mutations that arise each generation. For example, exposure of insect populations to the pesticide DDT (dichloro-diphenyl-trichloroethane) caused a widespread increase in the presence of various DDT-resistant mechanisms, including changes in the permeability of the insect to absorption of DDT, and increased frequency

of those enzymes that break down DDT into relatively less toxic products. It is therefore not surprising that insecticide resistance is associated with numerous genes, and that genetic changes have arisen in response to insecticide exposure (Figure 13.2).

Continuous Variation

From Charles Darwin onward, many evolutionary biologists have suggested that rather small heritable changes provide most of the variation on which natural selection acts. In Darwin's words from *The Origin of Species,*

> Extremely slight modifications in the structure and habits of one species would often give it an advantage over others; and still further modifications of the same kind would often still further increase the advantage . . . Under nature, the slightest differences of structure or constitution may well turn the nicely balanced scale in the struggle for life, and so be preserved.

For many measurable traits such as crop size and yield, animal height/length or body weight, researchers usually focus on small changes or **continuous variation**, evidenced in characters for which the plots of the distribution of the variation form a bell-shaped curve or normal distribution, as seen, for example for human height (FIGURE 13.4). From an evolutionary view, it is clear that many genetic differences of small phenotypic

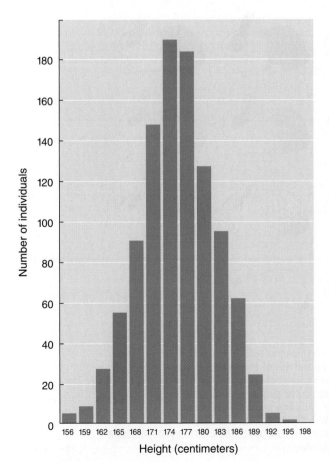

FIGURE 13.4 Distribution of the heights of 1,000 Harvard College students aged 18 to 25. (Adapted from Castle, W. E., 1932. *Genetics and Eugenics, Fourth edition*. Harvard University Press, 1932.)

effect can accumulate through selection to give large quantitative differences. Selecting for the presence or absence of white spotting in Dutch rabbits can lead to completely colored or completely white strains (FIGURE 13.5), a result that illustrates the presence of previously hidden genetic variation. The genetic causes for these changes are twofold but related. Genes with small phenotypic effect are important contributors to continuous variation. Such genes are often located in large regions of chromosomes known as **quantitative trait loci (QTLs)**. Analyses of QTLs provide excellent demonstrations of the integration of genetic variation in individuals and in populations.

Quantitative Trait Loci

The term *quantitative trait loci (QTL)* is shorthand for all of the genes (alleles) in a particular region of a chromosome that affect a quantitative aspect of the phenotype,

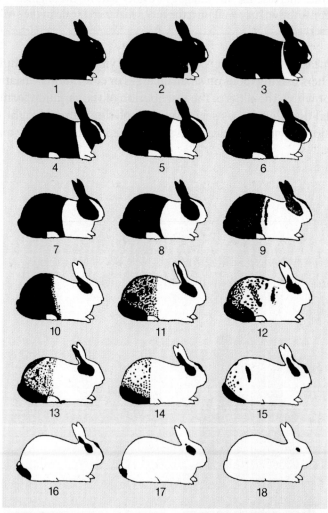

FIGURE 13.5 Selection for white spotting in Dutch rabbits, beginning with white only on the tips of the forepaws (grade 1) results in all ranges of spotting to animals that are totally white (grade 18). (Reprinted by permission of the publishers from *Genetics and Eugenics: A Textbook for Students of Biology and a Reference Book for Animal and Plant Breeders* by W.E. Castle, p. 281, Cambridge, Mass.: Harvard University Press, Copyright © 1916, 1920, 1924, 1930 by the President and Fellows of Harvard College.)

such as size or shape, or processes such as growth and morphogenesis that influence size and shape. Individual QTLs may affect the entire individual (as in growth) or may affect different regions of the organisms (tail growth versus leg growth) or different characters (liver versus kidney).

QTL analysis enables us to isolate suites of genes acting on particular parts of the phenotype at particular stages in ontogeny and to determine their relative affects. For instance, in recent studies with selected inbred lines of rats, four QTLs were shown to influence tail growth and body weight. One QTL has a substantial effect on both tail growth and body weight throughout development. A second has a large effect late in development but only a small effect early. A third shows the reverse pattern, while the fourth has minor effects at one or two discrete developmental stages only. Analysis of these QTLs has identified genomic regions whose influence varies by feature and over developmental time.

A second example comes from natural populations of a fish, the threespine stickleback, *Gasterosteus aculeatus* in lakes in British Columbia. Sticklebacks have complex sets of skeletal elements including spines, bony plates in the skin (a dermal armor), a three-part pelvic girdle, and bony gill rakers in the gills. Illustrated in FIGURE 13.6 is the variation expressed under different conditions in the length of the dorsal spines, the number of bony plates, and the number of gill rakers, each of which is influenced by QTLs. One QTL explains much of the variation in plate number, two QTLs explain 66 percent of the variation in gill raker number.

In the past, it was assumed that the cumulative actions of many genes of small effect were responsible for the entire *QTL effect* (as it is known). More recent analyses in *Drosophila*, mice, and fish have revealed that a small number of genes within the

(a)

(b)

(c) 10 mm

FIGURE 13.6 Individual threespine sticklebacks, *Gasterosteus aculeatus,* cleared and stained to reveal the skeleton. From **a–c** note the reduction in the pelvic girdle to a nubbin in **c**, and the reduction of the dermal plates to six and five in **b** and **c**, respectively, compared with the complete set of plates in **a**.

QTL may contribute disproportionately to the QTL effect; the QTL shows us the region of the chromosome with many genes of small effect, but may also alert us to a gene or genes of large effect located in the same region of the chromosome.

To continue with the stickleback example, freshwater sticklebacks arose from marine populations. An ancestral allele (*complete*) of the gene *Ectodysplasin* in marine sticklebacks is associated with maintaining bony plates. The derived allele *low* found in freshwater fish is associated with (causes?) plate loss. *low* has been fixed independently in different freshwater populations, and may well be a major gene in the QTL responsive to selection resulting in reducing plate number.

You can see that QTL analysis enables us to match environmental conditions to evolution quite precisely. Sorting out the various influences of environmental conditions, the genetic response to different environmental conditions, and gene by environment interactions is a major preoccupation of many evolutionary biologists and quantitative geneticists.

Population Genetics and Gene (Allele) Frequencies in Populations

Darwin proposed that natural selection operates on small, continuous, hereditary variations. His cousin, Francis Galton and others accepted evolution, but maintained that variations are sharp and discontinuous. The controversy was resolved when it was shown that several genes, each with small effect (polygenes) can have a large effect when they influence the expression of a single phenotypic trait. By the 1930s, it became accepted through genetics that

■ The union of population genetics and evolution as the neo-Darwinian theory of evolution (*neo-Darwinism*) is not the same as the *Modern Synthesis* of evolution in which systematics and paleontology were added to population genetics as components of a theory of evolution.

1. evolution is a population phenomenon that
2. can be represented as a change in gene (now allele) frequencies because of the action of various natural forces such as mutation, selection and genetic drift (see Chapter 15), and that
3. these changes can lead to differences among populations, species, and higher clades. This population genetics view of evolution became known as the neo-Darwinian theory with its emphasis on the frequency of genes in populations as the basis of evolutionary change.

Population genetics deals with genes as alleles and gene frequencies as allele frequencies. Allele frequencies — the frequencies of individual alleles — and the gene pool — all the alleles of all individuals in the population — are two major attributes of a population, the latter being defined as a group of potentially interbreeding organisms (see below and also Chapter 14). The gene pool represents all the variation available in the current generation, and, setting mutation aside, all the variation that can contribute to the next generation.

Population genetics is sophisticated statistically, mathematically and conceptually. Below you will find a very abbreviated introduction to population thinking, the concept of population frequencies, the gene pool, and the genetic attributes of populations.

According to a principle devised independently by Geoffrey Hardy (1877–1947) in England and Wilhelm Weinberg (1863–1937) in Germany and known as the **Hardy–Weinberg principle**, in a population of randomly mating individuals, allele frequencies are conserved and in equilibrium unless external forces act on them (FIGURE 13.7). Because mutation rates are usually observed to be of the order of 5×10^{-5} or less, the

ASSUMPTIONS **STEPS**

1. Parents represent a random sample of the
 gene frequencies in the population.

A. Provides gene frequency in parents

2. Genes segregate normally into gametes
 (heterozygotes for any gene pair produce their
 two kinds of gametes in equal frequencies).

3. Parents are equally fertile (gametes are
 produced according to the frequency of
 the parents).

B. Provides gene frequency in gametes

4. The gametes are equally fertile (all have
 an equal chance of becoming a zygote).

5. The population is very large (all the possible
 kinds of zygotes will be formed in frequencies
 determined by the gametic frequencies).

C. Provides gene frequency in the
 gametes that form the zygote

6. Mating between parents is random (not
 determined by any preferences associated
 with specific genotypes).

7. Gene frequencies are the same in both
 male and female parents.

D. Provides genotype frequencies in the
 zygotes

8. All genotypes have equal reproductive ability.

E. Provides genotype frequencies in
 adult progeny produced by zygotes

F. Repeat of steps A, B, C, D, E, etc.

FIGURE 13.7 Assumptions and steps in the Hardy–Weinberg equilibrium. (Adapted from Falconer, D. S. and
T. F. C. Mackay. *Introduction to Quantitative Genetics, Fourth edition*. Longman, 1996.)

shift toward equilibrium is slow, so slow that mutational equilibrium is rarely if ever reached, especially because mutation rates are not constant. The Hardy–Weinberg principle is the founding theorem of population genetics. Assumptions underlying this principle (random mating in large populations, equilibrium allele frequencies, absence of gene flow into the population), and stepwise changes in gene frequencies as the principle is acted out in populations are outlined in Figure 13.7.

In natural populations, we can determine genotype frequencies and Hardy–Weinberg equilibrium for single genes far apart on the chromosome, and if the number of alleles is limited, ideally to two. The more gene pairs, the longer it takes to achieve equilibrium.

Populations, Allele Frequencies and the Gene Pool

Geneticists define a **population** as a group of sexually interbreeding or potentially interbreeding individuals. The size of the interbreeding population may vary, but is usually taken to be a local group (also called a **deme**) in which random mating occurs (BOX 13.1). Structural features of the environment — a river that individuals cannot cross, or a landslip — reduce effective deme size (see Chapter 14). As a consequence, variation

BOX 13.1

Demes: Reproductive Units Within Populations

All members of a species share a common gene pool, although populations may vary genetically from one another. We expect widely separated populations to have less opportunity to share gene pools than those closer together. Consequently, many species consist of genetically diverse local populations known as *demes*.

A deme is a local population of organisms of one species that interbreed with one another and share a distinct gene pool. A species therefore may consist of many demes (local populations) that do not exchange alleles. Demes can be differentiated from one another on the basis of specific gene frequencies. Demes could contribute to the gene pool of the entire species, but are sufficiently separated geographically to maintain specific gene frequencies. Selection may take demes in different directions, potentially leading to the isolation of one or more groups.

Because the forces acting on demes may change among localities, differences among populations arise and are maintained. A transect across central California shows that populations of the yarrow plant, *Achillea,* differ significantly in such traits as height and growing season (FIGURE B13.1). The adaptive nature of such traits is evidenced by the differential responses of populations when moved to different localities. Coastal plants are weak when grown at higher altitudes; high-altitude plants grow poorly at lower altitudes (FIGURE B13.2). The adaptive features of many such plant populations reflect the accumulated genetic response of a population to a particular ecological habitat.

The considerable genetic variability in human groups shows that populations of modern humans are demes. On the basis of blood types (A, B, O, Rh+, Rh−) five groups of humans can be distinguished: African, Caucasian, Greater Asian, Amerindian and Australoid. Members of each group are not genetically pure in the sense of sharing a uniform genetic identity, although differences among these groups have not reached the point where each population is fixed for a different allele. Rather, an allele fixed in one population is usually polymorphic in other populations. Genetic uniformity does not even apply to members of the same family or related individuals. About 84 percent of the genetic variation among humans comes from differences among individuals and populations of the same group. Only 16 percent comes from differences among groups.

■ Rh is shorthand for Rhesus-factor. Rh-positive blood type is the blood group (approximately 85% of people) whose red cells have the Rh antigen.

BOX 13.1

Demes: Reproductive Units Within Populations (*continued*)

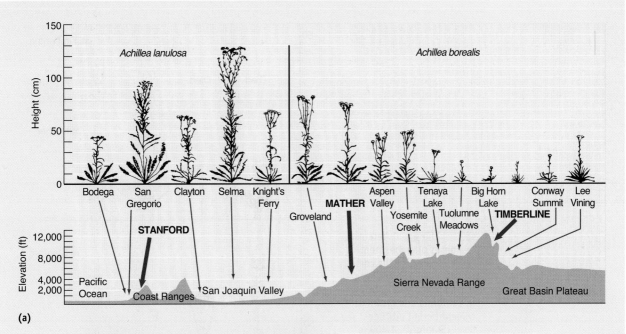

(a)

(b) (c)

FIGURE B13.1 Representative plants from different populations of the common yarrow, *Achillea*, gathered from locations across central California and grown under uniform conditions in Stanford, California. The differences in plant size, leaf shape and other characteristics, and the transition between *A. lanulosa* and *A. borealis* shown in (a) relate to the locations of the populations and indicate that genetic differences have evolved among them. Examples of *Achillea borealis* and *A. lanulosa* are shown in (b) and (c), respectively. (a Adapted from Clausen, J. D., D. Keck, and W. M. Hiesey, *Carnegie Inst. Wash. Publ.*, **No. 581**, 1948: 1–219.)

(*continued*)

BOX 13.1
Demes: Reproductive Units Within Populations (*continued*)

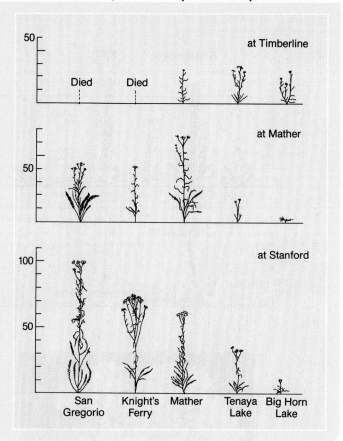

FIGURE B13.2 Differential responses of clones from representative of the common yarrow, *Achillea*, originating from five localities in California and grown at three different altitudes: sea level (Stanford), 1,400 m (Mather) and 3,050 m (Timberline). For localities, see Figure B13.1a. (Adapted from Clausen, J. D., D. Keck, and W. M. Hiesey, *Carnegie Inst. Wash. Publ.* **No. 581**, 1948: 1–219.)

in "the population" is effectively variation in each deme. Furthermore, selection may take demes in different directions in such populations, potentially leading to isolation of one or more groups.

Measuring Allele Frequencies

Allele frequencies as a measure of evolution were discovered soon after Mendelian genetics was discovered in 1900. This measure arose directly from Mendel's 3:1 ratio of dominant to recessive phenotypes (see Chapter 7). Dominant alleles, so the argument went, would reach a stable equilibrium frequency of three dominant individuals to one recessive.

Hardy and Weinberg independently demonstrated that allele frequencies do not depend upon dominance or recessiveness (that is, on genotype frequencies) but remain essentially unchanged from one generation to the next, *provided that* mating is random and all genotypes are equally viable. Therefore, by confining our attention to alleles rather than genotypes, we can predict allele and genotype frequencies in future generations, and after the first generation, the genotype frequencies will remain at Hardy–Weinberg equilibrium, all else being equal. Traits do not become

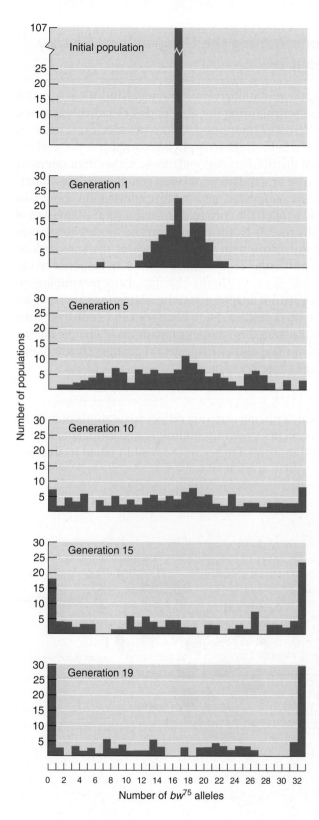

FIGURE 13.8 Distributions of the numbers of *brown* (bw^{75}) alleles in 107 populations of *D. melanogaster,* each with an initial frequency of 0.5 bw^{75} (top panel). The populations were continued for 19 generations (bottom panel), using 16 parents to start each generation. By generation 19, the bw^{75} allele had been eliminated (0 alleles) from 30 populations and had been fixed (32 alleles) in 28 populations. (Data from Buri, P., *Evolution,* **10**, 1956: 367–402.)

Phylogeography: Reconstructing the Geographical History of a Lineage Using Genetics

A recent approach to understanding the evolutionary processes regulating the geographic distributions of groups is based on restructuring the genealogies of individual genes, groups of genes or populations. Differences among populations of a species are detected by differences in gene frequencies. The field is known as **phylogeography** — the working out of the basis for the geographical distribution of populations. Rather than sampling a single population, as is often the case in standard population genetics analyses, different populations within a species are examined. Consequently, information about past patterns of migration can be used to explain the current distribution and subdivision of species into groups. Phylogeography reveals group, population and species histories as illustrated by the current geographical distribution of strains of wheat, discussed below.

Migration to remain within a particular ecosystem as it shifts in response to climate change (behavioral adaptation) can occur in the absence of any morphological changes. It appears as if the species are not adapting, but the adaptations are revealed using large-scale temporal and spatial studies. Such changes can occur over long periods of a species' existence, as illustrated by studies on patterns of change among African antelopes as they tracked climate and environmental changes during half a million years during the last ice age some 2.5 Mya.

Spread of Wheat
Strains of the Asian common wheat, *Triticum aestivum* reflect adaptations of a single species to a large number of environments, growing conditions, and artificial selection pressures across Asia from the Ukraine and Turkey in the west to Japan in the east.

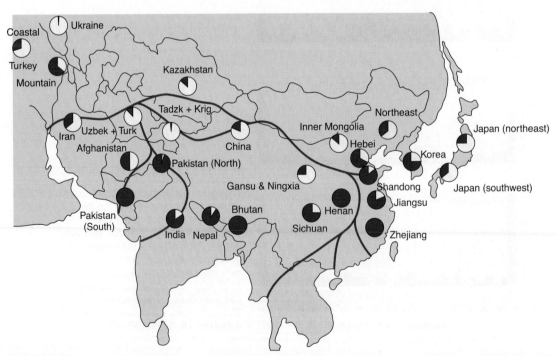

FIGURE 13.9 Phylogeographic analysis of Asian common wheat, *Triticum aestivum,* based on isozyme analysis. The collection sites of the populations used in the study range across all of Asia. The major trade route (the "Silk Road" is shown in brown. (Adapted from Ghimire, S. K., Y. Akashi, C. Maitani, M. Nakanishi, and K. Kato, *Breeding Sci.,* **55**, 2005: 75–185.)

Using the argument that the Tibetan plateau and large areas of desert in west China form effective geographical barriers to migration, three routes for the spread of wheat along trade routes to East Asia have been proposed (FIGURE 13.9).

- From Turkey to Sichuan China along an ancient Myanmar route;
- Along the Silk Road; and
- From the coastal area of China and Korea.

This interpretation of the distribution of the strains of Asian common wheat based on geographical barriers and trade routes developed and maintained by humans is supported by two studies based on the distribution of genes for two enzymes. Six hundred and forty-eight races of wheat revealed 33 populations that could be grouped into six clusters originated from three lineages (FIGURE 13.10). This study revealed

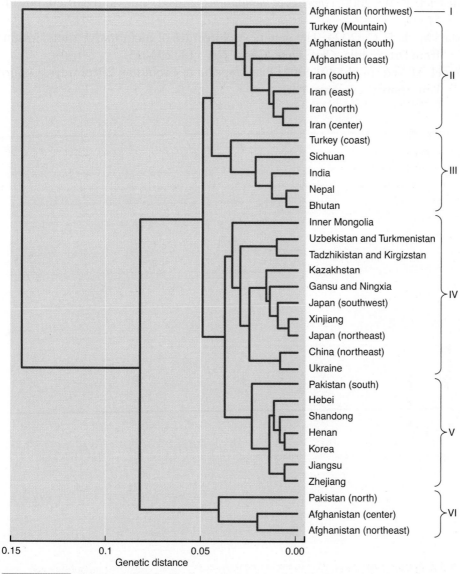

FIGURE 13.10 The 33 populations of Asian common wheat shown in Figure 13.9 were resolved into six clusters (I–VI) derived from three lineages.

genetic and geographical differentiation of Asian wheat and allowed the possible sources of origin and routes of migration of populations to be verified. This is phylogeography at its best.

Recommended Reading

Crow, J. F., 1988. Eighty years ago: the beginnings of population genetics. *Genetics*, **119**, 473–476.

Dobzhansky, Th., 1937, 1941, 1951. *Genetics and the Origin of Species* (three eds.). Columbia University Press, New York.

Gillespie, J. H., 2004. *Population Genetics: A Concise Guide*, 2nd ed. Johns Hopkins University Press, Baltimore, MD.

Hedrick, P. W., 2000. *Genetics of Populations*, 2nd ed. Jones and Bartlett, Boston, MA.

Mackay, T. F. C., 2004. The genetic architecture of quantitative traits: lessons from Drosophila. *Curr. Opin. Genet. Dev.*, **14**, 253–257.

Pagel, M. (ed. in chief), 2002. *Encyclopedia of Evolution*, 2 Volumes. Oxford University Press, New York.

The Biology of Populations

<div style="font-size:2em">14</div>

KEY CONCEPTS

- Individuals interact within populations and with elements and other organisms in their environment.
- Populations modify their physical and biological environments.
- The effective unit within a population is an interbreeding group.
- Competition and predation bind species into coevolving units.
- Reproductive strategies result in the production of many or few offspring, depending on the species.
- Coevolution as seen between parasite and host or pollinating insect and host plant is common in nature.
- Many species can change their morphology in response to signals received from prey, predators or the environment in a process known as phenotypic plasticity.

Above: A population contains animals of different generations.

Overview

Central to evolution for almost a century has been analysis of interactions among individuals and between individuals and their environment, interactions that occur in populations. Biological, ecological and genetic interactions are three broad and overlapping types of interactions affecting populations. Adaptation to local environments establishes one of the first barriers that can lead to reproductive isolation and speciation.

Population interactions such as competition, predation and coevolution influence the maintenance and selection of particular traits. Competition results in niche and character distinctions between demes, in some cases eliminating one group entirely. Predation has complex effects on the size and structure of the populations of both prey and predator. Coevolution of species from different lineages, including different kingdoms, is illustrated using predation and insect-host plant interactions.

The structures of populations are complex. One characteristic that can be measured is population growth. Unlimited population growth becomes exponential, as Malthus, Darwin and Wallace were well aware (see Chapter 7). As the environment imposes restrictions, a population will tend to stabilize at a size (number of individuals) called the carrying capacity (K).

In nature, the structure and relationships of populations depart from many of the ideal conditions that would make their evolutionary behavior simple to understand. Populations are neither of constant size nor uniformly distributed in space and are subject to variable degrees of mutation, migration and selection (see Chapters 12 and 15). Moreover, neither the physical nor the biological environment — temperature, amount of light, prey, predators — remain constant over time. Furthermore, and importantly, populations are not passive recipients of environmental or ecological information. A population modifies its physical and biological environment in ways that can diminish or enhance its own resources and those of other populations. Because of all these interactions, it has been said that populations "must continue running in order to keep in the same place" (the Red Queen hypothesis; see Chapter 15).

In general, the biological and ecological approaches briefly reviewed in the first two sections of this chapter emphasize how populations respond to their environment and to other species as assessed by the numbers of individuals (BOX 14.1, and see Chapter 1 for individuals versus populations). Animal and plant populations share common features in their response to inbreeding, selection, mutation, migration and genetic drift. In the overwhelming majority of populations, however, it is difficult to discover the genetic information that could explain how such changes correlate with the evolutionary mechanisms of mutation, selection and genetic drift (see Chapters 12 and 15).

Biological Interactions Within Populations

Fecundity is the term used for the number of offspring produced by an individual, usually a female.

In terms of survival, growth and/or fecundity, a slight increase in the numbers of individuals in one species may cause an increase, decrease, or have no discernible effect on numbers of another species. Interactions between two species can be classified according to the terminology set out in TABLE 14.1 and illustrated in FIGURE 14.1, with categories ranging from neutral interaction with no affect on either species to competition with maximal effects on both species. The essential features of three of

BOX 14.1
Can Individuals Exist Outside a Population?

In Chapter 1, I made the claim that "In most species of uni- and multicellular organisms, individuals exist in **populations** that inhabit discrete ecological niches" (Box 1.3). Exceptions that occurred to me were few: The many parasites that live their lives as individuals; organisms with both sexes in the one individual (hermaphrodites), many of which live, reproduce and die alone (for example, solitary wasps).

However, I continued to think of examples of species that appear to exist as individuals living alone and so began to wonder whether most species do exist as populations rather than as individuals. The notion of a sexually reproducing species mostly consisting of individuals does not seem sustainable, and so I asked several of my evolutionary biology colleagues for their views. Their responses were varied at one level but uniform on another: **species must exist as populations of individuals**. I think this a sufficiently interesting issue that I have outlined some of the responses below; this might make an interesting discussion topic.

Hermaphrodite fish exist as populations, with the possible exception of the 75-mm-long mangrove killifish, *Rivulus marmoratus,* which is endemic to the east coast of North, Central and South America, from Florida to Brazil. According to the textbooks this is the only fish, and possibly the only vertebrate, with internal self-fertilization. However, it turns out that self-fertilization is not obligatory in *R. marmoratus;* separate sexes can occur. A colleague with a strong ecological and natural history bent observed, "even hermaphrodites probably make decisions about their movements with respect to others of the species, e.g., not to feed or lay eggs where some conspecific (species in the same genus) has done so."

An evolutionary botanist colleague kindly pointed out that my statement was "a little too facile," and reminded me that most vascular plants are hermaphrodite but have evolved mechanisms to prevent self-fertilization (see Chapter 10). Therefore, individuals are members of a population. This conclusion was echoed by another evolutionary botanist, who said, "A great many plants (possibly the majority) are hermaphrodites. Many do not self fertilize at all, and I can't think of any that self-fertilize all the time." He also reinforced the statement that species must exist as populations. "I think this is a necessary condition of evolution, since individuals don't evolve—populations do." He related the problem to the many definitions that can be applied to the concepts "population" and "species," something we saw also to be so for genes (see Box 1.2):

> I think you've highlighted one of the maddening facts of biology, namely that many of our most useful (or used) terms cannot be consistently defined. This is true of "species" where no single definition really works, so in practice we use different conceptions for different cases. The term "population" is even worse as it is a wholly human construct, defined for the question at hand. We usually delimit populations by geography, and then ask questions about phenotypic and genetic differentiation, gene flow, etc., but we could equally define population based on phenotypic and genetic differentiation, gene flow, etc., and ask how geography maps onto that. My short answer is that, yes, all organisms live in populations, but the average exposure time to another individual of the same species varies greatly.

A marine ecologist echoed these sentiments; "I also believe individuals of any species (hermaphrodites and parasites included) exist as populations at some level—it's a matter of scale."

TABLE

14.1 Potential Interactions That Can Occur Between Populations of Two Species

Type	Nature of Interaction
Neutralism	Neither population affects the other.
Commensalism	One species is unaffected while the second (commensal) species benefits.
Mutualism	Both species benefit, as seen in Müllerian mimicry (see Figure 15.11).
Predator-prey-induced polymorphism	Both species benefit, as seen in phenotypic plasticity (Figure 14.5 and see Figure 13.6)
Coevolution	Both species benefit.
Predation	The predator benefits at the expense of the prey.
Parasitism	The parasite benefits at the expense of the host.
Competition	Each species inhibits the other.

these interactions are discussed below, the three being **competition, predation** and **coevolution**.

Each interaction may be seen only in one combination of species, reflecting the past evolutionary history of the populations involved, spatial limitations, climatic conditions, soil nutrients, and the effects of other species in the community. Each interaction must be disentangled from others and explored separately. In their various forms and through interactions between them, these processes determine the dynamics of demes, populations and species.

Competition

Competition arises when two groups depend on the same limited environmental resource(s) so that each group causes a reduction in the other's numbers, often with important ecological or behavioral consequences. Three consequences of competition are outlined below.

Resource Partitioning

Competition often leads to ecological diversity. It can be to the advantage of competing groups to minimize the harmful effects of direct competition by using different aspects of their common environmental resources. Among the many examples of such **resource partitioning** is the one illustrated in Figure 14.1a for five species of warblers, each using different parts of their spruce tree habitat. Should such resource partitioning be disrupted and habitats overlap, competition will lower the fitness of competing groups. A study showing that nest predation increases when nesting sites overlap between different species supports resource partitioning.

Different closely related (conspecific) species may occupy different parts of the habitat as illustrated by species of Anolis lizards in trees on Caribbean Islands. Each of the three species (ecomorph species) shown in Figure 14.1b occupies a different

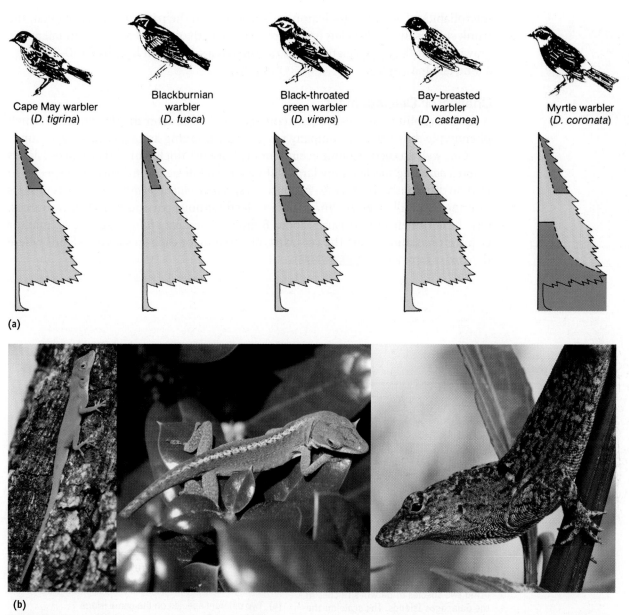

(a)

(b)

FIGURE 14.1 (**a**) Resource partitioning as demonstrated by the most common feeding zones (dark green) in spruce trees for five species of northeastern U.S. warblers of the genus *Dendroica*. The Cape May warbler is quite rare unless there is a large outbreak of insects. The Myrtle Warbler is both the most abundant and the least specialized warbler in most coniferous forests. The more common warblers — Blackburnian, Bay-Breasted and Black-Throated Green — are different enough in feeding zone preferences to explain their coexistence. (Adapted from Krebs, C. J., *Ecology: The Experimental Analysis of Distribution and Abundance, Fifth edition*. Benjamin Cummings, 2001.) (**b**) Ecomorphs of Anolis lizards inhabit different parts of the habitat (tree, bush, grass) and have distinctive morphologies.

microhabitat. Each displays features associated with the particular microhabitat: the trunk species *Anolis distichus* has a snout-vent length of 40–58 mm, a short tail and is gray in color; the crown giant, *Anolis ricordii* has a snout-vent length of 130–190 mm, a long tail and is green.

Character Displacement

A further evolutionary response to competition is **character displacement** in which phenotypic differences accompany resource partitioning among coexisting groups.

One well-studied example occurs among Darwin's finches in the Galapagos Islands where *coexisting species* show large differences in bill sizes, enabling each species to feed on differently sized seeds. In contrast, *species isolated on different islands* possess intermediate bill sizes enabling them to feed without partitioning seed resources. One beak dimension measures about 8 mm for *Geospiza fuliginosa* and 12 mm for *G. fortis* on islands where they co-exist, but 10 mm for each species on islands where they exist separately (FIGURE 14.2).

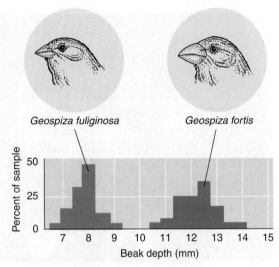

(a) Two different species on the same island (Abingdon, Bindloe, James, Jervis Isls)

FIGURE 14.2 Character displacement illustrated using beak depth (mm) using two species of Darwin's finches in the Galapagos Islands. The scale on the X-axis of (**a**) applies to all three graphs. (**a**) *Geospiza fuliginosa* (brown) and *G. fortis* (blue) coexisting on four islands show large differences in beak sizes (8 versus 12.5 mm), enabling each to feed on different sized seeds. However, when either species exists alone on different islands (**b, c**), beak size is intermediate (about 10 mm, although note the bimodal distribution in **c**) enabling feeding without partitioning seed resources. (Adapted from Givnish, T. J. and K. J. Sytsma (eds.). *Molecular Evolution and Adaptive Radiation*. Cambridge University Press, 1997; based on Grant, P. R. *Ecology and Evolution of Darwin's Fiches*. Princeton University Press, 1986.)

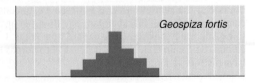

(b) A single species on an island (Daphne)

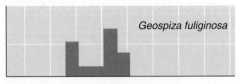

(c) A single species on an island (Crossman)

The advances that can come from close and continuous observations of natural populations are demonstrated by the 40-year-long and ongoing studies on Darwin's finches by Peter and Rosemary Grant of Princeton University. Among numerous discoveries, the Grants documented a series of processes beginning with competitive interaction and character displacement, leading to evolutionary change in one of the participating species (BOX 14.2).

BOX 14.2
Darwin's Finches and the Jaws of Fish

Darwin's finches were introduced in Box 5.4 as an important cluster of species used by Charles Darwin as evidence for species change (see Figure 14.2 and Figures 5.5, 5.10 and 6.10).

The importance of this cluster of 13 species to our understanding of evolution has been greatly enhanced by the work of Peter and Rosemary Grant, who have studied Darwin's finches continuously since 1973. Their work has documented evolution in action by relating environmental changes leading to evolutionary changes in beak size and body size over the span of a few finch generations. An early study by Peter Grant and colleagues published in 1976 confirmed a prediction made by Leigh Van Valen that the degree of variation in important environmental factors should be reflected in the amount of variation in the morphological traits relevant to those factors. Thus, species encountering a food that varies greatly in size and hardness should show higher variation in beak dimensions. Within each species, individual birds choose food that is of the appropriate size and hardness for their beaks.

As their work progressed and the longitudinal data sets built up, the Grants showed that natural selection can produce evolutionary change very rapidly. They also showed, surprisingly, that the direction of selection in particular populations can change frequently and unpredictably as environmental conditions fluctuate (see Figure 14.2). By documenting evolution in action in natural populations, this work has made profound contributions to our understanding of the evolutionary process.

A question raised by the results of the Grants' study of evolution in Darwin's finches is how, in genetic and developmental terms, evolution can proceed so rapidly. If changes to beak size and shape required selection to alter the expression or function of many genes in some coordinated fashion, it is unlikely that evolutionary changes could happen as quickly as they clearly have over the past 30 years in these finches.

An insight into this question has come from a 2004 study by Arhhat Abzhanov and his colleagues on the developmental–genetic basis for variation in beak size and shape in Darwin's finches. Remarkably, the amount and area of expression of a single gene, *Bmp-4*, is correlated with variation in beak size and shape across species of Darwin's finches (FIGURE B14.1). In a fascinating parallel, increased levels of *Bmp-4* in a cichlid fish from Lake Malawi are associated with biting and crushing feeding on algae (see Figure B17.3). This can be compared with lower levels of *Bmp-4* in another cichlid species from the same lake that feeds by suction feeding on plankton in the water column.

Feeding is effected through jaws and teeth. Although expression of *Bmp-4* is conserved in tooth-forming regions across three species of fish (zebrafish, Japanese medaka, and Mexican tetra), *Bmp* genes are not expressed in toothless regions of the jaws, implicating *Bmp* in evolutionary tooth loss, and linking genetic regulation of teeth and jaws.

(continued)

■ This cluster of finches on the Galapagos Islands was first referred to as Darwin's finches by David Lack in 1936 (see Box 5.4).

■ In their recent book *How and Why Species Multiply: The Radiation of Darwin's Finches* (2008), the Grants provide an outstanding overview of speciation.

BOX 14.2

Darwin's Finches and the Jaws of Fish (*continued*)

These results suggest that the rapid evolution of the beak in Darwin's finches (and perhaps in many other lineages of birds), and jaw and tooth morphology associated with feeding in fish, is based on a relatively simple developmental-genetic mechanism that can produce a coordinated pattern of variation in beak, jaw and/or tooth sizes and shapes. Other similar genetic and developmental mechanisms underlying speciation are explored in Chapter 17.

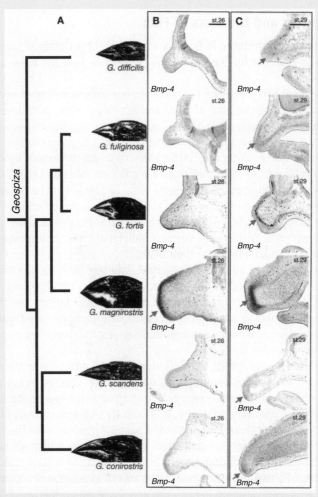

FIGURE B14.1 Relationship between evolutionary relationship (**A**), beak morphology and variation in *Bmp-4* expression (**B, C**) in six species of Darwin's finches. The histological sections are from the developing beak at two embryonic stages, 26 (**B**) and 29 (**C**). Darkly stained areas indicate high levels of *Bmp-4* expression. (Reproduced from Abzhanov, A., Protas, M., Grant, B. R., Grant, P. R., and Tabin, C. J. 2004. *Bmp4* and morphological variation of beaks in Darwin's finches. *Science,* **305**, 1462–1465. Reprinted with permission from AAAS.)

Competitive Exclusion

The many variants of the definition of **niche** outlined in Box 1.3 capture the concept that a species' niche includes all the environmental resources used by the species as well as the ways in which the species exploits those resources. The American evolutionary biologist Richard Lewontin captured the essence of the niche and of the relationships of an organism to its (self-made) environment in his 2000 book, *The Triple Helix: Gene, Organism, and Environment.* "Niches do not preexist organisms but come into existence as a consequence of the nature of the organisms themselves . . ."

In laboratory experiments in which two species occupy the same niche, one species commonly dies out (FIGURE 14.3), supporting the principle of **competitive exclusion:**

■ When competition is not checked by partitioning or fluctuation of resources and when two competing species use exactly the same resources in the same environment, they occupy the same niche.

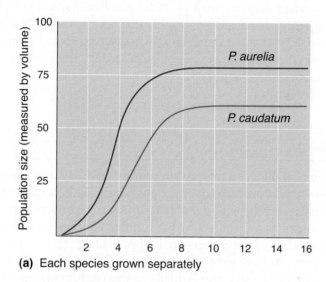

(a) Each species grown separately

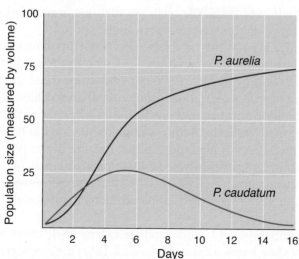

(b) The two species grown together

FIGURE 14.3 Growth of two species of *Paramecium* (**a**) in separate cultures, and (**b**) in mixed cultures. Although *P. aurelia* replaces *P. caudatum* (**b**), in some mixed culture conditions *P. caudatum* multiplies faster than *P. aurelia.* (Adapted from Gause, G. F. *The Struggle for Existence.* Williams & Wilkins, 1934.)

two species cannot continue to coexist in the same environment if they use it in the same way. This principle was foreshadowed by Darwin in *On the Origin of Species*.

> Owing to the high geometrical rate of increase of all organic beings, each area is already fully stocked with inhabitants; and it follows from this, that as the favored forms increase in number, so generally will the less favored decrease and become rare.

Competitive exclusion also may be sensitive to external environmental factors. In the example shown in Figure 14.3, one paramecium, *Paramecium caudatum*, is more sensitive to metabolic by-products than is the other species, *P. aurelia*.

Evolutionary biologists have seriously questioned whether competitive exclusion alone accounts for the differences observed among coexisting species — invoking for example, character displacement — or for the finding that closely related and potentially competitive species often occupy different habitats. Morphological differences among species that coexist currently may have evolved in the past in places where these species did not compete; related species occupying different habitats may not have diverged because of competition but because of different food preferences, nesting sites, and so on.

Predation

Predation is an important factor diminishing competitive exclusion among competing coexisting species; predators can reduce the possibility of a dominant species reaching its carrying capacity, thereby making room for other competitors. A predator entirely or partially consumes its prey, affecting the numbers of those organisms it feeds on. The intimate dependence of predators on their prey often leads to a coupling of their relative abundances; increase in numbers of prey is followed by an increase in predators, which, in turn, can reduce prey, which can then reduce predator numbers — an arms race. As shown in FIGURE 14.4, cyclic oscillations in population numbers correlate, especially for predators that prey on a single species.

Such simplicities are not the rule, however. Some predators don't cause a substantial drop in the abundance of their prey. As seen in the predation of wolves on caribou herds, wolves prey mostly on young caribou or those weakened by age or disease (and which have less reproductive potential than do older, fitter individuals).

Moreover, predator numbers may be buffered when more than one species of prey is being exploited. Buffering reduces large oscillations in predator population size and spreads the effects of predation so that the size of the prey population also remains fairly stable. Compensatory or additive mortality may be seen when a new predator is introduced into an ecosystem; when wolves or lions were introduced into populations of mule deer, *Odocoileus hemionus*, lions but not wolves influenced the mortality rate of the deer.

Different combinations of factors produce extensive differences in populations of Canadian snowshoe hares, *Lepus americanus*. Indeed, the differences are greater than we might have expected from the actions of each factor alone. Reducing predation *or* increasing the food supply *each* cause a 2- to 3-fold increase in snowshoe hare numbers, that is, *a cumulative 4- to 6-fold* increase. In combination, reduced predation and increased food supply caused *an 11-fold* increase in numbers. Obviously, different interactions occur when predation is low and food is plentiful than when food is plentiful and predators are prevalent, or when predation is low and food is meager.

(a)

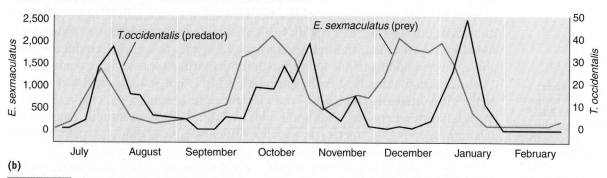

(b)

FIGURE 14.4 Three cycles of oscillating population numbers for two species of mites in a defined, controlled environment. One species, the western predatory mite *Typhlodromus occidentalis* (**a**) is the predator. All stages in the life cycle of a herbivorous species, the six-spotted mite, *Eotetranychus sexmaculatus* (**b**) are the prey. For each cycle, as numbers of prey increase (green) predator numbers follow (gray) causing a "crash" in the prey population, followed by a crash in the predator population. (Adapted from Huffaker, C. B., *Hilgardia,* **27**, 1958: 343–383.)

Coevolution

Group interactions have various consequences, one of which, **coevolution**, is common. The intimate ecological relationships among many species derive from coevolutionary events in which adaptive changes in one species follow adaptive changes in another(s). What begin as casual interactions between different species develop into obligatory coevolving associations.

Paleontological evidence points strongly to competition between herbivorous and carnivorous mammals during which size and speed seemed to increase sequentially in various members of both groups. Similar evolutionary changes in size, speed and protective devices occurred among dinosaurs (see Chapter 18), have been a consistent trend in many prey-predator groups, and are a requirement when coevolution involves pollination of a plant by an insect (see Box 10.3) or parasitism (below).

Parasitism

Notable examples of coevolution are between parasites or pathogens and their hosts. Twenty-seven genes in the flax plant, *Linum usitatissium* confer resistance against a fungal rust pathogen. The pathogen, in turn, has a similar number of genes allowing

it to overcome the resistance these host genes confer. In such cases and in others, one can reasonably claim that an increased frequency of a resistant mutation in the host will be followed by selection for an increased frequency of one or more mutant genes in the parasite or pathogen that overcome resistance.

Not unexpectedly, coevolutionary events between host and parasite can be complex and may reduce the virulence of the parasite. One prominent example is a myxoma virus imported from South America to Australia to control the phenomenal population growth of the European rabbit, *Oryctolagus cuniculus*. Although the virus caused only a mild disease in native South American cottontail rabbits, *Sylvilagus brasilensis*, it acted as a highly lethal pathogen among the Australian rabbits, being transmitted primarily by mosquitoes. Viral-based lethality in 1950–1951 was as high as 99 percent among infected rabbits. It seemed as though the Australian rabbit population either would be eradicated by the virus or persist only at low numbers. This expectation was not fulfilled, however. Although some Australian rabbits became increasingly resistant to the virus, the virus itself became less virulent. Because mosquitoes feed only on live rabbits, the rate at which they transmit the virus falls if the virus kills its immediate host too quickly. Highly virulent, rapidly replicating strains of virus were therefore selected against. Strains with reduced virulence were selected for — they allow infected rabbits to live long enough for the virus to spread more readily.

■ Some small DNA viruses are genetically stable compared to rapidly evolving RNA viruses, and can persist and coevolve with their hosts, causing little disease.

Interestingly, this host-pathogen relationship is not static; as rabbit resistance to the virus increases, increased viral virulence can become a selected trait. Thus, the relationship between virus and rabbit in Australia may eventually emulate the relationship in South America, in which a virulent virus has only a partially harmful effect on its native host. Such evolutionary outcomes are not unusual.

Reduced parasitic virulence, however, is not always a successful option for competing parasitic strains. As just discussed, parasitic infection and parasitic virulence are positively correlated. Success among competing parasitic strains depends on their ability to replicate rapidly; rapid parasitic replication is a major factor in causing virulence. Viruses that incorporate into host chromosomes and are transmitted between generations by vertical transmission have been selected for reduced virulence because they depend on host reproduction for survival. On the other hand, viruses that enter organisms exclusively through infection from other hosts (horizontal transmission) are selected to increase infectivity by replicating rapidly, and thus almost always cause host destruction.

Insects and Host Plants

The many instances of specificity between an insect, moth or bee species and its host plant are usually presented as examples of coevolution. Many do represent coevolution but some do not, as revealed by comparing phylogenies of insects and plants.

Nuclear DNA was used to construct a molecular phylogeny of 100 species of leaf-mining moths to obtain estimates of the time of origin of the moth species. Comparison with the distributional ages in the fossil record of the host plants of these moths led to the conclusion that the main radiation of the moth genus (*Phyllonorycter*) took place 27.3 to 50.8 Mya, which was well after the radiation of their host plants, which was 84 to 90 Mya. The moth radiation appears not to be an example of coevolution (in the strict sense of that word) but of delayed colonization of host plants: "follow-along evolution."

Extent of Coevolution

Although competition, predation and parasitism reflect the popular concept of "nature red in tooth and claw," instances of cooperation and mutualism modify its impact. Among these examples are (a) cooperative relationships discussed under group selec-

tion (see Chapter 15) and phenotypic plasticity (this chapter), and (b) symbiotic relationships, including those between cellular organelles and their eukaryotic hosts (see Chapter 9), algae and fungi, and cellulose-digesting protozoans in termite intestinal tracts. Long periods of coevolution made many such associations obligatory, but others such as between some pollinators and plants may be facultative, neither species relying on the other for survival. Others have been forced upon species by our methods of cultivation; the dwarf palm, *Chamaerops humilis* has been forced into a mutualism (in this case, forced pollination) by being cultivated in botanical gardens with the weevil, *Derelomus chamaeropsis*.

Mutualism refers to a mutually beneficial relationship between two species.

Depending on how one defines the concept, many other examples of coevolution abound. They include mimics that evolve in step with the evolution of their models, ants that cospeciate with fungi they "farm" for food, and competing species that evolve changes between them to reduce competition (for example, character displacement), and many others.

Phenotypic Plasticity

Species with more than one stage in their life history were introduced in Chapter 11 as examples of the concept that one genotype can give rise to more than one phenotype. The caterpillar/butterfly and tadpole/frog are familiar examples of species in which formation of two life history forms is independent of population or environmental cues (see Chapters 12 and 13).

A challenge arises at the population level because of the production of alternate body forms (morphs) within a single species. Morphs arise in response to external cues received from a predator, prey, population density or environmental signals. Production of such morphs shows how the evolution and ecology of populations is built upon genetic and developmental changes in individuals. The external signal may not necessarily be present in every generation. If individuals are not exposed to the environmental signal in a given generation, the environmentally induced phenotype will not appear, but the ability to respond to the signal is passed on to the next generation. We refer to this facultative ability of a single genotype to produce more than one phenotype as **phenotypic plasticity**.

It is important to remember that natural selection takes place within a generation and affects individuals of that generation. Response to natural selection happens between generations, and only for those features or aspects of features that are heritable.

The large number of examples testifies to the ubiquity and utility of phenotypic plasticity, a topic that has been the subject of many studies in many fields, especially ecology and **life history theory**. Indeed, life history theory, which is normally studied by ecologists, has become of major interest to evolutionary, developmental and molecular biologists who seek to understand how phenotypes arise and are maintained. An even greater challenge is integrating the evolution and ecology of populations with phenotypic plasticity.

Some species in aquatic communities are held in balance because of chemical interactions between individuals of predator and prey species. This fascinating phenomenon has been studied especially thoroughly in two situations

- interactions between species of predatory and prey rotifers (FIGURE 14.5a), and
- interactions between the water flea *Daphnia pulex* (a microscopic crustacean) and its predator, the larvae of a midge (Figure 14.5b).

In both situations, the predator releases a chemical that *acts on some but not all* the eggs of the prey species. Rotifers that develop from those eggs produce an extra set of spines that prevent them from being eaten by the predator (Figure 14.5a). Water

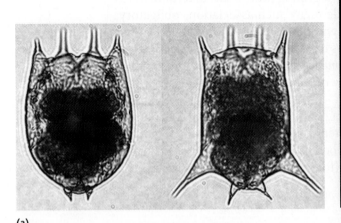

(a) (b)

FIGURE 14.5 Two examples of phenotypic plasticity in response to chemical signals released by predators (see text for details). **(a)** A rotifer (*Brachionus calyciflorus*) in the absence (left) or presence (right) of a predator to show the additional spines formed in the presence of the predator. **(b)** A water flea (*Daphnia*) in the presence (left) or absence (right) of a predator to show the helmet that forms in the presence of the predator.

fleas that develop from those eggs produce enlarged spines and an enlarged helmet (Figure 14.5b) that prevent the *Daphnia* morphs from being eaten by the predator(s).

Four other examples of phenotypic plasticity in aquatic organisms or aquatic life history stages of various types are briefly introduced below.

1. In the presence of predatory dragonfly larvae, tadpoles of the Australian brown striped marsh frog, *Limnodynastes peroni,* develop increasing tail height and muscle mass, features that enhance their chance of escaping predation.

2. When food supply is restricted or population numbers high, some anuran and salamander tadpoles display modified development; they become cannibalistic with much larger heads and more massive jaws than their non-cannibalistic brothers and sisters, which they eat (FIGURE 14.6).

3. In a bony fish, the threespine stickleback, *Gasterosteus aculeatus,* one, two, or three of the bones of the pelvic girdle are reduced in response to a combination of a *biotic factor* — absence of predatory fish — and an *abiotic factor* — the level of dissolved calcium in the lake (see Figure 13.6).

4. Plasticity in body forms in the pumpkinseed sunfish, *Lepomis gibbosus* is reflected in inshore and open ocean morphs which diverged from one another in response to predation by walleye, *Sander vitreus.* The conclusion is that this diversification is driven by selection related to predation rather than the use of available resources by the two morphs; that is, selection for predator-induced polymorphism drove and drives divergence.

Phenotypic plasticity is not confined to aquatic communities. In the life cycle of the North American moth, *Nemoria arizonaria,* caterpillars take on a morphology depending on whether they hatch out as *twig morphs* on oak trees when catkins are absent (summer) or as *catkin morphs* when catkins are present (spring), a seasonal polymorphism that is triggered by the levels of tannin in their diet.

An example of the role that phenotypic plasticity plays in allowing birds to adjust to changing environments (including food abundance) is a more than 30 year study on phenotypic *and* genetic variation in a small Mediterranean passerine bird, the

(a)

(b)

FIGURE 14.6 Cannibalistic tadpoles of Couch's spadefoot toad, *Scaphiopus couchii*, from the side to show the body form (**a**) and from the front to show the massive head and tooth rows (**b**).

blue tit, *Cyanistes caeruleus*. Comparisons have been made between deciduous and evergreen forest environments, both on the mainland and on islands. Extensive data are available on population structure, food abundance, breeding performance, and morphology. Phenotypic plasticity emerged as a function of the distances over which birds dispersed. Short-distance dispersal results in local specializations, longer-distance dispersals in phenotypic plasticity, which, when local selection regimes opposed gene flow, led to local adaptations out of proportion to the geographical separation between the populations. TABLE 14.2 summarizes some of the habitat-specific differences recorded on the mainland and on two of the deciduous island sites. Jacques Blondel and his colleagues distinguished deciduous oak and evergreen oak types but found maladapted populations in which, for example, breeding time did not coincide with peak food (caterpillar) supply. They concluded, "wherever habitat patchiness is a mixture of deciduous and evergreen patches, the resulting reaction norm includes local specializations, phenotypic plasticity, or local maladaptation, depending on the size of habitat patches relative to the average dispersal range of the birds."

> ■ A reaction norm, a way to measure the response of a phenotype to different environments, is an index of organismal-environmental interaction on the one hand (this chapter) and of flexibility/plasticity-canalization/constraint on the other (see Chapter 15).

Population Growth

Although the capacity for reproduction is counterpoised against selection, the evolutionary potential of the **reproductive power** of a population can be determined.

Reproductive modes differ among organisms. Many annual plants and insects breed only once during their lifetimes. Many perennial plants and vertebrates breed repeatedly, although not necessarily annually. The number of offspring a female produces at any one time varies significantly between species, ranging from a single offspring

TABLE

14.2 **Characteristics of Blue Tit (*Cyanistes caeruleus*) Populations from One Mainland and Two Island Habitats in the Mediterranean**

Trait	Mainland	Island 1	Island 2
Blue tit population density (pairs/hectare)	1.0	1.28	0.35
Prey (caterpillar) abundance (mg/m²/day)	23	493	—
Blue tit body weight (g)			
Males	11.2	9.9	9.4
Females	10.7	9.7	9.4
Clutch size	9.8	8.5	7.2
Number of fledglings	7.5	7.3	5.0
Fledgling weight (g)	10.7	10.4	9.3

Source: Blondel, J., et al., *BioScience*, **56**, 2006: 661–673.

in many larger mammals to thousands of eggs produced by a salmon. Furthermore, because an individual may die before reaching reproductive age, the sooner reproduction begins the greater the chances of producing offspring. In some organisms, reproductive stages follow soon after hatching.

The consequence of birth rate minus death rate, which is the primary factor determining population size, is normally presented as the **rate of increase** (*r*). In most species, natural selection favors genotypes that confer survival to reproductive age, with little or no selection for genotypes to survive longer. As Richard Dawkins pointed out, "we inherit whatever it takes to be young, but not necessarily whatever it takes to be old."

Under ideal conditions the early stages of population growth in an environment well supplied with resources can occur geometrically in each generation ($2 \rightarrow 4 \rightarrow 8 \rightarrow 16$, and so on), closely following the growth curve shown in FIGURE 14.7a , and as known since Malthus (see Chapter 7). Such a pattern of population growth would be limitless if reproduction rate was unchecked and if space and resources were limitless.

Environmental resources and space are not limitless, however. Nor is population size, which may stabilize at some near-constant value or may "crash." Malthus' popularization of the idea that war, famine, and disease held in check exponential growth of human population led Darwin and Wallace independently to the concepts of the struggle for existence and natural selection (see Chapter 6). To evaluate such impacts on a population we need to determine how many individuals the environment can support. This value, **K**, is the **carrying capacity**. Using the population of yeast cells shown in Figure 14.7b, we see that the population levels off at a plateau of about 665 individuals, a rate of increase displaying an S-shaped growth curve. In reality, populations rarely follow such smooth growth curves.

The size of a population can change dramatically over a short period of time as shown in FIGURE 14.8 for a bird, the great tit, *Parus major* in Holland. Many parameters affect such fluctuations in numbers within a population. *Density independent* agents such as climate are independent of population size and crowding. *Density dependent*

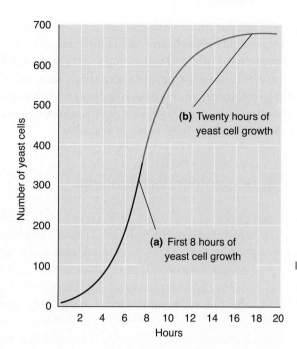

(b) Twenty hours of yeast cell growth

(a) First 8 hours of yeast cell growth

FIGURE 14.7 Numbers of cells of the budding yeast, *Saccharomyces cerevisiae*, in a defined volume of culture medium for two growth periods, beginning with approximately 10 cells per volume. **(a)** Geometric growth during the first 8 hours. **(b)** S-shaped growth curve over the 20-hour growth period. (Data from Carlson, T., *Biochemische Zietschrift* **57**, 1913: 313–334.)

Saccharomyces cerevisiae (sugar mold of beer), sometimes known as Brewer's yeast, has been used for thousands of years to make bread and brew beer.

agents depend on a combination of population size and space available. A sample of examples of adaptations to climate is outlined in Box 1.1.

Organisms such as bacteria and plant weeds display rapid population growth in the face of fluctuating environments generally and have a rapid rate of increase (*r*) at low population densities. As stated by Darwin, "A large number of eggs is of some importance to those species, which depend on a rapidly fluctuating amount of food, for it allows them rapidly to increase in numbers." Other organisms, such as large vertebrates, face more uniform or predictable environments with population sizes that are close to carrying capacity (K). The selective advantages of this pattern of reproduction are increased efficiency in resource use and a greater likelihood that offspring can be raised to the stage when they themselves can compete.

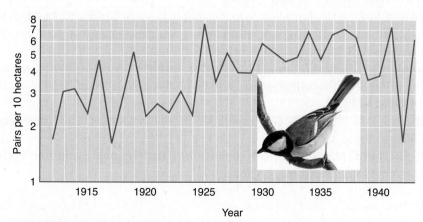

FIGURE 14.8 Fluctuations in population size of the great tit, *Parus major*, in Holland between 1912 and 1943. (Adapted from Begon, M. and M. Mortimer. *Population Ecology, Second edition*, Blackwell, 1986.)

Population density also can modify population growth. In natural populations of *Drosophila melanogaster,* allele differences in a foraging gene that produces a protein kinase used in signal transduction pathways (see Chapter 13) affect feeding behavior; "rovers" move longer distances than "sitters." These differences in foraging activity are selected by differences in population density; rovers are adaptive at high density where larvae must travel longer distances to obtain food; sitters are adaptive at lower densities where food is more available.

Recommended Reading

Abzhanov, A., M. Protas, P. R. Grant, and C. J. Tabin, 2004. *Bmp4* and morphological variation of breaks in Darwin's finches. *Science,* **305**, 1462–1465.

Albertson, R. C., J. T. Streelman, and T. D. Kocher, 2000. The beak of the fish: genetic basis of adaptive shape differences among Lake Malawi cichlid fishes. Assessing morphological differences in an adaptive trait: a landmark-based morphometric approach. *J. Exp. Zool.,* **289**, 385–403.

Fox, C. W., D. A. Roff, and D. J. Fairbairn (eds.), 2001. *Evolutionary Ecology. Concepts and Case Studies.* Oxford University Press, New York.

Grant, P. R., 1986. *Ecology and Evolution of Darwin's Finches.* Princeton University Press, Princeton, NJ. [Reissued in 1999 with a Foreword by J. Weiner.]

Grant, P. R., and B. R. Grant, 2008. *How and Why Species Multiply: The Radiation of Darwin's Finches.* Princeton University Press, Princeton, NJ.

Hall, B. K., 1999. *Evolutionary Developmental Biology,* 2nd ed. Kluwer Academic Publishers, Netherlands.

Hall, B. K., and W. M. Olson (eds.), 2003. *Keywords and Concepts in Evolutionary Developmental Biology.* Harvard University Press, Cambridge, MA.

Lewontin, R., 2000. *The Triple Helix: Gene, Organism, and Environment.* Harvard University Press, Cambridge, MA.

Losos, J. B., 2009. *Lizards in an Evolutionary Tree. Ecology and Adaptive Radiation of Anoles.* University of California Press, Berkeley, CA.

Petrusek, A., R. Tollrian, K. Schwenk, et al., 2009. A "crown of thorns" is an inducible defense that protects *Daphnia* against an ancient predator. *Proc. Natl. Acad. Sci. U.S.A.* **106**, 2248–2252.

Rockwood, L. L., 2006. *Introduction to Population Ecology.* Blackwell, Malden, MA.

Rose, M. R., and L. D. Mueller, 2006. *Evolution and Ecology of the Organism.* Prentice-Hall, Upper Saddle River, NJ.

Tatarenkov, A., S. M. Q. Lima, D. S. Taylor, and J. C. Avise, 2009. Long-term retention of self-fertilization in a fish clade. *Proc. Natl. Acad. Sci. U.S.A.* **106**, 14456–14459.

Thompson, J. N., 2005. *The Geographic Mosaic of Coevolution.* The University of Chicago Press, Chicago.

West-Eberhard, M. J., 2003. *Developmental Plasticity and Evolution.* Oxford University Press, Oxford, England.

Natural Selection

15

- Natural selection acts on the phenotype: structural, physiological and behavioral aspects of the phenotype.
- Response to selection can be tracked in nature and under experimental conditions.
- Selection may stabilize or change a phenotype, depending in part on patterns of change in the organisms' environment.
- Changes in the genotype can tract phenotypic selection leading to heritable changes in the genotype.
- Selection can act on any stage in the life cycle from gametes onward.
- Because it leads to differential reproduction, selection has the greatest affect when it operates before organisms reproduce.
- Selection on one species can influence the evolution of other species, especially evident in mimicry but not limited to mimicry.
- Selective change is not unbounded but is influenced by constraints from genetic and developmental processes.

Above: Natural selection results in conformity and in variation.

- Selection is facilitated by the large amount of allele variation in populations, variation itself being subject to selection.
- A population's response to selection can be modeled using concepts such as adaptive landscapes.

Overview

Selection, which acts on the fitness differences among different traits of the **phenotype**, can operate on a population in at least four ways: (1) stabilizing selection, which maintains phenotypes close to a mean, (2) directional selection if an extreme phenotype has adaptive value, or to eliminate deleterious mutations rapidly, (3) disruptive (diversifying) selection when different phenotypes are better adjusted to particular environments, and (4) sexual selection, when selection operates differently on males and females of the same species.

With systematic and persistent selection on the phenotype from generation to generation allele frequencies change, that is, the **genotype** changes. Genetic variation in populations facilitates the genotypic response to selection. Indeed, selection acts on variation itself. Selection can preserve genetic variation if the heterozygous genotype is more fit than either homozygote, as in sickle cell anemia, or when the frequency of an allele affects fitness. Allele frequency may vary in different environments, as in populations of the British peppered moth, *Biston betularia*, in England or the rock pocket mouse, *Chaetodipus intermedius* in the American southwest. Advantageous genotypes are said to occupy adaptive peaks of varying value according to their degree of fitness.

Species are bound together as predators and preys, through mimicry and by other ecological interactions (see Chapter 14). Consequently, selection on one species can, and almost inevitably does, influence other species. Selection is not limitless, however. Dynamics and constraints of genetic and developmental pathways work with selection to canalize evolutionary change. This interaction can be observed in nature and mimicked in laboratory experiments in which an organism's response to environmental change is measured as a reaction norm; the greater the number of reaction norms, the more responsive the organism is to the environment. Although the issue is still unresolved, in some instances, for example, in a population with sexual reproduction and altruistic social behaviors, selection seems to occur for the benefit of the group, even though it may harm the genetic future of the individual.

For populations to evolve — that is, to change their gene frequencies — mutation must introduce nucleotide differences in the alleles of genes of individuals within the population. The mere appearance of new alleles, however, is no guarantee that they will persist or prevail over others. Explanations for the persistence of many mutations and their increase in frequency are sought elsewhere than in the original mutational event.

Although evolution is marked by phenotypic changes, the transmission of such changes between generations is genotypic. The genotype need not only be that of the individual under selection; genotypic inheritance is both direct and indirect. Genes of the parent (maternal gene effects) and epigenetic influences from the gene products of

other organisms or from the environment (see Chapter 13) transmit change between generations. The ability to learn is transmitted both genetically and culturally (see Chapter 20). Finally, selection is not only on the adult phenotype. In multicellular organisms, selection can elicit changes in gametes, embryos and/or larval stages, depending on the life history stage at which selection influences survival or fertility (FIGURE 15.1).

■ In most animals and plants selection takes place primarily on the diploid zygote and on post-zygotic embryonic, larval and adult stages.

Survival of the Fittest

Evolution is described as the **survival of the fittest** because of the operation of natural selection. The term, coined by Herbert Spencer, was used by Wallace (and by Darwin in later editions of *The Origin of Species,* although Darwin thought struggle for reproduction was a more apt term for evolution by natural selection).

Darwin's concern has been raised by others. The argument is that, because it defines those that survive as the fittest and the fittest as those that survive, "survival of the fittest" is a circular, tautological, or unprovable statement, and so should be abandoned. However, the term and concept is appropriate for at least three important reasons.

1. Selection does change the range of phenotypes that survive in a single generation.
2. Natural selection on a phenotypic character can be a cause for change in gene frequency but is *not the same as a change* in gene frequency.
3. Most genes and phenotypes require multiple generations of selection in the same direction before their proportions change.

The relative fitness of a genotype is certainly not the only factor contributing to its survival. For example, if there is heterozygote advantage, the frequency of a harmful recessive allele may remain stable despite selection against homozygotes for that gene. Evolutionary theory includes more than survival of the fittest. R. A. Fisher begins his classic 1930 treatise, *The Genetical Theory of Natural Selection,* with the statement "Natural Selection is not Evolution." Natural selection is the umbrella term for a range of evolutionary processes, the outcome of which is differential survival and reproduction.

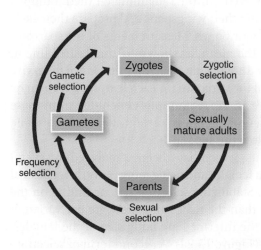

FIGURE 15.1 Simplified diagram of kinds of selection acting on the life stages of a multicellular organism during a single generation, from zygote to zygote. (Adapted from Christiansen, F. B. The definition and measurement of fitness. In *Evolutionary Ecology,* B. Shorrocks (ed.), Blackwell, 1984.)

Kinds of Selection

It is commonly assumed that selection is always for the optimum, weeding out the strong and eliminating the weak. As a broad generalization this is so; selection is associated with survival of the fittest. Selection, however, is the name given to a variety of processes that affect phenotypes in at least four ways, none of which are mutually exclusive.

- **Stabilizing selection** maintains phenotypes close to a mean.
- **Directional selection** operates when an extreme phenotype has adaptive value, or to eliminate deleterious mutations.
- **Disruptive (diversifying) selection** enables different phenotypes to do better in particular environments.
- **Sexual selection** operates differently on males and females of the same species.

Because environmental conditions are often variable, even on short time scales, these different types of selection do not remain separate, but combine in different ways.

A further and more controversial form of selection, **group selection**, has been proposed for selection on traits that benefit a group (usually a population) at the expense of individuals within the group. Situations in which an individual "sacrifices" themselves for the population (altruism) are perhaps the clearest example of the operation of group selection. The first four modes of selection are outlined below. Group selection is discussed later in the chapter in the context of adaptive landscapes.

Stabilizing Selection

When selection has operated on particular characters over many generations, most populations achieve phenotypes that are well if not optimally adapted to their surroundings; many phenotypes cluster around some value at which fitness is highest. Individuals that depart from these optimum phenotypes are expected to show less fitness than those closer to the optimal values.

A classic study documenting stabilizing selection in the wild came about as the result of a major storm, followed by a very cold first of February morning in 1898 in Providence, Rhode Island. Hermon Bumpus (1862–1943), a comparative anatomist at Brown University, found 136 dead or stunned house sparrows (*Passer domesticus*) on the ground. Seventy-two sparrows survived but the remaining 64 died. Bumpus had the foresight to weigh the birds, measure their wingspans and prepare their skeletons to measure lengths of some of the major bones. Measurements of the sparrows that *survived* the storm clustered around the mean, as shown by the dashed line in FIGURE 15.2a . Measurements of the same features on the sparrows that *failed to survive* the storm clustered around the extremes of the variation for each feature, as shown by the shaded portions of the curve in Figure 15.2c. As Bumpus concluded from these data, "it is quite as dangerous to be conspicuously above a certain standard of organic excellence as it is to be conspicuously below the standard."

Bumpus retained all the measurements, which have been reanalyzed in several independent studies, separating males from females and using more powerful statistical analyses that were available in the 1890s. Stabilizing selection was operating on the females, which were, on average, smaller than the males and show greater variation in body size; stabilizing selection favors the intermediate body size of the females, maintaining body size closer to the norm (Figure 15.2a). A second form of selection,

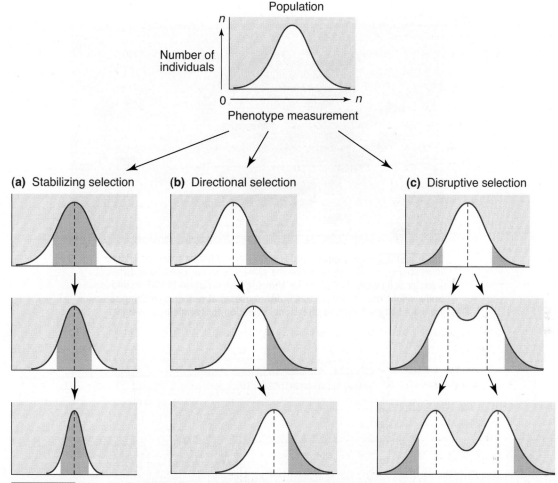

FIGURE 15.2 Three kinds of selection and their effects on the mean (dashed lines) and variation of a normally distributed quantitative character. The horizontal axis of each bell-shaped curve represents measurements of the character. The vertical axis represents the number of individuals found at each measurement. Shaded areas represent individuals selected as parents in experiments to demonstrate each type of selection.

directional selection (below) was operating on the males, which are larger and show less variation in body size (Figure 15.2b). The sensitivity of these two modes of selection is reinforced when the size differences between males and females in this sample are considered. Male weights are 29.41 ± 0.86 g and females 28.49 ± 1.72 g, differences that are indistinguishable to our eyes (**FIGURE 15.3**) but discriminated by modes of selection.

Directional Selection

Not all character selection is stabilizing. Selection can result in an extreme phenotype as organisms move in one or the other direction of a phenotypic distribution (Figure 15.2b). Animal and plant breeders, who select for extremes of yield, productivity or resistance to disease, commonly practice such directional selection (**FIGURE 15.4**). Its role in evolution is especially important when the environment of a population is changing and when one or only few phenotypes are adapted to new conditions (see Chapter 17).

Directional and stabilizing selection can operate sequentially, as has been shown in situations where different genotypes produce a single phenotype. **FIGURE 15.5**

FIGURE 15.3 Male (upper) and female (lower) house sparrows (*Passer domesticus*), which appear to be the same size to our eyes, have different average body weights with males showing less variation (±3%) around mean body weight than females (±6%), differences that reflect stabilizing selection for male body weight and directional selection for female body weight.

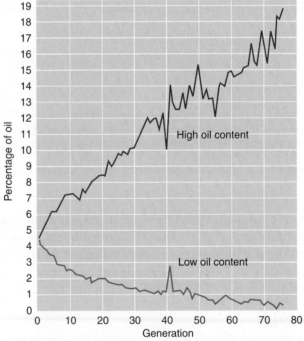

FIGURE 15.4 Results of selection for high and low oil content in corn kernels in an experiment begun in 1896 at the University of Illinois and continuing for 80 years. Selection for high oil content still continues to yield increases, whereas the effect of selection for low oil content tapered off on reaching the 0 percent lower limit. (Adapted from Dudley, J. W. "Seventy-six generations of selection for oil and protein percentages in maize." In *Proceedings of the International Conference on Quantitative Genetics,* E. Pollack et al. (eds.), Iowa State University Press, Ames, 1977.)

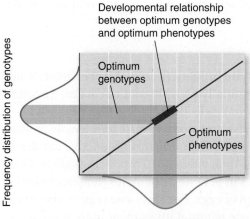

(a) Before canalizing selection

(b) Selection for canalization

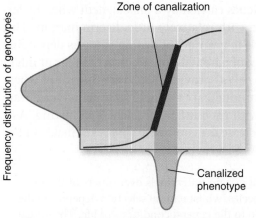

(c) After canalizing selection is completed

FIGURE 15.5 Sequence of selection for a canalized phenotype. The diagonal, running from lower left to upper right, represents the developmental relationship between genotype (*vertical axis*) and phenotype (*horizontal axis*); the steeper this developmental curve, the greater the number of different genotypes that can produce the same phenotype. **(a)** Before canalizing selection only a small section of the genotype produces the optimum phenotype. **(b** and **c)** As selection for canalization proceeds, the developmental curve assumes more of an "S"-shape as larger portions of genotypic variation produce the optimum phenotype, shown as a zone of canalization in **(c)**. (Adapted from Strickberger, M. W. *Genetics, Third edition*, Macmillan, 1985.)

diagrams such a scheme devised by the English-born but Australian-based geneticist James M. Rendel (1915–2001) in 1967. In a *Drosophila* population selected for a change in bristle number caused by a mutation in the *scute* gene, directional selection leads to alteration in bristle number. Stabilizing selection then maintains the new bristle number (Figure 15.2). Selection has affected both variation of the trait and the pattern of variability of the trait.

Disruptive Selection

Selection, whether stabilizing or directional, may act in a constant fashion if the selective environment is uniform. However, when conditions are changing, a population may be subjected to divergent or cyclically changing (oscillating) environments to which different genotypes among its members are most suited. Such selection is known as disruptive or diversifying because it establishes different optima within a population (Figure 15.2c).

Because of environmental variability or other changing biological conditions, one type of selection may cease and be followed by or overlap another. Disruptive

selection may be followed by directional selection, which may yield to stabilizing selection.

Sexual Selection

Mating systems are among the most important influences on species behavior. When selection operates preferentially and differentially on males or on females, we follow Darwin in referring to such selection as sexual selection.

Although both sexes benefit if their offspring survive, male and female animals often differ in the cost of reproduction; females begin their reproductive careers by investing more resources in producing eggs than males do in producing sperm. Thus, a female's genes benefit if she discriminates in her choice of mates (female choice) to protect her relatively expensive production of eggs. For this same reason, there is an advantage for females to seek increased male parental investment in their offspring, a strategy especially notable among mammals and birds, where reproductive success can depend on a relatively long-term commitment to their progeny.

Males, in contrast, may be more extravagant in disposing of their relatively inexpensive and more plentiful sperm. Genes carried by a male benefit when he fertilizes as many females as possible, often with relatively little discrimination. This conflict of interest between the sexes leads to a variety of mating patterns depending on a variety of factors, including the degree of parental care necessary for egg or infant survival, and which sex (or whether both sexes) provides such care. In groups where females are primarily responsible for parental care — including many vertebrate species — males are likely to compete with each other for success in mating. As a result, selection can occur for traits that increase the combative abilities of males and/or that increase their attraction to females. As Darwin put it,

> Sexual selection depends on the success of certain individuals over others of the same sex, in relation to the propagation of the species; whilst natural selection depends on the success of both sexes, at all ages, in relation to the general conditions of life. The sexual struggle is of two kinds; in the one it is between the individuals of the same sex, generally the males, in order to drive away or kill their rivals, the females remaining passive; whilst in the other, the struggle is likewise between the individuals of the same sex, in order to excite or charm those of the opposite sex, generally the females, which no longer remain passive, but select the more agreeable partners.

Much evidence exists for competition between males in those species where one male mates with many females. Males may have special competitive armaments, such as horns and antlers, in order to gain access to females, which can lead to considerable sexual dimorphism (FIGURE 15.6a). Longer horns in male dung beetles are a well-studied example (FIGURE 15.7). Males are generally larger than females in such groups; male elephant seals (*Mirounga leonina*) are eight times larger than females and male northern fur seals (*Callorhinus ursinus*) are seven times heavier than females (FIGURE 15.8).

Because of female choice, male ornaments can become quite conspicuous, as in the dramatic plumage of peacocks, birds of paradise and hummingbirds (Figure 15.6b). Such exaggerated traits, like the peacock's tail, are adaptive and selected because they signal to others that the carrier is in sufficiently good condition to expend the extra energy to produce and maintain the trait. However, although male decorative fea-

and that canalization can be selected for, there must be **canalized traits** and **canalizing selection**.

> . . . canalizing selection limits the expression of a trait, presumably by eliminating genotypes that would broaden phenotypic responsiveness to the environment and so broaden expression of the character, which could occur by eliminating genotypes that would permit an individual to respond to any genetic, epigenetic or environmental influences that would lead to greater variability in the phenotypic expression of the trait.

Among Waddington's experiments on canalization were demonstrations in *Drosophila* that phenotypically uniform expression of the normal *Ultrabithorax* gene in the face of environmental stress depends on other "background" genes. The genetic basis for such constant expression became apparent when specific genetic loci that support *Ultrabithorax* transcriptional stability were identified.

Traits elicited by environmental stimuli, such as *Ultrabithorax* or *Crossveinless* wings in *Drosophila*, can be genetically incorporated and appear developmentally without the environmental stimulus in an evolutionary process known as **genetic assimilation**. This is not a Lamarckian process of direct instruction by the environment, but occurs because of selection for genotypes capable of such response. The close fit between organismal flexibility and environmental change is a product of underlying genetic components. This was demonstrated beautifully using color polymorphism in the hornworm *Manduca quinquemaculata*. Color polymorphism is sensitive to environmental signals as a consequence of regulation of juvenile hormone. A polymorphic line, initiated by heat shock and then subjected to selection over multiple generations, produced the environmentally sensitive color morph in the *absence of heat shock*. The environmentally induced morphology had been genetically assimilated, which means that after generations of selection the morphology can arise in the absence of the environmental signal.

Selection and Fitness

Although hereditary transmission is through the genotype, selection is on the phenotype. To the extent that phenotypic differences are due to genetic variation, phenotypes differ in viability and fertility and so influence the frequencies of their genotypes.

The consequences of selection on a particular trait in a population are measured in terms of reproductive success or **fitness**. In their simplest form, fitness and selection are measured by the number of descendants produced by one phenotype relative to the number produced by another. Sustained selection on the phenotype can result in change in gene frequencies from generation to generation. Effects of selection are recorded as **selection coefficients** (s), giving the greatest affect on fitness a value of 1. The lowest selection coefficient would be 0, indicating no affect on fitness, an unlikely possibility.

As long as the selection coefficient does not change, equilibrium between two alleles is impossible without new mutations. This is not so if the heterozygote has superior reproductive fitness to both homozygotes, a situation known as **heterozygous advantage**, **heterosis** or **hybrid vigor**. Because heterozygous advantage is often expressed in characters that affect fitness — longevity, number of offspring produced, resistance to disease — it has been studied in some detail. An oft-cited example is the dramatic increase in agricultural yield of hybrid corn achieved by crossing selected

inbred lines. Beginning with an average yield of about 25 bushels per acre in the 1920s, hybrid corn enabled increases to as much as 140 bushels per acre.

An example of the interplay of factors affecting fitness is that fitness is often **frequency dependent**, especially in situations where a variant allele is either common or rare. Frequency dependent fitness has the advantage of allowing single evolutionary events — the origin and spread of a single mutation — to be modeled and compared with the evolutionary stable strategies discussed later in the chapter. Understanding frequency-dependent fitness therefore has the potential to inform us about branching points in evolution. So too does an approach that considers selection and fitness in terms of populations occupying adaptive peaks or valleys in an **adaptive landscape**.

■ Analysis of *frequency-dependent fitness* or *adaptive dynamics* originated in game theory. It is not for the mathematically faint-of-heart.

■ Adaptive Landscapes and the Shifting Balance Theory

Sewall Wright's shifting balance theory for maintaining alleles in a population was outlined in Chapter 13. According to this theory each genotype at a locus occupies an adaptive peak in an adaptive landscape, an adaptive peak being a position of high fitness associated with a specific environment (FIGURE 15.9). As long as no other factors change the fitness of these genotypes, each of the six peaks shown in Figure 15.9 will be of equal height. A population consisting entirely of any one of these genotypes would achieve maximum fitness for this phenotype.

The number of adaptive peaks increases astronomically with the number of loci and alleles; one locus with four alleles gives 10 gene combinations; 100 loci with four alleles each gives 10,100 gene combinations. Thus, even if only a small portion of gene combinations is adaptive, there are many more possible adaptive peaks than a species can occupy at any one time. To move from peak to peak until a population finds the

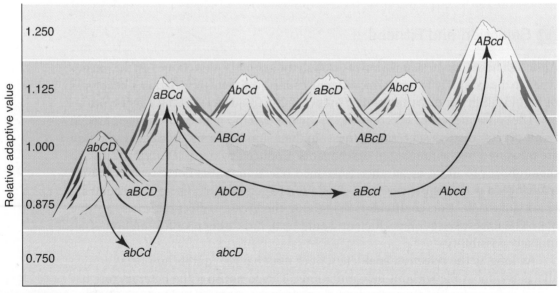

FIGURE 15.9 Adaptive landscape for homozygous genotypes at four loci in which six genotypes homozygous for capital letter alleles at two loci (AbcD, AbCd, aBcD, and so on) attain relatively high relative fitness (adaptive) values (peaks). Further differences among them are caused by the fitnesses of different alleles at the Aa and Bb loci. Genotypes bearing more capital letter alleles at these loci have higher relative fitness values than those that do not (for example, aBCd, abCD). (Adapted from Wright, S. Genic interaction. In *Methodology in Mammalian Genetics*, W. J. Burdette (ed.). Holden-Day, 1963.)

highest one demands that the population travel through genotypes that occupy the lower adaptive valleys of this adaptive landscape. Arrows in Figure 15.9 indicate such reductions in fitness, showing the general route a population with genotype abCD might take to reach the highest peak at genotype ABcd.

Movement of a population from peak to peak depends on various factors, especially the size of the population. Small populations are more subject to random fluctuations in gene frequencies (genetic drift; see Chapter 12) so their frequencies vary more easily than in large populations. As the environment changes, the relative fitness of genotypes changes. As new alleles are introduced, interaction with other alleles (epistasis) changes their relative fitness. What is a "valley" at one time need not be a valley at another, and a population's position on the landscape can fluctuate accordingly. Once such an adaptive landscape has evolved, further evolution will depend on the origin of a new selective environment and new adaptive peaks. However, if conditions are not changing rapidly, the same set of adaptive peaks may remain for long periods of time.

A potentially high adaptive peak for a population need not coincide with a high selective peak for a genotype within the population. This discrepancy arises because the selective values of genotypes are based on competition with other genotypes but may not indicate their effect on the population. Altruists (**BOX 15.1**) who sacrifice themselves for the benefit of shared genotypes may have low selective value as individuals, although a population bearing such altruistic genotypes may have higher reproductive values than one without them. Sterile castes among insects that forgo reproduction represent an example of altruism. Situations in which the altruist is not disadvantaged also have been proposed. When alarmed, many species of antelope "bounce" up and down with the joints of the legs locked and the legs in an extended posture. This behavior, called stotting or pronking, warns other individuals of danger, but, it has been argued, also alerts the predator that the individual "pronker" is aware of his/her presence, and so may not be an easy target for predation.

How can a population located on a relatively low adaptive peak evolve so that it occupies the highest or near-highest peaks on the adaptive landscape? To answer this problem, Sewall Wright proposed that many populations break into small groups of subpopulations. These demes are small enough to differ genetically because of random fluctuations in gene frequencies (genetic drift; see Chapter 12) but are not so widely separate as to completely prevent gene exchange and the introduction of new genetic variation. The adaptive landscape is therefore occupied by a network of demes, some at higher peaks than others. Selection takes place not only between genotypes competing within demes but also between demes competing within a general environment. Sewall Wright called the kaleidoscopic pattern of evolutionary forces acting on these demes the shifting balance process, outlined in its simplest form in Box 13.1. By contrast, Ronald Fisher suggested that most populations are large and fairly homogeneous, and that selection tests each new allele independently in competition with all other alleles in the population. This large population primarily increases its fitness by small, incremental ("additive") selective steps.

Investigation into the theories proposed by Fisher and Wright continues. Strong arguments against Wright's shifting balance theory point to its dependence on unsupported assumptions of genetic drift, selection and gene flow (see Chapter 12). Others have presented examples of deme structure and intergroup selection in support of Wright's theory. Until further evidence appears, perhaps both Fisherian and Wrightian populations should be assumed to exist, large and homogeneous at one time and

BOX 15.1

Altruism, Kin Selection and Reciprocal Altruism

Altruism

Social interactions can produce results that are positive or negative for individuals involved in the interactions, ranging from a mutual benefit, to a positive advantage to the performer but not to the recipient, to the negative effects of altruistic sacrifice to the performer but advantage to the other individual(s). The fitness effect of **altruism** on the altruist may be negative compared to selfishness, but it will have a positive effect on others in the population.

On the genetic level, an altruist really benefits any genes it shares with relatives, although possibly sacrificing its own. Thus, individuals in a socially interacting group containing many altruists are better off (achieve higher fitness) because the effect on others is more positive than in a group with fewer altruists. So, the success of a group depends on a *group property* (the frequency of individuals expressing certain behaviors) rather than on characters confined to only single individuals. As Darwin (1871) put it for humans

> It must not be forgotten that although a high standard of morality gives but a slight or no advantage to each individual man and his children over the other men of the same tribe, yet that an increase in the number of well-endowed men and an advancement in the standard of morality will certainly give an immense advantage to one tribe over another.

Information provided by an individual affecting the survival of other members of the group is dramatically revealed when a monkey encounters a leopard and reacts with a loud scream, signaling other monkeys in the group to take refuge. Although this warning signal may call the predator's attention to "the screamer" and diminish its chance for survival, the effect can help preserve other members of the group to which it is related.

Although unresolved, in some instances, for example, in a population with sexual reproduction and altruistic social behaviors, selection seems to occur for the benefit of the group. This is so even though altruism may harm an individual's own genetic future. Since the early 1930s, population geneticists have proposed that there are genetic advantages in altruistic behavior, in which individuals endanger their own survival for those who carry closely related genotypes. In the flour beetle, *Tribolium confusum,* egg-eating cannibalism by larvae declines in groups in which larvae only feed on genetically related eggs. This altruistic behavior of refraining from cannibalism is selected because it enhances the survival of related individuals that would be considered prey in the absence of altruism.

In 1964, William Hamilton (1936–2000) popularized this cooperative process under the name **kin selection**, and provided mathematical formulae by which some of its benefits could be evaluated. As John Maynard Smith (1920–2004) pointed out two decades later, "the main reason for thinking that kin selection has been an important mechanism in the evolution of cooperation is that most animal societies are in fact composed of relatives" (see Chapter 18).

Kin Selection

Advantages of kin selection are seen in the alarm calls of vervet monkeys (*Chlorocebus pygerythrus*) and Belding's ground squirrel (*Spermophilus bedingi*). Although these cries enhance the caller's danger, they provide the individual with indirect benefits by helping its genetic relatives.

BOX 15.1

Altruism, Kin Selection and Reciprocal Altruism (*continued*)

Because selection seems to operate on the populational level in some organisms, kin selection is used to explain social behavior in some lineages of hymenopteran insects: ants, bees and wasps. These social insects have haploid males and diploid females, a situation known as haplodiploidy (see Box 18.4). All females (sisters) derived from a single pair of parents are more closely related than are mothers to their own daughters. The genotype of the sisters as a group (*kin*) benefits from female workers who sacrifice their own reproductive ability. Instead of reproducing, they rely on their mother's reproductive ability by helping raise sisters rather than producing their own daughters.

Reciprocal Altruism

The situation in which individuals cooperate only with those who also cooperate is known as **reciprocal altruism**, a concept introduced by Robert Trivers in 1985.

Trivers hypothesized that altruism can become established in a group where the frequency of interaction among individuals is high and the life span sufficiently long to enable recipients of altruistic acts to return favors to the altruists. Because altruism may have significant benefits to the altruist when it is reciprocated, the benefits to individuals who partake in reciprocal altruism can far outweigh the costs. Cooperative defense roles taken on by members of the group, whether lions, primates, cattle, birds, and so forth, involve shared genetic survival. Fish aggregate in schools, the schools offering increased defense against predators compared to isolated individuals.

A similar type of selection probably occurs in distasteful prey species in which individuals with an **aposematic warning pattern arise**. In some species, all individuals are aposematic: yellow and black striped wasps, or brightly colored poisonous frogs and snakes (FIGURE B15.1). The warning pattern makes them more susceptible to predation, but through their death they protect related genotypes that carry the same pattern.

■ *Aposematic* is the term used for the warning colors that make a poisonous or distasteful animal conspicuous and recognizable to a predator.

(a) (b)

FIGURE B15.1 Warning (aposematic) coloration as seen in yellow and black striped wasps (**a**), and yellow-banded poison dart frog, *Dendrobates leucomelas* (**b**).

subdivided at another. Perhaps, as Wright stated, different aspects of populations demand different theoretical approaches.

One important consequence of Wright's shifting balance theory was to emphasize differences in survival or extinction among populations rather than only among individuals (see Table 1.1), an emphasis that seems to require selection on the group (group selection) rather than on the individual.

■ The various ways in which individuals can be differentiated from populations are outlined in Chapter 1.

Group Selection

An important consequence of Wright's shifting balance theory discussed above and outlined in Box 13.1 was to emphasize differences in survival or extinction *among populations rather than only among individuals*. Selection among individuals in a population is a conservative force that pushes the population up a single adaptive peak (Figure 15.9). Selection among populations (accompanied by genetic drift) leads to the occupation of higher adaptive peaks and the replacement, extinction, or colonization of populations at lower adaptive peaks.

Evolutionary biologists have long considered the extent to which group selection occurs, or whether it occurs at all. Among the chief argument against group selection is the claim that many so-called group adaptations arise from selection among individuals. Mathematical models show that selection among groups necessitates high extinction rates and practically no gene flow. Nevertheless, most would agree that group selection is at least theoretically possible, and from the examples below and others, group selection is receiving more attention than in the past. In part, this is because the process has been broadened to consider selection at levels other than individuals or populations, including selection between species (*species selection*) or accidental factors such as catastrophes that cause survival differences among species (*species sorting*). Other mass interactions, such as susceptibility or resistance to parasitic infection, also can involve group survival or extinction.

Three major situations susceptible to explanation by group selection have been considered.

1. **Sexual reproduction** may involve hazards as well as significant expenditure of resources with the consequence that individuals often incur considerable disadvantages. Benefits in the evolution of sexual reproduction therefore seem most likely related to variability conferred on the sexual population as a whole.
2. **Kin selection**, which is used by population biologists to explain social behavior in some lineages of ants, bees and wasps (hymenopteran insects; Box 15.1 and see Box 18.4).
3. The third, **altruism** and **reciprocal altruism**, are situations in which individuals cooperate only with those who also cooperate with them (Box 15.1).

Although the importance of group selection is unresolved, in a population with sexual reproduction and altruistic social behaviors, selection seems to occur for the benefit of the group, even though it may harm an individual's own genetic future. Such situations remind us that fitness can be affected by behavior in various ways: courtship display and mating behaviors, feeding and care of offspring by individuals other than the parents (as occurs in birds), altruism, behavioral mimicry, schooling or swarming and aggression are some examples. Such relationships often are gathered under the concept of **evolutionary stable strategies**.

Evolutionarily Stable Strategies

Social relationships in which members of a group or species interact for breeding, feeding and defense, are aspects of behavioral evolution given much attention. Behaviors involved in social interactions among primates are discussed in Chapter 20, but many behaviors, such as cooperation and dominance relations, apply to other social

groups. Because the evolutionarily stable investment in competition between groups increases as competition within groups increases, evolutionary stable strategies can lead to a sophisticated level of interactions that allows, for example, social insects to be modeled as superorganisms.

In evaluating social interactions between individuals, various models dealing with the advantages of cooperation as opposed to the advantages of individual action have been proposed. For example, when two individuals confront each other in a social group there may be conflicting interests between cooperation (in which each individual gains some advantage) and individual action (in which one individual gains greater immediate advantage than it could by cooperation). Cooperation can become an especially effective strategy, even between strangers.

Among evolutionarily stable strategies is one called "**tit for tat**," in which an individual behaves cooperatively in a "game's" first move or interaction and then repeats its opponent's previous move. Thus, an opponent acting selfishly is punished by a selfish response, while cooperative behavior is rewarded by a cooperative response. As pointed out in 1993, "the advantage of tit for tat lies in it being quick to retaliate and quick to forgive." A danger arises in tit for tat strategy, however. Incorrectly evaluating an opponent's response as selfish can cause a cycle of retaliatory moves until a random correction occurs that reestablishes cooperation. One solution is called "*generous tit for tat,*" in which the probability of correcting such mistakes is greater than chance, because opponents occasionally overlook selfish responses. It has also been argued that "tit for tat" is highly susceptible to any errors in implementing the strategy; one mistake made by either participant and the strategy will fail.

When these concepts are extended to repetitive interactions among more than one pair of individuals, further strategies may develop, such as reciprocal altruism, in which individuals cooperate only with those who cooperate with them (Box 15.1).

■ Theorists have hotly evaluated the relationship of social (sociobiological) determinants to human behavior and culture, a topic discussed in Chapter 20.

Selection and Balanced Polymorphism

Polymorphism was introduced in Chapter 12. The persistence of different genotypes through heterozygote advantage is an example of **balanced polymorphism**, a term coined by the Oxford ecological geneticist E. B. Ford (1901–1988) to describe the preservation of genetic variation through selection. A gene locus is polymorphic if at least two alleles are present, with a frequency of at least one percent for the second allele (of the least frequent allele when more than two alleles are present). As discussed in Chapter 12, polymorphisms are ubiquitous in practically all populations examined.

One well-known example of polymorphism is the *sickle cell* gene in humans, where heterozygotes survive the malarial parasite more successfully than do either of the homozygotes (see Chapter 12). As shown in FIGURE 15.10, this gene — and others that appear to offer protection against malaria (also shown in Figure 15.10) — persists in notable frequencies in geographical areas where malaria is endemic. Of course, the malarial parasite has evolved multiple genes to evade the immune response mounted by the host. Individual *Plasmodium falciparum* parasites contain some 60 genes for one such protein. Different individuals contain different variants of the genes, providing an impressive arsenal mounted by the parasite in the parasite-host arms race.

A second example at the phenotypic level is selection of cryptic moth prey by blue jays (*Cyanocitta cristata*). Jays capture and consume more of the abundant prey

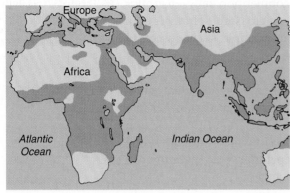

(a) Falciparum malaria

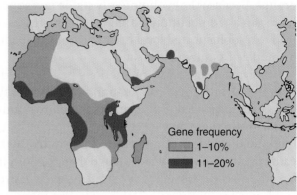

Gene frequency
1–10%
11–20%

(b) Sickle cell anemia

(c) β-Thalassemia

(d) G-6-pd deficiency

FIGURE 15.10 Relationship between the geographic distributions of the most dangerous form of malaria, falciparum malaria, caused by the protozoan parasite *Plasmodium falciparum* and genes that confer resistance against malaria. (**a**) Distribution of falciparum malaria in Eurasia and Africa before 1930. (**b**) Distribution of the gene for sickle cell anemia (*Hbˢ*). (**c**) Distribution of the gene for β-thalassemia. (**d**) Distribution of the sex-linked gene for glucose-6 phosphate dehydrogenase deficiency in males in frequencies above 2 percent. (Adapted from Strickberger, M. W. *Genetics, Third edition*. Macmillan, 1985.)

types than of the rare types, switching capture strategy in response to changes in abundance of prey types.

Selection and balanced polymorphism can be demonstrated in laboratory experiments. When two types of *E. coli* are co-cultured in glucose-limited minimal medium, the relative frequency of each is frequency dependent, each type having an advantage when rare, maintaining a balanced polymorphism in the population.

It has been proposed that not all polymorphism is caused by selection. In 1979 Motoo Kimura (1924–1994) proposed that the rate of evolution at the molecular level is far more rapid than been thought. Because of the presumably associated high cost of selection, Kimura proposed that most amino acid changes are neutral in effect. Kimura argued that as selection does not act on neutral mutations, the fixation of such alleles incurs no genetic load and depends only on their mutation rates and on random changes in gene frequency.

This **neutralist hypothesis** — called by some non-Darwinian evolution because of its dependence on mutation and not selection — seemed supported by the extensive

degrees of enzyme and protein polymorphisms that demonstrate that allele differences persist at many thousands of loci in many species. Such polymorphism must consist of neutral mutations, according to Kimura and his followers. High frequencies of selectively neutral mutations in a rapidly mutating strain of *Escherichia coli* support the neutralist position.

However, recent extensive studies demonstrate functional consequences of much polymorphism: A survey of 35 species from Israel, including insects, mollusks, vertebrates, and plants, concluded that the amount of allozyme polymorphism in a species correlates to factors such as life habit and climate. Perhaps even more striking is the finding that clones of the bacterium *E. coli* isolated from the intestinal tracts of animals as diverse as lizards and humans, and from localities as widespread as New Guinea and Iowa, share common allozyme frequencies. For each of five different enzymes, one particular electrophoretic band was frequent in almost all samples. Because other allozymes of these proteins exist, the finding of such a narrow distribution of allozymes is difficult to explain on any basis other than selection.

The discovery that many genes do not code for proteins is taken by most evolutionary biologists to further undercut the neutralist hypothesis. In general, nucleotide sequence analyses show that polymorphisms are significantly greater in those DNA sequences that do not determine amino acid sequences than in those DNA sequences that do. This finding suggests that selection reduces the variability in amino acid coding regions because such sequences have a greater effect on the phenotype than do noncoding regions. For the same region, more polymorphism occur in introns (intervening sequences; see Chapter 9), which do not code for amino acids, than in exons (expressed sequences), which do. Both these expectations are fulfilled for the *D. melanogaster* alcohol dehydrogenase gene, which has six percent polymorphism among introns, seven percent polymorphism among third-codon positions but almost no polymorphism in exons. Because random events can hardly explain such pronounced differences in polymorphism, these findings strongly suggest that selection must be the discriminating agent that determines which nucleotide base substitutions will become established in functionally different DNA sequences.

Selection and Mimicry

Other conditions responsible for polymorphism include a change in selection coefficients so that genes detrimental at one time are advantageous at another. Selection against a gene may depend on its frequency and may change or even be reversed when the gene is at low frequency, before it can be eliminated. An example of such **frequency-dependent selection** is **Batesian mimicry**, a mechanism proposed by the explorer Henry Bates (see Chapter 6) in which palatable species that mimic distasteful models are protected against predators (FIGURE 15.11a).

In general, the more frequent the mimic and the less frequent its model, the greater the chances that predators will attack the mimic. Conversely, the less frequent the mimic compared to the model, the greater the chances that the mimic will be protected. Mimicry among insects evolved early, and, as you might expect if mimicry is effective, has been remarkably stable. A leaf-mimicking stick insect, *Eophyllium messelensis,* from the Eocene, 47 Mya, closely resembles extant insect mimics. Furthermore, this insect mimicked the leaves from several types of plants (myrtle and laurel trees, and alfalfa), indicating that sophisticated forms of mimicry were established in the Eocene.

■ Strictly selection does not act on mutations, whether neutral or not, but acts on the phenotypes that result from the incorporation of mutations. In this sense, Kimura's hypothesis can be said not to depend on selection.

■ Allozymes are different forms of the same enzyme coded for by different alleles of the same gene.

■ Because of their composition, particular DNA sequences — for example, guanine-cytosine rich sequences that replicate early during DNA synthesis — may mutate at different rates and in different directions from other sequences.

(a) Batesian mimicry

Danais plexippus
monarch butterfly
(unpalatable model)

Limenitis archippus
viceroy butterfly
(palatable mimic)

(b) Müllerian mimicry

Heliconius eucrate
(unpalatable)

Lycorea halia
(unpalatable)

FIGURE 15.11 Mimicry in different species of butterflies. (**a**) Batesian mimicry by a North American species, in which the more palatable viceroy butterfly (*right*) mimics the more unpalatable monarch (*left*). Resemblance between two South American unpalatable species in (**b**) provides a common warning pattern to predators and helps protect both prey species (Müllerian mimicry). Whether Batesian or Müllerian, mimicry is one of the most obvious examples of convergent evolution.

As shown in Figure 15.11b, mimicry between different species — **Müllerian mimicry** — benefits both species by enabling predators to learn a single warning pattern that applies to all these potential but distasteful prey. When rare, conspicuous warning patterns on unpalatable individuals probably offer little protection because predators have few chances to learn their distastefulness. Distinctive patterns, however, offer greater protection to unpalatable individuals when they are at higher densities, as shown by the evolutionary ecologist, Gregory Sword, who discovered that when a Texan grasshopper (*Schistocerca emarginata*) grows up alone it develops colors that blend in with the vegetation. When the grasshoppers grow up in groups, they develop yellow and black warning colors if they consume a plant that is toxic to predators.

Selection and Industrial Melanism

A population sufficiently widespread to occupy many environments may maintain a variety of genotypes, each of which is superior in a particular habitat. Spatial and temporal organization of the environment may significantly affect the extent to which a population will rely on genetic polymorphism as an adaptive strategy. Coarse-grained environments, in which different individuals in a population are exposed to different experiences, promote greater genetic polymorphism than fine-grained environments, in which all individuals experience few environmental differences. A prominent example is polymorphism associated with **industrial melanism**. In the 19th century, certain moths and butterflies showed increased frequency of a dark-colored (melanic) morph (**FIGURE 15.12**), a change that occurred in industrial

(a) **(b)**

FIGURE 15.12 Dark-colored and light-colored tree trunks, one with a melanic **(a)** and one with a non-melanic **(b)** peppered moth (*Biston betularia*) placed to show the contrast between them. The light-colored trunks derive their appearance from lichens. Although trees are commonly shown as resting sites for these moths, their actual resting habits are unknown.

areas of England where air pollution darkened vegetation because of coal smoke deposits.

In the early 1970s in the industrial city of Birmingham, Bernard Kettlewell (1907–1979) and others released known numbers of light and melanic morphs of the British peppered moth, *Biston betularia* into the countryside. When they recovered the moths, they recaptured a significantly greater proportion of melanic than light forms. Their data suggested that sooty areas offer greater protection to melanic than to light-colored morphs — the light forms are at a selective advantage — so that more melanic than light morphs survive to be recaptured. The relative fitness of the melanic morphs may lie, at least partly, in their ability to remain concealed from bird predators on darkened twigs or tree trunks (Figure 15.12). In non-industrial areas, in contrast, trees covered with normal gray lichens offer decided advantages to the light-colored moths.

Whether because of environmental camouflage or unknown factors, English populations of the peppered moth show various degrees of polymorphism, ranging from high frequencies of the melanic form in industrial areas to almost zero in many rural areas. Based on the dates when specimens were collected by amateurs, and on records in museum collections, and assuming one generation per year, it is estimated to have taken about 40 generations during the 19th century for the frequency of non-melanic phenotypes to decrease in some industrial areas from about 98 percent to 5 percent. The selection coefficient against a non-melanic gene in some of these industrial areas was fairly intense, about .2 or somewhat greater. Passage of clean air legislation in Britain in 1956 reduced industrial smoke in many formerly polluted areas. This reduction in pollution altered selection and the frequency of melanic forms of the peppered moth and of other insects declined dramatically.

A parallel situation is seen in light and dark forms of the rock pocket mouse, *Chaetodipus intermedius* in the American southwest. Mice with light sandy colored fur are found on sandy substrates, mice with dark (melanic) fur on black lava flows (FIGURE 15.13), adaptations that reduce predation from predation such as owls, other raptors and mammals that hunt using vision. Ability to maintain the two fur colors has been traced to the gene *MC1R* that codes for the melanocortin-1-receptor. MC1R plays a key role in whether melanin granules will be deposited into the fur in rock pocket mice and in many other animals. Four amino acid differences separate the *MC1R* gene in

■ Although some of the "classic" images of this example of industrial melanism were "staged" — moths were placed on trees and photographed — the evolutionary mechanism is real and substantiated with numerous examples.

(a)

(b)

FIGURE 15.13 Rock pocket mice (*Chaetodipus intermedius*) that forage on sandy substrates have sandy colored fur (**a**), and those that forage on black lava flows have black fur (**b**).

light and dark rock pocket mice, 63 differences separate wild mice (*Mus musculus*) from rock pocket mice. A single amino acid substitution in the beach mouse, *Peromyscus polionatus* is associated with adaptation to a changing environment. We can conclude that small genetic changes can lead to significant phenotypic change, illustrating selection and evolution in action.

Recommended Reading

Endler, J. A., 1986. *Natural Selection in the Wild.* Princeton University Press, Princeton, NJ.

Fisher, R. A., 1930. *The Genetical Theory of Natural Selection.* Clarendon Press, Oxford, England (2nd ed., 1958, Dover, New York).

Gillespie, J. H., 2004. *Population Genetics: A Concise Guide,* 2nd ed. Johns Hopkins University Press, Baltimore, MD.

Grant, P. R., and B. R. Grant, 2008. *How and Why Species Multiply: The Radiation of Darwin's Finches.* Princeton University Press, Princeton, NJ.

Hallgrímsson, B., and B. K. Hall (eds.), 2003. *Variation: A Central Concept in Biology.* Elsevier/Academic Press, Burlington, MA.

Kettlewell, H. B. D., 1973. *The Evolution of Melanism.* Clarendon Press, Oxford, England.

Majerus, M. E. N., 1998. *Melanism: Evolution in Action.* Oxford University Press, Oxford, England.

Williams, G. C., 1996. *Adaptation and Natural Selection.* Princeton University Press, Princeton, NJ.

On the Nature and Origin of Species

16 Species and Similarity: On Being the Same Yet Different

KEY CONCEPTS

- Understanding the nature and reality of species has been a long-standing issue.
- Species are biological units; classification is our way of relating species one to another.
- Species are classified on morphological grounds but increasingly are compared to other species using both morphological and molecular data.
- Biological and evolutionary species concepts define species as interbreeding or evolutionarily isolated units, respectively.
- Any comparison between any two biological units is an implicit statement about their relationships.
- Homology is similarity resulting from recent common ancestry.
- Parallelism is the evolution of similar features in closely related lineages.
- Convergence is when similar features evolve by different means in independent lineages.
- Identifying and separating these different categories of similarity is critical when determining evolutionary origins and relationships and when constructing phylogenetic trees.

Above: Species are distinct and distinctive as this blue-footed booby (*Sula hebouxis*), named for the Spanish *bobo* for stupid fellow, seems to realize.

Overview

It comes as a surprise to many people that the question of the reality of species has been an issue for millennia. Before Darwin, classification involved observations of similarities and differences among organisms without regard to the origins of the organisms. Since Darwin, however, species have been arranged in ways that reflect similarity and evolutionary relationships.

Different species concepts reflect emphases on different aspects of species (and of speciation). For multicellular organisms (plants, fungi, animals), morphological characters are used to define morphological species (morphospecies). Since the advent of molecular phylogeny, species and lineages are usually arranged into phylogenetic trees using morphological and molecular criteria, although the identification of species remains centered on morphological characters. The ability of populations to interbreed and produce viable offspring is the primary criterion for determining species boundaries using the biological species concept. The evolutionary species concept defines species on the basis of their evolutionary isolation from one another using morphological, genetic, molecular, behavioral and ecological data. As with any species concepts, the evolutionary species does not resolve all the problems intrinsic to identification and relationships; for example, not all traits evolve at the same rate or at the same time (mosaic evolution).

Central to any comparisons in biology is determining whether similar appearance is a consequence of shared evolutionary ancestry. Phenotypes may be similar because of recent shared ancestry (homology), because closely related lineages develop similar features (parallelism) or because similar features evolve by different means in independent lineages (convergence).

Documenting, recording and comparing similarities and differences in morphology and/or function among and between organisms is the traditional province of **classification** (see Chapter 6), which has two roles

- to identify and describe the basic unit of classification, the **species** and
- to order species into **classificatory systems** that, ideally, connect them into **phylogenetic trees**.

This chapter presents different concepts of "what is a species?" and how relationships and phylogenies are determined.

To ensure we are comparing apples with apples and frogs with frogs, the concept of **homology** is discussed in some depth. Homology allows us to identify granny smith apples as homologous with golden delicious and as not homologous with the green toad. How we do this is not as easy as this trivial example makes it out to be.

■ *Systematics, classification, and taxonomy* are often used interchangeably.

■ Whenever any two biological objects are compared, we make an explicit assumption about their essential similarity, that is, their homology. Homology is central: apples are apples and oranges are oranges, but a bird is not a bat.

Classification and the Reality of Species

Considerable difficulties may be associated with how species are distinguished one from the other and placed into groups that reflect their most significant features and relationships. Without a rational system of classification, evolutionary relationships between most species would have been impossible to establish. Nevertheless, recognition of the biological importance of species took time. The centuries-old problem of changing views on the nature of species was introduced in Boxes 5.2 and 5.3; species

as imperfect representations of the true essence (Plato) or immutable members of a Great Chain of Being (Aristotle).

In Europe during the middle Ages, species were collected and described on the basis of their culinary or medical properties. The discovery of new lands, floras, faunas and species of plants and animals as a result of the expansion of worldwide exploration and trade in the 16th and 17th centuries greatly increased the known diversity of organisms but raised issues about relationships. As Thomas Moufet (1553–1604), a prominent sixteenth century English entomologist, described grasshoppers and locusts

> Some are green, some black, some blue. Some fly with one pair of wings, others with more; those that have no wings they leap, those that cannot either fly or leap, they walk; some have longer shanks, some shorter. Some there are that sing, others are silent. And as there are many kinds of them in nature, so their names were almost infinite, which through the neglect of naturalists are grown out of use.

■ Carl von Linné, usually known by the Latinized form of his name, Carolus Linnaeus, inherited his love of plants and their names from his father, Nils Ingemarsson Linnaeus, a Lutheran pastor.

The method of classification devised by Carl Linnaeus (1707–1778), the founder of systematics, began with a precise description of each species, which were then grouped by their morphology into a **genus** (plural **genera**). Linnaeus grouped related genera into **orders**, orders into **classes**, and established a system of **binomial nomenclature** in which each species name defines its membership in a genus and provides it with a unique species name and identity, for example, *Homo sapiens*, modern humans. Designating the species as the basic unit of classification enabled Linnaeus to arrive at groupings far more natural in their interrelationships than many of the schemes proposed before. Even though his system treated species as ideal forms (see Box 5.2), his classification was a crucial step on the path that revealed natural evolutionary relationships between organisms (see Chapter 6).

■ Less well known is that Linnaeus spent much of his life attempting to organize the economy of Sweden according to scientific principles, to adapt crops such as rice and tea to grow in the Arctic tundra, and to domesticate elk, buffalo and guinea pigs as farm animals.

Late in life Linnaeus toyed with the concept of **transitions between species**. For much of his career, however, he conceived of species as fixed entities, although he was aware that varieties within a species could show considerable differences. One example is his subdivision in 1758 of modern humans into four races: Asiatic, American, European and African. A second example is his designation in 1753 of the species *Beta vulgaris* for beets, whose cultivated varieties such as spinach beet, chard, beetroot, fodder beet and sugar beet, were given varietal names; *Beta vulgaris perennis* for sea beet, for example.

Under Linnaeus, classification developed rapidly as many species were described (mainly on the basis of their reproductive parts) and classified into groupings. In accord with idealist concepts, each species was believed to possess a unique "essence" that determined its specific characters (see Box 5.2). This essentialist (typological) view was reinforced by taxonomists who deposited type specimens in museums or in herbaria to be used as the standards (types) for classifying further specimens.

Although Linnaeus placed special emphasis on the species as the practical unit of classification, Buffon (see Chapter 2) codified the notion that species are the only biological units that have a natural existence. Buffon also introduced the (modern) idea that species distinctions should be based on whether there were reproductive barriers to interbreeding (evidenced by whether fertile or sterile hybrids were produced).

> We should regard two animals as belonging to the same species if, by means of copulation, they can perpetuate themselves and preserve the likeness of the species; and we should regard them as belonging to different species if they are incapable of producing progeny by the same means.

To Buffon, considerable variation could occur between individuals of a species, perhaps eventually even producing completely new varieties. Despite such variation, a species remained permanently distinguished from other species.

Strangely enough, the eighteenth-century barrier to the acceptance of evolution seemed to rest mostly on the **reality of species**. If species were indeed real, they seemed inevitably fixed. How could new species arise? Buffon, who had proposed evolutionary events on cosmological and geological scales (see Chapter 2), established three basic arguments *against* biological evolution, arguments that were used by antievolutionists well into the nineteenth century.

1. New species have not appeared during recorded history.
2. Although mating between different species fails to produce offspring or results only in sterile hybrids, this mechanism could certainly not apply to mating between individuals of the same species. How could individuals of a single species be separated from others of the same kind and become transformed into a new species?
3. Where are all the missing links between existing species if transformation from one to the other has taken place?

These arguments were not refuted until after Darwin. Thus, it is no surprise that one of the first serious pre-Darwinian proponents of biological evolution, Jean-Baptiste Lamarck proposed that one must do away with the concept of the fixity of species to establish the possibility of evolution. The differences between species, genera, families and so on were only apparent, not real. All intermediate forms existed someplace on Earth, although they were not necessarily easy to discover. Lamarck shared the concept that species do not become extinct (as embodied in the Great Chain of Being; see Box 5.3) but did not believe that species were separately created, proposing rather that they had evolved from each other. His branching classification of animals (FIGURE 16.1) introduced a direct challenge to the layered Scale of Nature (see Figure 6.1), which goes in only one direction, from imperfect to perfect (see Box 5.3).

Aware that species were not immutable, Lamarck evaded the problem of stasis by proposing that species are arbitrary units and so not "real," and that there must be forms intermediate between species. Lamarck's view of a species was very different from Charles Darwin's. If species were arbitrary (Lamarck), then species never went extinct but instead evolve into other "species." Before Darwin, classification involved observations of similarities and differences among organisms, without regard to their origins. Since Darwin, species have been arranged in ways that reflect evolutionary relationships.

Classification and Phylogenetic Relationships

Natural historians — which is what biologists were called before the words biology and biologist came into common usage in the early twentieth century — formulated techniques of classifying organisms much earlier than they accepted the concept that the similarities and differences among species arose from evolutionary causes. It is, in fact, fairly easy to classify organisms in ways that distinguish among them but do not necessarily reveal their common origins. For example, fish and whales can be classified in one group, butterflies and birds in another and squirrels and

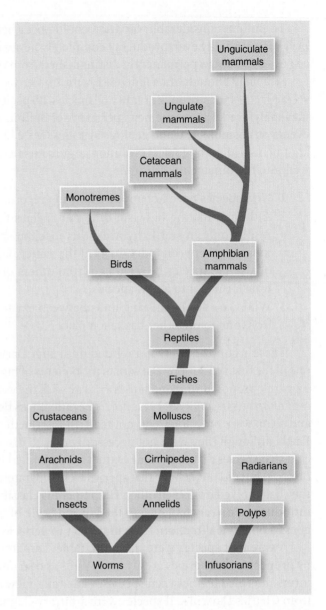

FIGURE 16.1 Evolutionary relationships among animals according to the view developed by Jean-Baptiste Lamarck in the late 18th and early 19th centuries.

monkeys in a third, using the criteria swim, fly and possess hair and live in trees, respectively.

With the advent of Darwin's theory that all organisms represent a single tree of life (see Chapter 7), natural historians had to decide whether they could or should use classification to reflect evolutionary relationships. There were strong indications that many organisms, grouped together because they possessed a large number of similar features, could also be said to have descended from a common ancestor. But such determinations revealed only one side of evolutionary classification. Another was to discern **lines of descent** among groups that perhaps shared only a few features. As Darwin put it, "Our classification will come to be, as far as they can be so made, genealogies . . . we have to discover and trace the many diverging lines of descent in our natural genealogies, by characters of any kind which have long been inherited."

One reason for the difficulty paleontologists have in determining phylogenetic relationships using morphological data is the difficulty in finding an unbroken line of ancestors connecting different groups. The fossil record may be fairly complete for some groups but far less complete for others. A classic and complete record is 60 million years of evolution of horses (BOX 16.1).

BOX 16.1
Horse Evolution

One year after the publication of *Origin of Species,* Richard Owen described the earliest known horse-like fossil, *Hyracotherium,* often referred to as *Eohippus* (the dawn horse).

Hyracotherium was some 50 cm high (the height of a good-sized dog), weighed about 23 kg, had four toes on its front legs and three on its hind legs (modern horses only have one toe on each foot), and simple teeth adapted for browsing on soft vegetation (FIGURES B16.1 and B16.2). Later fossil finds indicate that *Hyracotherium* was present from North America to Europe as a number of herbivorous species, some no larger than an average-sized modern-day fox.

In the approximately 60 My after *Hyracotherium* arose, horses changed radically. They now run on hard ground, chew tough, silica-containing grasses, and show special adaptations for this particular environment. Their elongated legs are built for speed, bearing most of the limb muscles in the upper part of the legs, enabling a powerful, rapid swing. Today, horses are the only land vertebrates with a single toe on each foot (Figure B16.2). This arrangement, coupled with a special set of ligaments, provides horses with a pogo-stick-like springing action while running on hard ground. Their teeth also show exaggerated qualities adapted for chewing tough, abrasive grasses, being much longer than the teeth of other grazers.

By the 1870s, paleontologists such as America's Othniel C. Marsh (1831–1899) were able to use fossils of North American and European horses to present the first classic example of a stepwise evolutionary tree among vertebrates, showing various transitional stages (Figures B16.1 and B16.2). Remarkably, almost all the intermediate stages between *Hyracotherium* and the modern horse, *Equus* are known, including transitions from low- to high-crowned teeth, browsers to grazers, pad-footed to spring-footed, and small- to large-brained. As shown in Figure B16.1, evolutionary changes among these forms did not proceed in a single direction, being better represented as a "bushy" family tree. Horses evolved adaptations for their habitats in different ways with some lineages maintaining distinct structures until they went extinct.

Although all occupied the same general area, horses made use of different environmental resources (*resource partitioning*). Some species became browsers, feeding on shrubs and trees. Others remained grazers, feeding on grasses. Still others grazed and browsed; reversion from grazing to browsing occurred in some Florida species. Differences in feeding habits can be deduced from dental scratches caused by grazing and dental pits caused by browsing, and from differences in the carbon isotope ratios ($^{12}C/^{13}C$) of grasses and shrubs; different diets produced different $^{12}C/^{13}C$ ratios in teeth.

The rate of evolution for any particular trait among the various branches was not constant. Size, for example, underwent relatively few changes for the first 30 million and the last few million years of horse evolution. Even when evolution was proceeding rapidly, as it did during the Miocene, both small- and large-sized species evolved. The finely detailed phylogeny of horse evolution encompasses hundreds of fossil species and is one of the best illustrations of some of the realities and complexities of adaptive radiation in evolution.

(continued)

■ *Eohippus* is an alternate but later genus name for *Hyracotherium*. Priority goes to *Hyracotherium*.

■ Analysis of mtDNA from 22 fossil horses by Orlando et al. (2009) has revealed two new species of horses and revised the patterns of relationships known previously only from the fossil record.

BOX 16.1
Horse Evolution (*continued*)

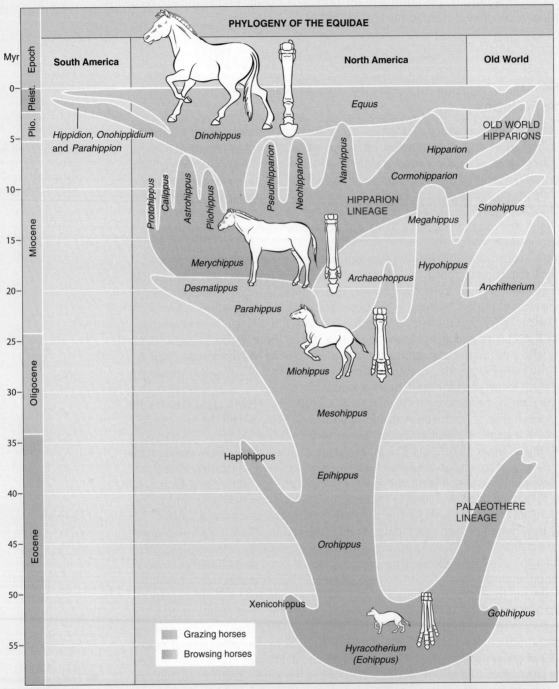

FIGURE B16.1 Evolutionary relationships among various lineages of horses, with emphasis on North American and Old World groups. Sample reconstruction of the digits ("toes") of the hind feet of some fossil horses and of the extant horse *Equus* are shown. The number of digits declined from four to one during evolution of the lineage. These horse lineages show both branching and non-branching patterns of evolution. See also Figure B16.2. (Adapted from MacFadden, B. J. *Fossil Horses: Systematics, Paleobiology and Evolution of the Family Equidae.* Cambridge University Press, 1992.)

BOX 16.1
Horse Evolution (*continued*)

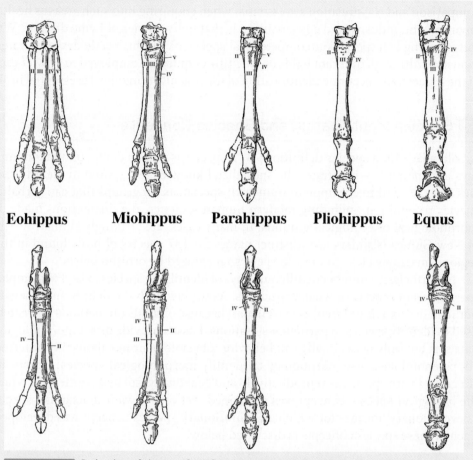

Eohippus Miohippus Parahippus Pliohippus Equus

FIGURE B16.2 Reduction of the toes from four to one in both forelimbs (top row) and hind limbs (bottom row) in horses from the Eocene "dawn horse," *Eohippus* (*Hyracotherium*), to the modern horse genus *Equus,* which appeared in the Pleistocene and has persisted to today. Digit III is retained in all, digits II and IV are reduced to splint bones in *Parahippus* and *Equus*, while digits I and V (the outer digits) were lost as early as *Miohippus*. See also Figure B16.1. (Adapted from Gregory, W. K., 1951. *Evolution Emerging. A Survey of Changing Patterns from Primeval Life to Man*. Two Volumes. The Macmillan Company. New York.)

Many extant species are separated by barriers that are seasonal or behavioral. Therefore, we cannot necessarily be confident that a species identified in the fossil record was not also subdivided into subspecies by seasonal or behavioral barriers with consequent lack of reproduction across the entire species range. This is where molecular data from extant genera/species can be used to construct a tree of phylogenetic relationships against which fossils can be compared. Fossils serve several important roles, including providing the physical record of organismal evolution, the historical order in which characters were acquired or lost, and a basis for the calibration of molecular-based trees.

Because knowledge is never stagnant, classifications of groups of organisms change over time. The single genus *Rubus* (blackberries, raspberries, loganberries) has been

divided into 500 species by one botanist, 200 by another and 25 by a third. While this means that you have to run to keep up with the latest work, it undercuts neither the validity nor the importance of taxonomic analyses. Taxonomic identification, comparison and classification are essential when enquiring into such areas as health, environment, industry and conservation. Is that tick a carrier of Lyme disease? Will this invading fish species outcompete local species? Will this beetle destroy a forest industry? Although different fields of scientific enquiry are employed to answer each of these questions, accurate identification of the organisms involved is essential in all.

Species Identification and Species Concepts

An old adage for a species definition — "a species is what a competent taxonomist says is a species" — illustrates the specialized knowledge required to identify species and the need to continue to train such specialists. In groups that can verbalize species recognition, including modern human societies, the distinctions made on morphological or ecological grounds, in many cases, are strikingly similar. A tribe of New Guinea islanders uses distinct names for 137 species of birds found in this region, amazingly close to the 138 species recognized by ornithologists.

Nevertheless, a number of different ways of identifying species exist. Each emphasizes different aspects of what a species is. As we saw with the definitions of a gene outlined in Box 1.2, differences exist in part because of limits on methods that can be used to identify species in particular situations: Fossils provide direct morphological data sets but only occasionally can behavior, physiology or metabolism be inferred. So, paleontologists use morphology to identify **morphological species**. Others are concerned with species as reproductively and genetically isolated entities, for which the **biological species** concept was developed. Yet others want to examine species as evolutionary lineages for which an **evolutionary species** concept was developed. Each of these species concepts is discussed below.

Morphological Species

For multicellular organisms in all kingdoms, morphological characters of adults and other life history stages have long been used to define species; Linnaeus delineated species using morphological characters. Such species are now referred to as *morphological species* or *morphospecies*, a morphological species being a population of individuals that share more characters (features) with one another than they do with any other organism. Comparative anatomy and comparative embryology are central to the identification of morphospecies.

Comparative Anatomy. Comparative anatomy is the study of comparative relationships among anatomical structures in the adults of different species.

Comparative anatomists followed the logic that organisms with shared homologous structures shared a common ancestor. (Homology is discussed below.) Organisms with nonhomologous or unlike structures represented divergent evolutionary pathways. A search for evolutionary relationships made it possible to trace many stepwise changes in tissues and organs (see Chapters 7 and 11). Such studies made clear that as each species evolved, previously inherited structures were modified, sometimes in entirely new ways.

Careful anatomical dissections and comparisons provided the criteria for constructing detailed evolutionary trees, an activity that preoccupied late nineteenth-century biologists as they sought evidence for Darwin's theory of the relatedness of all

organisms and for descent with modification. Such comparisons were complicated by any change in form and/or function of organs within a group (the flippers of whales, for example), by loss of a feature/function over time (the loss of limbs in snakes, for example), and by the persistence of vestigial organs (BOX 16.2). Darwin used all three types of evidence to support his theory of evolutionary change.

Another class of practitioners of comparative anatomy were/are paleontologists, for whom anatomy is often the only evidence on which species can be erected, lineages recognized and evolutionary trends identified. Interestingly, some ancient lineages have persisted with minimal morphological changes to the present day. They are known as "living fossils" (BOX 16.3). Aside from such rare living relics, however, fossils in almost every instance differ from present-day forms.

Comparative Embryology. The third major practitioners of comparative anatomy were *embryologists* (*developmental biologists*). For some decades after Darwin, comparative anatomy and especially **comparative** (evolutionary) **embryology** — the study of comparative relationships among anatomical structures in the embryos of different species — became the most popular and the most successful ways to study evolution.

BOX 16.2
Vestigial Organs

Comparative anatomy and embryology come together in various ways, one of which is the study of **vestiges**. Of particular interest are structures that seem to have lost some or all of their ancestral functions (FIGURE B16.3).

From an evolutionary viewpoint, biologists explain rudimentary or **vestigial organs** as arising because an organism adapting to a new environment usually carries along some previously evolved structures. As evolution continues, obsolete structures would tend to diminish, showing only traces of their former size and function. Examples are the rudiments of hind limb and pelvic girdle bones in some species of whales and snakes (FIGURE B16.4), hind limbs and the pelvic girdle having been lost in both groups when they diverged from limbed ancestors. Although to speak of

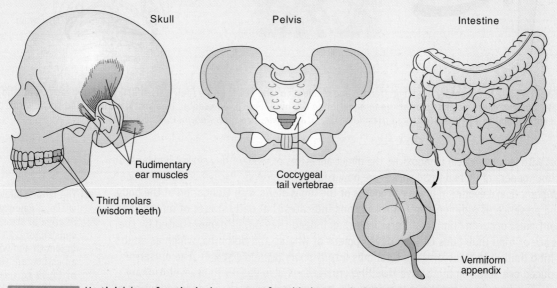

FIGURE B16.3 Vestigial (non-functioning) structures found in humans include the third set of molar (wisdom) teeth, muscles that move the ears in other mammals, ear muscles, tail vertebrae, and the appendix. (Adapted from Romanes, G. J. *Darwin, and After Darwin.* Open Court, 1910.)

(continued)

BOX 16.2
Vestigial Organs (*continued*)

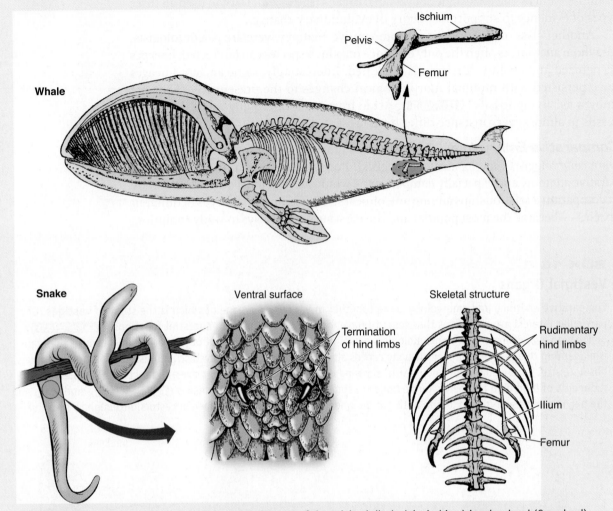

FIGURE B16.4 Rudimentary hind limb (femur) and elements of the pelvic girdle (pelvis, ischium) in a bowhead (Greenland) whale, *Balaena mysticus,* and spurs at the termination of the hind limbs,) represented by a vestigial femur and pelvic girdle elements (ilium) in a python. (Adapted from Romanes, G. J. *Darwin, and After Darwin*. Open Court, 1910.)

snakes with legs may seem paradoxical, the direct ancestors of modern-day snakes had legs; at least four genera of limbed fossil snakes are now known: *Haasiophis, Pachyrhachis, Eupodophis* and *Najash*. Snake evolution involved the loss of the limbs along with elongation of the body.

The presence of rudiments of limb skeletons indicates that early stages of limb development must occur in "limbless" vertebrates, as indeed they do, as demonstrated by the presence of hind limb buds early in development of whales and dolphins, which as adults lack hind limbs and transform the forelimbs into flippers (**FIGURE B16.5**). The presence of reduced eye stalks in blind cave-dwelling crustaceans also speaks to an evolutionary process by which obsolete structures gradually became rudimentary. So does reduction of the eyes in blind Mexican cave fish as other sensory organs assume a greater role in the dark caves (see Figures 6.6 and 13.4).

■ The largest known snake, a Paleocene relative of the boa constrictors, 13 m in length and with an estimated weight of 1,135 kg was described in 2009 by J.J. Head and colleagues.

BOX 16.2
Vestigial Organs (*continued*)

(a)

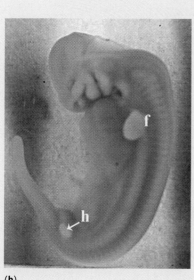

(b)

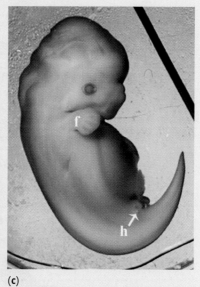

(c)

FIGURE B16.5 (a) Hector's dolphin, *Cephalorhynchus hectori*, showing the well-developed flippers and absence of hind limbs. Embryos of the spotted dolphin, *Stenella attenuata* at 24 (b) and 48 (c) days of gestation to show the well-developed flipper buds (f) and the rudimentary hind limb buds (h). At 48 days (c), the primordia for the digits can be seen in the flipper bud (f) but the hind limb bud (h) has almost disappeared.

Early in the nineteenth century, Karl von Baer (1792–1876) discovered remarkable similarities among the embryos of vertebrates that differ from one another as adults. von Baer generalized his findings into a "law:" early embryos of related species bear more common features than do later, more specialized developmental stages. von Baer's views were comparative and taxonomic but not evolutionary; he used classes of embryos to erect a scheme of classification. Darwin on the other hand asserted the importance of comparative embryology for understanding evolutionary change (see

BOX 16.3
Living Fossils

■ A living fossil is not necessarily a missing link but may represent the end of the line (a crown taxon) for their lineage.

Interestingly, some ancient lineages have persisted to the present day with minimal morphological changes. Such species are so similar to organisms believed to have become extinct many ages ago that they are called **living fossils**. Sturgeons, lungfish, coelacanths, horseshoe crabs and ginkgo trees are examples of living fossils. For example, coelacanths are ancient lobe-finned fishes (FIGURE B16.6) related to those that evolved into terrestrial vertebrates about 200 Mya. Although the fossil record of coelacanths (which began in the Devonian about 380 Mya) ended 80 to 100 Mya, fishermen find live coelacanths (*Latimeria chalumnae*) in deep waters off the eastern coast of South Africa. Similarly, an ancient form of segmental mollusk (*Neopilina*), believed extinct since the Devonian, has been found in deep-sea trenches off Costa Rica, Peru, and southern California.

These findings show the lack of completeness of the paleontological record. Disappearance of fossils of a particular type from the record does not necessarily mean that this type went extinct; "absence of evidence is not evidence of absence."

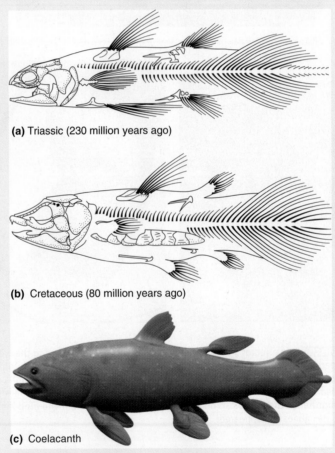

(a) Triassic (230 million years ago)

(b) Cretaceous (80 million years ago)

(c) Coelacanth

FIGURE B16.6 Coelacanths demonstrate retention of major features throughout a long evolutionary history resulting in the extant species being regarded as a living fossil. (**a**) A fossil lobe-finned coelacanth, *Laugia groenlandica,* from the Early Triassic, about 230 Mya. (**b**) A coelacanth, *Macropoma mantelli,* from the Late Cretaceous about 80 Mya. (**c**) The extant coelacanth, *Latimeria chalumnae,* found off the eastern coast of South Africa. (**a** and **b** Adapted from P. L. Forey, *Nature,* **336**, 1988: 727–732.)

Chapter 7). In 1861 and many later publications and lectures, the German embryologist, naturalist, philosopher and artist, Ernst Haeckel (1834–1919) hypothesized that embryos during their development repeat the evolutionary history of the groups to which they belong. Consequently, embryos during development were equated with stages of evolution (FIGURE 16.2). Haeckel built on von Baer's law in considering the early stages of development more conservative or evolutionarily stable than adult stages. Consequently, this *building on* represented a major shift to an evolutionary explanation. As Darwin stated

> In two groups of animals, however much they may at present differ from each other in structure and habits, if they pass through the same or similar embryonic stages, we may feel assured that they have both descended from the same or nearly similar parents, and are therefore in that degree closely related. Thus, community in embryonic structure reveals community of descent.

Haeckel, who was a major proponent of evolution in Germany, took comparative embryology into evolution in 1866 with what became known as the **biogenetic law**.

> Ontogeny [development of the individual] is a short rapid recapitulation of phylogeny [the ancestral sequence] . . . The organic individual repeats during the swift brief course of its individual development the most important of the form-changes which its ancestors traversed during the slow protracted course of their paleontological evolution according to the laws of heredity and adaptation.

To Haeckel, this meant that early stages of embryonic development recapitulate *adult ancestral forms,* that is, the tadpole of a living frog recapitulates a *tailed ancestor.* However, early stages of embryonic development recapitulate only *early ancestral developmental stages,* not ancestral adults. The discovery that juvenile stages of ancestral organisms can be retained in the adult forms of their descendants, as, for example in the preservation in adult humans of features found in juvenile apes (see Chapter 19), directly contradicts the Haeckelian notion that descendants retain ancestral adult features. Rather, organisms that share common descent make use of common underlying embryological patterns. Closely related organisms use shared genes and gene networks to produce characteristic developmental stages that have persisted for tens of millions of years. Evidence from genetics, molecular and developmental biology and from the integration of evolutionary and developmental biology (evo-devo; see Chapter 13) provided strong support for this view.

Biological Species

The morphological species concept says nothing about whether individuals so identified *interbreed* or have the potential to interbreed. Many biologists therefore adopt a **biological species concept** initially formulated by Ernst Mayr. A biological species is a sexually interbreeding or potentially interbreeding group of individuals normally separated from other species by the absence of genetic exchange through reproductive and other barriers. The concept of a species as an interbreeding group of individuals distinct from other such groups is based on the knowledge that such groups exist in nature and are separated by gaps across which interbreeding does not occur.

Whether a population fits the definition of a biological species can be tested. One means is to determine whether individuals in populations in the same locality normally interbreed. Inability to interbreed indicates that the populations are biological species. But the block to interbreeding may occur later in the reproductive cycle as a

Fish Salamander Tortoise Chicken Pig Cow Rabbit Human

Pharyngeal
(gill) arches

Vertebral
column

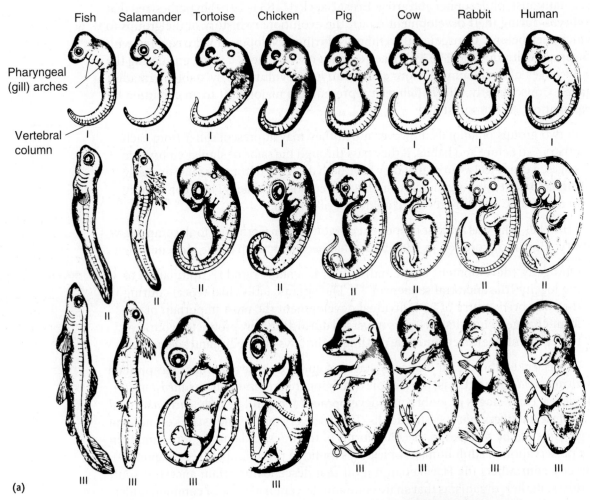

(a)

FIGURE 16.2 (**a**) One of Haeckel's classic nineteenth century illustration of different vertebrate embryos at comparable stages of development. Although Haeckel took some liberties in drawing these figures, the earlier stages are more similar to one another than later stages are to one another. Embryos in the different groups have been scaled to the same approximate size so that comparisons can be made among them. (**b**) Early embryonic stages are more alike than are adults, depicted here as shark, chicken and human embryos and adults. (**a** Adapted from Romanes, G. J. *Darwin, and After Darwin*. Open Court, 1910; **b** Adapted from de Beer, G. R. *Atlas of Embryology*. Nelson, 1964.)

(b)

post-mating barrier. Should interbreeding occur in nature, we can ask whether embryos develop and/or whether the progeny are viable and fertile. If the answer to this question is no, we consider the organisms as species separated by a reproductive barrier(s).

Applying the biological species concept has allowed taxonomists to make species distinctions between similar-appearing populations that cannot be separated on the basis of their morphology (**sibling species**). Two leafy-stemmed sibling species in a genus of the phlox family, *Gilia tricolor* and *G. angelensis* are an example from the plant kingdom. European short-toed and common (Eurasian) tree creepers (*Certhia brachydactyla* and *C. familiaris*) differ morphologically only in the size of the third toe and the patterning of feathers on the wing (FIGURE 16.3a) but their distribution, behavior and ecology are sufficiently distinctive to prevent interbreeding (Figure 16.3b). The fruit fly species *Drosophila pseudoobscura* and *D. persimilis* are almost identical in appearance and cannot be distinguished on the basis of their morphology. They do interbreed but only the female offspring are fertile.

Applying the biological species concept also has allowed taxonomists to unify different groups into single species that had been separated into distinct species on the basis of morphological and/or geographical criteria. One example is the union of several species of North American sparrows into a single **polytypic** species, the song sparrow, *Passarella melodia*, consisting of multiple geographic subspecies. Similarly, groups of yarrows (*Achillea* sp.), plants that show distinct ecological adaptations restricting their growth to particular environments (see Figures 13.9 and 13.10), are

■ A polytypic species is defined as a species with two or more subspecies that normally do not interbreed because they are geographically separated, but which can interbreed and where they do in nature, often create a narrow hybridization zone at a geographical boundary.

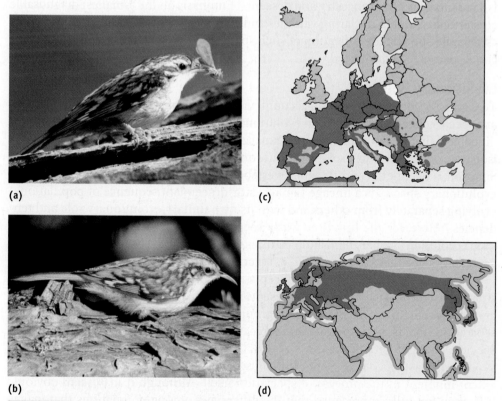

FIGURE 16.3 A sibling species pair, the European short-toed and common (Eurasian) tree creepers (**a**) *Certhia brachydactyla* and (**b**) *C. familiaris* are almost indistinguishable morphologically. Distinctive geographical distributions (**c, d**) and behaviors reinforce their inability to interbreed.

generally identified as varieties because they could exchange genes if they came into contact.

Three caveats prevent the universal application of the biological species concept.

- As it cannot be determined whether they could interbreed or not, fossil species are defined as morphological species.
- The biological species concept may not apply to animals with asexual reproduction. Each clone of individuals is genetically identical but genetically isolated from every other clone, yet few, if any, biologists would consider describing each clone as a separate species.
- Species that have long been isolated geographically may still be able to interbreed if brought into contact, usually in a nursery or a zoo. Examples include two widely separate populations of trees occupying similar habitats, one, the Chinese catalpa, *Catalpa ovata* and the other, found in the eastern United States, the Indian bean tree, *Catalpa bignoides*. Although they will cross in a nursery, and produce hybrids that are as viable and fertile as the parents, these populations have been separate for many millions of years, and are identified as separate species.

Defining taxa strictly according to cladistic relationships (see Chapter 6) has led some who use cladistics to replace the morphological and biological species with a **phylogenetic species concept**. Ancestor-descendant relationships rather than reproductive isolation define a phylogenetic species, which is a monophyletic group composed of "the smallest diagnosable cluster of individual organisms within which there is a parental pattern of ancestry and descent." Emphasis on the "smallest diagnosable cluster" can result in declaring varieties and even smaller genetically distinct groups as separate species. Identification of such species has allowed what might be endangered species to be protected.

Evolutionary Species

To explicitly take evolution into account, various specialists proposed an **evolutionary species** concept; species as an evolutionary entity. Species are defined in terms of differences that are not dependent on sexual isolation but rather on their "evolutionary" isolation, of which sexual isolation is only one aspect. In the words of George Gaylord Simpson, a twentieth century expert vertebrate paleontologist, writing in 1961, "an evolutionary species is a lineage (an ancestor-descendant sequence of populations) evolving separately from others and with its own unitary evolutionary role and tendencies." Here, for the first time was a species concept incorporating change over time (evolution) rather than static features, a concept that laid the groundwork for considering changes resulting from competition and interaction among species.

Of course, as with all species concepts, the evolutionary concept does not fit all situations. One particular problem presents itself. Because speciation is an evolutionary process, defining the stage when groups of organisms have reached complete separation (that is, have speciated) is subjective. And, applying the concept to extant species is fraught with problems.

As will have become evident, difficulties in species classification are, to a large extent inherent in the process of speciation itself. Although it may seem obvious, it is often not fully appreciated that the differences among populations that makes some of them hard to classify arise from the fact that evolutionary change differs in intensity and duration, over time, and from place to place when environments differ

■ As Simpson was the paleontologist who deciphered the evolutionary lineage of horses (Box 16.1), it is most appropriate that his initials are G.G.

over long periods of time. Neither phenotypes nor genotypes evolve uniformly among groups, resulting in different degrees of reproductive isolation, and morphological or genetic distinctiveness. Members of a species are identified by their similarity but their relationships derive from a shared history, which rarely results in all individuals in a species being identical. Classification and evolution inevitably emphasize different aspects of organisms.

A species name indicates singularity, but the individual members of a species display variation (see Chapter 12). A central issue, therefore, is the basis on which similarity is determined. This issue, which is fundamental to all biology, occupies the remainder of the chapter.

■ The **type specimen** defines a species and is housed in a museum for any and all to check.

■ Similarity: Knowing When Characters Are the Same or Different

In general, the more similar features shared by a group, the more likely the group is to have descended from a common ancestor. A classic example of discovering the evolutionary history of a character and of the lineage of organisms in which that character is expressed is reduction in the number of toes in the evolution of horses over some 60 My (Box 16.1, Figures B16.1 and B16.2). We can readily recognize and equate the parts of the skeleton at the ends of the feet as toes, even when the number decreases over time from five to one as it does during horse evolution. The fossil record is detailed, enabling the evolutionary changes to be reconstructed, lineages that became extinct to be recognized, and the lineage that led to the modern horse to be identified.

The evolutionary record is rarely as complete as it is for horse evolution. Greater difficulties in interpretation occur when similar characters arise in different lineages. This problem arises in organisms that have evolved to mimic another species in their environment. An example is a palatable insect that mimics a poisonous species in the same environment (see Figure 15.11). The species are similar, and may even have co-evolved, but their similarity is not based on shared ancestry. A second and frequent situation is when organisms in far-flung parts of the globe evolve in parallel, even though they have never come in contact. Examples include placental and marsupial "tigers" or "wolves" (FIGURE 16.4), marsupial, placental and monotreme "anteaters" (FIGURE 16.5) and African euphorbs and American cacti (see Figure 10.14).

These two examples bring us face to face with the **"apples and oranges"** problem raised in the introduction. Sometimes an apple looks like an apple because it is an apple. However, an orange may look like an apple but is really an orange, and an apple may look like an orange but it really is an apple. How do we know when the piece of fruit before us is an apple?

You will immediately see the problems that arise when attempting to determine whether features are the same and whether the similarities reflect evolutionary origin from a common ancestor or independent evolution. When does similarity mean sameness and when does similarity mean very close resemblance? Phenotypes may be similar because (i) of recent shared ancestry, (ii) because similar characters arose in groups with a more distant shared ancestor; or (iii) because similar evolutionary patterns arose independently in different lineages.

(a) Borhyaenid marsupial (Miocene, Argentina)

(b) Marsupial Tasmanian wolf (Tasmania, Australia)

(c) Placental wolf (North America)

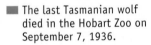
The last Tasmanian wolf died in the Hobart Zoo on September 7, 1936.

FIGURE 16.4 Striking examples of parallel evolution involving independent evolution of the "wolf" phenotype on three continents. (a) *Prothylacynus patagonicus,* a marsupial from the early Miocene in southern Argentina. (b) *Thylacinus cynocephalus,* the recently extinct marsupial Tasmanian wolf. (c) *Canis lupus,* the placental North American wolf. (Adapted from Marshall, L. Q. Marsupial paleogeography. In L. L. Jacob (ed.), *Aspects of Vertebrate History.* Museum of Northern Arizona Press, 1980.)

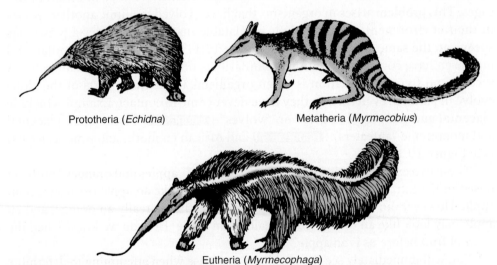

Prototheria (*Echidna*)

Metatheria (*Myrmecobius*)

Eutheria (*Myrmecophaga*)

FIGURE 16.5 Similar phenotypic features — long snout, long tongue, powerful claws — among anteaters that evolved independently within each of the three major groups of extant mammals. Depending on how their distance from a common ancestor is estimated, these forms can be considered either parallel (sharing mammalian ancestry) or convergent (descended from different mammalian groups).

Three concepts and terms arose in the mid-nineteenth century to deal with this problem of assessing similarity/sameness. Although differences between the three are outlined, you should be aware that this is a contentious area in evolutionary and developmental biology. Scientists regard themselves as objective. Assessing similarity, however, raises similar problems as defining a species. So be warned (FIGURE 16.6). This caution notwithstanding, these are absolutely central concepts, terms and challenges in evolutionary biology. First, the three concepts/terms are defined. Each is then considered in more detail.

1. When Similarity Means Shared Ancestry: Homology

 When similarity of a feature arises because of shared ancestry of the organisms with the feature, the features are considered **homologous.** Homology indicates that the same feature occurs in different species because of derivation from a common ancestor that bore the same characteristic (Figure 16.6).

(a) Homology: two species bearing the same phenotype caused by common ancestry for the same genotype

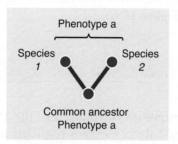

(b) Parallelism: two species with the same phenotype descended from a common ancestor with a different phenotype and genotype

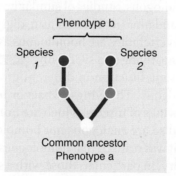

(c) Convergence: two species with the same phenotype whose common ancestor is very far in the distant past

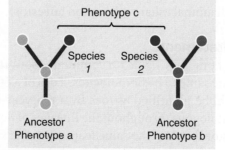

FIGURE 16.6 Homology, parallelism, and convergence diagrammed for two species (*1, 2*) that share a similar phenotypic character (phenotype a, b, c). Note that the distinction between parallelism and convergence may be subjective, because there are no rules that specify how far in the past one can establish a common ancestor for parallel evolution, and even convergent lineages have common, albeit distant, ancestors.

2. When Similarity Means Parallel Evolution: Parallelism

When similar features arise in closely related lineages that do not share a common ancestor, the features are considered to have evolved in **parallel**. Parallelism is a statement that similar features can occur in different lineages because of parallel evolution from a different ancestor. Because of shared earlier evolutionary history the two lineages, homologous genetic or developmental pathways, may form the features (Figure 16.6).

3. When Similarity Means Similar Solutions in Independent Lineages: Convergence

When a feature arises independently in unrelated organisms because of similar solutions to the same selective pressures, we regard the features as **convergent**. Convergence is a statement that similar features can arise in unrelated lineages as a common response to a similar environment and selection pressure. Because the lineages are independent, **different (non-homologous)** the features may arise using different genetic or developmental pathways (Figure 16.6).

Homology

A Brief History

> The term **homoplasy** is often used in the literature for features not sharing an evolutionary history in a lineage, without stating whether the underlying mechanism is parallelism or convergence.

Derived from terminology introduced in the 1840s by the comparative anatomist Richard Owen (1804–1892), organs regarded as the same, though serving different functions, are homologous. Studies of similar bones or muscles — the humerus in the upper arm, radius and ulna in the lower arm, digits — demonstrated that the forelimbs of widely different vertebrates are homologous as forelimbs, albeit with such different functions as walking, flying and swimming (FIGURES 16.7 and 16.8). In contrast, organs that perform the same function in different groups but do not share a similarity of structure are analogous. The wings of bats or birds, which are built around a bony skeleton, and the wings of insects, which are not, do not show a common underlying structural plan and so are analogous not homologous.

After Darwin, it was realized that homologues are found in organisms with a shared evolutionary history, in particular, those with a shared common ancestor. Analogues are found in animals without a common ancestor (although, going far enough back in evolution all animals share a common ancestor).

Homology Statements

Statements of homology make no comment about features having to be identical or even to look the same to be homologues. Toes of a 60-My-old horse are recognizable toes and would be identified as toes by a ten-year-old child. However, the toes of horses are homologous throughout the lineages of horse evolution because they arose from an ancestor that had the same features in the same position. The many similar features in the forelimb skeleton of the different vertebrates shown in Figure 16.7 are homologous because they arose from the same feature in a common ancestor, even though the features have changed in appearance with the evolution of wings in birds, flippers in seals and so forth (Figures 16.7 and 16.8). As Charles Darwin noted, "What can be more curious than that the hand of man formed for grasping, that of a mole, for digging, the leg of a horse, the paddle of a porpoise and the wing

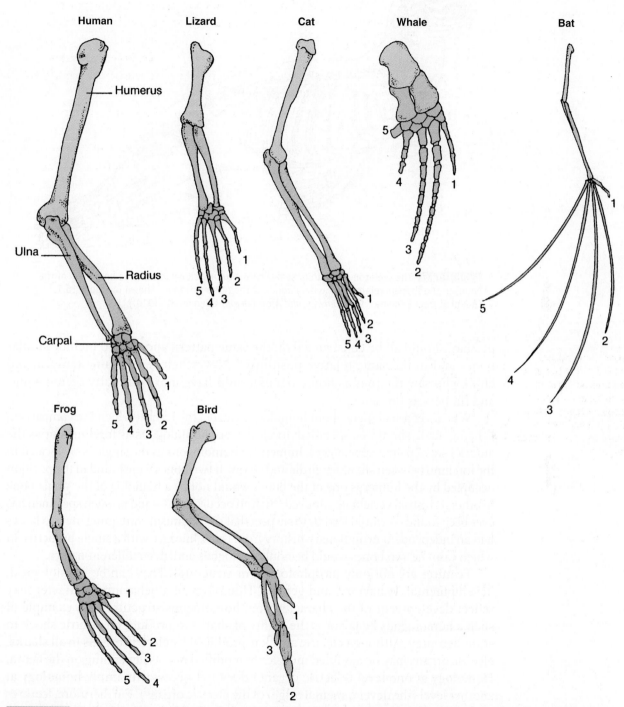

FIGURE 16.7 Skeletal structures of the forelimbs of representative terrestrial vertebrates to show the homology among bones at different levels in the limbs. Note that homology is preserved despite evolutionary changes in size, shape and function of the forelimbs. Compare, for example, the humerus in beige in each limb, or the digits in dark green. Even when digit number or length varies, homology of the digits remains: shown as digits numbers 1–5 in all except the bird where the number is reduced to three, and (for growth) in the expanded lengths of digits 2 and 3 in the whale, digits 3–5 in the bat and digit 2 in the bird.

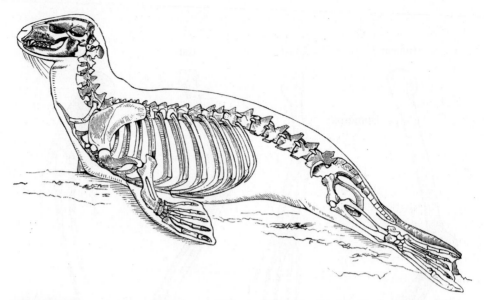

FIGURE 16.8 The skeleton of a harbor seal, *Phoca vitulina*, showing how the homology of the bones in the flippers relates to the limb bones of the other tetrapods shown in Figure 16.7. (Adapted from Romanes, G. J. *Darwin, and After Darwin*. Open Court, 1910.)

of a bat, should all be constructed on the same pattern and should include similar bones and in the same relative positions?" Nevertheless, the same ten-year-old child who saw the toes as homologous would have more difficulty seeing wings and flippers as the same.

Whenever a statement about homology is made, the level at which the comparison is being made should be specified: forelimbs are homologous as forelimbs or as the anterior set of paired appendages; humeri are homologous as the single bone located in the forelimb between shoulder girdle and elbow. If two bones were found in the position occupied by the humerus one of the bones would not be a homolog of the single bone in other terrestrial vertebrates. Indeed, such an occurrence — and no such specimen has ever been found — would lead us to suspect that the organism containing the two bones had an independent evolutionary history from those lineages with a single humerus, in which case the two bones would be nonhomologous and given different names.

Features are not only morphological or structural. They can be physiological, developmental, behavioral and genetic. Homology of a behavioral character may reflect development of the character from homologous structures. An example of such a homologous behavior is the ability of sharks to produce an electric shock to stun their prey. Although electric shock is produced by electric organs in all sharks, electric organs may be modified muscles or modified nerves depending on the taxon. Homology at one level (electric organs) does not necessarily imply homology at another level (the developmental origin of the electric organs). Furthermore, features of the phenotype not homologous as structures (limbs and genitalia in terrestrial vertebrates, for example) may share genes or gene pathways that are homologous. Homology is a hierarchical concept, reflecting the fact that evolutionary change at different levels (genes, development, structures) need not occur in tandem.

As molecular sequences became available and were compared, the term homology was extended from features of the phenotype to molecular sequence similarity. As introduced in Box 11.4, to enable comparisons of genes between different organ-

isms and to differentiate a gene duplicated within a species from a gene(s) shared with other species (see Box 12.1) the terms paralogous and orthologous genes were coined, **orthologous genes** (orthologues) for genes *shared between species* because of shared species ancestry, and **paralogous genes** (paralogues) for genes *duplicated within a species*; that is, they are extra copies of the gene.

Parts repeated in an individual are eminently liable to vary in number, structure, and/or function in response to natural selection. **Serial** (iterative) **homology** is used for similarities among parts of the same individual or organism, for example, different types of vertebrae (neck, thorax, tail as in Figure 16.8) or hemoglobin molecules (α, β, γ chains). Serial homology often reflects the duplication of a gene responsible for producing or affecting a particular structure; duplication of globin genes is a good example. Serial homology also can reflect duplication of a particular structure; duplication of vertebrae is a good example. Such duplicates originally may have had similar features, but evolved differently from each other, a point made by Darwin (although not in the context of genes).

Parallelism

Parallelism is the evolution of similar features in closely related lineages that do not share a most recent common ancestor. Examples include marsupial and placental "tigers" and "wolves" (Figure 16.4), the anteater-like features in several lines of mammals, each of which descended from non-anteater mammalian groups (Figure 16.5) and the similar leaf form and habit of African euphorbs and American cacti (see Figure 10.16). As with homology and convergence, parallelism can occur at molecular or other levels of the phenotype.

Moreover, as indicated above and in Chapter 1, organisms are organized hierarchically, with information building upon the level(s) below. New emergent properties arise at each level and, just as importantly, cannot be predicted from the properties of the level below (see Chapter 1). If evolution is constrained, as, for example, by stabilizing selection (see Chapter 15), we might expect little evolutionary change in the genetic or developmental processes from which parallel features are constructed.

Convergence

The examples of parallel evolution of the "wolf" phenotype on three continents shown in Figure 16-4 nicely illustrates the need for knowledge of evolutionary relationships to separate parallelism from convergence. Parallelism is the evolution of similar features in closely related lineages. Convergence (convergent evolution) is the evolution by different means of similar features in independent lineages. The existence of convergence is evidence that selection to meet similar problems in different evolutionary lineages can, and often does, lead to functionally similar anatomical structures (Figure 16.5).

Convergence derives from exposure of different lineages to similar environmental factors leading to selection for similar features. The wings of birds and bats are homologous as limbs with digits because they share an ancestor that possessed limbs with digits built using similar regulatory processes. Neither bat nor bird wings are homologous to insect wings as no common ancestor has a feature from which both types of wings can be derived. The evolution of wings in insects and in vertebrates is an example of convergence — two independent lineages of animals responding to selection for flight.

▮ Recommended Reading

Felsenstein, J., 2004. *Inferring Phylogenies*. Sinauer Associates, Sunderland, MA.

Hall, B. K. (ed.), 1994. *Homology: The Hierarchical Basis of Comparative Biology*. Academic Press, San Diego, CA. [paperback issued 2001]

Hall, B. K., 1999. *Evolutionary Developmental Biology*, 2nd ed. Kluwer Academic Publishers, Dordrecht, Netherlands.

Hall, B. K., and W. M. Olson (eds.), 2003. *Keywords & Concepts in Evolutionary Developmental Biology*. Harvard University Press, Cambridge, MA.

Head, J. J., J. I. Bloch, A. K. Hastings, et al., 2009. Giant boid snake from the Paleocene neotropics reveals hotter past equatorial temperatures. *Nature* **457**, 715–718.

Lee, M. S. Y., 2003. Species concepts and species reality: salvaging a Linnean rank. *J. Evol. Biol.*, **16**, 179–188.

Mayr, E., 1982. *The Growth of Biological Thought. Diversity, Evolution, and Inheritance*. Harvard University Press, Cambridge, MA.

Orlando, L., J. L. Metcalf, M. T. Alberdi, et al., 2009, Revising the recent evolutionary history of equids using ancient DNA, *Proc. Natl. Acad. Sci. U.S.A.*, **106**, 21754–21759.

Valentine, J. W., 2004. *On the Origin of Phyla*. The University of Chicago Press, Chicago.

Wilson, R. A. (ed.), 1999. *Species. New Interdisciplinary Essays*. MIT Press, Cambridge, MA.

Origin of Species

17

KEY CONCEPTS

- Although many species persist for long periods of time both genotypic and phenotypic changes occur over time.
- Even if change is occurring, individuals that interbreed within a population(s) maintain the status of the population(s) as a species.
- Conditions for speciation can be initiated when a population subdivides or when a few individuals found a separate population in a similar or different environment.
- Without the separated population occupying a different environment, speciation will be slow, on the order of millions of years in many groups.
- Isolation of a group within a population also can lead to speciation, which can be much faster than speciation in geographically isolated populations.
- In an isolated population, natural selection will lead to adaptation to the new environment as the isolate begins to differentiate from the original population.
- Behavioral, seasonal or habitat differences can facilitate reproductive isolation of the isolate or of a group within a single population.

Above: Speciation in all its splendor: a monkey flower (*Mimulus* sp.).

325

- Subsequent reproductive isolation provides a barrier to interbreeding with the original population.
- Reproductive isolation becomes complete and the population is reproductively isolated as a new species.

Overview

Given that new species evolve from existing species, the question of the origin of species is the question of how new species arise. Change within a lineage (micro-evolution) may result in the formation of subspecies or varieties but not the formation of a new species. Darwin mostly devoted himself to explaining how beginning with a single species a lineage of organisms could change through time in response to natural selection to produce a succession of species over time (see Chapter 6). This mode of speciation, known as adaptive radiation, reflects the origin of differentiation among populations. Species diversification occurs if splits and divisions within an ancestral line result in the emergence of more than one species or a cluster of species (a *clade*).

Speciation can be initiated when genetic exchange within or among populations is impeded in processes known as *sympatric* and *allopatric speciation,* respectively. The initiating event in allopatric speciation is almost always geographical splitting of a population or the isolation of a few individuals, the latter known as the founder effect. If the environment occupied by the isolate differs from that experienced by the original population the isolate is likely to adapt through natural selection to the differing environmental conditions. Over time, the isolate may begin to differentiate phenotypically from the original species. Concurrently or subsequently, reproductive isolating mechanisms may arise and provide a barrier to gene exchange with the original population and/or with other populations in the new environment. Isolating mechanisms that prevent interbreeding may be seasonal, behavioral or based on habitat preference or changes in reproductive biology that prevent interbreeding, fertilization or embryonic development. A similar series of changes occurs in sympatric speciation but in the absence of geographical subdivision of the original population.

Species Can Change Without Speciation Being Initiated

Much of evolution is about populations maintaining the integrated features that enabled them to speciate in the first place. Even so, long-term change within a species need not initiate speciation. In large part, this is because speciation involves more than phenotypic change, although such change is often the first sign that a species is changing. Speciation requires reproductive isolation of previously interbreeding demes.

As discussed in Chapter 14, most species consist of more than one population and many species consist of many populations. The local environments inhabited by separate populations of a single species may differ. As a consequence local adaptations may occur among the populations of a species. Such local changes are recognizable by differences in morphology, physiology and/or behavior. If such local adaptations do not affect the reproductive biology of the populations, individuals from different populations will still be able to interbreed should the populations recombine; the populations remain as a single species, although subspecies or varieties may be recognizable.

■ It should be clear that the discussion of speciation in this chapter defines species on the basis of their reproductive isolation from other species, that is, the biological species concept is being used (see Chapter 16).

Adjustment within a species to environmental change is one form of microevolution (see Chapters 6 and 11).

Even when a species consists of many populations, gene flow between populations may slow or even inhibit local specializations and so promote continuance and stability of the species. The greater the gene flow, the fewer the differences that develop over time and the more stable the species. Mechanisms that reduce gene exchange between populations accelerate the formation of distinctive groups. If such groups can still interbreed, they are often referred to as varieties, subspecies or sibling species (see Chapter 16). As one example, gene flow is greater between populations of many non-migratory species of birds than it is between populations of migratory species; migratory species have, on average, less than half the number of varieties found in non-migratory species. Male parental care in birds also is seen more often in species with more varieties than found in species where the female cares for the young. The implication is that behavioral and reproductive patterns such as migration or the sex providing parental care influence gene flow to the extent of influencing differentiation within a species.

The remainder of the chapter is devoted to discussion of how speciation is initiated and of examples that demonstrate speciation in action. Two major modes of speciation are recognized. One, initiated between populations following geographical isolation is known as **allopatric speciation**. The other, initiated within a population, is known as **sympatric speciation**. Allopatric speciation is usually followed by sympatric speciation within the isolated populations but sympatric speciation can be initiated in the absence of geographical isolation.

Adaptation and Differentiation

Groups of organisms that are evolutionarily related are often but not always connected geographically. Large geographical barriers such as oceans and mountain ranges isolate groups from one another and can establish conditions that facilitate the development of considerable differences among the separated groups. Colonizers capable of transcending geographical barriers and moving into new environments often become the ancestors of entirely new groups, as demonstrated most graphically in the wide evolutionary radiation of species that descended from finches reaching the Galapagos Islands. Beginning with what was probably an ordinary mainland finch, new species of finches evolved in the unoccupied niches of the Galapagos Islands (see Figure 6.10 and Box 14.2). The name given to this process, **adaptive radiation**, signifies the rapid evolution of one or a few forms into many different species occupying a variety of habitats within a new geographical area. The radiation of marsupial mammals in Australia (see Figure 6.11) shows how protection from competition by the isolation of a continent (in this case, the absence of placental mammals from Australia) can lead to an array of species with widely divergent features and functions, ranging from herbivores to carnivores, and paralleling placental mammals on other continents.

Speciation Initiated by Geographical Isolation

The set of processes regarded as the most common mode of speciation (allopatry or allopatric speciation) is initiated only after a species has separated into two or more populations. As discussed later in the chapter, speciation also can be initiated within a population (sympatry or sympatric speciation).

As introduced in Box 6.6, the concept of geographical isolation was developed in the 19th century by John Gulick to explain speciation of land snails on Hawaii.

Although **geographical isolation** is the first step in allopatric speciation, speciation is by no means an automatic consequence of geographical isolation of a subset of the species. Conditions encountered by the isolate may be identical, in which case it could remain as a population of the original species. As discussed above, if the environments differ, populations can adapt to changing conditions without becoming reproductively isolated from other populations of the same species. Speciation is rare, we might even say unusual.

For speciation to be initiated, a geographically isolated population has to be exposed to differential selection in the new environment and has to adapt phenotypically in response to that selection. **Differential selection and adaptation** are therefore the next steps in allopatric speciation. However, without a further step, **reproductive isolation**, the isolated populations would remain members of the same species, albeit phenotypically distinct. These three steps — isolation, local adaptation, and reproductive isolation — and the consequent prevention of interbreeding with the parent or other species, are the essence of allopatric speciation.

Forms of Geographical Isolation

The potential for reduced gene exchange and reproductive isolation occurs when populations subdivide to occupy different areas or habitats (FIGURE 17.1). Subdivision may (a) split a population in two, as when a new river cuts through a region, (b) isolate a small portion of the original population or (c) be initiated if a few individuals leave or become isolated from the original population (which is how many new species have arisen on islands). These three modes of separation differ with respect to the

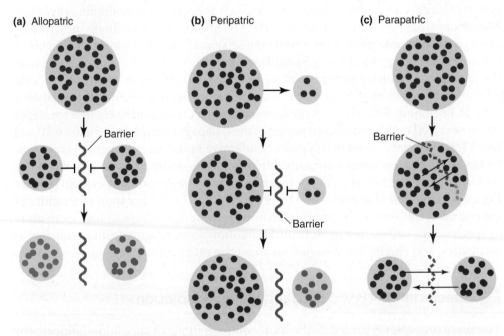

(a) Allopatric **(b)** Peripatric **(c)** Parapatric

FIGURE 17.1 Comparison between three patterns of speciation: (**a**) Allopatric, where a barrier divides a population into two, preventing gene flow (—|); (**b**) Peripatric, where a small population is isolated and forms a founder population, again with no gene flow (—|) between the populations; (**c**) Parapatric, where a barrier isolates a group but where gene flow between the two groups still occurs (⇄).

degree of geographical isolation and so in the potential for continued gene exchange (Figure 17.1). Evolutionary biologists speak of

- **Allopatric speciation**, when a population is divided because of a natural physical barrier or because intervening geographical populations become extinct;
- **Peripatric speciation**, when a population is divided because of the budding off of a small completely isolated founder colony from a larger population; and
- **Parapatric speciation**, when a population at the periphery of a species adapts to different environments but remains contiguous with its parent so that gene flow is possible between them (Figure 17.1).

As introduced above and discussed below, when gene flow is interrupted within a population without geographical isolation, sympatric speciation can be initiated.

The isolated population could be small. In theory, a single female carrying fertilized eggs/ovules or a single pair of individuals in a species with separate sexes could establish a new population that could form the basis of a new species. This step in speciation is known as the **founder effect**, a term coined by the enormously influential evolutionary biologist, Ernst Mayr, for a process discovered by John Gulick in the mid-19th century (see Box 6.6). A classic example of the founder effect discussed in **BOX 17.1** is speciation on the Hawaiian Islands of fruit flies in the genus *Drosophila*. Radiation and speciation of Darwin's finches on the Galapagos Islands also was initiated by the founder effect (see Box 14.2).

■ Theoretically, one individual is sufficient to found a new population if that individual is hermaphrodite, as are land snails (see Box 6.6).

Speciation can proceed quite rapidly under circumstances in which a peripheral population is genetically isolated from its parent (peripatric speciation; Figure 17.1). A small, isolated population would be subject to genetic drift and inbreeding and exposed to new adaptive landscapes and strong selection pressure (see Chapter 12). The combined effect of such forces can result in novel, coadapted gene combinations that affect behavioral, morphological and physiological traits, a necessary prelude to reproductive isolation from neighboring and/or ancestral populations.

Once isolated geographically, a population can become further isolated by processes that are seasonal, behavioral or habitat related, then by further restricted gene flow and reproductive isolation, which is the final phase in any speciation process. These changes of adaptation and differentiation, which occur **within a population**, are elements of sympatric speciation. As discussed above, such changes within a population are secondary steps in speciation initiated by geographical isolation. An important debate in evolutionary biology has been whether speciation can be initiated sympatrically by mechanisms that reduce gene flow within a population in the absence of initiating geographical isolation. Speciation initiated without geographical isolation is discussed in the following section before discussing mechanisms that facilitate reproductive isolation.

Speciation without Geographical Isolation

The sequence of evolutionary events in allopatric speciation begins with the formation of distinctive populations and ends with reproductive isolation. Under these conditions, populations can only accumulate genetic differences when they are sufficiently spatially separated to prevent gene exchange that could and probably would eradicate those differences. Alternatively, a population in a single locality selected for adaptation to different habitats within that locality could accumulate increased genetic variability

BOX 17.1

Speciation of Fruit Flies in Hawaii

Tectonic events (see Chapter 2) formed and continue to form the Hawaiian Islands. A localized "hot spot" in Earth's mantle under what is now Hawaii pierced the lithosphere and produced volcanic eruptions that formed a succession of islands as the Pacific plate moved northwestward. Hawaii, which is larger than all the other Hawaiian Islands combined, was built from five volcanoes. In time various older islands eroded, first becoming atolls, then seamounts (submerged volcanoes) to form the existing series of islands that extend from Hawaii to Midway to a point near the far-western Aleutians.

The Hawaiian Islands are home to more than 800 native species of *Drosophila,* the most ancient of which arose over 30 Mya. Among these are 100 species in the "picture-winged" species group, so named because of their large decorated wings (FIGURE B17.1). Careful analysis of salivary chromosome banding patterns has shown that these 100 species were derived from founder events in which each Hawaiian island was settled by relatively few individuals whose descendants evolved into different species (FIGURE B17.2). Each successful founder is presumed to have been a fertilized female.

The oldest Hawaiian island, Kauai is 5.6 My old. Ten of the 12 species of picture-winged *Drosophila* on Kauai are the most ancient species on any Hawaiian island. Forty-one species endemic to the Maui island complex (Figure B17.2) derive from only 12 founders, ten from Oahu and two from Kauai. The youngest island, Hawaii, was colonized entirely by founders from the older islands, 26 species evolving in what may have been less than 500,000 years.

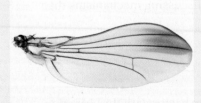

(a)

(b)

(c)

FIGURE B17.1 Wing patterns of picture-winged *Drosophila* from Hawaii. (**a**) Wing of *Adiastola.* (**b**) Wing of *Planitibia.* (**c**) Wing of *Antopocerus.*

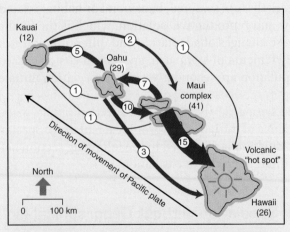

FIGURE B17.2 Colonization pattern showing the founder events associated with speciation of the "picture-winged" group of *Drosophila* in the Hawaiian Islands. The width of each arrow is proportional to the number of founders (circled). The number of *Drosophila* species present on each island in 1992 is given in parentheses. (Adapted from Carson, H. L. *Drosophila Inversion Polymorphism.* CRC Press, 1992.)

through disruptive selection (see Chapter 15), which could lead to different forms (*polymorphism*) arising within the species. Examples include polymorphism in the British peppered moth *Biston betularia* (see Chapter 15), mimicry in the butterfly *Papilio dardanus,* the evolution of distinct populations or varieties via mimicry in the butterfly *Heliconius erato,* and Batesian mimicry by an insect, the apple maggot *Rhagoletis pomonella* of the forelegs and feeding appendages of a jumping spider.

Although formation of morphs within a species is not speciation, you can see that, as with geographical isolation, the presence of morphs could establish conditions under which reproductive isolation could arise and speciation could be initiated. **Phenotypic plasticity** (which underlies polymorphism) has emerged as an important mode of incipient speciation, especially in situations in which changes in morphology are accompanied by changes in behavior, both of which can serve as mechanisms isolating individuals or groups within a population. Such conditions are most likely to be met in animal or plant species whose development is plastic (allowing phenotypic variation to arise), where the environment is subdivided into microhabitats, and, for animals, where behaviors such as aggression, territoriality or defending breeding sites provide effective reproductive isolation.

Cichlid Fishes

Sympatric speciation could occur under such conditions and has been shown to have occurred in several lineages, of which the cichlid fishes in small crater lakes in the African Rift Lakes are perhaps the most well studied example. Having originated no later than about 3 to 4 Mya, cichlids represent a prime vertebrate example of explosive radiation and sympatric speciation. Perhaps the most dramatic of such speciation events occurred in Lake Victoria, the youngest of the East African lakes. Lake Victoria is known to have dried up completely during the last Ice Age 12,400 years ago, yet 500 new cichlid species have arisen since that time.

Lakes Victoria, Malawi and Tanganyika in Africa contain 300, 200 and 125 endemic species of cichlids, respectively, all of which appear to have arisen rapidly and by sympatric speciation. Lake Victoria is no more than 750,000 years old, yet contains 300 species found nowhere else. Indeed, if Lake Victoria dried up between 12,500 and 14,000 years ago as geological evidence suggests, these species have arisen amazingly rapidly. If recent findings for the South American cichlid genus *Apistogramma* hold true for African cichlids, these species numbers may be considerably underestimated. Individuals from the same populations of *Apistogramma caetei* that *differ only in color* and so have been regarded as the *same species,* fail to interbreed because of female mate choice, that is, they are good biological species.

Sadly, decline in Lake Victoria over the past four decades has been the most rapid and perhaps the most drastic of any lake on Earth. The deadly combination of over-fishing, the introduction of exotic species — especially the Nile tilapia, *Oreochromis niloticus,* and the Nile perch, *Lates niloticus* — pollution and poor management of the surrounding land, all contributed to depleting oxygen levels and "choking" of the lake with algae; more than half of the indigenous cichlid species are now extinct. Cichlids are not the only species at risk; 30 million people depend on Lake Victoria to make a living, and ultimately to survive. A crater of speciation has become a crater of doom.

Cichlids are diverse ecologically and behaviorally. The combination of phenotypic plasticity (FIGURE 17.2, and see Chapter 14), aggressive behavior, territoriality, ability to consume specialized diets and the availability of many microhabitats in the lakes

FIGURE 17.2 A sample of the diversity of cichlid fish body form, coloration and patterning.

provide "textbook" conditions for sympatric speciation. An important specialization is the evolution of a second set of jaws, known as **pharyngeal jaws**, which develop by modification of one or more of the cartilages supporting the gills (BOX 17.2). With two sets of jaws, the pharyngeal jaws take over the function of processing prey, freeing the mandibular jaws to specialize for prey capture. The ability to capture specialized diets has evolved to a spectacular degree. One clade of seven species of cichlids in Lake Tanganyika has asymmetrical jaws. With this asymmetry comes extreme specialization to the point of feeding behavior in which some species feed by scraping scales from the left sides of prey fishes, and other species by scraping scales from the right side of prey fish.

How do we know that these cichlids speciated sympatrically? One class of evidence is mtDNA sequence analysis, which indicates that the many morphologically diverse species of cichlid fish in East Africa are monophyletic (have a single common ancestor). Phenotypic and behavioral similarities in fish in different lakes arose by parallelism or convergence through sympatric speciation (FIGURE 17.3).

Ecological heterogeneity can reinforce genetic diversity and contribute to reproductive isolation, as shown in a *Drosophila melanogaster* population selected for radically different experimental habitats. Other examples of sympatric speciation among

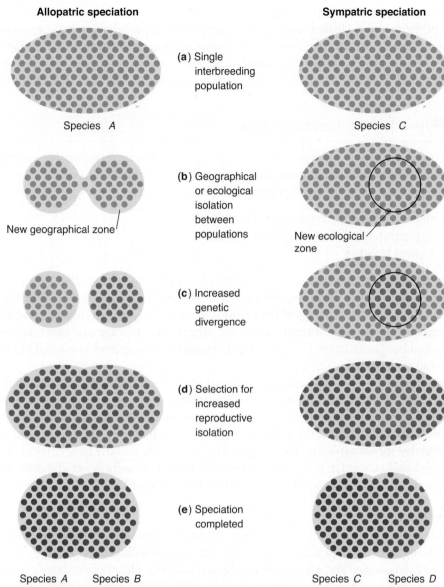

FIGURE 17.4 A simplified diagram of two modes of speciation, distinguished from one another by whether geographical isolation occurs. In allopatric speciation, a population of Species A (**a**) splits into one or more new geographical zones (**b**) allowing genetic differentiation to occur (**c**). Subsequently, mixture of the groups (**d**) can result in selection for increased reproductive isolation mechanisms among them. Speciation is complete (**e**) when gene flow between the groups can no longer occur, even when the two species (A and B) occupy the same locality. In sympatric speciation, a population of Species C (**a**) splits into one or more groups that occupy different ecological zones, such as special habitats or food sources, within a single geographical locality (**b**). Increased genetic differentiation between the groups (**c**) permits selection for reproductive isolation mechanisms (**d**) that eventually lead to complete speciation (**e**), resulting in species C and D. (Adapted from Strickberger, M. W. *Genetics, Third edition.* Macmillan, 1985.)

Although the distribution ranges of these two species overlap throughout large areas of the western United States, these three isolating mechanisms are enough to maintain them as separate species. Chromosome inversions that render male offspring sterile reinforce the reproductive isolation; to date, only a few females resulting from interbreeding have been found in nature among many thousands of flies examined.

Reproductive Incompatibility

Isolation by season, habitat or behavior may, but need not, lead to **reproductive incompatibility**. The three most common mechanisms by which reproductive incompatibility functions to isolate previously joined populations (which otherwise could form a hybrid) are

- **death** of gametes, zygote or early embryo,
- formation of a **hybrid with low viability**, or
- **hybrid sterility**.

Pollen grains in plants, for example, may be unable to grow pollen tubes in the styles of other species, with the consequence that development cannot be initiated. Crosses among 12 frog species of the genus *Rana* revealed a wide range of inviability. In some crosses, no egg cleavage occurs. In others, the cleavage and blastula stages are normal but gastrulation fails. In still others, early development is normal but later stages fail to develop.

In some *Drosophila* crosses, an insemination reaction in the vagina of the female causes swelling and prevents successful fertilization of the egg. As discussed above, *Drosophila pseudoobscura* and *D. persimilis* are virtually completely isolated reproductively. Even when interbreeding occurs gene exchange is impeded because of reproductive incompatibility; the F_1 hybrid male is sterile and the progenies of fertile F_1 females backcrossed to males of either species show markedly lower viabilities than the parental stocks (a process known as hybrid breakdown).

Hybrids

Hybrid sterility, long thought to be important in isolation, has received much attention. The most familiar example is probably the mule, which is the progeny of a male donkey (*Equus asinus*) and a female horse (*Equus caballus*). The opposite cross produces a hinny. Sterility comes from chromosomal incompatibility. The horse has 64 pairs of chromosomes, the donkey 62, and the mule 63.

In 1889, Alfred Wallace proposed that natural selection might favor the establishment of mating barriers among populations if the **hybrids were adaptively inferior (hybrid inferiority)**. According to this hypothesis, subsequently supported with experimental evidence, selection for sexual isolation arises because most groups and species are strongly adapted to specific environments. Maintenance of speciation enables populations to preserve their adaptive advantages from gene flow from non-adapted groups. Hybrids between two highly adapted populations represent a genetic dilution of their parental gene complexes that can be of great disadvantage in the original environments. In this case, hybrid inferiority would reinforce species separation. The likely sequence of evolutionary changes would be the evolution of genetic differentiation between allopatric populations of a single species, overlap of these differentiated populations in a sympatric area, intensification of sexual isolating mechanisms through selection and selection against the hybrid (Figure 17.4, left column). However, and as discussed below under *Speciation and Hybridization Zones*, hybrids can represent new

■ Because the mule is sterile it is not given a species name. Neither are most domesticated species.

gene combinations that may be better adapted to changed conditions than those of either parent species, in which case hybridization would reinforce speciation.

Sexual Isolation in Sympatric and Allopatric Populations

Because they are geographically close enough to produce deleterious hybrids, sexual isolation should be strongest among **sympatric populations** of related species. Because they are too distantly separated to produce such hybrids, sexual isolation should be weakest among **allopatric populations**. Evolutionary biologists have sought evidence for these hypotheses by comparing the degree of sexual isolation among different sympatric and allopatric populations, either in the wild or under experimental conditions. Some of the best information has come from studies on fruit flies.

One experiment tested the degree of sexual isolation between two sibling species, *Drosophila arizonensis* and *D. mojavensis*, by attempting crosses in which the species strains were collected, either from connected (sympatric) or separated (allopatric) populations. When the strains came from separated population, the interspecific cross *arizonensis* × *mojavensis* occurred only in 4% of matings. This contrasted with 25% of matings between connected populations. The same has been shown experimentally in plants. The most difficult to cross of nine species in the annual herb *Gilia* are the allopatric species. Sympatric species, by contrast, show no barriers against intercrossing even though all F_1 hybrids are sterile.

To these observations can be added data from hundreds of moth species, which show that male scent-emitting organs used to attract females are significantly more common among species associated with the same host plant than among species associated with different host plants. Because these organs produce species-specific courtship pheromones, these sexual-isolating mechanisms are more frequent when moths use the same host plant (sympatry) than when moths use different host plants (allopatry).

Normally isolated sibling species *D. pseudoobscura* and *D. persimilis* were used in an experimental demonstration that prezygotic isolating mechanisms can operate in sympatric populations. Although sexual isolation exists between these two species in nature and at normal temperatures in the laboratory, cold temperatures can cause a significant increase in interspecific mating. By identifying each of the two species on the basis of different homozygous recessive alleles, hybrids formed under these low-temperature conditions could be recognized and removed from interspecific population cages. When this operation was performed each generation, fewer and fewer hybrids appeared. For example, after five generations, the frequency of hybrids in the mixed populations had fallen to five percent from values that initially were as high as 50 percent. This is striking evidence that **selection against the hybrid** caused rapid selection for sexual isolation and reduction in hybrid formation. A somewhat similar experiment involved planting a mixture of yellow sweet and white flint strains of corn. By eliminating plants that produced the greatest proportion of heterozygotes, intercrossing was reduced from about 40 percent to less than five percent in five generations.

As the fruit fly species were already reproductively isolated — they are separate species — the increased sexual isolation observed reinforced past speciation and cannot therefore be used as evidence for a mechanism of future speciation. Nevertheless, the fact that prezygotic isolation can be increased experimentally by selecting against hybrids indicates that this isolating mechanism may well function in cases where hybrid fitness declines. Mating tests between *D. pseudoobscura* females and *D. persimilis* males taken from natural populations show that sexual isolation is increased in areas where populations overlap compared to areas where *D. persimilis* is absent.

■ A **prezygotic** isolating mechanism is one in which gametes fail to form if species are crossed, in contrast to a **postzygotic** isolating mechanism in which mating occurs but embryos fail to develop.

■ Sweet corn, the corn on the cob sold in supermarkets, is yellow because of the high content of vitamin A in the kernels. Flint (Indian) corn has hard kernels that can be ground into meal and used to make flat cakes or added to stews and soups.

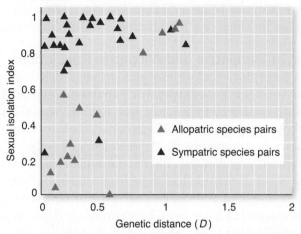

FIGURE 17.5 Measurements of the degree of sexual isolation for pairs of allopatric and sympatric *Drosophila* species plotted against genetic distance. A sexual isolation index of 1 = complete isolation, 0 = no isolation. As the figure shows, when species in a pair are closely related — genetic distance between them is small, for example, 0.5 — they are more isolated from each other when they are sympatric than when they are allopatric. Allopatric populations achieve reproductive isolation at a much greater genetic distance than do sympatric populations; D = 0.54 and 0.04, respectively. As an approximation, species of *Drosophila* took 200,000 years to speciate sympatrically but 2.7 My to speciate allopatrically. (Adapted from Coyne, J. A., N. H. Barton, and M. Turelli, *Evolution,* **51,** 1999: 643–671.)

An extensive survey of populations in nature also shows that sexual isolation between pairs of *Drosophila* species is greater for sympatric species than for allopatric species pairs (FIGURE 17.5). The basis for the analysis is **genetic distance**, which is a measure of the evolutionary divergence between two species, usually determined on the basis of allele frequencies. The lower the genetic distance, the more closely related the species. As shown in Figure 17.5, when species in a pair are closely related — genetic distance between them is small, for example, 0.5 — they are more isolated from each other when sympatric than when allopatric. Furthermore, allopatric populations require much greater genetic isolation (greater genetic distance) to achieve reproductive isolation than do sympatric populations. For the species in Figure 17.5, genetic distance is 0.54 and 0.04, respectively.

Speciation and Hybridization Zones

Where species barriers break down to produce viable and fertile hybrids, as often occurs in plants, **zones of hybridization** or **hybrid swarms** can develop in which genotypes and phenotypes differ from both parental species. In some cases, fertile hybrids can act as intermediaries, introducing genes from one species into the other, enhancing a species' ecological range and evolutionary flexibility. If a habitat exists to which the hybrids are better adapted than are the parents, the new population may eventually become isolated, a process that occurs in plants and in animals.

Plants
A contribution of hybridization to speciation is supported by detailed demonstrations of changes in chromosome number (polyploidy), especially in plants (see Chapter 12). Between 40 and 70% of all plant species are polyploid and so could have arisen by hybridization, although chromosome number can change within a single species without hybridization, and hybridization can occur without polyploidy.

(a) (b) (c)

FIGURE 17.6 Three western United States species show rapid evolution of a hybrid *Helianthus anomalus* (**a**) initially formed from a cross between *H. annuus* (**b**) and *H. petiolaris* (**c**) less than 60 generations ago.

One well-investigated example between plant species occurs in sunflowers of the genus, *Helianthus*. Three western United States species show rapid evolution of a hybrid *Helianthus anomalus* initially formed from a cross between *H. annuus* and *H. petiolaris* less than 60 generations ago (**FIGURE 17.6**). Interestingly, "synthetic hybrids" can be made experimentally by crossing the two parental sunflower species and then successively crossing and backcrossing them. Synthetic hybrids acquire genomes similar to the natural *H. anomalus* hybrid, incorporating similar parental genes and excluding others. Once an initial hybrid is formed, selection becomes an important factor allowing genes that further develop the genetic architecture of the hybrid to accumulate.

Animals

Although less common, hybridization has been associated with speciation in animals.

As discussed earlier in the chapter, mating in two of Darwin's finches — the medium ground finch and the cactus finch on Daphne Island — is based on recognition of song type. Because this cross-species learning does not provide complete reproductive isolation, interbreeding can lead to hybridization, although hybridization is rare, occurring in less than one percent of breeding pairs. Hybrids have beak sizes intermediate between their parental species but only survive when seeds of the appropriate size are present. Hybrid reproduction and survival is under environmental control.

After a year of exceptionally heavy rains, the ecological condition of Daphne Island changed and an abundance of small soft seeds were produced. Under these new conditions, hybrids with intermediate beak size enabling them to eat these seeds survived and had the highest breeding success. However, hybrids did not mate with each other (probably because there were so few of them). Rather, hybrids backcrossed to one or other of their parental species **according to their imprinted song type**, allowing genes to flow from one species into the other. Despite a low rate of gene flow but because of high survival of several backcross generations, over 30 percent of individual cactus finch had some genes from the medium ground finch. Interbreeding significantly increased the genetic and phenotypic variation of the cactus finch population.

Frequency and Impact of Hybridization in Nature

As this example from Darwin's finches shows, the evolutionary consequence of episodic hybridization is to increase the genetic variation on which selection can later act. If the environment subsequently changes, rapid evolutionary change may ensue (see Box 5.1). Episodic genetic exchange of this type between closely related species is

widespread in nature; examples are found in insects, cichlid fish, sticklebacks, many species of plants and birds and mammals with male parental care, and could be an important contributor to rapid adaptive radiation.

It has been difficult to document how often new hybrid species occur. In part this is because of natural variation in the **frequency of hybridization**, which, in vascular plants, varies between families. Some plant families show hardly any hybrid species while in others, 50 percent or more of the species arose by hybridization. With these caveats in mind it has been estimated that some 25% of plant and 10% of animal species can hybridize with one or more other species. Ability to hybridize is most commonly seen in groups undergoing rapid radiation where species are more likely to be closely related than in lineages that diverged longer ago.

The **impact of hybrids** on evolution may be more significant than their frequency suggests, however. Some plant hybrids may be the ancestors of entire lineages comprising many species and occupying many habitats. Of course, as discussed earlier, hybrid sterility normally is a barrier to further evolution, Even then, especially in plants, polyploidy may arise in a vegetatively propagating or self-fertilizing hybrid, enabling it to produce fertile gametes (see Figure 11.2). Because these gametes are diploid relative to the haploid gametes of the two parental species, a new species is born at one stroke, fertile with itself or other such polyploid hybrids but sterile in crosses with either parental species. A laboratory polyploid hybrid strain of silk worm, produced by crossing the domesticated silk worm *Bombyx mori* with the wild silk worm *B. mandarina* is viable because triploid gametes were used to backcross with haploid gametes to produce tetraploids that could interbreed. Relationships between hybridization and speciation remain an important topic for future analysis.

Genes and Speciation

Are specific genes responsible for speciation?

For many, the search for "speciation genes" focuses on sexual traits, which are often the traits involved in erecting the barriers that isolate species. Variation between sexual and nonsexual traits for a group of *Drosophila* species shows that sexual traits exhibit greater variation between species and less variation within species, consistent with selection acting differently on sexual traits at different times. Sexual traits undergo greater selection for differences between populations during speciation, and greater selection for uniformity within species after attaining speciation. It was proposed that *changes in sexual traits correlate with early speciation events*. Studies pointing to similar genetic roles for other sexual traits in various organisms include investigations of pheromonal differences between *Drosophila* species, mating preferences in *Heliconius* passion-vine butterflies, evolution of mating type genes in the unicellular green alga, *Chlamydomonas* and in sperm–egg fertilization interaction in animals.

Clearly, speciation events can occur in various ways and at various rates. In some groups, such as the picture-wing Hawaiian Drosophilidae, speciation has been dramatically rapid (Box 17.1) and may well involve fewer genes with greater phenotypic effects than in the slower speciation events in many other species. Relatively few mutations seem to account for the rapid transition from teosinte to maize (FIGURE 17.7) and from bee-pollinated to hummingbird-pollinated species of monkey flower, *Mimulus* sp. (FIGURE 17.8). In other groups allopatric speciation in the absence of

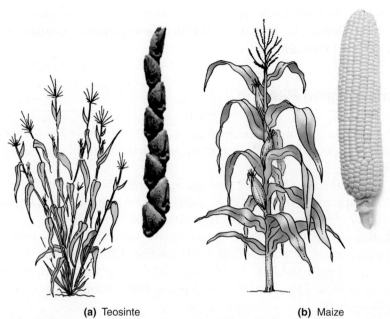

(a) Teosinte **(b)** Maize

FIGURE 17.7 (**a**) Teosinte (*Zea mays parviglumis*), the wild ancestor of cultivated maize, showing the mature plant and a kernel-bearing ear. (**b**) A mature plant and ear of its descendant, corn (*Zea mays mays*). Although strikingly different in plant and ear architecture, these two forms differ in relatively few genes.

(a) **(b)**

(c) **(d)**

FIGURE 17.8 Adaptation in flower structure (**a, b**) associated with the transition from bee-pollination (**c**) to hummingbird-pollination (**d**) of monkey flowers, *Mimulus* sp.

bottlenecks may have been more common. Such allopatric speciation modes, however, do not exclude selection for sexual isolation between sympatric populations because of hybrid sterility or inviability.

Recommended Reading

Barluenga, M., K. N. Stölting, W. Salzburger, M. Muschick, and A. Meyer, 2006. Sympatric speciation in Nicaraguan Crater Lake cichlid fish. *Nature,* **439,** 719–723.

Coyne, J. A., and H. A. Orr, 2004. *Speciation.* Sinauer, Sunderland, MA.

Grant, P. R., and B. R. Grant, 2008. *How and Why Species Multiply: The Radiation of Darwin's Finches.* Princeton University Press, Princeton, NJ.

Grant, P. R., and B. R. Grant, 2009. The secondary phase of allopatric speciation in Darwin's finches. *Proc. Natl. Acad. Sci. U.S.A.* (doi:10.1073/pnas.0911761106).

Hey, J., W. M. Fitch, and F. J. Ayala (eds.), 2005. *Systematics and the Origin of Species on Ernst Mayr's 100th Anniversary.* The National Academies Press, Washington, DC.

Huber, S. K., L. F. De León, A. P. Hendry, E. Bermingham, and J. Podos, 2007. Reproductive isolation of sympatric morphs in a population of Darwin's finches. *Proc. R. Soc. Lond. (B),* **27,** 1709–1714.

Ready, J. S., I. Sampaio, H. Schneider, C. Vinson, T. Dos Santos, and G. F. Turner, 2006. Color forms of Amazonian cichlid fish represent reproductively isolated species. *J. Evol. Biol.,* **19,** 1139–1148.

Rieseberg, L. H., B. Sinervo, C. R. Linder, M. C. Ungerer, and D. M. Arias, 1996. Role of gene interactions in hybrid speciation: Evidence from ancient and experimental hybrids. *Science,* **272,** 741–745.

Mass Extinctions, Opportunities and Adaptive Radiations

18

KEY CONCEPTS

- Extinctions may be small scale (single populations) or massive (entire biotas).
- Most dinosaurs and many other terrestrial and marine groups died out at the end of the Cretaceous Period 65 Mya.
- The Late-Cretaceous mass extinction is the only such event that can conclusively be associated with the impact of an asteroid.
- Stem mammals survived the mass extinction that removed most dinosaurs 65 Mya.
- Ecological factors — the opening up of niches vacated by the dinosaurs and the origination of new ecosystems as a consequence of continental drift — facilitated mammalian evolution and adaptive radiation.
- Mammalian evolution teaches us that dominance at one time is not a necessary recipe for continued dominance; long-term evolutionary success is unpredictable.
- Although most dinosaurs became extinct at the end of the Cretaceous, the lineage that gave rise to birds persisted.

Above: Fossils, such as this dinosaur, document extinction and, therefore, past life.

345

- Pterosaurs, pterodactyls and birds independently evolved the ability to fly.
- Feathers evolved only in birds and in avian ancestors, the feathered dinosaurs.
- Because birds are reptiles (living dinosaurs) we now speak of avian reptiles and non-avian reptiles.
- Insects underwent the most explosive radiations of any animals, diversifying into 900,000 extant described species and perhaps as many as eight million undescribed species.

Overview

At the end of the Cretaceous 65 Mya, most dinosaurs along with other large marine reptiles and various invertebrates died out. The most well supported theory to explain these extinctions involves the impact of an asteroid. The rare element iridium, often found in meteorites, is found worldwide in a stratum at the Cretaceous-Tertiary boundary, as are quartzes that only form as a result of high impact. The Chicxulub crater, a 193-km–diameter crater off the Yucatán Peninsula, is thought to be the site of impact.

Stem mammals survived the mass extinction that removed the nonavian dinosaurs. Opening up of niches vacated by the dinosaurs, the origination of new ecosystems as a consequence of earlier continental drift and the evolution of other groups such as land plants facilitated mammalian evolution and adaptive radiation. Mammals have much to teach us about survival, radiation, diversification, adaptation to new environments and parallel evolution.

Adaptations for sustained flight appeared independently in two lineages of reptiles: pterosaurs/pterodactyls and birds. Pterosaurs developed hollow bones and flight membranes between the body and the wings, and in some cases lost the tail and teeth, leaving the jaws as a beak. Birds arose from bipedal, ground- or (not tree-) dwelling small carnivorous dinosaurs. On the paleontological level, little was known of evolutionary relationships among birds until recent discoveries of fossil birds in China and elsewhere.

Feathers are homologues of and derived from reptilian scales, and likely evolved as an insulating mechanism and/or as devices for display.

With their ability to swim, fly and crawl, adapt to evolving land plants and develop social organization, insects have undergone perhaps the most extensive radiation of any animals since the Cambrian. Whether such radiations involve increase in complexity is considered.

Extinction

Extinction is the flip side of speciation; species arise and species disappear.

Extinction has appeared in several contexts through the text. As discussed in Chapter 2, once the realities of fossils and extinction were accepted, it was possible to conceive of a "law of succession" in which one form replaced another: evolution. The

"poster-child" for evolution from the fossil record, the evolution of horse lineages, was shown to involve many speciation and extinction events (see Box 16.1). Because species cannot always adapt to large or rapid environmental changes, extinction was seen to be common if not inevitable. Populations or subsets or populations can crash or become extinct because of environmental catastrophes; the Ediacaran Biota completely disappeared, as did a major group of Cambrian arthropods, the trilobites (see Chapter 11).

Extinction can conveniently be considered at several levels of increasing severity and impact.

- Extinction may be local, specific and not result in extinction of the species, as might occur if a flood or avalanche wiped out a population of a species but spared other populations. Genetic variation in a species might be reduced by such local events but the species would survive.
- Extinction may eliminate an entire species, as might occur if the species consisted of one or few local populations, all of which were wiped out by an environmental event (flood) or by a biological event such as elimination of a specific food item, or a species-wide disease.
- Extinction may eliminate all the species in a region or ecosystem, as might occur following a volcanic eruption.
- Extinction may be of much larger scale, eliminating most of the species on a continent or on Earth. Snowball Earth, discussed in Chapter 11, is an example. Elevation in CO_2 levels discussed in Chapter 8 is another. Collision between Earth and an asteroid (this chapter) is a third.

Our ability to project present-day mechanisms into the past does not mean following a philosophy restricted to gradual change, such as the one Lyell expounded to explain all events on Earth (see Chapters 1 and 2). Astronomers have demonstrated that the universe is a violent place. It had an abrupt birth. Its stars and galaxies were born in the midst of violent interactions. Its elements originated from the debris of many violent episodes, and violent impacts still occur — even in our small solar system. Whether from impacts, volcanism or plate tectonics (see Chapter 2), occasional catastrophes in Earth's history have had major effects on organisms and so on evolution. Effects include mass extinctions and the elimination of ecosystems. A mass extinction opens up niches and so can serve as an important factor facilitating the radiation of organisms that survived the mass extinction event.

Extraterrestrial Impacts

As a result of data gathered during the *Apollo* space program, a series of heavy extraterrestrial impacts are known to have battered the moon about 4 Bya. We infer that similar events occurred on Earth's surface. The sterilizing heat such impacts generated, whether caused by the collision of asteroids or comets with even larger planet-like bodies may be characteristic of the period after which life arose. Although their frequency greatly diminished, impacts persisted. Geologists have identified more than 100 craters on Earth. Ten meteors, each one km in diameter, are estimated to have each produced 20-km–wide craters at a frequency of one every 400,000 years. A 50-km–wide crater is produced every 12.5 My, a 150-km–wide crater every 100 My.

Depending on their size, objects causing such impacts may have had enormous and long-term environmental effects. The impact crater throws large numbers of particles into the atmosphere, producing dust clouds that interfere with photosynthesis, causing the food chain to collapse in various localities. Depending on whether the impact is on land or at sea, it can cause large climatic temperature changes, which may run through a cycle of an immediate "winter" from the dust thrown up, followed by a high-temperature "greenhouse effect" from the water vapor released. Heat generated by the object entering the atmosphere and the heated material it ejects on impact can spark raging forest fires and generate nitrous oxides that seed acid rains that destroy vegetation and marine organisms.

Given such potential effects, can asteroid impacts be shown to have initiated mass extinctions?

Mass Extinctions

Mass extinctions are not singular events. As far as can be concluded from fossil data, there have been five major mass extinctions since the Cambrian (TABLE 18.1), each marked by the relatively abrupt disappearance of at least 75 percent of marine animal species. Extinction events on the order of the type in Table 18.1 are estimated to have occurred on average about every 100 My (FIGURE 18.1).

Mass extinctions are allied to other less massive extinctions, perhaps caused by a series of impacts from extraterrestrial bodies from the Cambrian Period on. As might be expected, extinctions with less effect are more common, but the average interval between such major events is long enough that most marine invertebrate species could face extinction from such an event. For example, an event that eliminates five percent of species occurs on average once every million years. If the life spans of marine invertebrate species are of the order of four My, all marine invertebrate species would have experienced such an event.

What about land plants and animals? One major extinction event can be shown to have resulted from collision between Earth and an asteroid. That event, which

TABLE

18.1 Details of the Five Major Mass Extinction Events Since the Cambrian[a]

Extinction Period	Date (Mya)	Estimated % of Extinction of Marine Animals	
		Genera	Species
Late Ordovician	440	61	85
Late Devonian	365	55	82
Late Permian	245	84	96
Late Triassic	208	50	76
Late Cretaceous	65	50	76

[a]Each mass extinction was marked by the relatively abrupt disappearance of at least 75 percent of marine animal species.
Source: Raup, D.M. and J.J. Sepkoski, Jr., Science 231 (1986): 833–835.

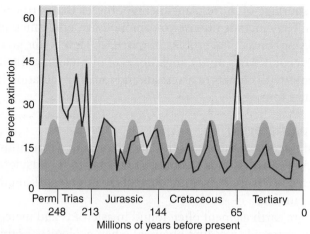

FIGURE 18.1 Percentage of marine animal extinctions during the 260-My interval between the Permian and the present (*solid line*). Areas shaded in green are based on a periodicity of an extinction event every 26 My. (Adapted from Fox, W. T., *Paleobiology* **13**, 1987: 257–271.)

occurred at the end of the Cretaceous Period, 65 Mya, resulted in the extinction of many animals (including most dinosaurs) and many plants, opening up niches for an extensive radiation of mammals; extinction of one group creates an opportunity for others. This mass extinction is often referred to as the **K–T extinction** event, K from the German word for the Cretaceous Period, *Kreidezeit,* and T from the Tertiary Period.

Dinosaurs and Late Cretaceous Mass Extinctions

No large dinosaurs survived beyond the end of the Cretaceous Period 65 Mya. Indeed, no land vertebrate larger than about 23 kg survived into the Tertiary Period. For perspective on size and extinction, a Labrador dog weighs around 25 kg, a horse, 500 kg. Other terrestrial and marine organisms became extinct at the same time; paleontologists estimate that more than half of all species of reptiles became extinct during a relatively short geological period at the end of the Cretaceous. For many groups of animals, species/generic diversity was no higher five to ten My after extinction than before, that is, many groups had not recovered and many did not recover from the extinction event.

Many possible causes have been proposed to account for such a wide spectrum of extinction. A once popular hypothesis, now abandoned, sought to explain the extinction of dinosaurs by internal rather than external causes. Just as individuals are born, grow old and die, this hypothesis suggested that species and other classification categories follow a similar life history driven by internal factors (**orthogenesis**) that cause evolution to proceed in a direction unrelated to selection and adaptation. Evidence used to support orthogenesis included the appearance of bizarre and what appeared to be nonadaptive characters. Some long considered the large, seemingly clumsy 3.3-m–wide antlers of the Irish elk *Megaloceros* (see Figure 15.5) to be a cause of its senescence and extinction. However, no evidence supports any biological mechanisms other than the inability to cope with changing environmental or competitive challenges as explaining the extinction and replacement of dinosaurs or any organism.

There is no shortage of more plausible hypotheses than orthogenesis for these mass extinctions. They include intense volcanic activity, epidemics, changes in plant composition, shifting continental profiles, elevated CO_2 levels (the greenhouse effect), changes in sea level or ocean salinity, high doses of ultraviolet radiation, dust clouds caused by collisions with comets or asteroids and ionizing radiation from supernova explosions or other sources.

Collision with an Asteroid

The most generally accepted theory for the late Cretaceous extinctions is the collision of an asteroid with Earth. The **Collision Theory** gathered considerable support in the 1980s after the discovery of iridium deposits in strata marking the Cretaceous–Tertiary (K–T) boundary.

Iridium is a rare earth element often found in asteroids and meteorites. Evidence linking iridium to asteroids was first obtained when geologists examined an area of destruction of all the trees within a 14.5-km radius of a small area of the remote Tunguska region of Russia, and the flattening of any trees within a 40-km radius. High levels of iridium in the soil were interpreted as meteoritic in origin, the result of a 30-m–diameter meteor that exploded 8 km above the ground in 1908. The worldwide presence of iridium in Cretaceous-Tertiary boundary strata, along with high-impact particles — glasslike spherules and shocked, fractured quartz (see Chapter 2) — strongly indicated collision with an extraterrestrial body. An anomalous iridium-rich layer at the K–T boundary, now identified at more than 100 different localities, lends credence to a large impact at the end of the Cretaceous. Such an explosion would have carried the iridium and high-velocity particles to the top of the atmosphere and spread them worldwide. Good evidence for a close association with a foreign body impact exists *only* in the case of the Late Cretaceous extinction.

Many paleontologists demonstrated, however, that dinosaurs and other animals had already declined in numbers or disappeared *before* these impact layers were deposited. Contrary to the immediate effects of an extraterrestrial impact, dinosaur extinction may have taken a million years or more; no unequivocal non-avian dinosaur bones are known after the K–T extinction event. Some paleontologists proposed a combination of stressful environments and an extraterrestrial impact, that is, **both gradual and catastrophic explanations.** Some species of other groups, including small mammals, both survived this major extinction event and expanded. Small size and efficient control of body temperature may have been crucial for the survival of terrestrial vertebrates in the Late Cretaceous. Traits that were of benefit during extinction might not have been the traits that were most advantageous before extinction.

Chicxulub Crater

Among the best candidates for a crater large enough to record the impact of an asteroid that could have produced such global effects in the Late Cretaceous is the **Chicxulub crater.** Discovered in 1993 off the coast of the Yucatán Peninsula in Mexico, Chicxulub has an outer diameter of about 195 km. According to its discovers, Virgil Sharpton and others, Chicxulub "records one of the largest collisions in the inner solar system since the end of the early period of heavy bombardment almost four billion years ago. . . . Earth probably has not experienced another impact of this magnitude since the development of multicellular life approximately a billion years ago."

Does this impact account for all the Late Cretaceous extinctions? Not necessarily. Paleogeologists have proposed that deposition of iridium globally in atmospheric

dust by the most violent volcanic eruptions and ash could have been a primary or auxiliary cause for these extinctions. An enormous outpouring of nearly 4.1 million km^3 of volcanic lava — the "Deccan Traps" — which covered one-third of present day India during the Late Cretaceous extinctions would support this hypothesis. Other evidence shows that the meteor fell between eruptions of plumes of deep mantle material. Both theories may be correct.

No firm evidence supports hypotheses that extinctions other than the Late Cretaceous mass extinction were initiated by meteor or asteroid impacts. Iridium deposits in strata associated with other extinctions are not great enough to assume extraterrestrial impact. Strata deposited after the greatest of all extinctions, the Permian-Triassic (P–T) extinction 251 Mya (Table 18.1) contain almost no iridium. The P–T extinction event is referred to as the Great Dying; as much as 96 percent of all marine species, 70 percent of terrestrial vertebrate species and 85 percent of terrestrial insects went extinct in what may have been three pulses of extinction over 165,000 years. Proposed mechanisms include release of methane from the sea floor, droughts of long duration and decrease in atmospheric oxygen associated with climate change.

■ Even when non-explosive, volcanic eruptions can affect world climate; the eruption of a chain of volcanoes in Iceland in 1873 produced a cloud of ash that stretched across Europe into parts of Asia and Africa, causing famines as far away as the Nile Valley in Egypt, where a sixth of the population died. Chemicals in the ash, especially sulfur dioxide and hydrogen chloride, produced acid rain that destroyed crops and killed farm animals.

Radiations of Mammals

Some species survived the asteroid impact, including small mammals, although we should recognize that traits such as small size that were of benefit during extinction might not have been the traits that were most advantageous before extinction, when large dinosaurs dominated.

Mammals not only survived the K–T boundary, they thrived, radiating into many forms and diversifying into many new habitats, both terrestrial and aquatic, making use of new adaptations such as specialized limbs. Extinction of the non-avian dinosaurs allowed mammals to invade niches (see Box 1.3) previously occupied by herbivorous and carnivorous dinosaurs.

An earlier significant stimulus for mammalian radiation was the breakup of the large Pangaea landmass that began in the Triassic Period, 225 Mya (BOX 18.1). Both the breakup of Pangaea and the movements of tectonic plates after the K–T extinctions established new land masses initially (225 Mya), the equivalent of Africa and South America as one land mass and India, Antarctica and Australia as a second. Within just over 80 My South America had separated from Africa and North America from what is now Greenland (Box 18.1). The evolving connections and separations of landmasses facilitated dispersal and/or isolation of major mammalian groups.

To these land movements with their marked effects on climate and environment, and their production of new geographical and ecological regions, can be added

■ You will appreciate that "North America" (or any other modern continent) did not exist in their current form in the Cretaceous Era. Names such as North America or Australia are used as a convenient shorthand for the location on the globe being referred to.

- uplifting of mountain systems that took place from the Cretaceous onward leading to chains such as the Rockies, Andes, Alps and Himalaya;
- submersions and regressions of shallow seas; and
- delineation of new shorelines.

Changes in vegetation, especially the diversification and radiation of ferns, flowering plants and plant-insect associations discussed in Chapter 10 took place during these periods leading to new landscapes of grasslands, savannas, and forests (FIGURE 18.2). This origination, modification and shaping of new and different habitats

BOX 18.1
Fragmentation of Pangaea

Drifting continents result from the action of plate tectonics (see Chapter 2).

Geography in the Devonian, 375 Mya, shows two major landmasses, the **Gondwana** group of continents and a North American-Eurasian group called **Laurasia** (see Figures 2.6 and 2.7). By the end of the Paleozoic, 100 My later, these two major continental groups had united to form the giant landmass **Pangaea** (see Figure 2.7). In turn, Pangaea began to break up during the Triassic, about 225 Mya.

Fragmentation of Pangaea began with the separation of Western Gondwana (South America and Africa) from Eastern Gondwana (India, Antarctica and Australia). By the Late Jurassic, 145 Mya, sea-floor spreading had begun to separate South America from Africa. By the Cretaceous, 142 Mya, North America had separated from Greenland and South America from Africa.

In the Western Hemisphere, the rapid drift of South America away from Africa, which began about 100 Mya, led eventually to a reunion with North America some 4 to 5 Mya. In the Southern Hemisphere, New Zealand had drifted away from the Australian-Antarctican-South American landmass before the end of the Cretaceous. The other Southern continents were joined until the beginning of the Tertiary about 65 Mya. However, by the Eocene, 20 My later, Australia had begun the northward journey that would eventually unite it with Asia.

had a major impact on mammalian adaptation, variation and distribution resulting in the evolution of some 4,000 genera of fossil mammals and 5,400 species of extant mammals.

Radiation into South America

A colony of North American marsupials reached Europe during the Early Eocene, 50 Mya. Other short-lived lineages made their way to Asia and Africa, probably through a North Atlantic–Greenland–Europe connection. Extinctions also played a role in these later stages of mammalian evolution. During the Miocene, marsupial populations of both northern continents became extinct. South and North America united in the Pliocene 5.3 Mya, re-establishing a North American–South American land bridge. An extensive interchange between the mammals of these two continents followed with marsupials invading North America from the south and placental mammals invading South America from the north (BOX 18.2). Many of the South American marsupials became extinct as invading North American placental mammals diversified rapidly and took their place. Here extinction resulted from competition for the same habitats. As elsewhere, extinction and radiation in South America seemed to go hand in hand, testifying again to the basic opportunism of evolutionary change. An important lesson is that the survival or extinction of any group or lineage may be closely connected to the survival or extinction of other groups or lineages; **coevolution and coextinction.**

Mammalian radiation continued through the Pleistocene (1.8 Mya to the present), an epoch that marked the appearance of many mammals in their current forms. Climatically, the Pleistocene also marks a period of at least **seven glaciations** (the Ice Ages), which at times covered one-third of the Earth's surface. Woolly mammoths and woolly rhinoceroses made their appearances in the northern continents during this interval, along with giant deer, giant cattle and large cave bears. Interestingly, these large mammals, in addition to horses, camels, ground sloths and various other groups,

■ New Zealand apparently separated from the Gondwana continents during the Cretaceous. Only native marsupial mammals and bats had been found there until recently when a mouse-size mammal was discovered in Miocene deposits, the age of which is consistent with this mammal having been present before the divergence of marsupial from placental mammals.

(a) Early Eocene—50 millions years ago

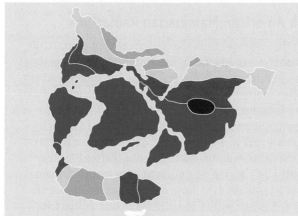

(b) Early Oligocene—32 millions years ago

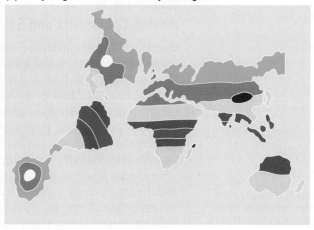

(c) Late Miocene—10 millions years ago

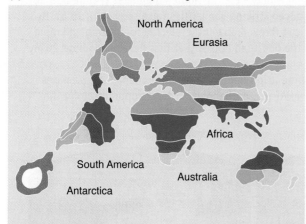

North America

Eurasia

Africa

South America

Australia

Antarctica

■ Tropical forest

■ Paratropical forest (with dry season)

☐ Subtropical woodland (broadleafed evergreen)

☐ Polar broadleafed deciduous forest

■ Woody savanna

■ Temperate woodland (broadleafed deciduous)

☐ Temperate woodland (mixed coniferous and deciduous)

☐ Grassland/open savanna

■ Mediterranean-type woodland/thorn scrub/chaparral

■ Tundra

☐ Ice

FIGURE 18.2 Positions of continental landmasses at three stages over 50 My during the Cenozoic Era, showing the distribution of different kinds of vegetation during the **(a)** Early Eocene, **(b)** Early Oligocene, and **(c)** Late Miocene. (Adapted from Janis, C. M. *Ann. Rev. Ecol. Syst.*, **24**, 1993: 467–500.)

all became extinct in North America about 11,000 years ago. Among approximately 79 mammalian species weighing more than 45 kg, 57 (72%) became extinct at that time; an Alaskan malamute dog weighs around 45 kg. Limited extinction occurred in Europe. Among possible explanations for these Late Pleistocene extinctions is rapid deterioration of the climatic advantages for large animals as the ice sheets retreated.

Another explanation is the predatory role of humans; stone-age hunters entered formerly glaciated areas of North America and Europe at this time (see Chapter 19). One hypothesis is that humans initiated the first "man-made extinctions" at this time by killing these mammals for food, as also occurred in Australia. A species of giant kangaroo, *Protemnodon anak*, 2 m tall and weighing 100 to 150 kg disappeared from Tasmania within a few thousand years of the arrivals of humans 41,000 to 43,000 years ago. Its loss was almost certainly due to hunting, although climate and/or vegetation change contributed to the disappearance of elements of the megafauna on mainland Australia. The invasion of Australia by humans, both during the Pleistocene and more recently, was accompanied by other placentals, including dogs, rabbits, sheep and rodents, all of which had an impact on the indigenous fauna.

BOX 18.2

Moving Continents and South American Mammalian Radiations

The picture of mammalian evolution that emerges from the pattern of continental drift, outlined in Box 18.1, is that early monotremes and marsupials entered southern parts of Pangaea by the Late Jurassic and Early Cretaceous (FIGURE B18.1a). The subsequent rifting of Australia isolated its mammalian fauna from later competition, with more derived eutherian groups evolving in western Pangaea during the Late Cretaceous and Early Tertiary (Figure B18.1b). In South America, marsupials and early placentals had replaced monotremes by the Early- and Mid-Tertiary, but by that time the South American continent had drifted considerably from Africa and separated from North America (Figure B18.1c).

The South American placentals, although beginning only with some ungulates and xenarthrans ("strange-jointed"), radiated perhaps even more rapidly than did marsupials on that isolated continent. By the Early Eocene, within 15 to 20 My of their initial Late Cretaceous colonization, placentals comprised 75 to 100 new genera in some 15 families. The xenarthrans produced a strange bestiary of armadillos, glyptodonts, sloths and anteaters (FIGURE B18.2a). Also radiating widely were the hoofed ungulates (Figure B18.2b). Convergent or parallel evolution produced striking similarities (also see Figure 16.6). By the Early Miocene some South American forms, apparently selected for grazing and rapid running, were remarkably similar to the one-toed horses that first developed about 20 My later in North America.

In the Oligocene, a similar rapid radiation began among the rodents and primates that had reached South America from Africa, probably by "island hopping" along the

■ Xenarthrans are also called edentates because of their reduced or suppressed teeth.

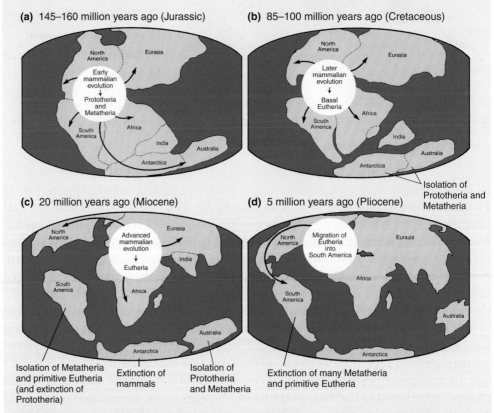

(a) 145–160 million years ago (Jurassic)

North America
Eurasia
Early mammalian evolution
↓
Prototheria and Metatheria
South America
Africa
India
Australia
Antarctica

(b) 85–100 million years ago (Cretaceous)

North America
Eurasia
Later mammalian evolution
↓
Basal Eutheria
South America
Africa
India
Australia
Antarctica

Isolation of Prototheria and Metatheria

(c) 20 million years ago (Miocene)

North America
Eurasia
Advanced mammalian evolution
↓
Eutheria
India
South America
Africa
Australia
Antarctica

Isolation of Metatheria and primitive Eutheria (and extinction of Prototheria) Extinction of mammals Isolation of Prototheria and Metatheria

(d) 5 million years ago (Pliocene)

North America
Migration of Eutheria into South America
Eurasia
Africa
South America
Australia
Antarctica

Extinction of many Metatheria and primitive Eutheria

FIGURE B18.1 Effect of continental drift on the dispersion and isolation of major lineages of mammals over the 140 My spanning the Jurassic to the Pliocene.

BOX 18.2

Moving Continents and South American Mammalian Radiations (*continued*)

island chains on the oceanic ridges cast up in the South Atlantic Ocean. Rodents produced a great diversity of caviomorphs (cavies) distinguished by special jaw muscle attachments. Primates, confined mostly to tropical areas, produced the wide array of New World monkeys (see Chapter 19).

The evolution of mammals on the isolated island of South America was largely independent of mammalian evolution elsewhere, until South America rejoined North America via the Panama Isthmus during the Pliocene (Figure B18.1d). During the Oligocene, however, some island hopping combined perhaps with transport on floating debris ("rafting") occurred as various monkeys and caviomorph rodents made their way to South America from Africa or North America.

By the Pliocene 5 Mya, considerable evolution toward more derived eutherian forms had occurred either in Africa or Laurasia (North America-Eurasia); most of the South American mammalian fauna showed far less change. When the Pleistocene began, massive invasions of northern eutherians south across the Central American land bridge resulted in the rapid extinction of many South American mammalian families. Only rarely, as with opossums, did early South American mammals manage to successfully invade North America. Supporting this view of plate tectonics and mammalian evolution is abundant fossil evidence of South American extinctions in the Pliocene and Pleistocene.

FIGURE B18.2 Reconstructions of early placental mammals of South America: (**a**) three lineages of xenarthrans (edentates) and (**b**) three lineages of ungulates that arose in the South American placental mammalian radiation. (Adapted from Steel, R., and A.P. Harvey. *The Encyclopaedia of Prehistoric Life*. Mitchell-Beazley, 1979.)

Invading the Air: Flying Reptiles

Escape from extinction is an important hallmark of biological survival, and may at times depend on the ability to move rapidly from threatened environments. Flight provides a number of advantages: rapid escape from terrestrial predators and menacing conditions; access to feeding and breeding grounds that would otherwise be difficult or impossible to reach; and relatively swift transit between localities.

Although forms capable of gliding for short distances arose in various vertebrate lineages (including "flying fish"; FIGURE 18.3), known adaptations for sustained powered flight have appeared only three times in the evolution of terrestrial vertebrates: in pterosaurs, birds and bats.

The two forms that arose within the reptiles — pterosaurs and birds — differ in respect to mechanisms of flight and the accompanying adaptations. In pterosaurs a flight membrane (patagium) of skin stretched between the trunk and wing. An elongated fourth finger was present on each hand. In birds, the flying surface consists of many stiff wing feathers that project posteriorly from the wing. Pterosaurs became extinct as a result of the K–T extinction event. Birds, which had originated before the mass extinction, flourished.

Pterosaurs and Pterodactyls

Early pterosaurs appear in the Late Triassic 220 Mya and birds in the Late Jurassic 135 Mya. Their skeletal features provide a number of clear homologies with other late Triassic bipedal archosaurs, including dinosaurs (FIGURE 18.4a).

A well-described Late Jurassic pterosaur, *Rhamphorhynchus* (Figure 18.4a), was about 0.6 m long. The tail was long with a small, rudder-like flap of skin at the end. The bones of *Rhamphorhynchus* and all other pterosaurs were light and hollow, and its elongated jaws were armed with strong, pointed teeth. The breast bone and its accessory bones provided sufficient surface for attachment of large flight muscles as is also seen in "flying birds." Although narrow at the tips, the wings broadened toward the trunk. It is not clear whether the wings attached nearer the waist or the ankle, and this may have varied in pterosaurs. However, their leg joints were structured like those of birds and other dinosaurs, so they would have walked with one

■ "Flying birds" because some birds have lost the ability to fly. Flightless birds are discussed at the end of this chapter.

FIGURE 18.3 A "flying" fish photographed from above showing adaptation of the fins for gliding through the air.

Bird Evolution

Within about 30 million years in the Early Cretaceous a range of aquatic and shore birds had originated, indicative of rapid speciation and habitat exploitation. Some of these fossils represent groups such as flamingos, loons, cormorants and sandpipers. Others still retained teeth indicative of their reptilian origins. Lineages that can be placed into most of the recognized extant orders of birds appear to have originated about 60 to 90 Mya, that is, just before the Late Cretaceous and into the Tertiary Period. Although bird fossils are not plentiful, there are fossils in sufficient numbers and kinds of Eocene, Oligocene and Cretaceous deposits in China to indicate that almost all the major clades of birds (recognized as orders) had evolved by then, some lineages having originated in the Late Cretaceous.

FIGURE 18.8 shows the phylogenetic relationships of birds and dinosaurs. *Archaeopteryx* is shown as a basal bird. *Confuciusornis* is known from 120-My-old (lower Cretaceous) deposits in China. Its beak was toothless, as in modern birds, indicating that teeth were lost early in avian evolution. Large claws on the forelimbs indicate that the several known species of *Confuciusornis* retained the claws seen on *Archaeopteryx*. The Enantiornithines (opposite birds), represented by *Nanantius eos*, comprised a lineage that had many of the features of extant birds — feet that allowed them to perch and well-developed flight — but they did not survive the mass extinction 65 Mya.

The oldest fossil bird with the greatest resemblance to modern birds is a loon-like shorebird, *Gansus yumenensis*, described on the basis of five specimens found in Cretaceous deposits in China. Analysis of *Gansus* and the oldest (already flightless) penguin, *Waimanu*, from the Paleocene on New Zealand, 60 to 62 Mya, when combined with analysis of mitochondrial DNA, have opened new windows on the origin and diversification of birds that could not have been imagined even a few years ago.

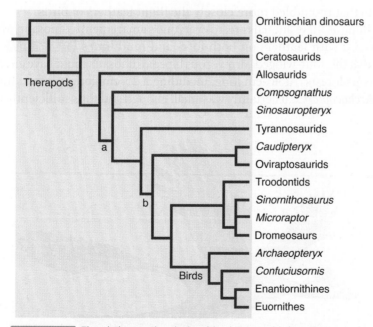

FIGURE 18.8 The phylogenetic relationships between birds and sauropod and ornithischian dinosaurs. Branch **a** indicates the presence of simple feathers; Branch **b** shows the presence of more complex (essentially modern) feathers.

Origin of Feathers

Until recently, the lack of any pre-*Archaeopteryx* fossils with preserved soft tissue kept alive the mystery of when and how feathers first evolved and left the origin of avian flight unresolved. Recent fossil discoveries lead us to ask whether any species with feathers should be called a bird, or whether some more derived dinosaurs also had feathers.

Two major hypotheses for the origin of feathers have been presented

- Feathers were an adaptation for insulating the (presumed) warm-blooded and ground-dwelling reptilian ancestors of birds. Fossils of **feathered dinosaurs** suggest — perhaps even demonstrate — an insulation/display function.
- Ancestral birds were tree-dwelling reptiles that used their developing wings to glide from branch to branch, or were ground-dwelling runners whose feathers formed planing surfaces enabling them to increase their speed.

Overwhelming weight of evidence supports the origin of birds from ground-dwelling dinosaurs.

In the mid 1970s the American paleontologist John Ostrom (1928–2005) presented the view that feathers evolved primarily as a means of controlling heat loss in warm-blooded dinosaurs. The report in 1998 of a mane of feathers down the back of a basal carnivorous bipedal dinosaur, *Compsognathus* (Figures 18.7 and 18.8) caused immediate discussion. A mane of feathers would provide strong support for Ostrom's notion of a feathered-dinosaur origin of birds; flightless birds have feathers with structures similar to those found in non-avian carnivorous dinosaurs.

In the last ten years, species of "feathered dinosaurs" belonging to 14 genera have been discovered. The most basal of these, *Sinosauropteryx* (Figure 18.8) from the Jurassic-Cretaceous boundary, some 150 to 120 Mya, had hollow, feather-like structures on its entire body. *Protarchaeopteryx* and *Caudipteryx,* from 135 to 121 Mya, had feathers that resembled more closely the flight feathers of birds.

Astonishing as these animals are, to everyone's amazement a dinosaur, *Microraptor,* was discovered with feathers on both fore- and hind limbs (**FIGURE 18.9**), strongly suggesting that this four-winged carnivorous bipedal dinosaur could have glided through the Cretaceous forests. At 40-cm long, including a 25-cm–long tail, and thus close to the size of *Archaeopteryx*, it certainly was small enough and thus sufficiently light to fly.

Considerable evidence supports the conclusion that some small dinosaurs could regulate their body temperatures without relying on behavior such as moving into the shade or facing their body profile away from the sun's rays.

FIGURE 18.9 Reconstruction of a four-winged dinosaur, *Microraptor.*

Even more astonishing is the publication by Hu and colleagues in September 2009 of a 35 million-year-older feathered dinosaur, *Anchiornis huxleyi* from the Jurassic Period. *Anchiornis* also had features on both fore- and hind limbs, raising the possibility that birds evolved from a "four-winged" ancestor with evolutionary change involving loss of the feathers from the legs.

Flightlessness

One dramatic adaptive opportunity opened up by the extinction of the dinosaurs was the availability of a large number of vacant terrestrial regions into which various large flightless ground birds evolved. Giant forms, such as the 2.1-m-tall *Diatryma gigantea* (FIGURE 18.10) and others that may have reached a height of 3 m or more were widely distributed until they became extinct later in the Cenozoic. One group of flightless birds, the "thunder birds" (Dromornithidae) from Australia, includes what was perhaps the largest bird ever, *Dromornis stirtoni*, estimated to have weighed 570 kg. Few flightless birds now survive. Ostriches of Africa, emus of Australia, rheas of South America and the smaller flightless species such as kiwis and island rails are mostly confined to diminishing habitats.

The evolution of flightlessness is assumed to have involved changes in selection pressure, caused either by an absence of predation or as a response to developing marine habitats. In the first case, in protected or island habitats where major carnivorous forms were absent, local birds could evolve to dominate the terrestrial food chain. Once assuming such roles, selection for reduction in former flight structures would have reduced their energy expenditure. In marine habitats, many birds "fly underwater," the major difference between flight in air and in water being the density of the medium and not the dynamics or mechanics of this form of locomotion. Some marine birds such as penguins and steamer ducks that spend little time flying in air responded to selection for wing modifications that enhanced underwater propulsion but reduced flight ability.

3.6 m

1.8 m

Diatryma gigantea

Aepyornis maximus
(elephant bird)

Dinornis maximus
(giant moa)

FIGURE 18.10 Reconstructions of some extinct large flightless birds showing their relative sizes. *Diatryma* was an early Cenozoic bird; the others date from the much later Pleistocene. (Adapted from Feduccia, A., *The Age of Birds*. Harvard University Press, 1980.)

Insects Conquer Land and Air

Jointed appendages, a hardened exoskeleton with inner projections for muscle attachments, and highly developed sensory structures on the head structures are among the many features that enable **arthropods** to exploit almost every conceivable ecological habitat. A hardened exoskeleton, along with waxy waterproofing acts as a barrier to desiccation, the ability to burrow into shoreline sand and soil, and protected gills that evolved into organs for terrestrial respiration appear to be major features accounting for the successful invasion of land.

Invasion of the land by arthropods resulted in the evolution of terrestrial spiders, some terrestrial crustaceans, insects, millipedes and centipedes and their relatives. **Insects** underwent what may be the most explosive radiation of any animals since the Cambrian, diversifying into 900,000 extant species and perhaps as many as eight million undescribed species. One group of insects, the beetles (coleopterans), contains 350,000 described species. Alfred Wallace alone collected 80,000 beetles in the Malay Archipelago, 80 of which are shown in FIGURE 18.11 .

Although some early insects reached relatively large sizes — some Carboniferous dragonflies measured 0.6 m between wing tips — insects have tended to remain small; there are limits to the volume of tissue in which insect respiratory tubules can effectively exchange gases. An even more limiting factor may be that the insect exoskeleton would have had to become much heavier and more unwieldy as insects became larger. Sea scorpions could get away with it in the ocean, but arthropods could not do the same on land.

FIGURE 18.11 A small fraction (0.1%) of the beetles collected by Alfred Russel Wallace in the Malay Archipelago.

The enormous diversification of insects has been attributed to the **modular** organization of the insect body in which antennae can evolve independently of wings (see Figure 11.8), mouthparts independently of legs, and so forth — a process known as **mosaic evolution** (see Chapter 15). Homeotic mutations (see Chapter 11), so common in insects, attest to the comparative ease with which individual segments can be altered in a body plan that consists of serially repetitive elements. Such partial independence of body parts or the embryonic units from which they arise is known as **modularity** (BOX 18.3, and see Chapter 13) and the speed with which gene regulation and development can be modified as **tinkering** (see Chapter 11).

The **coevolution** of flowering plants and insect pollinators discussed in Box 10.3 was a further major factor in the diversification of insects. So too was **life history evolution**; advantages accrue to those with specialized larval stages allowing living and feeding differently than adult forms. An aquatic larva with a terrestrial adult or a wingless larva and winged adult can exploit a wide range of environments and diets. Once such highly adaptive traits appeared, arthropods could enter and thrive in almost any part of the terrestrial environment.

Social Organization

Insects diversified into both **social** and **nonsocial forms**; the only **social organization** other than that seen in colonial animals and in vertebrates occurs in insects.

Social organization entails a **division of labor** among different members of a group, a phenomenon far beyond the kinds of simple aggregation exemplified by swarms of migrating locusts or the cooperative "tents" that some caterpillars construct (FIGURE 18.12a). Social insect societies include termites (Figure 18.12b) and various ants, bees, and wasps. The ability to develop different sexes or life history stages from fertilized and unfertilized eggs and the evolution of **kin selection** characterize some forms of social insects in which each colony has a single mated queen (BOX 18.4).

In social insects, different morphological types (castes) or age groups assume different functions. Each colony, for example, usually has only one fertile queen engaged in egg production, often many fertile males for egg fertilization, and one or more classes of sterile workers exclusively engaged in food gathering, cooperative brood care, nest maintenance, and defense of the colony. In worker honeybees, age-related division of labor ties younger bees to caretaking and maintenance, and older bees to food foraging and defense. Completion of the sequencing of the genome of the honeybee, *Apis mellifera,* has already revealed differences in patterns of gene expression between different castes and associated with foraging.

Hormones provide one of the mechanisms linking and integrating environmental changes to gene action and to changes in the phenotype. This is especially evident in organisms such as insects with one or more phases in the life cycle; behavioral differences between castes correlate with differences in juvenile hormone. Chemical influences on development and behavior also are responsible for determining who will be a queen and who the sterile workers. The queen produces a diffusible hormone (a pheromone) that suppresses fertility in workers, a suppression that can be overcome by various environmental influences including special foods such as royal jelly in honeybees.

■ Only one example of caste formation is known among vertebrates. It is in the naked mole rat *Heterocephalus glaber* from East Africa, in which one female produces all the young and in which there is complete sexual dimorphism between reproductive and helper females.

BOX 18.3
Modularity

Because of their multicomponent subunit structure, hierarchies are more stable and more easily constructed than a nonhierarchical system in which all parts must be simultaneously assembled and in which the absence of any single component could interfere with any kind of assembly at all. For example, a cell composed of multiple subsystems (modules) is less vulnerable to accident and more easily synthesized than a similar compartment that has no subsystems but can function only when all of its many chemical components match perfectly and aggregate simultaneously.

Modularity has emerged as an organizing principle at all levels from gene networks (see Chapter 13), through cell populations, to organ primordia. Modularity is being applied in developmental genetics, developmental biology, evolution, and in evolutionary-developmental biology. Because modules subdivide what would otherwise be large biological units, and because modules are subject to selection, the concept of modularity is fundamental to understanding how the phenotype arises; modules link the genotype with the phenotype.

Modules have a distinctive fate, internal integration, and the ability to interact with other modules. As with homology and reflecting the hierarchical organization of life, modularity is a hierarchical concept. A limb bud is a developmental module in a chicken embryo and forms either a wing or a leg, both of which are modular components of the adult. Anterior and posterior limb buds share some modular gene networks, that, for example, result in the differentiation of cartilage or muscle, but differ in other networks — those that specify whether a limb will be a wing or a leg, for example.

Modularity is not merely a theoretical construct; modules can be tested for. If a region of an early chicken embryo from where a wing is known to develop is grafted to another region of the embryonic body, the transplanted piece forms a wing in the ectopic position. The gene network for "wing" that is activated in the ectopic position, and which initiates wing formation, is acting as a modular gene network and conferring a higher modular property onto cells in the new location. Gene networks as one example of the modularity of development were discussed in Chapter 13 using the example of sensory modules in surface and blind Mexican cavefish. As the cavefish example illustrates, a gene network may operate in several tissue organs or even regions of an organ, change in one, but not in the others.

An example of modularity at cellular and genetic levels is illustrated in FIGURE B18.3 and B18.4 . What appears to be a single bone (the dentary) making up the mammalian lower jaw is comprised of six modules, each of which arises from a separate population of cells (Figure B18.3). Each module is subject to independent genetic control, as shown by the effects of knocking out single genes, in which case one module can be prevented from developing or diminished in size, while other modules develop normally (Figure B18.5). An example of the result of selection on such modules is shown in the extremely reduced dentary of a marsupial, the Western Australian honey possum *Tarsipes rostratus* (FIGURE B18.5). Honey possums feed by licking the nectar from flowers. Mice, other rodents and voles chew and gnaw their food. Comparison of the lower jaws of honey possums (Figure B18.4) with those of mice or voles (FIGURE B18.6) illustrates the extreme reduction of the honey possum dentary.

Both in development and in evolution, modularity provides a means for independent yet integrated development of embryos and for mosaic evolution in response to selection; modules are subject to natural selection and so evolve. Modules evolve in much the same ways as genes, namely by

- *duplication,* with one of the duplicates acquiring a new function;
- *dissociation,* in which a module separates either in space (homeotic mutations, for example) or in time and acquires a new function; and/or
- *co-option,* the incorporation of one module into another.

BOX 18.3
Modularity (*continued*)

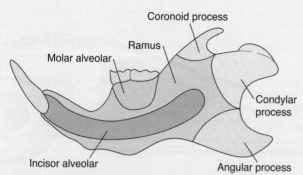

FIGURE B18.3 The single dentary bone of the rodent lower jaw is comprised of six modules that form a unified single, but complex element; see Figure B18.6 for the dentary in a meadow vole. The molar- and incisor-alveolar units are derived from the cell population that forms the molar and incisor teeth, respectively. The ramal population, a separate cell population not associated with tooth formation, forms the body of the dentary. The coronoid, condylar and angular processes each derive in part from the ramal population and in part from a cartilage-forming population of cells at the tip of each process.

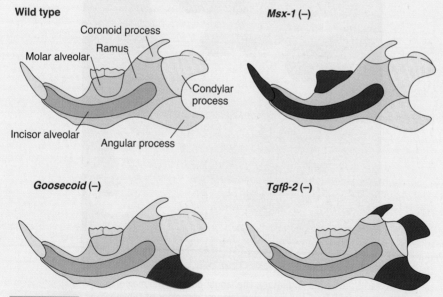

FIGURE B18.4 The modules in the single dentary bone of the rodent lower jaw are subject to independent genetic control as illustrated by the effect of knocking out individual genes. The wild type shows six modules (also see Figure B18.3). Modules that fail to develop or are underdeveloped in knockout mice are colored dark brown. Lack of the gene *Msx-1* [shown as *Msx-1* (−)] prevents the molar and incisor alveolar units from developing; other modules develop normally. Lack of *Goosecoid* results in smaller coronoid and angular units but does not affect the growth or shape of the other modules. Lack of *TGF β-2* results in underdevelopment of all three posterior processes.

(continued)

BOX 18.3
Modularity (*continued*)

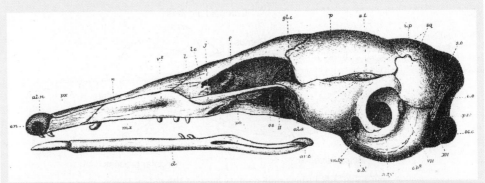

FIGURE B18.5 The Western Australian honey possum, *Tarsipes rostratus*, has perhaps the most reduced lower jaw and teeth of any mammal. The teeth consisting of peg-like molars situated midway along the dentary bone (**d**) and a forward-projecting incisor in the slender and much reduced dentary bone. Compare with the vole shown in Figure 18.6. (Reproduced from Parker, W. K. *Stud. Mus. Zool. Univ. Coll. Dundee* 1890: 79–83.)

■ The honey possum has the lowest weight at birth (<5 mg) and the longest sperm (360 mm) of any mammal.

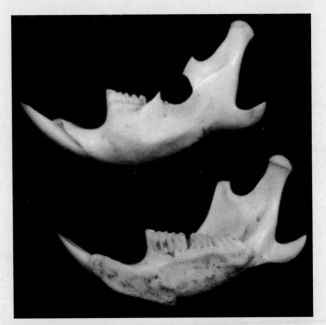

FIGURE B18.6 A dentary, the single bone of the mammalian lower jaw, from the Magdalen Island subspecies of the meadow vole, *Microtus pennsylvanicus magdalensis*, shown in lateral (upper) and medial (lower) views. Note the three prominent bony processes at the posterior end of the dentary (right). Compare with Figures B18.3 and B18.5.

(a) **(b)**

FIGURE 18.12 A swarm of caterpillars in and on the temporary "tent" they have constructed (**a**) and a termite mound in the Kakadu National Park in northern Australia (**b**).

BOX 18.4

Kin Selection in Social Insects

In termites, both sexes arise from fertilized eggs, but in bees, wasps and ants (hymenoptera) males develop from unfertilized (haploid) eggs and females from fertilized (diploid) eggs. This hymenopteran system of sex determination is especially prone to evolve sociality; the diploid female offspring of a queen share more genes with their sisters (75%) than they share with their own daughters (50%). Benefits of sociality include communal nesting, joint protection against predators and parasites, and sharing foraging for food.

Genetic relationship leads to a conflict of interest for a colony's sex ratio. Being more related to sisters, it is to the workers' genetic benefit that the colony supports more females and fewer distantly related males. Queens, by contrast, being related equally to sons and daughters, benefit genetically when males and females are produced in equal numbers. When the queen mates more than once, genetic relationships between female workers decline, but a worker is still more closely related to her mother's female offspring than to her sister's female offspring, which accounts for the destruction of workers' eggs by other workers.

This system of sex determination encourages what John Maynard Smith called *kin selection*, a concept developed by the British evolutionary biologist W. D. Hamilton and discussed in Box 15.2. It is to the genetic advantage of females to invest their energy in raising sisters — sisters are more closely related to them — than in producing daughters, who are more distantly related. In evolutionary terms, this means that the altruistic behavior (Box 15.2) of sterile female workers helping raise more sterile worker progeny for their mother (the queen) would be expected to arise repeatedly in hymenoptera. The independent evolution of altruism at least once in ants, eight times in bees and twice in wasps, bears this out.

Complexity

Extinction and replacement of species or biotas often raises the issue of whether **complexity** has increased during organismal evolution. Are mammals more complex than reptiles? Are organ systems composed of multiple modules more complex than those based on a single module? Are social insects more complex than nonsocial species? Complexity was introduced in Chapter 1, discussed in the context of the nature of life in Chapter 4, the evolution of complex organ systems in Chapter 6 (the eye; see Figure 6.4) and the prokaryote–eukaryote transition and the origin of multicellularity in Chapters 8 and 9. Historical context for the concept of complexity is provided by the concepts of "The Great Chain of Being" (see Box 5.3) and of progress toward perfection (see Chapter 1). The opportunity is now taken to consider complexity in the context of whether evidence can be obtained to support or refute the hypothesis that evolution is associated with increase in complexity. Such a hypothesis assumes that complexity is an attribute of organisms that can be measured.

Measuring Complexity

It seems self-evident that multicellular organisms (animals, plants, fungi) are more complex than the unicellular organisms discussed in Chapter 8. But are some multicellular organisms more complex than others? Are humans more complex than fungi? Are flies more complex than worms?

Evolutionary biologists shy away from the concept that evolution results in increasing complexity. To quote from *The Major Transitions in Evolution,* an influential 1995 book by Eörs Szathmáry and the late John Maynard Smith, "There is no theoretical reason to expect evolutionary lineages to increase in complexity with time, and no empirical evidence that they do so. Nevertheless, eukaryotic cells are more complex than prokaryotic cells, animals and plants are more complex than protists, and so on." Criteria used to measure complexity include

- **genome size** or the total number of genes in an organism;
- **gene (copy) number** or the number of copies of a gene in a given gene family resulting from gene duplication;
- **increase in the size** of organisms over the course of evolution;
- **the number of genes** that encode proteins;
- **the number of parts** or units in an organism (where parts might be segments, organs, tissues, and so forth);
- **the number of cell types** possessed by an organism;
- **increased compartmentalization**, specialization, or subdivision of function over the course of evolution;
- **the number of** gene, gene networks or cell-to-cell **interactions** required to form the parts of an organism; and/or
- **the number of interactions between the parts** of an organism, reflecting increasing functional complexity and/or integration over the course of evolution.

Several of these criteria — notably genome size, gene number, gene networks, and compartmentalization — are discussed in other chapters. Perhaps the most commonly used criteria — the number of cell types and increase in organismal size — are discussed below before evaluating life style evolution in the context of complexity.

Increase in Numbers of Types of Cells

Increase in complexity is perhaps most readily seen during animal embryonic development in the transformation from a single-celled zygote to a multicellular organism with as many as several hundred different cell types. Perhaps as a consequence of parallels between development and evolution (see Chapters 4 and 13), the most commonly used metric of complexity is the number of different types of cells possessed by an organism.

As illustrated in FIGURE 18.13, numbers of cell types have increased over the course of animal evolution. Estimates range from 6 to 12 cell types in sponges and cnidarians, 20 to 30 in flatworms, 50 to 55 in mollusks, arthropods, annelids and echinoderms to as high as 200 to 400 in humans (Figure 18.13). It is difficult to escape the conclusion that, as occurs during embryonic development and using the measure of numbers of cell types, complexity has increased during the evolutionary history of life on Earth.

Increase in Organismal Size

Embryonic development (and so increasing complexity) is accompanied by increasing size. As discussed in Chapters 8 and 9, early stages of the evolution of life were accompanied by an increase in organismal size. However, increase in organismal size over the evolution of a lineage is not routinely used as a criterion of complexity because no sustained size increase occurs within many lineages. Furthermore, evolution often leads to *decrease in size* and for very good reasons. Decrease in size is seen

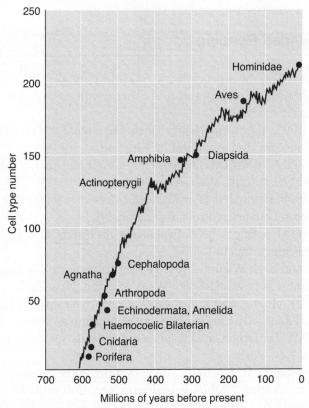

FIGURE 18.13 Time of origin of various animals (My before present) with the estimated numbers of cell types in each. The marked increase from low to high is considered by many evolutionary biologists to reflect an increase in complexity. (Adapted from Valentine, J. W., A. G. Collins, and C. P. Meyer, *Paleobiology* **20**, 1994: 131–142.)

in parasites in many different groups of animals, in organisms that live between the sand grains at high tide levels on beaches, or in large animals such as elephants after several generations of isolation on islands.

Life Style

When complexity is measured by criteria other than organism size different groups of organisms of the same size can show different levels of complexity.

Once relatively large potential host organisms evolved, **parasitism** became a successful way of life for many invertebrates. As parasites evolved, many of their organs become simplified or disappeared altogether. This is not the case for the **life cycle of a parasite**, which may become enormously complex and involve multiple intermediate hosts. Some tapeworms, for example, pass through a number of intermediate hosts ranging from arthropods to fish before the adult stage develops in the primary mammalian host. Such developmental and life history networks take advantage of opportunities provided by the evolution of other organisms with different life histories and which exploit different and often new environments. The adaptive advantage of such life cycle complexity is that a parasite can build up population numbers in intermediate hosts, enhancing the parasite's chance of infecting a primary host. In addition, spreading its early stages among intermediate hosts conserves the resources of the primary host, enabling the adult parasite to remain productive for relatively long periods; adult *Schistosoma* trematodes may live for 30 years in their human hosts. *Schistosoma* is the causative agent of the widespread tropical disease schistosomiasis.

■ Recommended Reading

Aguinaldo, A. M. A., J. M. Turbeville, L. S. Linford, et al., 1997. Evidence for a clade of nematodes, arthropods and other moulting animals. *Nature*, **387**, 489–493.

Benton, M. J., 2003. *When Life Nearly Died: The Greatest Mass Extinction of All Time.* Thames and Hudson, London, England.

Chiappe, L. M., 2007. *Glorified Dinosaurs: The Origin and Early Evolution of Birds.* University of New South Wales Press/John Wiley.

Erwin, D. H., 2006. *Extinction: How Life on Earth Nearly Ended 250 Million Years Ago.* Princeton University Press, Princeton, NJ.

Grimaldi, D., and M. S. Engel, 2005. *Evolution of the Insects.* Cambridge University Press, Cambridge, England.

Hölldobler, B., and E. O. Wilson, 2009. *The Superorganism: The Beauty, Elegance, and Strangeness of Insect Societies.* W. W. Norton, New York.

Hu, D., L., Hou, L. Zhang, and X. Xu, 2009. A pre-*Archaeopteryx* troodontid theropod from China with long feathers on the metatarsus. *Nature* **461**, 640–643.

Kemp, T. S., 2005. *The Origin and Evolution of Mammals.* Oxford University Press, Oxford and New York.

Kielan-Jaworowska, Z., R. L. Cifelli, and Z.-X. Luo, 2004. *Mammals from the Age of Dinosaurs. Origins, Evolution, and Structure.* Columbia University Press, New York.

Krogh, T. E., S. L. Kamo, V. L. Sharpton, et al., 1992. U-Pb ages of single shocked zircons linking distal K/T ejecta to the Chicxulub crater. *Nature*, **366**, 731–734.

Maynard Smith, J., and E. Szathmáry, 1955, *The Major Transitions in Evolution.* Freeman, Oxford, England.

Stanley, S. M., 2009. Evidence from ammonoids and conodonts for multiple Early Triassic extinctions. *Proc. Natl. Acad. Sci. U.S.A.* **106**, 15256–15259.

Turney, C. S. M., T. F. Flannery, R. G. Roberts, et al., 2008. Late-surviving megafauna in Tasmania, Australia, implicate human involvement in their extinction. *Proc. Natl. Acad. Sci. USA,* **105**, 12150–12153.

Vickaryous, M., and B. K. Hall, 2006. Human cell type diversity, evolution development classification with special reference to cells derived from the neural crest. *Biol. Rev. Camb. Philos. Soc.,* **81**, 425–455.

VII

Human Origins and Evolution

Human Origins and Evolution

19

- Humans, along with monkeys, apes and chimpanzees are primates, a lineage of mammals that arose 85 Mya in North America.
- Apes diverged from monkeys 29 to 34 Mya.
- Hominins (apes and humans) diverged from chimpanzees 6.3 Mya.
- Two (possibly three) species of bipedal stem hominins are known from Africa.
- The earliest fossils of the genus *Homo* are 2.4 My-old, four species having existed in Africa between 2.4 and 1.6 Mya.
- By half a million years ago, several additional species had arisen, again in Africa.
- The earliest anatomically modern humans (*Home sapiens*) are known from fossils in Ethiopia 160,000 years ago.
- *Homo sapiens* spread from Africa to Europe, Asia and then to the New World in multiple migrations.
- Chimpanzees, some monkeys and humans use and manipulate tools.
- The first tools associated with Homo (*H. habilis*) are from 2.5 Mya.

Above: Primates share structural and behavioral features.

375

Overview

Primates arose 85 Mya as tree-dwelling, large brained, social mammals. Apes diverged from monkeys in the early Oligocene (29–34 Mya), hominins and chimpanzees diverged 6.3 Mya or later (Figure 19.1). Chimpanzees, the sister group to modern humans, have many behavioral traits once thought unique to humans, including tool use and modification, planning, organized aggression, and complex social networks.

The first bipedal primates (and the earliest hominins) are represented by *Orrorin tugenensis* and *Ardipithecus kadabba,* both of which evolved in Africa. The slightly more recent species of Australopithecines ("southern apes") shows long bone changes indicative of predominant bipedal locomotion, and changes in the teeth and jaws indicative of changing diet. As the brain of australopithecines was no larger than the brain of the earliest hominins, it appears that increase in brain size *followed* the origin of bipedalism. Although this reads like an obvious conclusion, this issue has been contentious.

The earliest fossils of the genus *Homo* are 2.4 My-old with brains much larger than those of the australopithecines. Four African species existed between 2.4 and 1.6 Mya. *Homo ergaster* is regarded as having given rise to *H. erectus,* which was the dominant hominin species at the time and which may have persisted in South-East Asia until 600,000 to 700,000 years ago. By 500,000 years ago, sufficient evolutionary change had accumulated that several new species can be identified; *Homo heidelbergensis, H. helmei,* and *H. rhodesiensis.* These may be forerunners of species known as anatomically modern humans (*Home sapiens*) which first appear as fossils in Ethiopia 160,000 years ago. Populations of anatomically modern humans expanded rapidly, spreading in multiple migrations from Africa to Europe, Asia, and then to the New World: the "Out of Africa hypothesis." It is thought that *Homo heidelbergensis* gave rise to *Homo neanderthalensis* (Neanderthals) in Europe about 250,000 years ago. Neanderthals persisted in Spain and Gibraltar as late as 24,000 years ago, leading to suggestions that they could have coexisted with *Homo sapiens.*

Primates: A Rapid Survey

Primates arose some 85 Mya during the last third of the Cretaceous Period from a lineage of small, insect-eating, tree-dwelling shrew-like mammals (FIGURE 19.1). Abundant early primate fossils are known from the Middle Paleocene, 60 Mya. The structure of the ear region and the nature of the teeth allow these species to be identified as primates.

Primates are a successful group; some 400 species of living primates have been described within two monophyletic lineages.

- **Strepsirrhines** (Strepsirrhini, "wet-nosed primates") are small-bodied, nocturnal (active at night) lemurs of Madagascar and lorises of Africa and South East Asia.
- **Haplorhines** (Haplorhini, "dry-nosed primates") are larger-bodied, diurnal (active during the day) monkeys, apes and humans with shorter faces than strepsirrhines, forward-directed eyes and larger brains (FIGURE 19.2).

Molecular evidence suggests that strepsirrhines and haplorhines diverged about 77 Mya in the Late Cretaceous.

Primates are a group (order) of mammals comprising lemurs, tarsiers, monkeys, apes and humans. All primates share large brains, a grasping hand and stereoscopic vision. Body size ranges from 30 g (smallest lemur) to 227 kg (the largest gorillas).

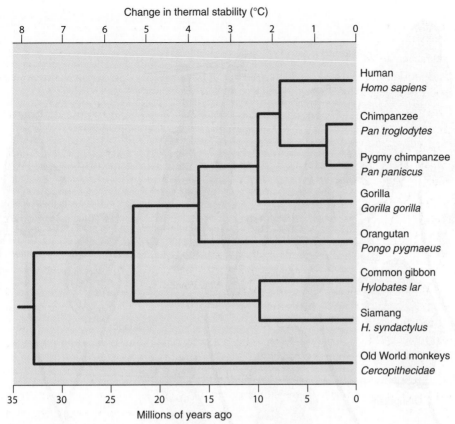

FIGURE 19.1 A phylogenetic tree and dates of divergence for humans, apes, and Old World monkeys based on DNA–DNA hybridization studies. (Adapted from Sibley, C. G., and J. E. Ahlquist, *J. Mol. Evol.*, **20**, 1984: 2–15.)

Three lineages of dry-nosed primates (haplorhines) — New World monkeys, Old World monkeys, and apes and humans — are recognized and discussed below.

Monkeys

Old and **New World monkeys** arose 35 to 40 Mya in what is now Africa.

Extant New World monkeys, which are found only in South and Central America, are tree dwellers with broad noses, widely spaced nostrils that face laterally, and prehensile tails. A third of all extant primate species are Old World monkeys, which are usually larger than New World monkeys, lack a prehensile tail (some are ground dwelling), and have close set nostrils that face downward. Although previously considered a trait unique to humans, both New World (capuchins) and Old World monkeys (chimps and baboons) are known to use tools.

An almost complete 47-million-year-old (mid-Eocene) skeleton unveiled in 2009 is a potentially important transitional stage in early haplorhine evolution. Launched with enormous fanfare, including a book, documentary film, and Web site, *Darwinius masillae* has been interpreted as a link between monkeys, apes and humans on one branch, and lemurs on the other (Figure 19.2a). *D. masillae*, which is a 5-cm–long juvenile female, has a lemur-like skeleton with such primate features as opposable thumbs, nails rather than claws on the fingers but with relatively shorter limbs than monkeys. (Interpretation of this species is already under discussion; an October 2009

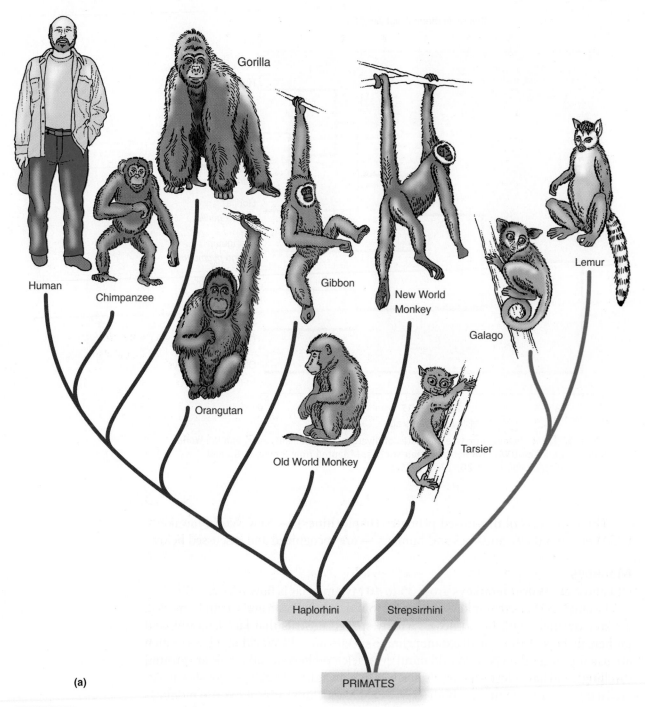

(a)

FIGURE 19.2 (a) Various living representatives of primates as traditionally represented.

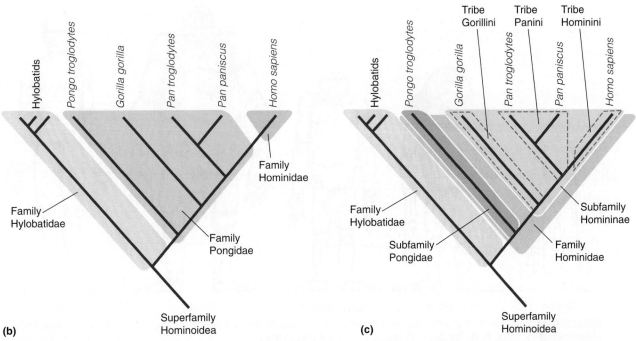

FIGURE 19.2 *(continued)* (**b**) Classification of the Hominoidea according to evolutionary systematics (**c**) and cladistics. (See text for details.)

publication of another species [*Afradapis longicristatus*] questions the evolutionary position of *D. masillae,* regarding such forms as convergent with the lineage from which humans eventually arose.)

Apes and Humans: Hominoids

In the conventional classification, the third group of haplorhines, which are **apes and humans** (**FIGURE 19.3**), are known collectively as **hominoids** (Hominoidea). The earliest hominoid fossils are 20 My-old Early Miocene forms such as the African genera *Proconsul,* and what may be the earliest ape, *Morotopithecus.* Molecular analysis places the divergence of great from lesser apes around 18 Mya (Figure 19.1).

Apes and humans are distinguished from monkeys by a distinctive pattern of cusps on the molar teeth. Many apes evolved a form of suspensory arboreal locomotion, hanging by the arms and swinging from branch to branch. Orangutans of Borneo and Sumatra (Figure 19.3) are large sexually dimorphic apes; males may weigh as much as 90 kg. Orangutans diverged from the lineage leading to the African apes and humans about 13 Mya. Humans are distinguished from other hominoids by adaptations for bipedal locomotion on the ground (shorter arms in relation to leg length; Figure 19.3), an enlarged brain and shortened fingers along with opposable thumbs.

Apes diversified rapidly during the Early and Middle Miocene, from which Period many genera have been described. Monkeys diversified a little later during the Late Miocene when most apes became restricted in distribution to wet forest habitats. As monkeys replaced apes in arboreal habitats during the last half of the Miocene 5 to 10 Mya, habitats opened up to those lineages that could adapt to ground dwelling.

The two extant species of **chimpanzees** from equatorial Africa are the common chimpanzee, *Pan troglodytes* and the more slender and lightly built bonobo, *Pan paniscus,* often and incorrectly called a pygmy chimpanzee. Individuals average 32 to 45 kg in

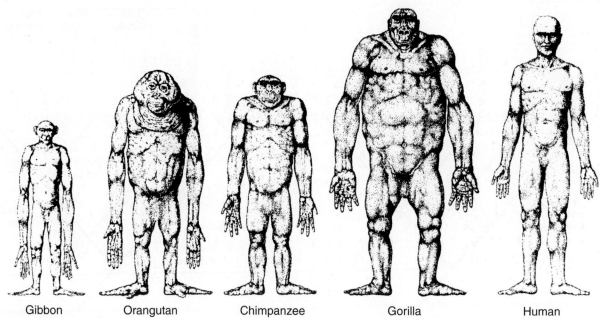

Gibbon Orangutan Chimpanzee Gorilla Human

FIGURE 19.3 Body contours and proportions of adult male apes and humans with all hair removed, drawn to the same scale. (Adapted from Schultz, A. H. *The Life of Primates.* Universe Books, 1969.)

weight depending on species and sex. Both species have complex social structures and hierarchies, reflected in living in large social groups of up to 120 individuals.

The largest of the apes (up to 227 kg) are the predominantly ground-dwelling lowland and mountain **gorillas** of equatorial Africa (Figure 19.3). A single dominant male or silverback usually controls a group of 5 to 30 individuals, usually all females and juvenile males.

African apes, particularly chimpanzees, share many features with humans and are now recognized as the sister group to humans. Unfortunately, in the process of sorting out hominoid relationships, older terms have been given new meanings, which is always confusing.

- Great apes and humans are now referred to as **hominids** (Family *Hominidae*; Figure 19.2c), a term previously used for extant humans and their closest extinct relatives.
- Humans and their closest extinct relatives are **hominins** in the tribe hominini (Figure 19.2c); see **BOX 19.1** for more on this change.

Among all the hominoids, evolutionary changes in humans allowed them to take *bipedal terrestrial locomotion* the furthest. You would therefore expect to find substantial genetic differences between, say humans and apes. Amazingly, this is not the case. The length of human hind limbs in comparison with the forelimbs (Figure 19.3), the larger brain, the absence of a protruding face, enlarged sexually dimorphic canine teeth, and sparse body hair — in short all the features that allow us to distinguish humans from apes — result from small changes at the molecular level.

Although a child can distinguish an ape or chimpanzee from a human, more than 96% of the DNA and protein sequences of chimps and humans are identical. An average chimpanzee and an average human differ genetically only by four percent (four base pairs out of every 100). By comparison, individual humans differ only in one base

■ Many explanations have been suggested for the evolution of bipedalism including advantages in foraging, evading predators, carrying provisions, using tools and manipulating weapons.

BOX 19.1

What's in a Name? Hominids and Hominins

It has become increasingly clear that humans and the African apes (particularly chimpanzees) are closely related, with chimps as the sister group to (closest relative of) humans. As the diversity of fossil species demonstrating the relationships between humans and the great apes increased, many paleoanthropologists saw that the traditional classification of humans and our closest relatives was inadequate and inconsistent. For this reason, a cladistic classification of hominoids has gained favor in recent years.

A consequence of this change is that humans and the great apes are now collectively referred to as *hominids,* a term previously used exclusively for humans and the fossil species most closely related to humans. In the cladistic classification, humans and our closest extinct relatives belong to the "tribe" *hominini* (Figure 19.2c). Tribe is a taxonomic level between subfamily and genus. Here is where it does become confusing and where you have to keep your wits about you. When paleoanthropologists use the term "hominin" in the recent literature they mean the same thing as "hominid" in older literature. At stake is more than terminology. This change reflects our changing view of the relationships within this portion of the Tree of Life.

■ Recall from Box 6.5, that, in cladistics, all groups include all descendents and the common ancestor of the group.

pair out of 1,000 or 0.001%. The current hypotheses, therefore, are that: (a) most of the differences between chimpanzees and humans were initiated by changes in gene regulation; (b) phenotypic change need not and often does not involve major genetic change; and (c) that small differences in gene function or pattern of gene expression can produce large changes in phenotype (for more of which, see Chapter 13).

Because chimpanzees are our closest relatives (the sister group to modern humans), much effort has gone into comparing the morphology, genetics, and behavior of humans and chimpanzees with the aim of determining when the two lineages diverged. The molecular data shown in Figure 19.1 place the divergence at 7.5 Mya. The fossil record places the divergence at 7 Mya. A recent analysis involving the alignment of 20 million base pairs of the genomes of five primate species places the divergence at 6 Mya.

Australopithecines: The Southern Apes of Africa

Exploration of human origins takes us into the transition from apes to humans and the group of apes known as australopithecines — the southern apes of Africa. Four species are of special interest.

Australopithecus africanus

A skull and lower jaw found in 1912 in England set the stage for what an early hominin should look like. But, as outlined in BOX 19.2, that specimen — **Piltdown Man** (*Eoanthropus dawsoni*) — had not been deposited by the usual modes of fossilization. It was a fake.

The real story of the discovery of early hominin fossils began in 1925 when the anatomist and anthropologist Raymond Dart (1893–1988) reported an early hominin fossil from a lime quarry at Taung in the Cape Province of South Africa. The fossil consisted of the front part of the skull and most of the lower jaw of a six-year-old (the "**Taung child**"). Dart named the new species *Australopithecus africanus*. Aspects of the teeth and brain were more like humans than they were like

BOX 19.2
Piltdown Man

At the time of the discovery of the early hominin fossil at Taung in South Africa, most anthropologists interpreted the available evidence as indicating that early humans had large brain cases, ape-like jaws, and large canine teeth. Evidence for this interpretation came from a fossil cranium and lower jaw found in 1912 at Piltdown, England, that showed such features (FIGURE B19.1c). Named *Eoanthropus dawsoni* (Dawson's dawn-man) after its discoverer, an amateur British archaeologist, Charles Dawson (1864–1916), it was regarded as the most ape-like human fossil.

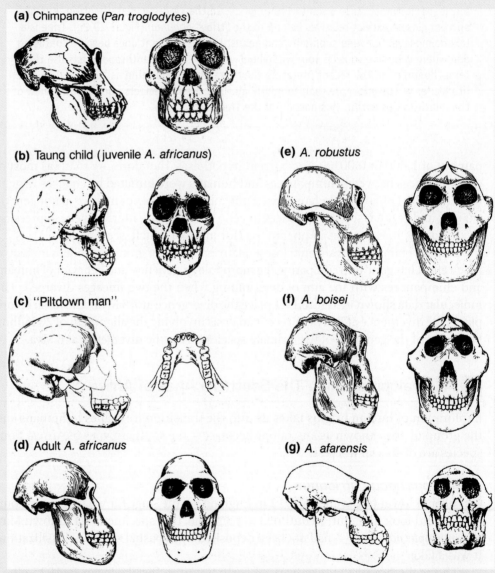

(a) Chimpanzee (*Pan troglodytes*)

(b) Taung child (juvenile *A. africanus*)

(c) "Piltdown man"

(d) Adult *A. africanus*

(e) *A. robustus*

(f) *A. boisei*

(g) *A. afarensis*

FIGURE B19.1 Skulls of a chimpanzee (**a**) and fossil hominids (**b–g**) in the genus *Australopithecus* (*A.* sp.) described in the text, including cranial and lower jaw fragments of the Piltdown forgery (**c**). (Adapted from Johanson, D. and M. Edey. *Lucy: The Beginnings of Mankind*. Simon and Schuster, 1981.)

BOX 19.2

Piltdown Man (*continued*)

Many anthropologists accepted the Piltdown fossil as valid, a situation that continued for about 40 years when it was shown that the entire skull was a hoax. The teeth had been artificially ground down, the cranium was of a different age than the jaw, artificial pigmentation had been used to color the bones, and the molar teeth had long roots like those of apes. Moreover, the associated animal fossils at the Piltdown site had a large accumulation of radioactive salts whose origin could be traced to a site in Tunisia. Piltdown man turned out to be a fabrication — a human cranium and the lower jaw of a female orangutan (and so should really be Piltdown female or Piltdown human) — a hoax perpetrated by someone who knew enough to destroy all obvious signs of the specimen's true origin by removing the jaw joint and modifying other features.

In a book on the specimen (*Piltdown: A Scientific Forgery*) published in 1990, Frank Spencer argued that the Piltdown forgery was perpetrated by Charles Dawson — the principal "discoverer" of the Piltdown fossils — in conspiracy with Arthur Keith, a leading British anatomist and physical anthropologist. Other historians of science suggest other suspects. Whoever perpetrated this hoax, the result for a time was the preservation of false views with false facts. The events that followed showed that false facts can be challenged in science and false views replaced.

apes. Adult australopithecine skulls, dated at 2.5 to 3 Mya, were discovered in the 1940s at *Sterkfontein*, not far from Taung. Further discoveries of postcranial skeletal material reinforced the human-like nature of the vertebral column and pelvic girdle.

Many more expeditions over the past 70 years revealed more and even older australopithecine species. These fossils tell us that considerable diversification of southern apes occurred over a two million-year period between 2.5 and 4.5 Mya. Four of those species are introduced briefly in **BOX 19.3** . These important finds and the conclusion that all known early australopithecines were bipedal, allow us to conclude that the human genus, *Homo,* had an immediate ancestor that was bipedal. The other features that characterize *Homo* evolved after bipedal locomotion arose. Lest these conclusions read as if they are final or as if all agree, the evolution of bipedal locomotion is a complex topic. Why? Because bipedality may have arisen more than once. Australopithecines are actively arboreal and terrestrial, a duality reflected in their bipedal locomotion, which differs functionally (and therefore, differs in origin?) from bipedality in humans.

Origins of Humans (Hominins)

Recall that humans are more closely related to the African apes than to orangutans and most closely related to chimpanzees, which are the sister group to (the closest relatives of) humans (Figure 19.1).

As human and chimpanzee lineages diverged around 6.3 Mya, the earliest species in the tribe to which humans belong — the Hominini (Box 19.1) — would be expected to have arisen shortly after 6.3 Mya, a date consistent with those outlined above. Further, DNA sequence analysis suggests that humanoids diverged from chimpanzees between 5.7 and 6.3 Mya. Many forms were competing for ground dwelling niches

BOX 19.3
Diversity of the Southern Apes of Africa

Australopithecines, the Southern apes of Africa radiated as at least four lineages between 2.5 and 4.5 Mya. One lineage, *Australopithecus africanus,* is discussed in the text. Four other species are briefly discussed here.

A bigger bodied South African hominin, with larger teeth and jaws, **Australopithecus robustus**, had a brain that was larger in absolute size than that of *A. africanus,* from which *A. robustus* may have arisen.

A Pliocene species from East Africa, **Australopithecus boisei** is the largest of the australopithecines. Its large molar teeth and powerful jaw muscles are interpreted as reflecting a diet of seeds and fruits with hard husks and pods.

Fossils from East African sites at Laetoli, Tanzania and the Afar (Hadar) region of Ethiopia, estimated as 3.9 to 3 My old, are classified as **Australopithecus afarensis**, a species with heavy brow ridges and a low forehead. Despite such ape-like features, *A. afarensis* displays a large number of important cranial, dental, and skeletal differences from apes. For example, although the canines are larger in *afarensis* males than in females, this sexual dimorphism is much less pronounced than in apes or early Miocene chimp, gorilla or orangutan-like fossils. These conclusions can be drawn because of quite a substantial record of *A. afarensis,* including the almost complete skeleton of a muscular, 1- to 1.2-m tall bipedal female known as "**Lucy.**" Furthermore, fossil footprints preserved under a layer of 3.7-My-old volcanic ash, record the bipedal gait of two individuals over a distance of some 21 meters.

The earliest australopithecine fossils currently known (mostly as teeth) and the most apelike hominin ancestor known are relics of an Ethiopian species (and genus) named **Ardipithecus ramidus**, dating between 4.3 and 4.5 Mya. *A ramidus* shares relationships with *Australopithecus afarensis* and with extant great apes, especially chimpanzees. In terms of the concept of transitional forms (**missing links** as they are often called), *A. ramidus* links apes and humans, a link established on teeth alone; no post-cranial fossils of *A. ramidus* are known.

■ Perhaps the most famous transitional form in human evolution, and perhaps the only species of any animal named before it was discovered, is discussed in Box 19.4.

and developing the features of bipedalism required for a ground dwelling existence. Two bipedal species are known.

- *Orrorin tugenensis* has features of the hind limbs indicating that it was bipedal when on the ground but features of the forelimbs indicating that it also was arboreal. With an estimated occurrence at 6.1 to 5.8 Mya, the chimpanzee-sized *Orronin* is the second oldest "putative" hominin.
- Dated at 5.8 to 5.2 Mya, the second species, *Ardipithecus kadabba*, is younger than *O. tugenensis;* you see why dating fossils (see Chapter 2) is so important when attempting to determine position within a lineage. *A. kadabba* was bipedal.

A third species, *Sahelanthropus tchadensis*, known from a cranium and fragments of jaws and teeth, had a brain estimated at 340 to 360 cm³. Whether this species was bipedal is unclear; lack of postcranial material makes comments on mode of locomotion premature. The large opening (foramen magnum) at the base of the skull is taken as indirect evidence for bipedalism. Dated 6.5 to 7.4 Mya, if bipedal, *S. tchadensis* would be the earliest known putative hominin.

■ Here, *putative* means that the fossil record is sparse, but given the evidence available, *Sahelanthropus tchadensis* is the earliest known hominin ancestor.

Of the three species *A. kadabba* has more features in common — shares more derived features — with later australopithecines. However, the presence of large canine

BOX 19.4
Pithecanthropus erectus

Although *Ardipithecus ramidus* may be the most deserving "missing link" between apes and humans, it is not the most famous. That prize goes to *Pithecanthropus erectus* (now *Homo erectus*), whose features were predicted by Ernst Haeckel in 1868. Haeckel even gave it a name, *Pithecanthropus alali* **before** a fossil matching his prediction was discovered in Java in 1891 by Eugène Dubois, a student of Haeckel's. This specimen became the center of discussion, confrontation and intrigue. Dubois's perseverance and truculence during two decades of withdrawal from social contact (1900–1920), during which he refused to allow others to see the specimens, polarized the scientific community, and, paradoxically, helped to found the field of Paleoanthropology. Pat Shipman's book *The Man Who Found the Missing Link* (2002) tells the tale.

teeth (which is a more basal feature) and its age of occurrence, makes *Ardipithecus* difficult to position in the hominin lineage (BOX 19.4).

Bipedalism and Brain Size: Hypotheses about Early Human Evolution

Before evaluating the changes that initiated the human lineage it will be helpful to outline the hypotheses that have been developed, especially concerning the relationship between the origin of the ability to walk upright on the hind limbs (bipedalism) and increase in brain size. The other major development, the origin of tool making and tool use is discussed in the following sections.

Charles Darwin said virtually nothing about human evolution in *The Origin of Species*. However, his second major book of evolution — *The Descent of Man* — published in 1871, dealt specifically with humans and their evolutionary origins. Ever since, bipedalism, increase in the size of the brain, the use of tools and language — the hallmarks of human evolution — have been regarded as a set of related changes, each reinforcing the others. The fossils record, however, informs us that bipedalism arose much earlier than did either of the other two traits. Therefore, at issue for some time has been whether the evolution of bipedal locomotion was an evolutionary innovation that facilitated changes in brain size and tool use.

We can reconstruct sufficient of their **environment** to know that early hominins lived in woodland, dry grassland and bush where food supplies would have been seasonal. This coupled with the patchy nature of the environment would have favored hominins who could adapt their diet to a wide range of foods (omnivores) and search for food over long distances. The hypothesis that standing upright and the ability to move rapidly enhanced this life style is easy to sustain; it can be seen in the fossil record. The combined advantages of an upright stance and faster locomotion on the ground while retaining the ability to climb would have enhanced avoidance of predators. Bipedalism would have fostered the use of weapons such as stick wielding and stone throwing, actions enabled by the freeing of the forelimbs from locomotion. The knowledge that chimpanzee use sticks as tools to obtain food (Figure B19.2) has been used to develop a scenario that tool making in *Homo*, perhaps also in Australopithecines, began with using unmodified sticks and stones from which the ability to modify sticks and stones arose, a topic discussed in BOX 19.5 .

■ A character such as bipedalism that ushers in other integrated changes related to a particular function(s) is known as a *key innovation*.

■ The anatomical structure of the hands and feet of the early australopithecines and early *Homo* (*H. habilis*) are consistent with the ability to climb trees.

BOX 19.5
Tool Use by Chimpanzees and Monkeys

Chimpanzee use sticks as tools to obtain food (FIGURE B19.2a). The scenario has therefore developed that tool making in *Homo*, perhaps also in Australopithecines, began with using unmodified sticks and stones from which the ability to modify sticks and stones arose.

In their natural habitat, imitation and the passing on of skills from one generation to the next through learning are obvious in chimpanzee tool making and tool use. This is especially so in the way chimpanzees employ twigs and vines in fishing for termites (Figure B19.2b). **Termite fishing** involves being aware of when to fish (October and November, although chimps probably know the season rather than the months), locating the sealed termite tunnels, often importing the necessary supply of tools from as far away as a kilometer, shaping some of the tools by removing leaves, biting off the ends of the tools to achieve an optimum length, inserting the tool with a proper twisting motion to follow the curves of the termite tunnel, vibrating the tool gently to "bait" the termite soldiers, and retracting it carefully to avoid dislodging the termites. Learning these tasks takes years; "even" an anthropologist who studied the technique for months was no better at it than a novice chimpanzee.

Similar demanding techniques used in cracking nuts require a chimpanzee to find a properly sized stone or hardwood club to be used as a "hammer," choose a well-shaped tree root as an "anvil," and precisely position each nut on the anvil; nuts from different species must be positioned differently. The hammer must be gripped in its most effective position, swung with the proper force, and aimed so that it hits the nut in exact locations to extract the maximum amount of nutmeat. Although young chimpanzees generally learn this technique from their mothers by imitation, incidents have been recorded in which mothers actively intervene in their offspring's unsuccessful nut-cracking attempts by taking the hammer, positioning the nut, and demonstrating the proper technique. We would call this active teaching if we saw it in humans.

(a) (b)

FIGURE B19.2 Tool use in chimpanzees (*Pan troglodytes*). (**a**) A stick fashioned for use in "fishing" termites from their nest (**b**). Young chimps learn tool use by observation and imitation.

BOX 19.5

Tool Use by Chimpanzees and Monkeys (*continued*)

Long thought limited to chimps and humans, monkeys have now been shown to use tools. Tufted capuchin monkeys (*Cebus apella*) that inhabit Brazilian savannah or forest environments use stones to crack open nuts during daily foraging for food. These monkeys also use tools to dig for tubers and to probe holes and crevices. Bearded capuchin monkeys (*Cebus libidinosus*) not only use stone tools, they select the most effective stone to crack open palm nuts, selecting heavier over lighter stones even when the heavier stones are smaller.

Origins of *Homo*

The genus *Homo*, which first appears in the fossil record in East Africa around 2.4 Mya, was similar in size to the australopithecines, but had a larger brain (600 to 700 cm^3) and smaller molar teeth. The prevailing trend in the evolution of *Homo*, as in mammalian evolution as a whole, is of parallel lineages arising contemporaneously or almost contemporaneously. Six species of *Homo* illustrate the adaptive radiation that took place over less than a million years between 1.6 and 2.4 Mya.

Homo rudolfensis and Homo habilis

Even though the numbers of known specimens is small they reveal that skeletal change occurred rapidly in the *Homo* lineage. Many experts assign early East Africa *Homo* fossils to two species that occupied similar geographical regions, *Homo rudolfensis* as early as 2.4 to 1.8 Mya, and *H. habilis* from 2.0 to 1.6 Mya. These are the most likely members of the lineage that gave rise to the next species of *Homo*, which appear in the fossil record 1.9 Mya, that is, no more than 500,000 years and perhaps as little as 100,000 years after *habilis* and *rudolfensis*. More than one species with similar features arose. One lived in Africa, the other in Asia. Both species were taller (1.7 m) and had larger brains (750 cm^3 or more) than their predecessors. The species are *H. ergaster* and *H. erectus*.

Homo ergaster and Homo erectus

The African species, *Homo ergaster* was the taller and leaner species with a brain capacity of 750 to 1,100 cm^3 and a larger body size than *H. habilis* or *H. rudolfensis*, either of which could have been its ancestor. *H. ergaster* used fire and large hand axes of the Acheulean type (Figure 19.4b, and see below). Whether these attributes indicate group living and hunting is hard to determine. As with Lucy — the almost complete skeleton of *A. afarensis* — the 1.6 My-old **Nariokotome skeleton** is the most complete fossil of a juvenile male *H. ergaster*.

A second species, *Homo erectus*, contemporaneous in time with *H. ergaster*, existed in Java. A thicker skull and heavier brow ridges along with the different geographical distributions allow the two species to be differentiated. Both had similar brain capacities (750 to greater than 1,200 cm^3 in *erectus*) and both are associated with Acheulean tools (Figure 19.4b). Later discoveries showed that *H. erectus* was present over a wide geographical area in Africa, China and Europe, and persisted until as late as 50,000 years ago in Java.

Fossils from the island of Flores in Indonesia discovered in 2004 (Figure B19.3) reveal what has been interpreted either as a relict population of dwarf *H. erectus* or as a new species, **Homo floresiensis** that existed as recently as 18,000 years ago. Such a recent existence of a human species other than our own is a remarkable suggestion and is highly controversial. Some experts claim that the best-preserved individual is more modern, a dwarf human (*H. sapiens*) but with signs of a congenital malformation, a topic taken up in BOX 19.6.

■ For the words highly controversial, you should read "has set expert against expert in an almighty set of arguments."

■ Much of the analysis has been conducted on the one fairly complete skull, LB1. Material from perhaps a dozen other individuals remains to be examined, so caution is required when drawing interpretations or conclusions.

BOX 19.6
Dwarf Hominins on the Island of Flores

Discovery of what has been interpreted as a dwarf (one meter tall) or pygmy population of *Homo erectus* on the island of Flores in Indonesia is so recent (2004) and the findings so unexpected that analysis and interpretation is ongoing at a fast pace. Especially controversial is whether these individuals represent a new species of *Homo* (*Homo floresiensis*), a relict population of *H. erectus* or a dwarf and/or a population of *H. sapiens* (FIGURE B19.3). Assignment to *H. erectus* would mean that a remnant of this species existed 18,000 years ago; specimens from the site date back to 95,000 years ago. Presence of more advanced tools than are known to have been used by *H. erectus* is evidence against assignment to *H. erectus*.

A more recent analysis of 140 cranial features led investigators to conclude that these individuals represent an early pygmy population of *Homo sapiens* with signs in the one complete skull (specimen LB1, an adult female) of a congenital malformation similar to a malformation known to occur in modern-day humans. The anomaly is **microcephaly**, a condition in which individuals have small brains and therefore small skulls. The one complete skull, which enclosed a brain of 400 cc^2, is comparable in volume to the brain of a chimpanzee and compares with microcephalic human skulls in various ways.

To explain the diminutive size, one study invoked mutation in the gene *Microcephalin (MCPH1)*, which is one of half a dozen genes in which mutation results in microcephaly; mutation of *MCPH1* results in *autosomal recessive primary microcephaly-1*. Enlargement of brain size during primate (including human) evolution has been associated with enhanced function of *MCPH* genes, which regulate the development of nerve cells in the cerebral cortex.

Another study proposed that a mutation in the gene for the growth hormone receptor made individuals insensitive to growth hormone. Disruption of growth hormone would disrupt the growth hormone–insulin-like growth factor 1 axis required to regulate growth. Consistent with this hypothesis, individual extant humans with the inherited disorder **Laron syndrome** (Laron-type dwarfism) have a mutation in the growth hormone receptor making them insensitive to growth hormone, resulting in short stature.

Cretinism also has been invoked to explain the small brain volume and features of the Flores skull. Both brain and skeletal size are reduced in individual modern humans with cretinism as a consequence of underactivity of the thyroid gland. Local high incidences of cretinism have been associated with iodine deficiency, which is a problem for individuals living on Flores today.

By definition, individuals with microcephaly have small skulls in relation to body size. More recent analyses of bones from seven individuals led to the conclusion that the size of the skull is proportional to the size of the postcranial skeletal elements, a finding that is inconsistent with microcephaly. The interpretation that this population

BOX 19.6
Dwarf Hominins on the Island of Flores (*continued*)

is "an island adapted population of *Homo sapiens* [hence the small size], perhaps with some individuals expressing congenital abnormalities" was reinforced by the discovery in March 2008 of the remains of a population of small-bodied (1.2 m tall) humans who lived in caves in the Western Caroline island of Micronesia 1,400 to 3,000 years ago. On the other hand, analysis of the feet leads some investigators to conclude that the feet are those of a basal biped, comparable to those known from 2- to 3-My-old human ancestors. The large, hairy feet of hobbits are invoked in comparison. With such divergent structures (mosaic evolution? pathology?) the jury remains deadlocked on the precise identification of this population.

(a)

(b)

FIGURE B19.3 (a) The skull of the most complete hominin from Flores dated at 18,000 before present (left), compared with a human skull on the right. (b) This individual is to have stood a little over 1 m in height, had long arms in relation to the length of the legs, unusual shoulders and a chinless mandible.

Homo sapiens

One of the reasons for *H. erectus* receiving so much attention and a partial explanation for the controversy over the Flores individuals is that our species, **Homo sapiens,** has been traced back to a lineage that contains *H. erectus.* Variation among specimens of *H. erectus* and the often subtle differences between *H. erectus* and *H. sapiens* led some experts to claim that *H. erectus* (or some variants within *H. erectus*) is (are) the first manifestation of *H. sapiens.*

Homo heidelbergensis

Part of the variation within *H. erectus* reflects individuals in different geographical location that would have existed as distinct and potentially non-interbreeding populations. Some conclude that the variation can in part be resolved by recognition of another species, **Homo heidelbergensis,** on the basis of fossils from Swanscombe in England and Steinheim in Germany and dated to about 200,000 years ago. These specimens have larger brain volumes than, and other features intermediate between, *H. erectus* and *H. sapiens.* To complicate things further, *H. heidelbergensis* may have arisen from a population of *H. ergaster* and not from the more geographically widespread *H. erectus,* which persisted in parts of the world long after *H. heidelbergensis* arose.

The tree of *Homo* evolution is much more like a bush or shrub with many branches than a tree with a single trunk. It is about to become even more bushy. Why? For two reasons: (a) another species, **Homo neanderthalensis,** discovered in the Neander valley, Germany, in 1856, arose from European populations of *H. heidelbergensis.* (b) Many paleoanthropologists have concluded that *H. sapiens* evolved from *H. heidelbergensis,* the separation of the two lines occurring in Africa around 200,000 years ago.

Neanderthals: Another Species of *Homo*

Neanderthals (*Homo neanderthalensis*) were widespread in Europe and Western Asia from 300,000 to 30,000 years ago. Shorter than modern humans, and with brains that were 10% larger, Neanderthals had distinctive brow ridges, large jaws, small chins, robust skeletons and other anatomical features not seen in humans.

Like humans, Neanderthals used stone tools to hunt large animals; bears and mammoths are shown in most reconstructions/dioramas. From flowers found placed on the graves of dead individuals it has been concluded that Neanderthals performed rituals. Other evidence from some 100,000 years ago indicates that Neanderthals practiced cannibalism. Because the Mousterian Culture (Figure 19.4c and see below) is associated with Neanderthals *and* with *H. sapiens,* some have concluded that there may have been cultural overlap between the two species.

The difficulty of assigning species status was discussed above in relation to *H. erectus* and the Flores specimens and is elaborated further in Box 19.6. Similar difficulties are encountered with Neanderthals. Some physical anthropologists have concluded that skeletal evidence — thicker skulls, differences in the pelvic girdles — and the presumed limited language ability of Neanderthals (see Chapter 20) support considering Neanderthals as a subspecies of *H. sapiens* designated *Homo sapiens neanderthalensis,* although Neanderthal faces and their large, deep-rooted teeth do not align them with any human populations.

Homo sapiens began to replace Neanderthals in various parts of the world about 40,000 years ago; Neanderthals are estimated to have died out 28,000 to 24,000 years ago. How is it known that Neanderthals died out rather than integrating with *H. sapiens*?

The two species certainly coexisted between 500,000 years and 24,000 ago, and both overlapped with *H. erectus* between 500,000 and 250,000 years ago.

Strong molecular evidence against integration comes from the recovery and analysis of sequences of *Neanderthal mitochondrial DNA* (mtDNA), which show little evidence of any contribution of Neanderthal to human mtDNA. (See Box 9.2 for ways of detecting ancient DNA.) mtDNA isolated from a 38,000 year-old Neanderthal skeleton is significantly more different from human mtDNA than expected if it were a sample of normal human variation. According to the data, Neanderthal and human mtDNA lineages diverged 660,000 ± 140,000 years ago, and "went extinct without contributing mtDNA to modern humans" (see Chapter 14). Even more recently in two studies published in 2006, sufficient Neanderthal DNA was isolated and sequenced that with sophisticated methods to estimate missing DNA, a reasonable first approximation of the Neanderthal genome was constructed. One study used one million base pairs of Neanderthal (humans have around 3.2 billion base pairs of DNA), the other 65,000 base pairs. These studies have been described as "perhaps the most significant contributions . . . since the discovery of Neanderthals 150 years ago." The one million base pair study produced an estimate of divergence of humans from Neanderthals 516,000 years ago. The 65,000 base pair study produced an estimated divergence 370,000 years ago. The weight of evidence supports Neanderthals and humans as separate species but we await even more detailed studies.

Tool Use by Species of *Homo*

Individuals in the genus *Homo* developed the capacity to make and use stone tools, the oldest of which are dated to 3.5 Mya (FIGURE 19.4a).

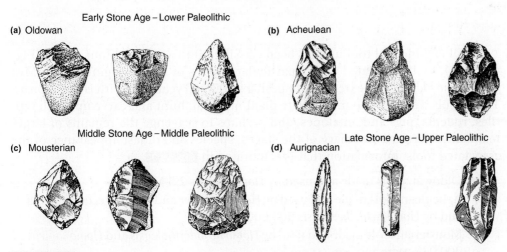

FIGURE 19.4 Stages in stone tool development beginning with the Oldowan stone industry (**a**) dating back at least 2.5 My and persisting for more than a million years. Some paleoanthropologists suggest that although Oldowan tools are traditionally ascribed to *Homo habilis,* they may have been used by earlier hominins. The cleavers and hand axes of the Acheulean stone industry (**b**) appear abruptly about 1.5 Mya and are assigned to *H. erectus.* Later stages were associated with other groups: (**c**) Mousterian (*H. sapiens neanderthalensis*), (**d**) Aurignacian and Upper Paleolithic (*H. sapiens sapiens*). These industries and cultures constitute the **Paleolithic Stone Age.** Following this period are the Mesolithic and Neolithic ages, the latter beginning about 10,000 years ago and marked by polished stone tools, pottery, domesticated animals, cultivated plants and woven cloth.

BOX 19.7

Stages in the Manufacture and Use of Tools by Species of *Homo*

As indicated in the text, tool making and tool use is classified into four stages and related to species of *Homo* that used each tool type, beginning 2.5 Mya with *Homo habilis*, and extending to *Homo sapiens*.

Oldowan tools (Figure 19.4a) consist of sharp-edged stones that could be made with a single blow of one rock against another. Such tools are sufficient to strip tough fibers from plants, break up animal carcasses, and expose the marrow cavities of long bones.

Acheulean tools (Figure 19.4b) consist of stones chipped from both sides using multiple strikes to create the cutting edge. They are most abundantly found 1.5 Mya associated with *Homo ergaster* and *Homo erectus*. Tools interpreted as hand axes are found about 0.5 By later. With the spread of *Homo heidelbergensis* through Europe, Acheulean tools are known from 500,000 years ago and were in use up until 200,000 years ago.

Mousterian tools (Figure 19.4c) require more complicated working and reworking (manufacturing?) than the other types of tools. Mousterian tools first appeared 200,000 years ago and are still found in 40,000-year-old deposits. They are associated with *Homo neanderthalensis* in Europe and with both Neanderthals and *Homo sapiens* elsewhere. The presence of spear points indicates tools made for hunting, with the implication that the spear points were affixed to long sticks (spears). The presence of scrapers indicates tools made for scraping animal carcasses, either for food or for skins for clothing.

Upper Paleolithic tools include hooks used for fishing and needles for sewing and are the dominant tools in Asia and in Africa between 90,000 and 12,000 years ago. Specialists divide the Upper Paleolithic tool period into temporal and species-specific subperiods (Figure 19.4d) reflecting increasing sophistication and diversified use of tools and other objects. Ivory beads and carved female Venus figurines are first found between 28,000 and 22,000 years ago, depending on geographical location. Barbed harpoons, spear throwers and the first rock paintings appear 18,000 to 12,000 years ago.

Tool making and tool use has been classified into a series of four stages (Figure 19.4), each associated with a particular species of *Homo*, beginning 2.5 Mya with *Homo habilis*, and extending to perhaps 200,000 years ago with *Homo sapiens*. These tools were used to manipulate plant material, hunt and carve up small reptiles, rodents, pigs and antelopes, and perhaps to scavenge the remains of larger mammals. Here is a summary of the stages, species and oldest appearance of the associated tools, information that is elaborated in BOX 19.7.

- Oldowan tools made and used by *Homo habilis* 2.5 Mya
- Acheulean tools made and used by *Homo ergaster* and *Homo erectus* 1.5 Mya and by *Homo heidelbergensis* 500,000 years ago
- Mousterian tools made and used by *Homo neanderthalensis* and *Homo sapiens* 200,000 years ago
- Upper Paleolithic tools made and used by *Homo sapiens* 90,000 years ago

Homo sapiens (Humans)

Many older texts used terms such as "primitive" or "advanced," "higher" or "lower" when referring to earlier and later forms within a lineage. Such terms are no longer regarded as appropriate, conveying as they do the notions of comparative evolutionary

progress on some subjective scale. The term **anatomically modern humans** is now used for extant and fossil individuals of the species *H. sapiens*. The earliest fossil evidence for *H. sapiens* is in Africa, a finding reflected in the "Out of Africa" hypothesis for the spread of *H. sapiens* around the world discussed in the following section.

Three individual specimens of anatomically modern humans discovered in 160,000 year-old deposits at Herto, Ethiopia are the geologically oldest members of *H. sapiens*. More recent fossils from Mount Carmel in Israel date to 90,000 years ago. Mousterian stone tools are associated with both sites. Even more recently — 35,000 years ago during the Upper Paleolithic in Europe — the features of human fossils and the tools present both had changed. Brow ridges decreased, skull vaults were higher (reflecting changes in brain size and form), faces were smaller and less protruding, and flaked flint tools of the Aurignacian Era (Figure 19.4d) were present. Humans were now representing their environment, especially the animals in it, as painting on cave walls and as sculptures in bone.

MtDNA and Human Migration

An important data set helping to resolve the immediate ancestor of *Homo sapiens* is mitochondrial DNA (mtDNA) from a large number of extant humans representing many populations and groups. One major conclusion from analyses is that all modern human mtDNA sequences originated with a single ancestral sequence in Africa between 140,000 and 290,000 years ago (FIGURE 19.5a), from which arose nine major descendant sequences (Fig. 19.5b–j) that went to other geographical regions through migration, further sequence changes occurring over time in populations in those regions.

Each geographical group outside Africa includes more than one mtDNA branch, that is, had more than one origin. Highland peoples of New Guinea have seven maternal origins, meaning that geographical regions were colonized more than once. Because the rate of mtDNA sequences divergence can be estimated (2–4% nucleotide change/My in vertebrates but slower in primates), timing of waves of colonization also can be estimated (Figure 19.5). Even though such estimates are influenced by the way in which fossil species are arranged phylogenetically, the hypothesis of a single African origin of *Homo sapiens* (elaborated below) remains the best explanation.

Out of Africa

There are two main views of the origin and subsequent dispersal of modern humans from a *Homo erectus* ancestor (FIGURE 19.6).

- The **single-origin hypothesis**, also called the **Out of Africa** or **Noah's Ark model**, proposes the origin of *Homo sapiens* in a single geographical region, Africa, followed by dispersal to other continents (Figure 19.6a).
- The **multiple-origin hypothesis**, also called the **Candelabra model** because of its shape (Figure 19.6b), proposes the parallel origin of *Homo sapiens* in different unconnected localities.

Accurate determination of the date of the last common ancestor of modern humans is a crucial piece of evidence in deciding between the two hypotheses. Existence of the last common ancestor one or more Mya would coincide with one of the dispersals of *Homo erectus* from Africa. It would indicate that modern humans found in different continents represent evolutionary lineages that each began with

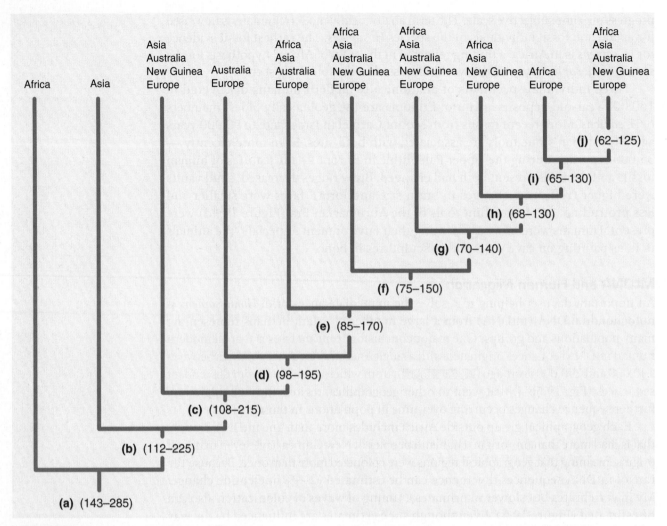

FIGURE 19.5 Phylogeny of mitochondrial DNA from 147 humans from five geographic regions. The ancestral sequence is designated as (**a**). Each node in the phylogeny (**b–j**) indicates a major descendant sequence. Estimated dates for each node (thousands of years ago) are given in parentheses. The areas colonized by each sequence are indicated at the top of the figure. (Based on data that Cann et al., 1987, obtained from surveying 370 restriction enzyme sites per individual, covering about 1,500 bases of the mitochondrial genome.)

H. erectus in geographically separated localities and so support the multiple origin hypothesis. If the last common *Homo sapiens* ancestor was much more recent, say 100,000 to 500,000 years old, then the dispersal of *Homo sapiens* would have occurred into areas occupied by *H. erectus* after a 1- or 2-My-old dispersal of *Homo erectus*, and so support the single-origin hypothesis.

Mitochondrial Eve

In 1987, on the basis of mtDNA nucleotide sequence analysis of geographically diverse human populations, Rebecca Cann and coworkers proposed a 200,000-year-old **common mitochondrial DNA ancestor** ("mitochondrial Eve") of modern humans, a finding consistent with the single-origin hypothesis. Subsequent studies in which calibration of the age of human origination was based on similar mtDNA sequence data, provided further support for the single origin hypothesis (although as discussed in Box 11.2, molecular clocks are difficult to calibrate). That being so, two further

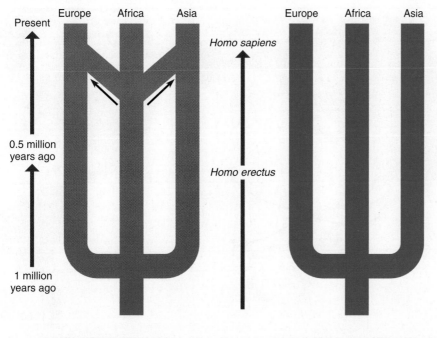

(a) "Out of Africa" model

Single origin of *Homo sapiens*
in Africa, and replacement of
Homo erectus in Europe, Africa,
and Asia

(b) "Candelabra" model

Multiple origins of *Homo sapiens*
from *Homo erectus* populations
in Europe, Africa, and Asia

FIGURE 19.6 Diagrammatic representation of two models for the origin of *Homo sapiens*. In (**a**) humans (blue) originated in one locality (Africa) and migrated to other continents where they replaced relict *H. erectus* populations (brown) that had entered these continents 1 My or more years ago. In (**b**) humans (blue) originated from *Homo erectus* in different localities.

findings are consistent with the single origin hypothesis: (a) all of the non-African mtDNA sequences are variants of the African sequence, and (b) most of the variability in mtDNA sequences occurs among members of African populations. Both these findings are consistent with the oldest mtDNA sequences being in African populations; multiple populations evolving in parallel would be expected to show similar amounts of sequence variability. More recent estimates place mitochondrial Eve as having existed 140,000 rather than 200,000 years ago.

Y-Chromosome Adam

MtDNA sequences are not complicated by the recombination of genes from the male and female parent; mtDNA sequences track the female genome back in time. Interestingly, perhaps surprisingly, the Y chromosome, which is inherited through the male line, also contains phylogenetic information, but of a different nature to that seen in mtDNA.

Ordinarily, changes in sequences in nuclear genes would be hard to differentiate into those resulting from recombination (and so potentially recent), and those that are much older. Fortuitously, a large proportion of the human Y chromosome does not participate in recombination. As a consequence, mutations build up in these regions over time. A phylogenetic analysis of more than 1,500 individual humans

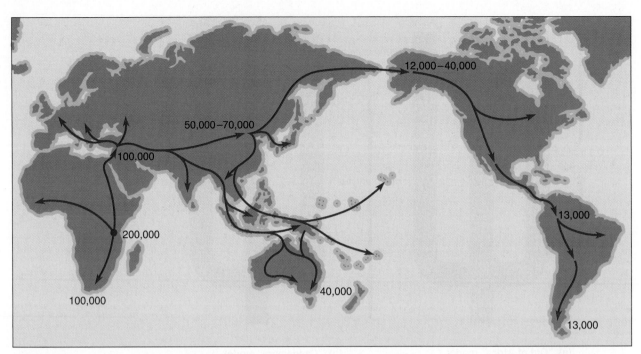

FIGURE 19.7 One scenario for the geographic distribution of *Homo sapiens* from their African origin. The numbers derive from a study of genetic distances between 26 human populations and represent estimated dates (thousands of years ago) when these populations reached their various destinations. These migrations correlate highly with many major patterns of linguistic evolution. (Adapted from Nei, M., and A. K. Roychoudhury, *Mol. Biol. Evol.*, **10**, 1993: 927–943.)

from all continents traced Y chromosomes to a common African ancestor living about 150,000 years ago. So both "Mitochondrial Eve" and "**Y-chromosome Adam**" confirm an African origin of humans. Given the margins of errors in such estimates, 150,000 and 140,000 years ago are respectably close.

Autosomal Human

A phylogenetic analysis of 29 autosomal genes — nuclear genes not located on the sex chromosomes — from 26 populations supports a single African origin of humans around 200,000 years ago. According to these data, dispersal resulted in humans reaching Eastern Europe 50,000 to 70,000 years ago, Australia 40,000 years ago, North America 12,000 to 40,000 years ago, and South America 13,000 years ago (**FIGURE 19.7**).

Humans did not undergo such vast migration without mastering their environment, developing social structures, and enhancing their ability to communicate, **three developments that occupy the balance of the chapter and lead us into a discussion of social and cultural evolution in the next chapter.**

Humans as Hunter-Gatherers

Jane Goodall listed more than 200 observed incidents where Gombe chimpanzees caught and/or ate colobus monkeys near Lake Tanganyika. The chimpanzee kill rate in Gombe is about 225 to 300 mammals a year.

Hunting in groups (packs) did not evolve with humans. Jackals and hyenas hunt in packs as their major means of capturing prey. Chimpanzees and baboons also occasionally form groups to hunt smaller animals; groups of two to five male chimpanzees will "tree" a monkey and cut off its escape by positioning themselves around it.

The earliest members of *H. sapiens* moved from forest to savanna as they roamed over large areas in search of food and prey items that were patchily distributed spatially and seasonally. Exposure to differing environments during migration and to the food types in those environments would have facilitated further structural and behavioral changes and adaptations; anatomical and inferred physiological evidence has been used to argue that adaptations for long-distance endurance running separate *Homo sapiens* from *Homo erectus*.

East African fossil sites indicate that scavenging was established 2 Mya. This early date probably indicates that humans were scavenging kills made by other large predators. A major evolutionary change was evolution of the ability to hunt, either as an isolated individual or in small groups. Evidence indicates that hunting and gathering arose much more recently than scavenging, around 55,000 years ago. A single kill of a large mammal would have provided food for more than one individual and for more than a day if kept away from other scavenging mammals. Such skills would have been passed on to the next generation by imitation, and later, as social living became more established, by active teaching of hunting skills and of the locations of seasonal populations of prey. A **hunter-gatherer society** would have begun.

The Rise of Agriculture

The changes in human societies produced by cultural change over the last 10,000 years (see Chapter 20) have been dramatic, as seen in the origin and evolution of agriculture, which was introduced in Chapter 12 in relation to selection of strains of wheat.

You may be unaware that the evolution of agriculture is not limited to human societies. Agriculture has evolved four times independently in four groups of animals: humans, termites, bark beetles, and ants. Ant agriculture arose when a lineage of ants began to cultivate a range of species of fungi some 50 Mya. Three novel systems of agriculture evolved in the last 30 My, each involving the cultivation of a separate species of fungus. A further system, the cultivation of a single species of fungus by leaf-cutter ants, evolved 8 to 12 Mya and is now the dominant form of ant agriculture in the New World tropics.

Somewhere during the Neolithic Age (Figure 19.4), the long-prevailing hominin lifestyle of hunting–gathering–fishing began to give way to the cultivation of food using domesticated plants and animals. Energies formerly expended in finding food were directed into methods of agriculture. Although originally developed in the Middle East, China, and Central America, such changes spread rapidly. Within 1,000 years or so, some form of agriculture had begun in many contiguous areas. Within 5,000 years agriculture and the technologies it stimulated extended widely (TABLE 19.1).

Perhaps the most immediate and far-reaching effect of agriculture was to increase food supply many-fold, a change permitting larger populations and a greater population density in agricultural communities. What matters here is the change in human lifestyle. Although there are uncertainties concerning whether sedentary communities preceded or followed agriculture, and whether agriculture was stimulated by climatic changes or by increased Neolithic social complexities, it did not take many generations for humans to settle arable areas on a permanent basis.

TABLE

19.1 Major Human Agricultural Expansions from the Neolithic Age Onward

Center of Origin	Area of Expansion	Time (Years Ago)	Technologies	Crop or Product
Middle East	Europe, North Africa, and Southwest Asia	10,000 to 5,000	Farming/ Domestication	Wheat, barley, goats, sheep, cattle
North China	North China	9,000 to 2,000	Farming/ Domestication	Millet, pigs
South China	Southeast Asia	8,000 to 3,000	Farming/ Domestication	Rice, pigs, water buffalo
Central America and North Andes	Americas	9,000 to 2,000	Farming	Corn, squash, beans
West Africa	Sub-Saharan Africa	4,000 to 300	Farming	Millet, sorghum, gourd
Eurasian steppes	Eurasia	5,000 to 300	Pastoral nomadism	Horses

Source: Cavalli-Sforza, L.L., et al., Science **259**, 1993: 639–646.

Humans as Genetic Engineers

Humans have been changing the genetic makeup of organisms for a long time. **Domestication** of plants and animals involved selection to promote some traits and eliminate others. In some species, the results have been dramatic. As discussed in Chapters 12 and 14, wheat is a good example. In the thousands of years since domestication began, humans have produced a world of agriculture, horticulture, and plantation forestry by manipulating genes and genomes, most commonly through selection.

The early 1970s, however, marked the beginning of a new era, the era of **genetic engineering**. Following the discovery that restriction enzymes can be used to cut sections of DNA at particular sequences (see Chapter 12), molecular biologists realized that they could splice together DNA fragments from different sources to form entirely new **recombinant DNA**, which could be used to introduce new traits to an organism.

A basic technique of genetic engineering is to insert a new or modified nucleic acid sequence into a virus or plasmid that can carry these novel sequences into host cells and thence host genomes where they can be incorporated and amplified many times over (FIGURE 19.8). This technology allows active intervention into the genetic material of any organism. Genomes of many organisms have been manipulated to affect such traits as protein production, resistance to infective agents, agricultural yield, nutritional value, toxic susceptibility, environmental stamina, tumor resistance, and so forth. Such manipulation of genetic traits allows us to move organismal evolution from the age-old province of random mutation and natural selection to **human-directed evolution**.

■ As discussed in Chapter 8, in nature, genes are transferred horizontally by hitchhiking on plasmids.

Food Crops and Animals

Genetic engineering is now big business, especially when it comes to agriculture. The first genetically modified (GM) organism, a tomato (FlavrSavr) was put on the

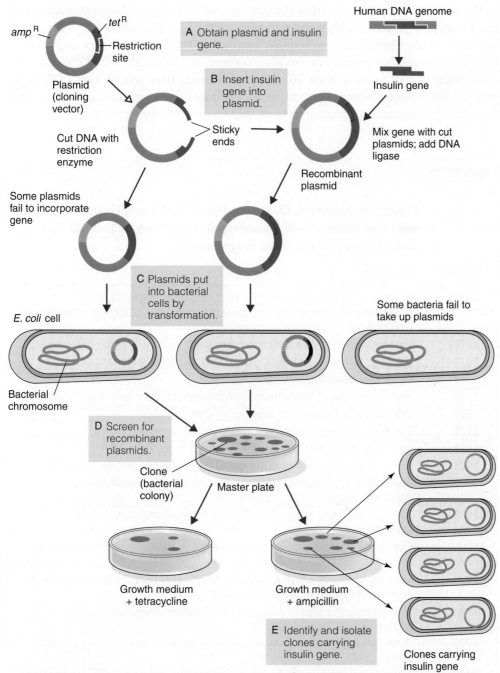

FIGURE 19.8 A general scheme for constructing a clone of recombinant DNA molecules using plasmids and restriction endonuclease enzymes. (**A**) A foreign DNA molecule (human insulin in this example) and a plasmid vector are selected, both carrying recognition sites that can be cleaved by a restriction enzyme. (**B**) Cleavage produces one or more fragments of the foreign DNA and opens the plasmid vector allowing the DNA to be inserted. (**C**) The plasmid carrying the foreign DNA is inserted into a host cell, in this example the bacterium *E. coli*. (**D**) Because some bacteria fail to take up the plasmid, the cells are screened for the presence of recombinant plasmids, a process that may involve growing the cells in the presence of specific molecules — tetracycline or ampicillin in this example. (**E**) Clones carrying the gene of interest are further replicated to obtain large quantities of the gene. (Adapted from Strickberger, M. W., *Genetics, Third edition*. Macmillan, 1985.)

U.S. market in 1994 (TABLE 19.2), but already by 2009, 85 percent of corn planted in the United States, 88 percent of the cotton, and 91 percent of the soybeans owed their origins to genetic technology, according to the U.S. Department of Agriculture statistics. A large proportion of food on North American supermarket shelves now contains at least some genetically modified content. Yet although by 2005 "GM crops" were being grown in more than 20 countries, to date they remain rare in Europe, where a huge public outcry against the technology on ethical, environmental, and food safety grounds continues to have an impact on government policy.

TABLE

19.2 **Major Discoveries Over the Past 40 Years Beginning with the Isolation of Restriction Enzymes in the 1970s and Leading to Genetic Engineering**

Year(s)	Advance
1970s	Isolation of a restriction enzyme that cuts DNA at specific sites Development of method to recombine the fragments, launching recombinant DNA technology (genetic engineering)
1972	First recombinant DNA molecules produced
1973	Introduction of recombinant plasmid vectors into bacteria for cloning
1975	DNA sequencing technology invented
1976	First clinical use of recombinant DNA for prenatal diagnosis of α-thalassemia
1977	Laboratory techniques for sequencing DNA invented
1978	Synthesis of the proinsulin peptide using recombinant DNA
1983	Discovery of polymerase chain reaction (PCR; see Figure 9.9)
1985	Polymerase chain reaction (PCR) used to amplify nucleotide sequences
1987	Production of a vaccine (anti-hepatitis B) based on recombinant DNA technology
1988	Incorporation and expression of genes inserted into mouse cells using retroviruses
1989	First human gene sequenced, and a defect in the gene product shown to "cause" cystic fibrosis
1990	Gene for defective enzyme (adenine deaminase) required for nucleic acid synthesis replaced in affected humans
1994	Marketing in the United States of FlavrSavr tomatoes, genetically engineered for longer shelf life
1996	Birth of a lamb named Dolly, first animal to result from cloning an adult somatic cell
1999	Sequencing of a human chromosome and of complete fruit fly (*Drosophila melanogaster*) genome
2001	First draft sequences of human genome completed independently by the Human Genome Project and Celera Genomics
2002	Mouse genome sequenced; birth of first genetically modified cat
2003	Sequencing of 99 percent of the human genome completed to 99.9 percent accuracy; marketing of GloFish, a pet fish to which a fluorescence gene was added
2006	Sequencing of the human genome completed

With their early emphasis on increased resistance to insect pests, diseases, and herbicides, the most common GM crops in use today were largely developed with financial profit in mind, but the potential is enormous. Improved nutrition (golden rice, for example, is rich in β-carotene, provitamin A) is just one promised benefit.

GM technology has been applied to many other organisms, including fish for enhanced growth, cold tolerance, and human insulin production; cows to produce human growth hormone and casein-enriched milk; pigs, for lower fat content, and mice to produce fish oils. Although most of this work has yet to be approved for commercial application, in 2006 approval for release of Atryn®, the first pharmaceutical drug generated in a mammal (a goat), was granted in Europe. Indeed, the pharmaceutical industry is investing heavily in research into this method of drug production.

■ Atryn®, an anti-clotting agent, is a recombinant form of human antithrombin designed to be used during surgery on people with antithrombin deficiency.

Microorganisms have been used to produce a wide array of human and other proteins, human growth hormone, and an essential blood-clotting factor (factor VIII) for hemophilia sufferers being two examples.

In summary, the major benefits conferred by GM technology are that in contrast to more traditional breeding techniques, which are time-consuming and inexact, GM is more efficient, more precise in its choice of genes, and allows genes from a far wider range of organisms to be introduced into the recipient organism's genome. When combined with cloning (see Chapter 20), its scope may be further enhanced.

Concerns About GM Technology

Perhaps the most fundamental charge leveled against genetic engineering is that humans are interfering with nature. But the physical/biological environment considered "natural" is not unchanging. Organisms have regularly affected and changed the environment, and continue to do so, while humans have long engaged in a succession of interventions into nature that have changed its form, function and substance in most localities and for many organisms.

An important feature of GM technology is that the genes employed usually come from a species other than the receiving organism. To many, this poses a substantial ethical problem. But as discussed in Chapter 8, gene flow via horizontal gene transfer between organisms occurs across the whole Tree of Life. Species barriers are continually being crossed by viruses, transposons (see Chapter 13), and plasmids, which carry genes from one species to another. Humans readily incorporate genes introduced by viruses and bacteria.

In agriculture, it has been standard practice for millennia to develop new varieties through hybridization between different species (see Chapter 12). At first, the species concerned were closely related. By the middle of the twentieth century, however, methods had been developed that allowed crosses between more distantly related species. For instance, triticale, a widely grown cereal, is a hybrid between wheat (*Triticum*) and rye (*Secale*; see Chapter 12). Protests mounted against the *green revolution*, a major achievement of this type of breeding, were on the basis of environmental and social concerns, rather than the ethics and safety of gene transfer between species. With genetic engineering, however, the genes transferred often come from a different phylum or even kingdom, which was not the case with earlier breeding practices.

Environmental concerns about GM crops are mostly focused on the perceived threats to local biodiversity and ecosystems, in particular the possibility that pollen from GM plants will fertilize closely related weeds (a problem if the crop is modified to resist pests or herbicides), closely related native plants (whose very existence might

then be threatened), or non-GM crops of the same species. The long-term impact of such crosses will depend, of course, on the relative fitness of the gene(s) in question. Again, this is not a new thing; gene flow from crops altered by non-GM methods has occurred for many decades. Still, the concern remains, and great care must be taken before introducing any new GM crop. After all, our experience with the technology is still relatively limited, and long-term effects can be difficult to envisage.

Recommended Reading

Brown, P., T. Sutikna, M. K. Morwood, et al., 2004. A new small-bodied hominin from the Late Pleistocene of Flores, Indonesia. *Nature,* 431, 1055–1061.

Cann, R. L., M. Stoneking, and A. C. Wilson, 1987. Mitochondrial DNA and human evolution. *Nature,* 325, 31– 36.

Crosby, A., 1986. *Ecological Imperialism: The Biological Expansion of Europe, 900–1900.* Cambridge University Press, Cambridge, England.

Diamond, J. M., 1997. *Guns, Germs, and Steel: The Fates of Human Societies.* Norton, New York.

Diamond, J. M., 2005. *Collapse: How Societies Choose to Fail or Succeed.* Viking Books, New York.

Fleagle, J. G., 1988. *Primate Adaptation and Evolution.* Academic Press, San Diego, CA.

Franzen, J. L., P. D. Gingerich, J. Habersetzer, J. H. Hurum, et al., 2009. Complete Primate Skeleton from the Middle Eocene of Messel in Germany: Morphology and Paleobiology. *PLoS ONE* 4(5): e5723. doi:10.1371/journal.pone.0005723.

Green, R. E., A. W. Briggs, J. Krause, et al., 2009. The Neanderthal genome and ancient DNA authenticity. *EMBO J.* 28, 2494–2502.

Groves, C. P., 2001. *Primate Taxonomy.* Smithsonian Institution Press, Washington, DC.

Hammer, M. F., T. Karafet, A. Rasanayagam, E. T. Wood, et al., 1998. Out of Africa and back again: Nested cladistic analysis of human Y chromosome variation. *Mol. Biol. Evol.,* 15, 427–441.

Harvati, K., and T. Harrison (eds.), 2006. *Neanderthals Revisited. New Approaches and Perspectives.* Springer, Netherlands.

Jungers, W. L., W. E. H. Harcourt-Smith, R. E. Wunderlich, et al., 2009. The foot of *Homo floresiensis. Nature* 459, 81–84.

Kingdon, J., 2004. *Lowly Origin: Where, When, and Why Our Ancestors First Stood Up.* Princeton University Press, Princeton, NJ.

Schulz, T. R., and S. G. Brady, 2008. Major evolutionary transitions in ant agriculture. *Proc. Natl. Acad. Sci. U.S.A.* 195, 5435–5449.

Seiffert, E. R., J. M. G. Perry, E. L. Simons, and D. M. Boyer, 2009. Convergent evolution of anthropoid-like adaptations in Eocene adapiform primates. *Nature,* 461, 1118–1122.

Shipman, P., 2002. *The Man Who Found the Missing Link: Eugene Dubois and His Lifelong Quest to Prove Darwin Right.* Harvard University Press, Cambridge, MA.

Spencer, F., 1990. *Piltdown: A Scientific Forgery.* Oxford University Press, Oxford, UK.

Wood, B., 2005. *Human Evolution. A Very Short Introduction.* Oxford University Press, Oxford, England.

Cultural and Social Evolution

20

- Unlike almost any other organisms, modern humans possess two modes of inheritance.
- One mode is biological inheritance through DNA.
- The second is cultural-social inheritance through experience and learning.
- The study of how these two modes of inheritance interact is socio-biology, which grew out of analyses of innate and learned behaviors.
- Although many animals communicate with sounds, only modern humans (and possibly Neanderthals) developed speech.
- Humans acquire and transmit culture through social exchanges using language.
- Cultural evolution has outpaced biological evolution for much of human history.
- The first recognition of social evolution came in the application of Darwinism to society: Social Darwinism.

Above: Survival is a family affair.

- Control over our own evolution permits the retention of deleterious alleles in particular groups or in the species as a whole.
- Movements to control evolution through selective breeding (eugenics) arose around the same time as did Social Darwinism.
- Eugenics continues to be practiced by individuals in their use of assisted methods of reproduction.
- Humans have been engineering the gene pool of other organisms since the origins of agriculture.
- Genetic technology is bringing us to the stage where human genes can be engineered.

Overview

Unlike most other animals, humans transfer information from generation to generation through genes and culture. The speed of human cultural evolution has accelerated so much more rapidly than our biological evolution that we each gather new experience at a rate many times that of our ancestors.

In the 19th century, proponents of Social Darwinism maintained that cultural differences evolved primarily by natural selection, as embodied in the concept of the survival of the fittest. This belief, which "justified" many social inequities, was based on the prevailing philosophy that society (which often incorporates nonbiological goals and value systems) is governed by the same laws as is biological evolution. Whether this is so remains an active area of investigation.

The term *biological evolution* for only those aspects of a species that are transmitted through the genome is an unfortunate one. Cultural evolution (cultural and social evolution, socio-cultural evolution) is no less biological than is inheritance of two arms and two legs, the ability to respond to the heat of a hot stove by rapidly withdrawing the hand or the ability of the tadpoles of some frog species to respond to environmental cues such as overcrowding or low levels of food by developing into cannibals. Biological and cultural evolution interact at several levels; many developments in sociobiology over the past quarter of a century are based on the proposition that there is a biological basis for much of human culture, and that genotypes predisposed toward cultural development accumulate through natural selection. Although biological factors are involved in most social patterns, they are often modified by cultural context.

Any desire to control our evolution is restricted by the potential incompatibility between cultural and biological fitness. Lengthening the life span may be a desired goal, but natural selection rarely influences the characters of post-reproductive individuals. Lowering fertility might be desirable, but biologically, high fertility has selective value. A large proportion of the human population is or will be affected by deleterious alleles that persist because of forces that maintain genetic variation or because the intervention of modern medicine allows individuals with genetic disorders to live longer and to reproduce. Three approaches to controlling our own evolution — eugenics, genetic engineering and cloning — are discussed, as are human efforts to engineer the genomes of other species, all of which raise ethical concerns.

BOX 20.1

Speech and Language (*continued*)

Broca's area on the left frontal lobe have difficulty speaking; this region coordinates vocal muscular movements. Aspects of language, such as grammatical structure and vocabulary, seem associated with neural circuits in the prefrontal cerebral cortex that lie somewhat forward of Broca's area. The disproportionate enlargement of the prefrontal cortex is an obvious feature of *Homo sapiens*.

It has been hypothesized that tool use and oral language share a common neurological evolutionary basis. Both language and manual tool manipulation are sequential processes. The appearance of stone tools coincident with the appearance of Broca's area in hominoids is consistent with a functional connection. Linkage between the two functions is consistent with their common localization in the left or dominant hemisphere of right-handed individuals. For those left-handed people whose manual control center lies in the right hemisphere, language control is often found there as well.

to population or from group to group (see Chapter 19). Thirty-nine behavior patterns, among them tool usage, grooming, and courtship behaviors, are present in some chimpanzee communities, even though genetic differences between the same communities are small. A recent cladistic analysis found cultural differences between populations of a single subspecies of the common chimpanzee *Pan troglodytes* to be greater than differences between three subspecies, consistent with cultural behavior being learned and not genetic.

Because of socially mediated transmission, cultural changes — unlike changes in genetic makeup — are not restricted to vertical transmission from one generation to another, but may be developed through interactions between *related and unrelated individuals*; the cultural "parents" of individuals need not be their biological parents. Nor need cultural parents derive from the same geographical area as their cultural offspring. Consequently, the kinds of isolation barriers that inhibit genetic exchange between biological species do not prevent transmission of culture between groups of modern humans. Humans therefore have two hereditary systems

- a **genetic system**, which transfers biological information from biological parent to offspring, and
- a **cultural system**, which transfers cultural information from speaker to listener, from writer to reader, from performer to spectator.

Importantly — indeed, very importantly — both systems are informational and heritable, the genetic system through the coding properties of DNA, and the cultural system through social interactions coded in language and custom, and embodied in records and traditions.

Cultural Evolution Outpaces Biological Evolution

One measure of how change continues to affect us is the time it takes to double our collective knowledge, a process that once took many thousands of years but now occurs in a mere handful. The generation time for cultural evolution is as rapid as communication methods can make it. Humans can now move from place to place

In 1963 D. J. da Solla Price estimated that of all the scientists who have ever lived, more than 90 percent were alive in 1963. Peter Gruss (President of the Max Planck Society for the Advancement of Science) made the same estimation in 2005.

faster than the speed of sound, and transfer ideas at the speed of electrons. In contrast, changes in human physical characters since agriculture arose seem relatively small, if detectable at all. Indeed, the most distinguished possession of *Homo sapiens,* the human brain, shows no change in size over the last 100,000 years (BOX 20.2). Why this difference in speed between cultural and biological evolution?

Oversimplifying a little, this contrast can be ascribed to differences between *two distinct types of evolution*: the mode of inheritance of learned characters associated with cultural evolution, and the mode of inheritance through natural selection involved in biological evolution. Evolution through experience is an extension of the method by

■ The large cranial volume of newborns — a cost of brain growth *in utero* — helps explain the difficulties faced by human females in giving birth.

BOX 20.2
Evolution of the Human Brain

The average human brain is 340% larger in volume and 290% larger when corrected for body weight than the brain of our closest sister group, chimpanzees. Human brains also differ significantly from those of gorillas. The average human brain is 267% larger in volume and 341% larger when corrected for body weight than the average gorilla brain. This difference represents evolution of the human and gorilla, both having undergone their own line of evolution.

The brains of humans become larger relative to body size because human brains follow a different pattern of development than the brains of other mammals or even other primates. In most mammals (including primates), brain growth is rapid relative to body growth during fetal stages but diminishes after birth. In humans, prenatal brain growth is also quite rapid relative to body growth, but this rate does not significantly diminish until infants are past one year of age. By adding 12 months of extrauterine development — the first year of life — such rapid early postnatal brain growth effectively extends the human gestation period from nine to 21 months.

Big brains relative to body size are metabolically expensive. Although the adult human brain represents only two percent of total body weight, it can consume as much as 20 percent of the energy budget. For this reason, strong selection is required to increase brain size relative to body size. Determining why, or even whether, bigger brains are better has preoccupied students of human evolution for centuries; the hypothesis is that changes in hominin brain size over the last 4 My must signify some crucial changes in mental capacity. Assumptions in the past were that humans are more complex than other organisms and so require more brain cells and so bigger brains, and/or that the ability to compartmentalize functions drove human brain evolution as it drove evolution at many other levels going back to the evolution of cellular organelles.

Comparisons between regions of the brain among closely related taxa can be used to correlate increases in specific regions with specific attributes of each taxon. When this was done for humans and great apes, much of the increase in human brain volume was found to be associated with increased area and thickness of the cerebral cortex. In particular there is a disproportionate increase in the neocortex, and increased connections between cerebellum and cortex. In apes, much of the increase is in the cerebellum, which is the region of the brain most involved in perception, coordination and motor control.

Another comparison is that the area of the cortex of a rat brain would cover a postage stamp, of a monkey a postcard, of a chimp a page, and of a human, four pages — all subjective measures but you get the point. Moving beneath the surface to genetic differences, a molecular study identified 49 regions of the genome that regulate neural development (some of which are expressed in the neocortex) and that are conserved in many mammals but that diverged rapidly after humans separated from chimpanzees.

which humans learn. It depends on conscious agents — humans with brains — who can modify cultural information in a direction that offers them greater adaptiveness or utility. Transmission occurs from mind to mind rather than through DNA. The information received from ancestors and contemporaries can be purposely changed to provide improved utility for ourselves, our offspring and others.

For reasons that should be apparent from earlier parts of this book, the high rate of cultural evolution is in striking contrast to the speed of evolution by natural selection. Biological (organic) evolution occurs through a process of selection (among other forces) on the phenotype. Much of genetic evolution is slow because it requires fortuitous changes in DNA sequences, their organization into the existing genome and selection of the resulting phenotype, before it can proceed. Each change may take many generations before it is incorporated into the population, although as discussed earlier, processes as diverse as horizontal gene transfer in prokaryotes (see Chapter 8), phenotypic plasticity in eukaryotes (see Chapter 12) and evolution of gene regulation in all organisms (see Chapter 13) greatly speed the rate(s) of biological evolution.

The incremental process of biological evolution is vastly different from the rapid, conscious selective process humans use to choose among behavioral alternatives, although the biological equipment needed to transmit and use cultural information (memory, perception, language ability) connects them both. Human minds have become agents of a novel selection mechanism by consciously choosing among alternatives because of their consequences. It is "human minds" and not "human mind" because cultural evolution relies on communication among individuals and on group interaction. Cultural evolution is vastly more than the sum of its parts, in that the products of a socially coordinated group of individuals — a city, a daily newspaper, an automobile factory, a cathedral, a film — are quantitatively and qualitatively more than such individuals could create alone.

Social Darwinism

Human culture has a biological foundation, and both culture and biology are based on informational systems that evolve over time. Awareness of these two facts prompted various writers in the late nineteenth and early twentieth centuries to suggest that biological concepts can be extended to society, and that nature and culture share similar evolutionary mechanisms, especially natural selection. These ideas were developed into a body of thought, later called **Social Darwinism** (see Chapter 1), the chief concepts of which are twofold.

- Differences among human individuals and groups arise through natural selection.
- Natural selection is the mechanism that led to social class structures and to national differences with respect to economic, military and social power.

In particular, early writers on the subject saw parallels in the role of competition in the biological and social spheres. Slogans such as "struggle for existence" and "survival of the fittest" (see Chapter 15), when applied to social traits, enabled English Social Darwinists, especially Herbert Spencer, to suggest that cultural evolution was proceeding inevitably toward social and moral perfection (and approaching its culmination in Victorian society in England!).

Spencer's writings, which ranged widely across biology, economics, philosophy and sociology, continued to exert their influence well into the twentieth century.

■ See for example, *The Factors of Organic Evolution* (Spencer, 1887) and *First Principles* (Spencer, 1897).

Interestingly, although Spencer coined the term "survival of the fittest," implying the action of natural selection, he remained a Lamarckian in respect to biological evolution until relatively late in his life (see Chapter 5). Whatever the mechanisms, Spencer conceived evolution as a powerful force that governed all spheres of existence and therefore justified social and economic policies that supported those who were most "morally fit." For many Protestant intellectuals, Spencer's belief in the cosmic power of evolution helped reconcile science to their religion and made his writings extremely popular.

In its harsher economic and social forms, the Spencerian approach became popular in various circles in the United States (where more than half a million copies of his books were sold), especially through the teachings of William Graham Sumner (1840–1910), the best-known American Social Darwinist. In 1883 Sumner wrote: "We cannot go outside of this alternative: liberty, inequality, survival of the fittest; not-liberty, equality, survival of the unfittest. The former carries society forward and favors all its best members; the latter carries society downwards and favors all its worst members." Not surprisingly, many wealthy capitalists found such views to their liking. John D. Rockefeller, Jr. (1874–1960), for example, whose father forged the gigantic Standard Oil Trust by destroying many smaller enterprises, justified such behavior with the observation that, "The growth of a large business is merely a survival of the fittest . . . the working out of a law of nature . . ."

Along with objections raised by others, including Thomas Huxley in his book *Evolution and Ethics,* the major problem with Social Darwinism was its assumption that society (economics, politics, government) operates through the same laws as biology and for the same goals. However, the laws of the inheritance of wealth and power in society are entirely man-made; the laws of biological inheritance are not. Because they can be consciously selected, social goals can be directed toward almost any objective: poverty or wealth, socialism or capitalism, for example.

Unfortunately, the possibility of implementing social goals attracted individuals and groups who aspired to occupy superior positions over other individuals or groups, and has been used to justify or reinforce racism, genocide, and social and national oppression. An extreme example is the role played during the 1930s and early 1940s by Adolf Hitler in the "racial health" movement — the purposeful destruction of millions of people because they were considered members of "inferior" racial groups. Even in the United States, with its more democratic social heritage, laws were passed during the 1920s restricting immigration from eastern and southern Europe because of their "inferior" or "undesirable" races. Immigration of virtually all Asians into the United States was halted in 1882 by the Chinese Exclusion Acts, which were not repealed until 1943. While it must be added that economic and political considerations played a large role in the push for immigration restrictions, racial consideration formed part of the social framework under which the laws were promulgated. The United States was far from alone. Australia enshrined what came to be known colloquially as the "White Australia Policy" in the second bill passed after confederation in 1901.

■ A wonderful opportunity to evaluate Thomas Huxley's development of evolutionary ethics is available. Huxley's 1894 book *Evolution and Ethics, and Other Essays,* contains the text of the Romanes Lectures (*Evolution and Genetics*) he delivered at the University of Oxford in 1893. Fifty years later his grandson, Julian, delivered the Romanes Lectures on *Evolutionary Ethics,* and published both sets of lectures with an extensive introduction and commentary (T. H. and J. S. Huxley, 1947). Both grandfather and grandson affirm their belief in evolution as a means to enhance our moral progress.

■ Sociobiology

Biology does exert influences on social interactions between people. The close mother-infant relationship is just one of many behaviors common to all human groups. But to what extent are such social behaviors genetically determined? How much is nature and how much nurture is how the question is often couched.

E. O. Wilson's book, *Sociobiology: The New Synthesis,* published in 1975, launched a new field of science, but evoked a firestorm of controversy. Defining **sociobiology** as "the systematic study of the biological basis of all social behavior," Wilson argued convincingly that animal behavior, like morphology, is shaped by natural selection: behavioral traits that maximize an individual's reproductive success are more likely to be carried into the next generation than those that do not. Even such social behaviors as **altruism** — a difficult problem for evolutionary biologists because an organism sacrificing itself for the greater good may not leave any descendants (see Box 15.1) — can be explained on the basis of genetic mechanisms. In altruism, selection acts not only on the fitness of an individual carrier of a favorable behavioral "gene," but also on that of the genetic relatives of such individuals ("kin selection" or "inclusive fitness"; see Box 15.1).

Drawing on methods used in population genetics, ethology, ecology and other disciplines, sociobiology provides a rationale for assessing how (and perhaps to what extent) biological causes account for social behaviors. In the years since Wilson's book, studies have been carried out on social behavior in a wide range of animals. For instance, why a new dominant male in a pride of lions kills off his rival's offspring, and why a female lion will nurse not only her own cubs (chapter opener photo), but those of her close genetic relatives.

■ The actions help to ensure survival of, respectively, his own and her close relatives' genes into the next generation.

Less successful has been the application of sociobiology to human behavior. Indeed, the major source of controversy in Wilson's landmark book was the last chapter, which extended the discussion to humans, as did a sequel, *On Human Nature* (1978). Because of interactions between genes and environment, Wilson said, "there is no reason to regard most forms of human social behavior as qualitatively different from physiological and non-social psychological traits." According to this view, human nature is determined as much by heredity as by culture, and there are limits to how much humans can change their behavior.

Sociobiology deals less with the immediate causes of a particular human behavior than with its "ultimate" underlying evolutionary function. Nevertheless, to its critics the discipline leads inevitably to the conclusion that most observed human social behaviors are biologically caused (with the potential although certainly not inevitable corollary that the status quo, including social inequities, is justified), a concept close to the views held by the Social Darwinists. But to the charge of *determinism* — to the implication that humans are slaves to biological destiny — Wilson countered,

> The moment has come to stress that there is a dangerous trap in sociobiology, one which can be avoided only by constant vigilance. The trap is the naturalistic fallacy of ethics, which uncritically concludes that what is, should be. The "what is" in human nature is to a large extent the heritage of a Pleistocene hunter-gatherer existence. When any genetic bias is demonstrated, it cannot be used to justify a continuing practice in present and future societies. ("Human Decency Is Animal." *New York Times Magazine*, Oct. 12, 1975.)

The challenge for sociobiology is to tease out the mix of biological and cultural evolution, an exceedingly difficult but important task.

Human Control over Our Own Evolution

The fact that cultural considerations can transcend biological considerations is apparent when **human control over evolution** is considered. In which direction is our evolution and the evolution of other species to be guided? What goals are to be set?

These questions arise not from any underlying biological laws, but from the conscious cultural realization that we would like to improve ourselves and the world in which we live, and from the availability of the technology to achieve such goals. This quest for human improvement, however, often comes face to face with our **biological limitations**, a few of which are now considered.

Although our lives have changed immeasurably as a result of our advanced technology, our genetic makeup has not. For example, many human occupations are sedentary, but our bodies still require exercise. Many who live in surplus societies tend to eat poorly, put on extra weight, and suffer from the accompanying ills, including heart disease and diabetes. The stresses of many aspects of modern life can enhance the affects of illnesses. In addition, pollution of various kinds caused by industry, automobiles, pesticides, and tobacco leads to a variety of diseases, ranging from induced cancer to emphysema and silicosis.

Perhaps one of the most important contrasts between what we are and what we would like to be lies in what might be called the difference between **biological and cultural efficiency.** *Biologically,* our *efficiency* begins slowly to decline in our 20s and 30s, ages that in the past coincided with reproduction or immediate post-reproduction. As measured by the contributions made to various professions and by increased life expectancy, however, *cultural efficiency* peaks decades later. Our biological heritage stresses early reproductive success, soon after which physical deterioration sets in. Human cultural development requires continued plasticity, adaptability and longevity. Human cultural development is limited by human biological decline.

In most organisms that attain reproductive maturity relatively early in their potential life span, survival after reproducing tends to be short-lived or non-existent. Individuals in many modern human societies enjoy a long period of **post-reproductive longevity.** Early in our own evolution, only about half the human population passed the age of 20 and probably not more than one in 10 lived beyond 40. These low longevity values extended into the time of the early Greeks and even later in some groups. Life expectancy remained between 20 and 30 years until the Middle Ages, then rose somewhat. Life expectancy has risen sharply among Europeans, Americans, Japanese and some others in the last century and a half: from about 40 years in 1850 to the present high 70s or low 80s. Less developed countries also have seen dramatic increases, the major exception being those countries in Africa where the AIDS epidemic has reduced already low life expectancies to as little as the low 30s.

These statistics are important. They indicate that as a result of improvements in sanitation, diet, and medical practice over the last century and a half, natural selection now exerts relatively little influence on our "fitness," that is, on the number of offspring produced, which more and more is a matter of personal choice. Further, most of the ills resulting from our changed lives — diabetes, heart disease and cancer — affect us mostly in our later, post-reproductive years, so that there is little or no chance for natural selection to weed out susceptibility to these ills. An individual who has produced three children and at the age of 50 develops cancer or other diseases with genetic components is no less reproductively successful than an individual of the same age who has produced three children but does not suffer from such diseases.

Our increasing longevity has had an even more dramatic impact, an impact that has grave implications not only for humanity but also for the world as a whole, and is discussed in the context of the population explosion in BOX 20.3 .

■ A DVD entitled "Evolution — Why Bother?", produced by the American Institute of Biological Sciences and the U.S. Biological Sciences Curriculum Study (available at http://www.aibs.org/bookstore), contains an up-to-date evaluation of the impacts of evolution on agriculture, human health, the development of new drugs, and other issues of the application of technology.

■ For a table of rankings of life expectancy by country, see *The World Factbook,* a publication of the CIA available online at https://www.cia.gov/library/publications/the-world-factbook/index.html. Swaziland, with an estimated 2007 life expectancy of 32.23, occupied the lowest ranking of any country.

BOX 20.3
Population Explosion

In most countries, infant mortality rates have declined markedly over the past century (FIGURE B20.3). From 15 or more percent of all births in 1900, infant mortality rates have fallen to less than one percent in more than 70 countries, a result of improved sanitation, nutrition and medical care, including control of infectious disease. Lower infant mortality comes at a cost, however. Defects such as cerebral palsy associated with premature birth cannot be brought down at the same rate as infant mortality and so increase in incidence.

In most countries, however, it was (and still is) usual for birth rates to remain high for many years, even decades, following such a decrease. When combined with increased longevity, the result has been an exponential growth of the human population, an explosion that at its height in the 1960s saw a doubling every 35 years, a rate many thousands of times greater than that experienced in Paleolithic societies. Since 1900, the number of people has risen from a billion to 6.8 billion (December 30, 2009). Although the overall growth rate has dropped (the doubling time is now 61 years), the absolute numbers continue to climb. Today about 210,000 people are added to the world population every 24 hours, by far the majority of them in developing nations. The U.S. Census Bureau predicts a world population of 9.3 billion by the year 2050.

This **population explosion** has grave implications not only for human beings but also for the planet and all its other inhabitants. The more of us there are, the more food, water, space, energy and other resources we need. The more resources are consumed, the more waste is produced. Our total impact on the environment, however, depends largely on how many resources each of us uses, and on how much waste each of us produces — our per capita **ecological footprint** — which varies immensely between countries. A person living in the United States today consumes far more resources, produces far more garbage, and emits far more carbon dioxide and other pollutants than does an inhabitant of, say, Nepal. Furthermore, per capita resource use in industrialized countries, which already far

■ The infant mortality rates are taken from estimates in the *2006 CIA World Factbook*. A recent analysis showed that a significant amount of the variation in human birth weight from region to region reflects adaptation to local selection pressures, especially parasitism.

■ Ecological footprint is a measure defined as the amount of land and water required to produce the resources consumed and to absorb the waste of an individual, population or lifestyle.

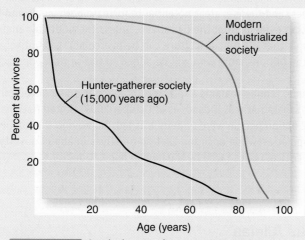

FIGURE B20.3 Survival curves for a population of hunter-gatherers who lived 15,000 years ago on the Mediterranean coast (based on skeletal remains) compared to a present-day population living in an industrialized society. (Adapted from May, R. M., *Nature,* **327**, 1987: 15–17.)

(continued)

BOX 20.3

Population Explosion (*continued*)

exceeds sustainable levels, continues to rise. At the same time, hundreds of millions of people in developing countries are responding to improved economic conditions by adopting the patterns of resource use characteristic of wealthier regions.

Here the complexities of just one concern associated with population growth are discussed: **access to food**. Current production, though rising, appears barely able to keep up with present requirements. What then of the future? This turns out to be a topic with many facets, some of them political. The list below is far from complete.

- The amount of arable land is finite, and the amount cultivated is at or approaching capacity in most countries.
- The average amount of arable land per person worldwide continues to shrink as populations expand and as land is taken out of production as a result of environmental problems and urban sprawl. (In 1950 there were 0.23 hectares per person, in 2000, 0.11 ha, a decline of more than 50 percent.)
- More and more land is used to grow food for livestock, a practice that ultimately results in production of far fewer food calories for humans per unit of land.
- Substantial amounts of food are lost through faulty distribution and storage systems, and through pests and disease.
- Extreme poverty means that a sizable fraction of the world's population cannot afford to buy nutritious food. At the same time, large surpluses exist elsewhere.
- Unwise subsidies and policies as well as increasing globalization and domination of agriculture by multinational corporations distort production and markets almost everywhere.

Other threats to the food supply locally as well as globally include soil erosion, fertilizer and pesticide contamination, water supply depletion, global climate change and extreme climatic events.

In terms of future food production, considerable amounts of arable land remain unused in several countries, and there is great potential to increase both yields and the nutritional value of crops. As is evident from the list above, unless the political, environmental, and other problems that plague agriculture can be overcome, it is unlikely that we will be able to feed the billions more who will be added to the world's population in the years to come.

In the meantime, famine remains a recurrent problem in some developing countries, especially in sub-Saharan Africa. Indeed, as of October 2009, United Nations statistics showed that 1.02 billion people were undernourished. Improving the quality of life for an increasing population while ensuring that no one lives in abject poverty are surely among the most critical issues faced by the human population.

Deleterious Alleles

Despite improvements in medical care, alleles that have obvious **deleterious effects** still affect human populations. Changes in the frequencies of such alleles are now considered.

The number of children born with marked physical and/or mental abnormalities is conservatively estimated at 20 to 25 per 1,000 births in the United States. The

mortality rate ascribed to congenital malformations is around 15 percent of all infant deaths. Many other defects, not immediately noted at birth, become apparent during childhood years and are more common than realized. In various studies, about 30 percent or more of hospital admissions for children and 50 percent of all childhood deaths are ascribed to birth defects or to complications such defects may have caused. Although not all birth defects are genetic, the proportion of those that are is high.

The incidence of each individual genetic disorder is low. Cystic fibrosis is 0.05% of live births. Tay-Sachs disease is 0.001% (see Table 12.1). Trisomy 21 is 0.13% and congenital heart defects 0.4%. Summing the incidence of genetic disorders in humans, whether autosomal, X-linked or chromosomal, gives a total frequency of around 2.6% of live births. If genetic defects such as muscular dystrophy that appear later in life are included, the frequency probably doubles to closer to 6%. If in addition, we include less obvious defects that nevertheless have strong genetic components — impaired resistance to stress and infection, and other physical and psychological weaknesses — the effects of harmful genes touch a significant proportion of the human population. As of January 5, 2010, the Human Gene Mutation Database listed 96,631 mutations in 3,611 genes, although most of these mutations are exceedingly rare.

■ The Human Mutation Database at the Institute of Medical Genetics in Cardiff, © 2007 Cardiff University. Available at http://www.hgmd.cf.ac.uk.

The ubiquity of deleterious alleles with lethal effect was dramatically demonstrated in studies made of the offspring of marriages between cousins. These studies show that apparently normal individuals in our society carry a genetic load equivalent to that of approximately one to eight deleterious lethal alleles, which if homozygous, would cause early death. Two important questions arise.

- What accounts for the prevalence of these harmful alleles?
- What, if anything, can be done to rid ourselves of them?

The reasons for the high frequency of deleterious alleles are not fully agreed upon although there is little question that they originally arose through mutation. In one hypothesis, held by the late Theodosius Dobzhansky and others, such alleles may offer considerable advantage to their heterozygous carriers by producing some form of hybrid vigor (see Chapter 15). According to this hypothesis, a gene will be maintained in the population although the homozygote produced by the gene is relatively inferior in fitness.

Another school, formerly headed by the late American geneticist Hermann Muller (1890–1967) was based on the hypothesis that such genes produce no advantage of any kind; their frequency is high because the usual effect of natural selection has been artificially reduced. According to Muller, genotypes that were formerly defective and would have been eliminated in earlier times are now retained in the population by medical techniques that enable those with these defective alleles to pass them on to their offspring. If not eliminated by selection, deleterious alleles will gradually increase in frequency in accordance with their mutation rate. Because the mutation rate is usually low, the frequency of any particular allele will increase rather slowly, but as there are many possible deleterious alleles, the allelic load will increase significantly. Nearsightedness is a trait whose frequency has most likely increased in recent periods, but can be corrected quite simply by an optometrist.

■ Hermann Muller received the 1946 Nobel Prize in Physiology or Medicine for the discovery of the production of mutations by X-ray irradiation.

That natural selection no longer operates to eliminate many genotypes is not necessarily an undesirable feature of modern life. Few individuals would argue today that fire and clothes should be abolished because they are artificial devices that circumvent natural selection by permitting hairless phenotypes to survive in cold climates. It would also be difficult for us to return to the "good old" prevaccination, presanitation

days of smallpox, diphtheria, typhus, cholera, and plague. What is a feature of modern life, however, is that despite advanced medical care, a great many deleterious alleles still exert an effect. Beginning in the early twentieth century, and detailed below, there have been a variety of attempts to improve human genetic quality. **Eugenics**, the earliest of these attempts, had a much broader focus than the individual allele, perhaps understandably given the state of genetics at the time.

Eugenics

In his dialogue *The Republic,* Plato suggested that humans could be improved through selective breeding. In his ideal philosopher-state only the most physically and mentally fit individuals would reproduce. Their offspring would be raised by the state and the governing class would be selected only "from the most superior."

In its modern form, suggestions for improving the human gene pool come under the name *eugenics,* a term coined by Francis Galton in 1883 from the Greek words *eu* (good) and *gen* (birth). After reading the works of Darwin, his first cousin, Galton became concerned with the heredity of quantitative characteristics, especially intelligence. From 1865 on, he promoted the idea that the evolution of human traits through natural selection could be substituted by their evolution through selective breeding: "What nature does blindly, slowly, and ruthlessly, man may do providently, quickly, and kindly. As it lies within his power, so it becomes his duty to work in that direction." Coming from a brilliant family himself, Galton was impressed by the way in which intellectual and personality traits tended to run in families. Convinced that such traits were inherited, and drawing on his knowledge of animal breeding, he concluded that "judicious marriages over several generations" could "produce a highly gifted race of men," and thus thwart the "reversion to mediocrity" he believed to be threatening society as a result of excessive breeding by those who were not superior. Although intelligence does have a large genetic component, much of it is remarkably plastic, being influenced by prenatal diet, maternal uterine environments, and cultural and socio-economic factors.

Galton's ideas were taken up and developed by many prominent thinkers. Like the Social Darwinists, however, early eugenicists reflected their own personal and cultural biases in deciding which traits they deemed undesirable. Strong racist and class-based biases played a part from the beginning. Charles B. Davenport (1866–1944), an influential leader of the early-twentieth-century eugenics movement in America, exemplified this racist approach by using New Englanders as the standard of comparison for all American social groups irrespective of their country of origin. According to Davenport and other members of his Eugenics Record Office — which from 1910 to 1944 collected an enormous database of information about American families — particular groups could be characterized by social characteristics with identifiable hereditary components: "Italian violence, Jewish mercantilism, Irish pauperism."

As now known, no evidence supports the biological superiority of any particular group in respect to social characteristics or intelligence. Many characteristics considered desirable — high intelligence, esthetic sensitivity, longevity and good physical health — are not caused by single genes that are easily identified but by complexes of many genes acting together and responding to different environments. As learned from domesticating animals (see Chapter 14), selective breeding to develop beneficial gene complexes is fraught with difficulty. The methods involve complicated schemes based on selection of parents along with testing of progeny under controlled environmental

conditions. Selective breeding enhances the selected characters but enhancement can be at the expense of the deterioration of other characters.

Eugenics Policies

Unlike Social Darwinists, eugenicists believed in active intervention in breeding. Given that some early proponents were visionaries, there was at least a possibility that the movement could have developed into a serious attempt to diminish human suffering and improve the human gene pool, if only it had been stripped of racism and provincial prejudice. Eugenics developed, however, in a society where those attitudes were commonplace. Two avenues of action were promoted and in some instances followed.

- **Positive eugenics** attempted to increase the frequency of beneficial genes.
- **Negative eugenics** attempted to decrease the frequency of harmful genes.

Francis Galton's approach of encouraging particular people to marry is an example of positive eugenics. Singapore's campaign in the 1990s, in which young graduates were offered inducements to produce children, is another. Unfortunately, however, negative policies dominated the vast majority of government programs instituted while eugenics held sway, roughly from the 1890s to the 1940s. So strong was the influence of the movement that almost every non-Catholic Western country was affected. Other areas included the southernmost Latin American states (where policies favored whiter complexions) and, since the early 1990s, China.

Viewed today through the prism of human rights, we are appalled at the legislation — some of it draconian — enacted and enforced in the name of eugenics. Methods employed included

- restrictions on immigration and marriage (although eugenics was only one of many issues in immigration, marriage, and segregation debates);
- racial segregation, including bans in the United States on marriage between whites and African Americans, overturned by the Supreme Court only in 1967;
- compulsory sterilization of the "feebleminded," certain criminals, and others deemed unfit;
- forced abortions; and, finally,
- in Germany under the Nazis, genocide of those (especially Jews) regarded as racially inferior and thus a threat to the "purity" of the Aryan race.

In the United States, home to the second largest eugenics movement (after Germany), marriage prohibitions were enacted in many states during the early decades of the twentieth century, and tens of thousands of individuals were sterilized, the last of them in the early 1960s.

Backing the eugenics movement was a large body of research. But even from the early years, some geneticists and members of the general public were sharply critical of the methods employed and of conclusions drawn from the findings. The desire to improve the human condition, however, continues in quite a different guise, that of **reproductive technology**.

Reproductive Technology and Eugenics

The search by today's parents for healthy children has many eugenic overtones. Genetic counseling before conception, artificial insemination, testing the fetus for inherited disorders, selective abortion of fetuses deemed defective, ultrasound

Recall from Chapter 13 that in thalassemia either α or β hemoglobin chains are either not produced or are produced in reduced numbers.

examination, and *in vitro* fertilization, including pre-implantation screening of embryos, are all in effect eugenic practices, albeit voluntary and on an individual basis. In countries where such techniques are widely used, the incidence of some congenital disorders has sharply declined in recent years. In Taiwan, for instance, the incidence of thalassemia amongst newborns dropped from 5.6 to 1.21 per 100,000 over eight years.

Because prenatal testing is expensive and often invasive, only ultrasound is carried out as a matter of routine (FIGURE 20.2); other tests are usually performed only when the family history (or mother's age, in the case of Down syndrome) indicates a clear risk that a particular mutation or chromosomal abnormality may be present. Because the likely consequence of such a testing regime is that the overall load of deleterious alleles in the newborn population will not decline significantly, some geneticists ask whether all fetuses should be screened for deleterious mutations. Armed with knowledge gained from sequencing the human genome (see below), it may not be long before screening tools can be designed to test for all known disease-causing mutations. Given the financial cost and risk to the fetus of prenatal screening, however, the most likely use of such tools will be in pre-implantation testing of embryos conceived via *in vitro* fertilization (IVF) (FIGURE 20.3). About one percent of Americans and four percent of Danes currently begin life by IVF.

Reproductive technologies, including genetic screening, evoke a host of ethical and social concerns, a few of which are listed below.

For a longer list of questions, see the Human Genome Web site, under Ethical Issues: http://www.ornl.gov/sci/techresources/Human_Genome/elsi/elsi.shtml.

- What constitutes a disease? If an individual carries a gene that increases their susceptibility to a condition later in life, do they have a disease? Sequencing the human genome may have narrowed even further the number of people considered normal.
- How do potential parents make an informed choice among our options?
- Can an individual live a fulfilling life if they inherit a particular condition? If yes, should that individual be denied that right?
- Who has the right to know about my genetic makeup — my school? My employer? My insurance company? My government? In fact, who owns this information?

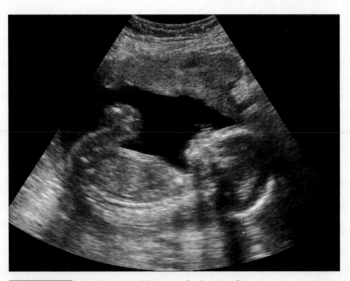

FIGURE 20.2 An ultrasound image of a human fetus.

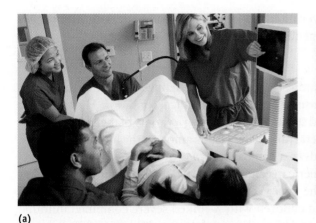

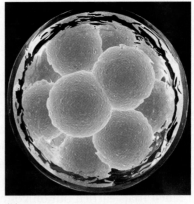

(a) (b)

FIGURE 20.3 *In vitro* fertilization. (**a**) A doctor, using ultrasound imaging, retrieves a woman's egg to be inseminated outside the body. (**b**) A 12-cell-stage intact embryo ready to be implanted into the uterus.

Despite the many ethical concerns, it is likely that the use of modern reproductive methods will continue to increase, and that over time they will have a larger and larger impact on our gene pool, at least in wealthy countries and well-off segments of others.

For now, however, and probably into the future, a far greater influence is likely to be modern medicine in general. A recent study documented that the incidence of severe congenital heart disease among adults in Quebec (Canada) climbed steeply between 1985 and 2000. Before open-heart surgery, few if any of these patients would have survived childhood. Less benignly, and despite laws banning the practice, the widespread use in India and China of ultrasound followed by abortion to ensure the birth of a male child is leading to significant imbalances in the male/female ratio of those societies.

So, if almost everyone now survives to have children, and if our children can be protected from natural selection so that they too go on to have children, **will natural selection continue to operate on humans?** The argument could be made that because only a handful of children now die of infectious disease in countries with robust health care systems, the door to a population with weaker average immune systems is being opened. But again, does it matter? If, as expected, uncontrollable pandemics appear, the answer could well be yes.

Genetic Technology

As discussed in Chapter 19, humans have been engineering the genetic makeup of organisms for a very long time. "Engineering" by manipulating genes and genomes, most commonly through selection, started with domestication of plants and animals and extended to agriculture, horticulture, and plantation forestry. Sequencing of the human genome, completed in 2006 after several drafts had been announced over the previous few years (see Table 19.2), provided a tremendous boost to both human genetics and the hope that humans may someday be able to control inherited diseases.

Two concurrent efforts were involved in this achievement, one carried out in laboratories around the world under the umbrella of the Human Genome Project and the other led by Craig Venter of Celera Genomics. Among the goals of the Human Genome Project were identification of what turned out to be about 23,000 protein-coding genes, determination of the sequences of the three billion chemical base pairs involved, and analysis of the ethical, legal, and social issues arising from the project.

In December 2006, Venter announced the sequencing of his own genome, the first for an individual person, and the first targeting all 46 chromosomes; the two human genome projects mapped only one of each pair. In October of the same year, a $10 million prize was offered to the first group to cheaply map the genomes of 100 people in 10 days.

A wide range of applications, covering many fields of human endeavor, is expected to flow from this work and from genomics in general.

Gene Therapy

An area of research evoking great interest is **gene therapy**, defined as the introduction of genetic material to treat or cure a disease or abnormal medical condition. Human trials, which began in 1990, have targeted several diseases using a variety of methods and vectors. Although there have been several promising results amongst the hundreds of trials, for example, in the treatment of advanced melanoma, a great many problems remain.

In 1999 an individual in an American gene therapy trial died, most likely as a result of a severe immune response to the viral vector. A further major setback occurred in French trials carried out on young children with X-linked severe combined immune deficiency ("bubble boy" syndrome). Although nine of the ten children in this trial were successfully treated, three subsequently developed cancer. In a study conducted in 2006, one-third of mice administered the same gene as was used in the latter trial developed lymphoma later in life. As of 2009, no gene therapy product had been approved for sale in the United States. In May 2007, Epeius Biotechnologies Corporation's Rexin-G™ was permitted by the United States Federal Drug Administration to move to phase I clinical trials as an orphan drug to treat pancreatic cancer.

■ Orphan drug is a seven-year right given by the U.S. FDA to a drug company to exclusively market a drug developed to treat a rare human disease, one affecting less than one out of every 200,000 individuals.

The impact of gene therapy on the gene pool and thus its ability to alter the path of evolution is limited by the fact that people who might otherwise die may now go on to produce offspring bearing the deleterious gene. However, there is a strong consensus among scientists and indeed the general public that the risks are so extraordinary that there should be no attempt to modify human germ plasm. It is also considered unethical to alter genes that would change or "enhance" nondisease characters such as an individual's appearance or height. An outline of how risks associated with such technologies are (should be) assessed is provided in BOX 20.4 .

Clones and Cloning

A **clone** is an organism descended from and genetically identical to another organism. Under this definition, all offspring produced by asexual means — a method of reproduction used by bacteria and many eukaryotes, especially plants — are clones. Clones provide genetic uniformity across the generations, an advantage when organisms face the same conditions for a long time, but clones lack genetic diversity. Sexual reproduction, on the other hand, has selective advantage for organisms facing changeable and unpredictable conditions, but is a comparatively expensive genetic gamble. The advantages and disadvantages of asexual versus sexual reproduction are discussed in Chapter 7.

Cloning, defined here as *the production of clones by artificial means,* has a long history in agriculture, horticulture, and forestry. Methods such as taking plant cuttings, grafting and layering have been employed for centuries if not thousands of years; familiar plants reproduced in this way include potatoes, pineapples, many horticultural and forestry species, and most trees bearing fruit or nuts.

Animals are cloned by subdividing an early-stage embryo and implanting each portion into a potential mother(s). Another method, *somatic cell nuclear transfer,* has been the subject of enormous controversy and media attention in recent years, and is what is usually referred to in discussions of cloning. In this method, a nucleus

BOX 20.4
Regulating Risk and Safety

From the first days of genetic modification technology, many have expressed concern over the potential dangers of recombinant DNA applications. It is a legitimate worry. How can we be sure that any type of genetic manipulation — indeed, any type of experimentation — does not incur unforeseen, detrimental consequences? How do we regulate risk and safety? Difficult as it may be to ensure against all risk, potential difficulties can be evaluated and precaution can be exercised.

It is important to realize that there are risks in almost any experimental procedure. For example, a drug that reduces the effects of the common cold may turn out to be a potent carcinogen. Or an experimental strain of wheat may have the potential to expand limitlessly, devastating other crops. At the same time, it should be kept in mind that "natural" catastrophes, so far, have exceeded those brought about by experimental procedures.

There is no way of proving the absence of risk in an experimental procedure. The ultimate effect of any activity, however trivial, is unpredictable; there is no way all future interactions can be anticipated. Should we then ban all beneficial technologies because there may be potential misuse or a possible unexpected deleterious effect in the future? Had experimentation been banned, even our moderate success in treating cancer, AIDS, and other serious diseases would not have been achieved. Moreover, it seems foolhardy to restrict the benefits of agricultural improvements, practically all of which have involved gene manipulation.

When risks are evaluated, it seems logical to base these risks on more immediate and foreseeable consequences. Again, the medical/agricultural model seems appropriate: restrict experiments to tissue culture, animal models, clinical trials, and/or enclosed agricultural plots. Such methods allow main as well as side effects to be closely observed before expanding the use of experimental treatments to the general public. One point cannot be emphasized enough: the importance of running trials for sufficient time so as to allow long-term side effects to become apparent.

For the most part, risk evaluations have functioned successfully. For example, most pharmaceutical drugs on the market are effective for the purposes for which they were designed, and their side effects can be taken into account during treatment. Keep in mind that strict regulation by politically independent governmental agencies has been the primary agent in averting pharmaceutical disasters. Because these new genetic technologies concern matters that affect everyone, decisions should be publicly based.

Widening the argument, many would agree that intervention in "nature" can be justified provided that the basic principle "above all, do no harm" is followed. Make sure our actions are beneficial or, at least, benign. Because intervention for beneficial goals is not always benign or risk-free, however, trade-offs are often necessary. In medicine, for example, some interventions involve surgery, radiation, chemotherapy, and other potentially damaging procedures. General agreement prevails, however, that medical intervention should take into account the risk of damage and the quality of life: in other words, to evaluate the means and not just the ends. The same is true for genetic modification in all its guises.

is removed from a body cell and inserted into an egg cell from which the nucleus has been removed. By the 1990s success had been achieved with a variety of mammal species, but always using nuclei from very early embryos (FIGURE 20.4). Nuclei from later embryos, and certainly from adults, appeared ineffective. The news in 1997 that Ian Wilmut and colleagues had succeeded in producing a sheep (named "Dolly") by transferring the nucleus of a mammary cell from a six-year-old ewe into one of its eggs, burst on the world like a thunderclap. If it could be done for sheep, why not for humans? Indeed, Wilmut told a British parliamentary committee that he expected a similar technique could be used to clone humans "within two years."

Successful cloning of mammals depends on "reprogramming" an adult nucleus to assume the functions of a zygote nucleus and thus direct the development of a new individual. This process is both expensive and highly inefficient — very few cells with transplanted nuclei lead to living offspring — and abnormalities are common in those offspring that do survive to birth. Still, the implications are unparalleled. You can now clone your pet cat or dog. What price for human cloning?

Reaction to the possibility of **human cloning**, most of it decidedly negative, was immediate. The media, representatives of medical, religious and other bodies, and members of the general public all voiced concerns about the ethics of the technology. Politicians in many jurisdictions acted swiftly. In the United States, for example, President Clinton banned all federally funded human cloning research, asked for a moratorium on non-federally funded research, and ordered the National Bioethics Advisory Commission to conduct hearings. Nevertheless, some individuals embarked on research in the field.

Little progress was made until 2004, when a South Korean biologist announced that he and his team had produced a human embryo by cloning; other advances in his lab were announced in quick succession. The aim of this work was not to generate an exact copy of a human being, but rather to produce a source of **stem cells** from which any cell type could be generated, opening the door to potential cures for a range of diseases. In late 2005 and early 2006, however, it became apparent that most of his results were fraudulent. Now regarded as one of the worst cases of fraud in the history of science, this case dealt a blow to stem cell research. For now, the many ethical and experimental challenges posed by human cloning are moot. As with gene therapy, the hoped-for benefits are huge, but it is not clear whether this line of research will yield fruitful results, let alone provide us with the tools to heal a large range of conditions.

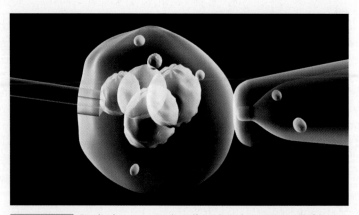

FIGURE 20.4 In cloning, a somatic cell nucleus in a pipette (left) is inserted into an early embryo held in place by the glass rod on the right.

Our Influence on the Evolution of Other Species and on Their (and Our) Environment

Until now this chapter has concentrated on the role of culture as a force in human evolution, and on human efforts to direct the evolution of our own and other species. Below is briefly addressed the extraordinary impact humans have had throughout history on the distribution and gene pools of organisms other than whose genomes humans have tried directly to change, impacts that have mostly been inadvertent.

Starting in the Paleolithic, when human actions such as hunting or the widespread use of fire may have led to the extinction of the megafaunas (very large animals) of Australia, the Americas and other parts of the world (see Chapter 19), humans have altered both the species mix and environment of almost the entire globe. A notable example today is the plight of the ocean's fish stocks, all of which are under threat from over-fishing. Populations of the northern cod, which live off Newfoundland and Labrador, have not recovered at all from a disastrous decline in the years around 1990.

Wherever humans have moved, especially since the development of agriculture, they have brought with them a suite of other organisms, both domesticated (cattle, goats, etc.) and opportunistic (rats, weeds, disease-causing microbes). The changes wrought on local environments took a quantum leap when Europeans began venturing across oceans. The impacts on native humans in newly discovered regions were calamitous, a story told by Jared Diamond in *Guns, Germs, and Steel* (1997).

In the decades following the arrival of Europeans in the Americas and the Pacific, infectious diseases such as smallpox, measles and tuberculosis decimated populations that up until then had been isolated. A similar fate befell a variety of animal and plant species, especially trees (for example, the American chestnut). Among the effects have been mass extinctions, especially on islands (Hawaii is a prime example), and displacement of native species by weedy ones (the list is long). Fueled by globalization, such impacts continue today. The agents of destruction can arrive from any quarter (for example, invasive species carried in the ballast water of ships; insect pests carried on lumber). Entire ecosystems are threatened or in some cases have disappeared.

In another book, *Collapse: How Societies Choose to Fail or Succeed,* Jared Diamond details the consequences of exceeding one's resource base. Using a number of case studies (the Maya of Central America, the Norse settlement in Greenland, and more), he shows how human societies in different times and different places have collapsed as a result of changes in their environment, changes in large part precipitated through their own actions, including faulty agricultural practices, deforestation, and water depletion, to name a few. Today our resource base is the whole world, and our impacts are accordingly global, extending even to climate change. Before us lie two very different worlds: a world where our ever-increasing numbers, our extravagant use of resources and the growing divide between rich and poor threaten our very survival, and a world where dazzling new discoveries in the realms of medicine and genetics hold out the possibility that one day we may be able to alter the path of our inheritance. Major genetic transformations, however, remain far in the future.

We turn in the next and last chapter to discussion of a very specific aspect of human culture: the ways in which humans have related and continue to relate their experience of nature to systems of belief, religion and evolution. Chapter 21 takes us back to historical perspectives introduced in Chapters 5 to 7. The traditional religious rationale for social and biological systems was that the universe followed a designed

order established by an intelligent deity or deities. By omitting God from the equation, evolutionary theory challenged that long-accepted worldview. Tension between religion and evolution remains, especially now that we can engineer other life forms, modify the environments of countless other species, and create changes that affect the entire globe.

Recommended Reading

Alcock, J., 2005. *Animal Behavior: An Evolutionary Approach*. Sinauer Associates, Sunderland, MA.

Diamond, J. M., 1997. *Guns, Germs, and Steel: The Fates of Human Societies*. Norton, New York.

Diamond, J. M., 2005. *Collapse: How Societies Choose to Fail or Succeed*. Viking Books, New York.

Huxley, T. H., 1894. *Evolution and Ethics, and Other Essays*. Macmillan, London.

Huxley, T. H., and J. S. Huxley, 1947. *Touchstone for Ethics 1894–1943*. Harper & Brothers, New York and London.

Kevles, D. J., 1995. *In the Name of Eugenics: Genetics and the Uses of Human Heredity*. Harvard University Press, Cambridge, MA.

Maienschein, J., 2003. *Whose View of Life? Embryos, Cloning, and Stem Cells*. Harvard University Press, Cambridge, MA.

Mindell, D. P., 2006. *The Evolving World: Evolution in Everyday Life*. Harvard University Press, Cambridge, MA.

Wilson, E. O., 1975. *Sociobiology: The New Synthesis*. Harvard University Press, Cambridge, MA.

Wilson, E. O., 1978. *On Human Nature*. Harvard University Press, Cambridge, MA.

Culture, Religion and Evolution

21

KEY CONCEPTS

- An important part of any culture is a system of beliefs by which individuals order their lives.
- All cultures developed belief systems in which the universe followed a designed order established by an intelligent deity.
- Discoveries that Earth was not the center of the universe called into question the necessity for a belief in a god(s) or deity.
- Discovery of physical mechanisms that governed the universe called into question the necessity of a creator.
- Darwin's theory of evolution provided a natural explanation and observable mechanisms for organismal origins, relationships and change.
- Form and function as adaptations to the environments by organisms with a heritable system and the ability to respond to changing circumstances replaced designs fixed by a creator.
- Replacement of a caring god by a process of chance and necessity produced enormous tensions within individuals, established religions and societies.
- Nevertheless, religious beliefs and scientific rationality coexist in many individuals and many sections of societies, a prominent exception being those who take records of creation as literal truth.

Above: The wonders of the organic and inorganic world are all around us, even in a drop of water.

427

◼ Overview

◼ Interactions between science and religion vary enormously from one religious/cultural tradition to another. For reasons of available space the discussion of evolution and religion is primarily illustrated using the Judeo-Christian religion.

◼ Philosophers (Herbert Spencer), revolutionaries (Karl Marx), psychologists (Sigmund Freud), and writers, playwrights and poets (Joseph Conrad, Thomas Hardy, Alfred Tennyson, George Eliot, George Bernard Shaw) are just a few of those who incorporated evolution into their studies, writings, politics and worldviews.

The present chapter continues the discussion of cultural and social evolution begun in Chapter 20 by exploring science in relation to culture and how the theory of evolution impacted on a very specific aspect of human culture — **systems of belief** — especially those associated with established Judeo-Christian religions. This chapter examines the roots of religion in a historical context, how the theory of evolution impacted on religion, and how creationists view evolution.

Science, society and culture are intimately interwoven (see Chapter 1). From a social point of view, the development of evolutionary theory in the late nineteenth century coincided with an all-pervasive political and economic revolution in social behavior. The economic challenges posed by capitalism and its new wealthy classes in Europe allowed — even encouraged — ideological challenges to the prevailing religious and philosophical systems that had supported the old social order, allowing European science to flourish. The influence of two of those challenges — Social Darwinism and the rise of eugenics — was introduced in Chapter 20.

As was discussed in Chapters 6 and 7, the impact of Darwin's theory on all aspects of Victorian and subsequent societies was profound. By proposing that the form and function of living organisms did not arise by creation but rather by natural processes, Darwin's theory made it clear that species were not fixed and unchangeable (see Chapter 16). These radical ideas, which revolutionized biology, also affected sociology, anthropology, economics, politics, women's rights, fiction, poetry, linguistics, philosophy and psychology.

As is true of any human activity, science is not independent of the society in which it is carried out. Indeed, whether and when science was conducted varied from society to society in the past. Examples are the much earlier discovery/invention and utilization of gunpowder, the compass, printing and many other items in China than in Europe.

The traditional Western religious rationale for social, cultural and biological systems for many centuries was that the universe followed a designed order established by an intelligent deity. Along with belief in a god who created and maintained the universe, religion provided a set of ethical and moral values upon which social systems were based.

Until Copernicus and Galileo in the sixteenth century, no one had seriously challenged the idea of a powerful deity controlling the physical universe. In a new worldview accepted by some after the discoveries of Copernicus, Galileo and others, God was seen as an initial creator who established the laws of nature by which the universe ran. The advent of Darwinism posed threats to Western religion by suggesting that humans had no special place in the universe and that their origin, features and relationships to the Great Apes could be explained by natural selection without the intervention of a god. The Darwinian view that evolution is a historical process — extant organisms were not created spontaneously but formed in a succession of past events as organisms adapted to changing environments — contradicted the common religious view of design by an intelligent designer. Given the great age of Earth and the power of natural selection, complexities that seem unlikely as singular spontaneous events become evolutionarily probable events. Consequently, for many, the reality of evolution challenged belief in a god or gods.

Charles Darwin and Society

Darwin's theory of evolution had a profound impact on biology and religion and on virtually all spheres of human society and culture. The ready acceptance of evolution in Europe owes much to the era in which it appeared, a time of social, economic and technological developments, the overthrow of old social orders, and the rise of capitalism. Acceptance of religious explanations eroded as more and more natural explanations for the origin and modification of Earth and its inhabitants were discovered, and as it was recognized that ethics and morality can differ between different human societies and that changes in such values need not depend on religious beliefs.

Nevertheless, some Judeo-Christian groups reject evolutionary explanations for biological events. Especially in the United States, but increasingly in Europe, fundamentalist religious groups who oppose evolution have attempted to prevent its teaching and to reduce or eliminate the subject of evolution in many biology textbooks. The "creation science" movement (which, despite its name, does not use the scientific method) and more recently the "intelligent design" movement are the latest attempts to deny evolutionary biology.

Science and Culture

Science and the scientific method often are presented as if they are immune or unconnected to the culture in which the science is conducted. A glance at any of the volumes in Joseph Needham's multi-volume opus on *Science and Civilization in China* dispels such a notion. Many discoveries and inventions were made in China centuries before they were made (independently) in the West; the circulation of the blood (2nd century BC versus 1616), printing and so the first printed book and the invention of the blast furnace (3rd century BC versus 1850s) are but three of dozens if not hundreds of examples (TABLE 21.1). China developed an advanced system of social development through its religious belief system and reverence for nature and because of a thirst for discovery and understanding the natural world and a quest to harness that understanding to benefit the Chinese people through application, that is, through technology.

The social context of science is amply illustrated when we see that the invention of gunpowder, the compass and printing in Europe between 900 and 1500 AD are rightly regarded as setting up the transformation of Europe during the scientific revolution in the 17th century. Francis Bacon (1561–1626), the philosopher, statesman and essayist — "knowledge is power" comes from one of his essays — whose promotion of the scientific method set the stage for the scientific revolution in 17th century Europe, saw that revolution as based on

> Printing, gunpowder and the compass: These three have changed the whole face and state of things throughout the world; the first in literature, the second in warfare, the third in navigation; whence have followed innumerable changes, in so much that no empire, no sect, no star seems to have exerted greater power and influence in human affairs than these mechanical discoveries. (*Novum Organum* [*New Instrument*], 1620).

However, printing, gunpowder and the compass were invented in China between 400 and 600 BC, one to two thousand years before their "invention" in Europe (Table 21.1; BOX 21.1). Science and cultural development go hand in hand. In the 5th century BC, the

The enormous contribution to human knowledge of Needham's project is admirably documented in the biography, *The Man Who Loved China* (2008) by the eminent biographer, Simon Winchester.

So far the 25 volumes encompass the history of scientific thought, mathematics, physics and physical technology/engineering, chemistry and chemical technology, biology and biological technology, and the social background to Chinese science.

21.1 Inventions and Discoveries in China

a. Inventions	Date
Abacus	190 AD
Ball bearings	2nd century BC
Blast furnace	3rd century BC
Cast iron	4th century BC
Crop rotation	6th century BC
Density of metals	3rd century BC
Deposition and erosion of sedimentary rocks	1070 AD
Gunpowder	9th century BC
Magnetic needle compass	1088 AD
Paper	200 BC
Porcelain	3rd century BC
Printed book (the *Diamond Sutra*)	868 AD
Rain gauge	1247 AD
Seismograph	132 AD
Spinning of silk	2850 BC

b. Discoveries	Date
Circulation of the blood	6th century BC
Estimation of Pi	3rd century AD
Inoculation against smallpox	10th century AD
Magnetic induction	1044 AD
Motion of the stars	725 AD
Negative numbers	1st century AD
Polar equatorial coordinates	1st century BC

c. Time of Inventions and Discoveries in China and in the West

China	The West	Invention/Discovery
14th century BC	900 AD	Decimal system
9th century BC	900 AD	Gunpowder
4th century BC	1100 AD	Compass
4th century BC	1300 AD	Cast iron
4th century BC	900 AD	First Law of Motion*
3rd century BC	1200 AD	Blast furnace
3rd century BC	900 AD	Estimation of Pi
3rd century BC	1400 AD	Porcelain
200 BC	1200 AD	Paper
2nd century BC	1200 AD	Circulation of the blood
1st century AD	1500 AD	Negative numbers
1st century AD	1900 AD	Suspension bridge
AD 132	1300 AD	Seismograph
AD 190		Abacus
AD 725		Motion of the stars
AD 868	1500 AD	Oldest printed book
10th century AD	1800 AD	Inoculation against smallpox
AD 1044	1700 AD	Magnetic induction
AD 1070		Deposition and erosion of sedimentary rocks
AD 1088	1700 AD	Magnetic needle compass
AD 1247		Rain gauge

*Isaac Newton's three Laws of Motion were published on July 5, 1687.

BOX 21.1

Printing, Gunpowder and the Compass in China

Paper and Printing

The first "paper" in China, was made from silk around 300 BC. It was expensive and limited in its usage. Paper derived from a variety of natural products — old clothing, bark from trees, stalks from threshed wheat — was first made early in the 2nd century BC. It was relatively cheap, light, thin, durable, and more suitable for writing on with a brush than was silk.

Block printing, in which words or images are engraved into a wooden board, which is then covered with ink and "printed" onto paper, one copy at a time, was established in China by 600 BC. Between 1041 and 1048, Bi Sheng invented a process of carving characters on a piece of clay, slowly baking the clay to harden it and so produce what is recognizable as movable "type"; that is, characters that could be arranged and reused multiple times for printing. Printing from movable metal type, which is how printing is usually categorized in the West, came much later.

Gunpowder

Gunpowder was invented in the 9th century BC when mixtures of saltpeter (potassium nitrate), sulfur and charcoal were discovered to be explosive when heated. Initially, the Chinese used gunpowder to make fireworks to celebrate and commemorate important events and festivals (FIGURE B21.1).

The Compass

The first magnetic compass, used by Chinese fortune-tellers and to plot direction on land, was invented in the 2nd century BC. It consisted of a spoon-shaped piece of lodestone (magnetite, a mineral oxide of iron) balanced on a bronze plate marked with stars and con-stellations (FIGURE B21.2). Invention of a magnetic compass with a magnetized needle that could be used for navigation followed the discovery that rubbing the tip of a needle with magnetite gave the needle the property of attracting iron particles that aligned along the North–South polar axis. The first such compass has been traced to the 8th century AD. By the 9th century AD magnetic compasses were installed on ships to aid navigation.

FIGURE B21.1 Fireworks over palaces in China celebrate New Year's Eve.

(continued)

BOX 21.1
Printing, Gunpowder and the Compass in China (*continued*)

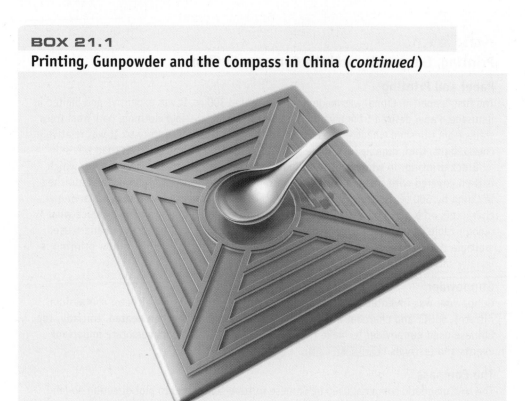

FIGURE B21.2 An early Chinese magnetic compass.

Chinese invented a process to cast iron. A palace tower 91 m high, weighing 1,300 tons, topped with a giant iron phoenix and covered with gold leaf, was built in the 7th century AD. This and a cast-iron lion 6 m high and weighting 40 tons, built in the 10th century BC and standing today, attest to the skill with which science was translated into technology and used to commemorate events in Chinese life and history. The only important aid to construction not invented in China appears to be the screw and screwdriver.

Science and Religion

As seen in the discussion of the early rise of science in China, the development of science is culturally dependent and our view of the world is influenced strongly by the culture in which we live. Different cultures place different emphases on how individuals perceive various events and relationships, and on how they explain these perceptions. What is considered as science is also culturally dependent; large differences exist between cultures as to whether or how they apply scientific concepts to events and experiences. Explanations that many of us accept as scientific — analyses based on the application to known and emerging knowledge of scientific principles and laws (see Chapter 1) — others do not accept, or accept only to varying degrees.

In Western European culture, nonliving phenomena generally have been considered more amenable to scientific analyses than matters that touch on life itself, and on human life in particular. Physics and chemistry were well established as sciences in Europe by the 19th century (even earlier in China; Box 21.1), before or during

which many physical laws were discovered. Biology, however, especially embryology and evolution, continued to be subject to interpretations according to which all living organisms were imbued with a vital force (vitalism, for which, see Chapter 5). Gaining freedom from such constraints was more difficult for biology (especially for evolution) than it was for physics or chemistry.

Bases of Religious Belief

Religious development in different cultures provides clues to the evolution of religion itself.

Religion first develops in a culture when, in an attempt to deal with aspects of experience that can neither be controlled nor understood, societies endow the forces of nature with the spirits of animals and/or supernatural powers. Ritual develops when ceremonies are repeated to help ensure their efficacy. Ritualized behavior seems to have become especially important in the transition from hunting to agricultural societies. Crops have to be planted and harvested at appropriate seasons each year. Individual efforts could be either rewarded or damned by mysterious external and uncontrollable forces such as droughts, floods, volcanic eruptions or plagues of insects (FIGURE 21.1).

Because societies saw the forces of nature as humanlike, religion sustained and encouraged the hope that those appeals that humans understood — gifts, sacrifice, obedience and loyalty — could appease nature's judgment and recrimination. Religion provided and preserved belief systems that helped explain and guide a societies' relationship to the world about it and provided answers to such questions as, "Where did we and our society come from, and why?" To these ends, the belief system of each culture usually offered comprehensive accounts of how and for what purpose the world was created. Accounts from 10 different cultures are outlined in BOX 21.2 . A theory of evolution by natural selection disrupted much more than beliefs in origins and design. It disrupted entire worldviews and codes of existence.

Recall from Chapter 19 that Neanderthals placed flowers on the graves of dead individuals in what is presumed to have been part of a ritual.

FIGURE 21.1 Plagues of locusts and other insects have devastated crops for millennia. Here desert locust swarm in Aleg, Mauritania.

BOX 21.2

Accounts of the Origin of the World from Ten Different Cultures

Egypt	God arose from the depths of the ocean, created dry land, and then created all creatures on the hill at Eliopoulos at the center of the universe.
Mesopotamia	God made sky and earth by splitting the powers of evil in half, and then produced humans for purposes of worship.
Iran	God created all that is good and struggled with an evil being that creates all that is bad. Each struggle lasts about 3,000 years and will continue until evil is vanquished, at which time creation will be complete and perfect.
Greece	God, a female, divided the sky from the sea, and produced a serpent with whom She copulated. She then laid a giant egg, out of which came the earth, its creatures, and all the heavenly bodies, as well as the subsidiary powers to rule these various entities.
India	God created himself from a golden egg, and from the various parts of his body everything was born. After a time, life is destroyed and the cycle begins again.
Israel	God created the universe in six days ending with the creation of humans, according to Genesis 1; or, God first created Adam in the Garden of Eden and then created animals and birds and eventually Eve, according to Genesis 2.
Benin, Africa	God was a woman who produced twins: the sun and the moon. During various eclipses, the twins came together to create the various gods and spirits of earth and sky that rule over humans.
Yucatán (Mexico)	God created the world in four distinct periods, each separated by a flood.
Crow Indians (USA)	God created the earth and its creatures from mud gathered in the webbed feet of ducks that swam on a primeval ocean.
China	The universe was originally in the shape of a hen's egg, out of which God emerged and chiseled its main physical features. After 18,000 years God died and the remainder of the world was derived from his body: the dome of the sky from his skull, rocks from his bones, soil from his flesh, rain from his sweat, plant life from his hair, and humans from his flesh.

■ See *The Evolution-Creation Struggle* by Michael Ruse (2005) and *Breaking the Spell: Religion as a Natural Phenomenon* by Daniel Dennett (2006) for how a philosopher of biology and a psychologist approach these issues.

In support of a supernatural view of events, religion has relied on two basic concepts that probably arose early in history, the *soul* and *God*. According to hypotheses proposed by psychologists and anthropologists, the idea that the soul is separate from the body probably originated in the separation of mind from reality in dreams. Theologians have often sought a "scientific" argument for the nonmaterial nature of the soul or the personality, claiming that intellectual processes cannot have a material origin. Today, such concerns are reflected in research whose goal is to understand **consciousness**, which embodies the awareness of a self that is aware of itself, looks out at a world that it perceives, and responds to it. Although the precise relationship between the cells and wiring of the brain and the thoughts and feelings it produces are not known, relationship exists and, as discussed in Chapter 19, there is every reason to believe that the complexities of thinking and feeling evolved like any other trait.

Darwinism and Religion

Darwinism had an especially dramatic impact on religion. To many of Darwin's religious contemporaries and to others since, *On the Origin of Species* and *The Descent of Man* raised issues of momentous importance: To the general public, Darwinism was at least as much a religious as a scientific issue. The stressful relationship between evolution and religion stemmed as much from the recognition that evolutionary concepts were not impregnable as from vulnerability of religion. Darwin was painfully aware of this and pointed out at least two discoveries that could refute his theory

- an inversion of the evolutionary sequence such as evidence of humans in the Paleozoic or Mesozoic Eras; or
- finding the same species in two separated geographical locations when their presence was not caused by migration between these areas.

No such evidence has been found in the century and a half since Darwin's theory was published in 1859.

Understanding the role that evolution plays in modern life requires understanding how religion and evolution interacted and interact. A confrontation between the two views soon after the publication of *The Origin of Species* was a debate on November 5, 1860 at Oxford University. Darwin's theory had been published just under a year earlier. As part of the ongoing public discussions generated by the book, Bishop Samuel Wilberforce (1805–1873) of the Church of England attacked Darwinian theory as incompatible with the Bible. Coached by the religious comparative anatomist Richard Owen, the Bishop attempted to destroy Darwin's theory through scientific arguments. Wilberforce aimed his final point directly at Thomas Huxley when he asked whether it was through his grandfather or grandmother that Huxley claimed descent from a monkey. The wit of Huxley's response, recounted a few months later in a letter to his friend Frederick Dyster, has often been quoted, as it is here again.

If, then, said I, the question is put to me would I rather have a miserable ape for a grandfather, or a man highly endowed by nature and possessed of great means and influence, and yet who employs these faculties and that influence for the mere purpose of introducing ridicule into a grave scientific discussion, I unhesitatingly affirm my preference for the ape.

Such religious attacks were worldwide, frequent, harsh, and almost always focused on the same points: religious opponents accused Darwinists of seeking "to do away with all ideas of God," "to produce in their readers a disbelief of the Bible," "to displace God by the unerring action of vagary," and to "destroy humanity's unique status."

In the most sensitive area of all — life itself — Darwinian evolution offered different answers to religion's claims of why life's important events occur. Darwin's works made clear that society no longer needed to believe that only the actions of a supernatural creator could explain biological relationships. Darwin presented a nature of continual change, unpredictable chance events and unrelenting competition for resources among living creatures with no obvious guidance. To religious believers natural selection substituted waste for economy by treating life as a continually expendable commodity. Darwin's theory replaced the creativity of a humanlike god with randomness and uncertainty. The fear that Darwinism was an attempt to displace

See http://www.hhmi.org/biointeractive/evolution/lectures.html#evodiscussion, a Web site maintained by the Howard Hughes Medical Institute for discussions of religion and science.

The suggestion that humans were related to the orangutan was raised in Chapter 19. Gorillas only became known after 1854 when the first bones were shipped from Africa, and in the early 1860s when a stuffed gorilla skin was taken along as a prop on lecture tours undertaken throughout England by an African explorer. "Victorians were horrified to think that these reputedly violent animals — distorted men in shape and size, representing the brutish, dark side of humanity — were possible ancestors" (Janet Browne, 2006).

God in the sphere of creation was quite justified. Consequently, for many, evolution presented a deep, abiding threat.

The transition from regarding nature as operating in understandable human terms to operating in terms of the evolutionary opportunism expressed by competition and reproductive success was difficult. The philosopher Emanuel Kant (1724–1804) who took an evolutionary approach to cosmology and to the origin of the solar system, found it at times abhorrent to admit that species could evolve, describing such notions as "ideas so monstrous that the reason shrinks before them." Darwin is reported to have felt quite uncomfortable about his role in proposing the evolution of species. In 1884, in a letter to Joseph Hooker, he wrote that "it is like confessing a murder." One source of ambivalence was the established belief that humans were created in the image of God and endowed to rule over other biological and social groups, Kinship to those below, whether ape or servant, was a repugnant idea. An oft-reported example of this repugnance is the response of the wife of the Bishop of Worcester when informed that Huxley had announced that man was descended from apes: "Descended from apes! My dear let us hope that it is not true, but if it is, let us pray that it will not become generally known."

Essential to the preservation of religion in the midst of the evolutionary bombardment was the place served by religion as the main repository for the ethics and morals of society (what is right and wrong), served to maintain confidence in the social order and to unify nationalistic and provincial sentiments ("God is on our side"). In outlining sociobiology thirty years ago, E. O. Wilson claimed that religion serves an important role; religion helps provide a common social identity to groups of individuals, and by increasing group power, confers "biological advantage" (see Chapter 20).

Evolution dealt with many basic questions of life that are of concern to religion, but as a science it did little to meet emotional needs. Society has held on to both evolution and religion because they serve different needs. With the exception of some fundamentalists, whose views are discussed in the following section, religion essentially withdrew from the domain of biological evolution, leaving both the origin of species and the origin of humans to evolutionary biologists and anthropologists.

The position of the Roman Catholic Church as enunciated by several recent Popes illustrates one type of accommodation between religion and evolution, namely acceptance of scientific findings when the evidence is incontrovertible. Although no supporter of evolution, Pope Pius XII (1876–1958) accepted the evidence that the universe was millions of years old. In 1993, in declaring that Galileo had not committed heresy in proposing that the solar system revolved around the sun, Pope John Paul II (1920–2006) used Galileo's achievements as an example of how scientific knowledge could lead the church to reinterpret the writings in the Bible. Three years later, he declared that scientific findings show that evolution is more than a hypothesis, it is "an effectively proven fact." Pope John Paul's acceptance was rational, predicated as it was on the basis that "truth cannot contradict truth."

The Question of Design

The first significant cracks in the theological armor of continued divine intervention in nature arose through the discoveries by Copernicus, Galileo and Kepler (FIGURE 21.2) of natural laws regulating the motion of the solar system. Their genius was that all

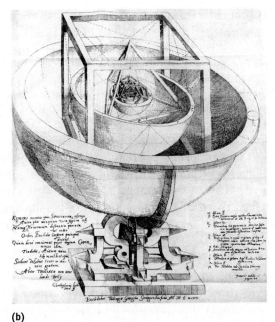

(a) **(b)**

FIGURE 21.2 Johannes Kepler **(a)** (1571–1630), a German astronomer and mathematician whose improved reflecting telescope and insights published in *Astronomia Nova* (1600) and other works demonstrated that the planets orbit the Sun, as shown in this historic illustration of Johannes Kepler's planetary system model **(b)**.

three provided simpler and more universal explanations of the workings of the physical universe than previously available (see Chapter 1). The elegance of their explanations made inevitable the rise of science and of the scientific method, for

> A momentous change had come about when what scientists did came to be taken for granted, even by those who understood little or nothing of it. The crucial change in the making of the modern mind was the widespread acceptance of the idea that the world is essentially rational and explicable, though very wonderful and complicated (J. M. Roberts, *The Triumph of the West*, 1985, p. 242).

These openings were widened considerably by the mechanistic explanations of the motion of the solar system through the force of gravity proposed by Isaac Newton (1642–1727), and by the hypothesis of an unbounded infinite universe in which our world was proportionately smaller than even a grain of sand. Later, geologists such as Charles Lyell (see Chapter 5) extended this mechanistic approach by proposing how natural forces could mold Earth's surface (see Chapter 2).

Although these scholars were not atheists, their findings about natural processes made through sciences such as mechanics, optics and chemistry, helped relegate God from a continually active, intervening agent to a prime force — a master artisan who has designed logical, self-functioning machines — and so led to a closer examination of the nature of God. If God functioned only as prime initiator of the universe, what was the need for God (and therefore for religion) at the self-functioning levels that followed the world's origin? Moreover, if the universe is logical, what was the logical purpose in creating it? Was the motivation to create perfection? If the world was not created perfect, what moral good could there have been in its creation? An example of how essential it was to believe that each design had a creative purpose is reflected

in the doctrine expounded by the English theologian, philosopher and advocate of Christian ethics, William Paley (1743–1805).

■ Paley's philosophy that Earth and its creatures were created, designed and maintained by God is contained in *Natural Theology; or, Evidences of the Existence and Attributes of the Deity* (1802), a treatise that formed the basis of the teaching of natural theology at Oxford and Cambridge for decades and which was last reissued by Oxford University Press in 2006.

> There cannot be design without a designer; contrivance without a contriver; order without choice; arrangement without anything capable of arranging. . . . Arrangement, disposition of parts, subservience of means to an end, relation of instruments to a use, imply the presence of intelligence and mind.

Simple common sense seemed to support supernatural design. Organismal design presupposes a designer, who by definition (but not based on any scientific evidence) is an intelligent supernatural being, a chain of reasoning that leads back to creation and a creator. Can there be a watch without a watchmaker, and by extension, can there be a person without a person-maker, laws without a lawmaker? So, it comes as no surprise to find that every culture/society has a creation story in its history (Box 21.2).

To evolutionary theory, the essential challenge that religion poses has always been, "How from the disorder of random variability can nature achieve the beauty of adaptation without intelligent intervention?" Darwin's fundamental contribution was to answer this question by means of a mechanism that no one had thoroughly explored before: natural selection. In providing his theory Darwin was explicit on the failure of Paley's arguments from design "now that the law of natural selection has been discovered." For Darwin there was "no more design in the variability of organic beings and in the action of natural selection than in the course which the wind blows." As he stated in his *Autobiography*

> The old argument of design in nature, as given by Paley, which formerly seemed to me so conclusive, falls, now that the law of natural selection has been discovered. We can no longer argue that, for instance, the beautiful hinge of a bivalve shell must have been made by an intelligent being, like the hinge of a door by a man.

The problem of design was never far from Darwin's thoughts. If plants and animals had been designed, the design failed to prevent the extinction of most species and suffering resulting from the laws of nature. In 1870 Darwin wrote: "My theology is a simple muddle. I cannot look at the universe as the result of blind chance, yet I can see no evidence of beneficent design, or indeed of design of any kind, in the details."

However much these philosophical questions of design and designer, watch and watchmaker challenged religion, they did not pose as big a danger as did Darwin and his theory. In the most sensitive area of all — life itself — Darwinian evolution offered different answers to religion's claims of why life's important events occur. By viewing life as a continually expendable commodity rather than a divinely premeditated and consecrated goal, Darwin replaced what many had seen as an understandable view of nature — the creativity of a humanlike God — with concepts of randomness and uncertainty and the fear that no one could understand the source and purpose of any natural event or design.

■ Creationism and Intelligent Design

Fundamentalist religious groups have generally not accepted the uneasy truce between evolution and religious institutions that exist in the Western world. In the United States, many individuals and groups who believe in the Judeo-Christian account of creation in Genesis have formed political pressure groups to insist that their beliefs be

treated as a scientific alternative to evolution in public education. According to a 1982 Gallup poll, 44 percent of Americans agreed with the statement that, "God created humankind in its present form almost 10,000 years ago." In 1999, the number had risen to 47 percent. A poll taken by the *New York Times* in November 2004 produced a statistic of 55 percent.

The origins of such groups within society date back at least to the early 1900s with strong roots among economically threatened tenant farmers and small landholders, especially in the U.S. South and Southwest. Sociologists have suggested that believing in the literal truth of the Bible, revivalism, and other aspects of fundamentalist religion, helped many of these rural groups defend their way of life against domination and control by the more intellectual but exploitative northern and eastern social and economic establishments.

Whatever their initial motivations, fundamentalist groups were successful in pursuing anti-evolution goals in the South and Southwest during the first few decades of the last century. By the end of the 1920s, fundamentalists had introduced anti-evolution bills into a majority of U.S. state legislatures, and had passed some in various southern states. Probably the most famous confrontation between evolution and biblical creationism during that period was the 1925 trial of a schoolteacher, John Scopes, who was convicted of ignoring the ban against teaching evolution in Tennessee schools (FIGURE 21.3).

Although many biologists felt that the Scopes trial essentially defeated the intellectual validity of the creationist position, creationists apparently lost little ground in these regions and had an impact on public education far beyond the South and Southwest. By influencing textbook adoption procedures in various local and state school boards in the United States, creationists successfully minimized evolutionary explanations in secondary school textbooks for a long time; by 1942, more than 50 percent of high school biology teachers throughout the United States excluded any discussion of evolution from their courses.

■ John Thomas Scopes (1900–1970) was a legally trained teacher in Dayton, Tennessee. William Jennings Bryan (1860–1925), a congressman from Nebraska who had run in three federal elections (1896, 1900, 1908) as the Democratic Party nominee for President, prosecuted the case against Scopes. Clarence Seward Darrow (1857–1938), a lawyer and leading member of the American Civil Liberties Union, defended Scopes. The trial, ended on July 21. Scopes was found guilty and fined $100. Bryan died of apoplexy five days later.

FIGURE 21.3 William Jennings Bryan (in the bow tie) being interrogated by the lawyer Clarence Seward Darrow, during the trial of the State of Tennessee vs. John Thomas Scopes, July 20, 1925. The trial had moved outdoors; the temperature in the courtroom exceeded 100°F.

The impetus for an increase of evolutionary teaching in U.S. secondary schools was the result of a movement to reform the science curriculum in the late 1950s and early 1960s, when it was realized that science education lagged behind that of other countries, specifically the Soviet Union, which in 1957 had launched the first space satellite, *Sputnik*. Among these innovations were new high school textbooks in both biological and social sciences (*Biological Science Curriculum Study; Man as a Course of Study*) that discussed evolution and analyzed changes in human social relationships. By the end of the 1960s, anti-evolution laws were either repealed or declared unconstitutional. Nevertheless, today only 90 minutes out of each teaching year is devoted to evolution in the average high school in the United States.

Within the last decade, a number of societies and institutes established by fundamentalists to promulgate creation science/intelligent design have entered the fray with the aim of including creationism in the science curriculum. Despite the name, there is little if any recognizable science in creation science. Although considerable literature deals with creationist attacks on evolution, the refusal of fundamentalist creationists to accept scientific evidence shows no promise of being resolved. By contrast, views in evolution, whether conforming or dissenting, are evaluated, as in all science, by evidence and not by faith or unverifiable decree.

Despite the overwhelming scientific evidence for evolution as a natural process, some religious groups adhering to creation have developed intelligent design as a purported scientific alternative to evolution. "Intelligent design" is latter-day creationism. Indeed, in a landmark December 2005 decision in a U.S. federal district court trial (*Kitzmiller v. Dover* in Dover, PA), intelligent design was ruled a form of religion and not science, and as such "cannot be taught alongside evolution in science classes in U.S. public schools."

Because intelligent design relies on supernatural explanations rather than natural causes, it is not science. Religious arguments have explanatory power with respect to belief systems, but they are not scientific explanations and should not be confused with, or regarded as, scientific explanations. Evidence that pertains to some of the more common creationist claims, all of which relate to topics discussed in earlier chapters, is outlined in **BOX 21.3**.

For the Web site containing the full decision in *Kitzmiller v. Dover* see: http://en.wikipedia.org/wiki/Kitzmiller_v._Dover_Area_School_District. Accessed October 8, 2009.

An extensive Internet source is the Talk Origins Archive at http://www.talkorigins.org. Two further sites are http://www.nap.edu/catalog/6024.html#toc (*Science and Creationism: A View from the National Academy of Sciences*, 2d ed. The National Academies Press, Washington, DC) and http://www.hhmi.org/biointeractive/evolution/lectures.html#evodiscussion, BioInteractive, a Web site maintained by the Howard Hughes Medical Institute with discussions of religion and science.

BOX 21.3
Evidence Refuting Common Claims Made by Creationists

Evidence countering some of the claims commonly made by creationists is outlined below, with cross-references to the chapters in which the evidence is discussed.

1. Analyses and explanations of historical events can be accepted even without the ability to provide experimental verification (see Chapters 1 and 2).
2. With respect to the origin of biotic organic molecules, the existence of abiotic synthesis of amino acids and other basic biological molecules indicates that chemical reactions are structured to produce such molecules, and that natural selection can operate to increase their complexity and organization (see Chapters 3 and 4).
3. The basic elements of organismal biology — distinctions between organisms, relationships of organisms to one another, and how those distinctions and relationships arose — can all be understood and explained by evolution (see Chapters 5–9).

BOX 21.3

Evidence Refuting Common Claims Made by Creationists (*continued*)

4. Large numbers of different kinds of hominid fossils are known, from the 4.4 My-old *Ardipithecus ramidus* to *Homo sapiens*. Humans are not a distinct creation (see Chapter 19).

5. The theory of evolution (genetic changes over time) explains the historical course of biology in terms of natural processes, such as mutation, selection, genetic drift and migration, providing explanations consistent with all observations to date and demonstrating the fact of evolution (see Chapter 15).

6. Mutations, even those not immediately adaptive, persist in populations through various mechanisms and provide the genetic variation allowing populations to change when environmental conditions change (see Chapter 17).

Specific examples often raised in contrary terms by creationists include

1. Geological strata are multilayered — some many kilometers thick — comprising sediments and fossils deposited over very long time periods (see Figure 5.3). There is no geological or paleontological support for a biblical one-year-long worldwide flood.

2. Fossils are not concentrated in any particular geological stratum and are found even in Precambrian deposits (see Figures 8.5 and 8.6).

3. Even when older geological strata have been elevated over younger strata (especially seen in association with mountain-building), the strata can be recognized as "older" by the kinds of fossils they contain, and by structural features that show continuity with non-overturned deposits elsewhere.

4. The demonstration of selection of resistance to insecticides in insects (see Figure 13.2) and for rapid adaptation of cichlid fishes to microhabitats in African lakes (see Box 14.2) provides direct evidence for evolution.

5. Selection for sight (see Figures 6.5 and 13.4) hearing, and vertebral support, include steps that are adaptive (see Figures 6.4 and 6.6), that is, "intermediate stages" are adaptive.

6. Cultivated corn evolved from wild grass in about 7,000 years because it was continuously selected for food by Central American Indians (see Figure 17.11), demonstrating the impact of humans on the evolution of other species.

7. The answer to why the emergence of new species is not commonly observed also is simple: the process of speciation takes time. In many groups, species formation is generally recognized when reproductive isolation occurs between two populations so that each can embark on its own distinctive evolutionary path (see Chapter 17). Because they involve many genetic and selective (adaptive) changes, such events can take many thousands of years. Intense selective differences have caused rapid genetic differentiation between populations (see previous point). Other selective effects, observed in both laboratory experiments and nature, include changes in temperature, food sources, habitats, and environmental toxins. Among such events investigated by population geneticists are fly populations that became resistant to insecticides, plants that developed new characters, *Drosophila* strains that developed specific sexual behaviors, cichlid fishes that rapidly adapted to different conditions in African lakes, and parasitic insects that adapted to different plant hosts.

Last Words

Many intellectual threads led to the modern theory of evolution, a theory that recognizes that Earth is ancient, that there is a common inheritance and a universal tree of life, and that natural events can be explained by discoverable natural laws. Essential to our understanding of evolution is that

- Groups of organisms are bound together by their common inheritance.
- The past has been long enough for inherited changes to accumulate, and, perhaps most essential of all, that
- Discoverable biological processes and natural relationships among organisms provide the evidence for the reality of evolution.

It took a long time to weave these threads into an evolutionary tapestry. It is hoped that this presentation of the evidence for and the fact of evolution will enable you to appreciate more clearly the richness and beauty of that evolutionary tapestry.

Recommended Reading

Appleman, P. (ed.), 2001. *A Norton Critical Edition*. Darwin: Texts, Commentary, 3rd ed. W. W. Norton, New York.

Ayala, F. J., 2007. Darwin's greatest discovery: Design without designer. *Proc. Natl. Acad. Sci. U.S.A.* **104**, 8567–8573.

Brockman, J., 2006. *Intelligent Thought: Science Versus the Intelligent Design Movement*. Vintage Books, New York.

Cracraft, J., and R. W. Bybee (eds.), 2005. *Evolutionary Science and Society: Educating a New Generation*. Biological Sciences Curriculum Study, Washington, DC.

Culotta, E., 2009. On the origin of religion. *Science*, **326**, 784–787.

Dennett, D. C., 2006. *Breaking the Spell: Religion as a Natural Phenomenon*. Allen Lane, London, England.

Reiss, J. O., 2009. *Not by Design: Retiring Darwin's Watchmaker*. University of California Press, Berkeley, CA.

Ruse, M., 2001. *The Evolution Wars: A Guide to the Debates*. Rutgers University Press, New Jersey and London.

Ruse, M., 2005. *The Evolution-Creation Struggle*. Harvard University Press, Cambridge, MA.

Scott, E. C., 2005. *Evolution vs. Creationism: An Introduction*. University of California Press, Berkeley, CA.

GLOSSARY

A

Abiotic Non-biological in origin; environments characterized by the absence of organisms.

Acheulean tools Large stone hand axes made and used by *Homo ergaster* and *Homo erectus* 1.5 Mya and by *Homo heidelbergensis* 500,000 years ago.

Adaptation The relationship between structure and/or function and an organism's environment that makes the structure or function suitable for ("adapted" to) life in that environment.

Adaptive landscape A model originally devised by Sewall Wright that describes a topography in which populations with high fitness occupy peaks and those with low fitness occupy valleys.

Adaptive peak *See* **Adaptive landscape.**

Adaptive radiation The diversification of a single species or group of related species into new ecological or geographical zones to produce a larger number of species.

Adaptive valley *See* **Adaptive landscape.**

Algae Photosynthetic multicellular organisms now divided among several super kingdoms of life.

Allele One of the alternative forms of a single gene.

Allele frequency The proportion of different alleles in a population.

Allopatric Species or populations in separate geographical locations.

Allopatric speciation Evolutionary change between populations initiated by geographical separation or because intervening geographical populations become extinct.

Alternation of generations A life cycle in plants in which a multicellular haploid stage alternates with a multicellular diploid stage.

Altruism Behavior that benefits the reproductive success of other individuals because of an actual or potential sacrifice of reproductive success by an individual (the altruist).

Amino acids Organic molecules from which peptides and proteins are made.

Amoebozoans Organisms including slime molds and many types of amoebae in the supergroup Unikonta.

Analogy The possession of a similar character by two or more species caused by factors other than common genetic ancestry.

Anatomically modern humans The name used for extant and fossil individuals of the species *Homo sapiens*.

Ancestral state The characteristic(s) of an organism that gave rise to related groups of organisms.

Ancient DNA The nucleic acid DNA extracted from dead of fossilized organisms.

Angiosperms Flowering plants with floral reproductive structures and encapsulated seeds.

Antennapedia complex A cluster of six *Antennapedia*-linked genes first discovered in fruit flies (*Drosophila* sp.) whose activation converts anterior into posterior body structures.

A-P axis The major axis of symmetry of bilateral animals where the head is anterior (A) and the tail posterior (P).

Apomorphic character A feature recently derived in evolution in contrast to an ancestral (plesiomorphic) character.

Archaebacteria Sulfur-dependent, and methane-producing organisms with cell walls and a single chromosome. Originated early in the history of life.

Archaeopteryx Perhaps the most famous fossil bird found 150 to 155 Mya.

Archaeplastida (also known as **Plantae**) A supergroup of eukaryotic organisms including red and green algae, land plants and Charophyta (the latter the ancestors of plants).

Archean Eon A Precambrian geological eon that lasted from 3.8 to 2.5 Bya.

Archetype A body plan ("*Bauplan*") characteristic of clades of organisms, usually phyla, but can characterize lower taxonomic levels.

Arms race In biology, used to describe the situation where two different organisms, for example a parasite and its host, or an animal and a bacterium, evolve to counter changes in the other organisms.

Arthropods The clade of animals (phylum) that includes insects, spiders and crustaceans with hard exoskeletons, a segmented body plan, and jointed appendages.

Artificial selection The process of selection of organisms by humans to change one or more characters.

Asexual reproduction Offspring produced by one parent in the absence of sexual fertilization or in the absence of gamete formation.

Atmosphere The gaseous envelope surrounding a celestial body such as Earth.

Autosomes Chromosomes other than sex chromosomes.

Axes of symmetry For organisms the major alignment of the body of animals or the stem of plants. Axes may be anterior-posterior, left-right, stem-root.

B

Balanced polymorphism The persistence of two or more different genetic forms through selection.

Baryonic matter Material composed largely of baryons, including all atoms and thus almost all matter we experience in everyday life.

Basalt A fine-grained igneous rock found in oceanic crust and produced in lava flows.

Base (chemistry) The nitrogen-containing component of the nucleotide unit in nucleic acid.

Base pairs *See* **Base, Base substitutions, Complementary base pairs, DNA (deoxyribonucleic acid), Homeobox, Homeotic genes, Nucleic acid, Nucleotide, RNA (ribonucleic acid)**.

Base substitutions The substitution of one base in a nucleotide, constituting perhaps the simplest type of mutation.

Batesian mimicry The similarity in appearance of a harmless species (the mimic) to a species that is harmful or distasteful to predators (the model), maintained because of selective advantage to the relatively rare mimic.

Bauplan (plural, *Baupläne*) The structural body plan that characterizes a group of organisms such as a phylum.

Behavior *See* **Altruism, Dominance (social), Homology, Imprinting, Instinct, Learning, Phenotype, Reciprocal altruism, Sociobiology**.

Big Bang theory A theory that the universe (time and space) expanded rapidly from an extremely hot, dense state about 13.7 Bya and that in the years since it has continued to cool and expand.

Bilateral symmetry The condition in which the left and right halves of an organism are equivalent so that the organism can be divided in half along a single plane.

Binomial nomenclature Part of the system of classification established by Carl Linnaeus in which each species name defines its membership in a genus and provides it with a unique species name and identity, for example, *Homo sapiens,* modern humans.

Biogenetic law The theory proposed by Ernst Haeckel that stages in the development of an individual (ontogeny) recapitulate the evolutionary history (phylogeny) of the clades to which the organism belongs.

Biogeography The study of the spatial (geographical) distributions of organisms.

Biological species concept The thesis that the primary criterion for separating one species from another is reproductive isolation.

Biosphere That part of Earth containing all living organisms.

Biota All the fauna and flora of a given region or time period.

Biotic Relating to or produced by biological organisms.

Bithorax complex A cluster of three genes first discovered in fruit flies (*Drosophila* sp.) and that regulates development of posterior body structures.

Blending inheritance The now-abandoned theory that offspring inherit a dilution, or blend of parental traits, rather than particles (genes) that determine those traits.

Body plan *See Bauplan*

Bottleneck effect A form of genetic drift that occurs when a population is reduced in size and later expands in numbers.

Box (Boxes) In molecular genetics, a DNA sequence specifying a protein that binds to DNA to act as a transcription factor activating or repressing other sequences of DNA. Families of such transcription factor genes share the same box.

Bricolage *See* **Tinkering**.

Bryophytes Mosses and liverworts.

Burgess shale fossils A 545-My-old assemblage of superbly preserved fossils located in British Columbia, Canada, and which record the Cambrian Explosion of animal body forms.

C

Cambrian explosion The appearance of fossils of many different forms of animals in the Cambrian Period, interpreted as a rapid diversification (explosion) of body plans.

Cambrian period The interval between about 545 and 495 My before the present, marking the appearance of many fossilized organisms (the Cambrian explosion). It is considered the beginning of the Phanerozoic time scale (Eon) and is the first period in the Paleozoic Era.

Canalization Restriction of phenotypic variation by intrinsic genetic or developmental mechanisms.

Canalizing selection Artificial or natural selection that results in reduced variation in expression of a phenotypic trait.

Carbonaceous Composed of or containing carbon.

Carbonaceous meteorites Extraterrestrial objects containing carbon compounds; also known as chondrites.

Carnivores Flesh eaters; organisms (almost entirely animal, but some plants) that feed on animals.

Carrying capacity (K) The number of individuals in a population that a given environment can sustain.

Catastrophism (Earth sciences) The eighteenth- and nineteenth-century theory that fossilized organisms and changes in geological strata were produced by periodic, violent and widespread catastrophic events rather than by naturally explainable events based on laws that act uniformly through time.

Cell wall The rigid or semi rigid extracellular envelope (outside the plasma membrane) that gives shape to plant, algal, fungal, and bacterial cells.

Celsius scale (°C) A scale of temperature in which the melting point of ice is taken as 0° and the boiling point of water as 100°, measured at 1 atmosphere of pressure.

Cenozoic (Caenozoic) Era The period from 65 Mya to the present, marked by reduction in dinosaur diversity and radiation of mammals. The third and most recent era of the Phanerozoic Eon, divided into two major periods, the Tertiary and Quaternary.

Central dogma (molecular biology) The tenet that the direction of flow of genetic information is from DNA → RNA → proteins.

Character A feature, trait, or property of an organism or population.

Character displacement An evolutionary response to competition in which phenotypic differences accompany resource partitioning among coexisting groups. Differences in bill sizes enabling each species of Darwin's finches to feed on differently sized seeds are a classic example.

Character state The designation of a character into (usually) binary states used to code characters in phylogenetic analysis.

Chengjiang fauna A 555-My-old assemblage of superbly preserved fossils located in Chengjiang, China, which record the Cambrian Explosion of animal body forms.

Chlorophyll A green pigment found in the chloroplasts of plants that collects light used in photosynthesis.

Chloroplast A chlorophyll-containing, membrane-bound organelle that is the site of photosynthesis in the cells of plants and some protistans.

Choanoflagellates Single-celled unikonts, the earliest forms of which may have been the ancestors of animals (or they and animals descended from a common ancestor).

Chondrites *See* **Carbonaceous meteorites.**

Chordates Members of the Phylum Chordata (which includes the vertebrates), united by the presence of a notochord and a dorsal nerve cord.

Chromalveolata A supergroup of organisms including various types of algae (kelp, dinoflagellates, diatoms), non-photosynthetic ciliates, and parasites such as *Plasmodium falciparum,* which causes malaria.

Chromosomes Elongate, threadlike structures in the nuclei of eukaryotes containing nuclear genes.

Cis-regulation The process by which an element of DNA that is adjacent to a gene interacts with the promoter of that gene to control binding sites for transcriptional activators or repressors.

Clade A natural grouping of taxa (all the descendants) derived from a single common ancestor.

Cladistics A mode of classification in which taxa are principally grouped on the basis of their shared possession of similar ("derived") characters that differ from the ancestral condition.

Cladogenesis Branching evolution involving the splitting and divergence of a lineage into two or more lineages.

Cladogram A tree diagram representing phylogenetic relationships among taxa.

Class A taxonomic rank that stands between phylum and order. A phylum may include one or more classes, and a class may include one or more orders.

Classification The grouping of organisms into a hierarchy of categories commonly ranging from species through genera, families, orders, classes, phyla and kingdom, each category reflecting one or more significant characters.

Cleavage The process of division by which a single-celled zygote becomes a multicellular animal.

Clone A group of organisms derived by asexual reproduction from a single individual; a group of cells that are identical because they arose from a single cell.

Cloning (genetics) Techniques for producing identical copies of a section of genetic material by inserting a DNA sequence into a cell, such as a bacterium, where it can be replicated.

Cnidaria Metazoans, commonly known as corals, jellyfish and hydrozoans.

Coacervate An aggregation of colloidal particles in liquid phase that persists for a period of time as suspended membranous droplets.

Codon The triplet of adjacent nucleotides in messenger RNA that codes for a specific amino acid carried by a specific transfer RNA or that codes for termination of translation (STOP codons).

Coelom (Coelome) The internal body cavity of eucoelomate animals (true coelomates) formed within mesoderm.

Coevolution Evolutionary changes in one or more genes, developmental processes, organs or species in response to changes in genes, developmental processes, organs or species.

Coextinction When two or more species, lineages or groups become extinct at the same time.

Cohort Individuals of a population all of which are of the same age.

Colinearity Applied to Hox genes to signify a correspondence between the positions of the genes on a chromosome and the antero-posterior parts of the body whose development they control.

Collision theory Proposed in 1749 by Georges Louis Leclerc, Comte de Buffon, the idea that a comet (or star) struck the sun and broke off fragments that formed the planets.

Commensalism An association between organisms of different species in which one species benefits from the relationship and the other species is not affected significantly.

Competition Relationship between organismal units (for example, individuals, groups, species) attempting

to exploit a limited common resource in which each unit inhibits, to varying degrees, the survival or reproduction of another unit by means other than predation.

Competitive exclusion The principle that two species cannot continue to coexist in the same environment (niche) if they use it in the same way.

Complementary base pairs Nucleotides on one strand of a nucleic acid that hydrogen bond with nucleotides on another strand according to the rule that pairing between purine and pyrimidine bases is restricted to certain combinations.

Complexity A state of intricate organization caused by arrangement or interaction among different component parts or processes: the greater the number of interacting parts, the greater the complexity.

Condensation theory The theory that the planets arose by the collision and accretion of planetesimals.

Consciousness The ability of an organisms to have an awareness of itself, to perceive the world around it and to respond to it.

Constraint Used in biology to describe factors that limit character variation or evolutionary direction.

Continents Seven large bodies of land (North America, South America, Asia, Europe, Africa, Antarctica, Australia) surrounded by water.

Continental drift Hypothesis proposed in 1912 by Alfred Wegener, according to which large landmasses move relative to each other across Earth's surface.

Continuous variation Character variations (such as height in humans), the distribution of which follows a series of small, non-discrete quantitative steps.

Convergence (convergent evolution) The evolution of similar characters (genes or morphologies) in genetically unrelated or distantly related species (homoplasy), mostly because they have been subjected to similar environmental selective pressures.

Core (Earth sciences) The innermost layer of Earth, composed mostly of iron with some nickel, and consisting of a solid inner core and an outer, liquid core.

Cosmic microwave background (CMB) A form of electromagnetic radiation that fills the entire universe and is a relic of the Big Bang. Tiny differences in the temperature of the CMB, which averages 2.7 K, reflect minute fluctuations in the matter of the universe and provide a snapshot of the early universe.

Cosmological constant The smooth energy density of the vacuum, introduced by Albert Einstein in his theory of general relativity to represent the inbuilt tendency of space-time to expand; a constant required to explain the universe as stationary.

Cosmological redshift Reflects the fact that light (photons) from distant galaxies is proportional to their distance from the observer and has been stretched so that the wavelength moves towards the red end of the spectrum; hence redshifted.

Cosmology Study of the structure and evolution of the universe.

Creationism The belief that each different kind of organism was individually created by a supernatural force.

Critical density (cosmology) The density of atoms — 10^{-29}g/cm^3 or six atoms of hydrogen/m^3 — that determines the geometry of the universe as flat. A higher density of atoms and the universe would collapse on itself; a lower density and the universe would continue to expand forever.

Crown Group The last common ancestor of the extant members of a clade and all its descendants.

Crust (Earth sciences) The exterior portion of the earth floating on the mantle, divided into continental (older) and oceanic (younger) crust, and composed of igneous, sedimentary, and metamorphic rocks.

Ctenophora Metazoans commonly known as comb jellies.

Culture The learned behaviors and practices common to a social group.

Cyanobacteria Photosynthetic organisms possessing chlorophyll a but not chlorophyll b. Formerly called blue-green algae, their color is caused by a bluish pigment masking the chlorophyll.

Cytology The study of cells — their structures, functions, components and life histories.

Cytoplasm All cellular material within the plasma membrane, excluding the nucleus and nuclear membrane.

Cytoplasmic inheritance *See* **Extranuclear inheritance, Inheritance, Maternal inheritance, Uniparental inheritance.**

D

Dark ages (cosmology) The time during the early life of the universe before the evolution of stars, when ordinary matter in the universe consisted of neutral hydrogen and helium, with a little lithium, and the universe was extremely dense, hot, and dark.

Dark energy A force that does not interact with light but is detectable through its gravitational effects, which accounts for 74% of the density of the universe, and which is thought to cause the acceleration of the expansion of the universe.

Dark matter That portion of the universe (22%) that neither emits nor absorbs light, which is known only from its gravitational effects on itself and on baryons, and whose composition is unknown.

Darwinism The theory, proposed by Charles Darwin, that biological evolution has led to the many different highly adapted species through natural selection act-

ing on hereditary variations in populations to give rise to descent with modification.

Degenerate (redundant) code Part of the genetic code for which there is more than one triplet codon for a particular amino acid but where a specific codon cannot code for more than one amino acid.

Deleterious allele An allele whose effect reduces the adaptive value of its carrier, either when present in homozygous condition (recessive allele) or in heterozygous condition (dominant or partially dominant allele).

Deletion A genetic change in which a section of DNA or chromosome has been lost.

Deme A local population of a species; in sexual forms, a local interbreeding group.

Descent with modification Darwin's phrase for biological evolution as change from generation to generation.

Design The concept that the complexity of organisms requires belief in a supernatural power (designer) to explain the diversity of life.

Development The progression from egg to adult in multicellular organisms.

Developmental control gene A regulatory gene high in a gene network that controls other (often many other) genes.

Dicotyledons Flowering plants (angiosperms) in which the embryo bears two seed leaves (cotyledons).

Dinosaurs Extinct terrestrial carnivorous or herbivorous reptiles that existed during the Mesozoic Era; members of the clades Saurischia and Ornithischia.

Diploblastic An animal whose embryos comprise only two major germ layers: ectoderm and endoderm (for example, Cnidaria).

Diploid An organism whose somatic cell nuclei possess two sets of chromosomes ($2n$), usually one from the male and one from the female parent, providing two different (heterozygous) or similar (homozygous) alleles for each gene.

Directional selection A form of selection resulting in change in a character or in the phenotype of a character in one direction.

Discontinuous variation Character variations that are sufficiently different from each other that they fall into non-overlapping classes.

Disruptive selection Change that favors the survival of organisms in a population that are at opposite phenotypic extremes for a particular character and eliminates individuals with intermediate values.

Division of labor *See* **Social Organization.**

DNA (deoxyribonucleic acid) A nucleic acid that serves as the genetic material of all cells and many viruses; composed of nucleotides that are usually polymerized into long double-stranded chains, each nucleotide characterized by the presence of a deoxyribose sugar.

DNA virus A small intracellular infectious agent, often parasitic, composed of DNA in a protein coat.

DNA world The hypothesis that the first chemicals of life to evolve were DNA.

Domestication The modification of plants or animals to suite human needs, often effected through artificial selection.

Dominance (social) The result of behavioral interactions between individuals in a group in which one or more individuals, sustained by aggression or other behaviors, rank higher than others in controlling the conduct of group members.

Dominant allele The allele that determines the phenotype in heterozygotes.

Dorsal The back side or upper surface of an animal; opposite of ventral. (In vertebrates, the surface defined by the location of the spinal column.)

Dosage compensation A mechanism that compensates for difference in the number of X (or Z) chromosomes between males and females so that the effects of their X-linked genes are equalized.

Double fertilization A distinctive feature of angiosperm plants in which two nuclei from a male pollen tube fertilize the female gametophyte, one producing a diploid embryo and the other producing polyploid (usually triploid) nutritional endosperm.

Downstream gene A gene(s) controlled by (downstream of) another gene, usually a regulatory gene.

Drift *See* **Genetic drift.**

Duplication Instances in which a particular section of DNA, chromosome segment, chromosome or entire genome occurs has been doubled.

Dwarf planets A class of planets in our solar system based on size, established in 2006 to accommodate Pluto, Ceres and UB313. Unlike planets, dwarf planets do not dominate their neighborhoods.

E

Ecdysoza One of two groups (the other being Lophotrochozoa) of protostomes, Ecdysoza being animals such as arthropods and nematodes with a molt in the life cycle.

Ecological footprint The amount of land and water required to produce the resources consumed and to absorb the waste of an individual, population, or lifestyle.

Ecology The study of the relations between organisms and their environment, in terms of their numbers, distributions and life cycles.

Ectoderm The outermost layer of cells of early animal embryos, from which nerve tissues and epidermal tissues are derived.

Ediacaran fauna/biota Soft-bodied fossils found in South Australia and other places, dating to a Precambrian

period lasting 60 or more My and with unknown relationships to other organisms.

Embryology The study of the development of organisms from their inception to birth/hatching and often into later life history stages such as larvae.

Enation hypothesis The earliest leaves to evolve; thin microphylls, evolved from extensions of tissues along the stem and not from small branches.

Endemic A species or population that is specific (indigenous) to a particular geographic region.

Endocytosis Cellular engulfment of outside material, followed by its transfer into the cellular interior encapsulated in a membrane.

Endoderm The layer of cells that lines the embryonic gut (archenteron) during the early stages of development in animals, and which later forms the epithelial lining of the intestinal tract and internal organs such as the liver, lung and urinary bladder.

Endosymbiosis A relationship between two different organisms, in which one (the endosymbiont) lives within the tissues or cell of the other, benefiting either or both. Eukaryotic organelles, such as mitochondria and chloroplasts, had an endosymbiotic prokaryotic origin.

Endothermic A body temperature maintained by internal physiological mechanisms at a level independent of the ambient (environmental) temperature.

Enhancer A region of the DNA (nucleotide sequence) of a gene that binds to another nucleotide sequence within the gene so that the promoter sequence can be transcribed. Individual genes may contain more than one enhancer.

Environment The complex of external conditions, abiotic and biotic, that affects and interacts with organisms or populations to provide the facilities and resources that enable hereditary data (genotypes) to produce organismal characters (phenotypes).

Enzyme A protein that catalyzes chemical reactions.

Eon A major division of the geological time scale, often divided into two eons beginning from the origin of Earth 4.5 Bya: the Precambrian or Cryptozoic (rarity of life forms) and the Phanerozoic (abundance of life forms).

Epigenesis The theory that tissues and organs are formed by interaction between cells and substances that appear during development, rather than being preformed in the zygote.

Epigenetic The sum of the genetic and non-genetic factors that influence gene action.

Epistasis Interaction between non-allelic genes.

Epithelium (plural, **epithelia**) One of the two fundamental types of cellular organization in animals, consisting of a sheet of laterally connected and interacting cells resting on an extracellular matrix — the basement membrane — that is produced by the epithelium.

Epoch One of the categories into which geological time is divided. Periods are often divided into three epochs: Early, Middle, and Late; for example, Early Cambrian.

Equilibrium (genetics) The persistence of the same allelic frequencies over a series of generations.

Era A division of geological time that stands between the Eon and the Period. The Phanerozoic Eon is divided into Paleozoic, Mesozoic and Cenozoic Eras; and each era is divided into two or more periods.

Eubacteria Prokaryotes, other than archaebacteria.

Eugenics The belief that humanity can be improved by selective breeding.

Eukaryotic cells Cells that contain nuclear membranes, mitochondrial organelles and other characteristics that distinguish them from prokaryotic cells.

Evo-devo *See* **Evolutionary developmental biology.**

Evolution Changes in organisms through time that lead to differences among them.

Evolutionary developmental biology The field in biology in which developmental processes are studied for their insights into evolutionary processes.

Evolutionarily stable strategies The result of a balance between cooperation and individual action among interacting individuals in a social group.

Evolutionary species A definition of a species based on isolation from other species, often identified from ancestor–descendant populations.

Evolutionary trees Arrangements of the evolutionary history of one or more groups of organisms presented in a tree-like branching pattern.

Evolutionary universe (cosmology) A more formal title for the Big Bang hypothesis to describe the origin and evolution of the universe.

Evolvability The concept that the genotypes of some organisms evolve more rapidly/readily than do the genotypes of other organisms.

Exaptation A character that was adaptive under a prior set of conditions and later provides the initial stage (is "co-opted") for the evolution of a new adaptation under a different set of conditions.

Excavata A supergroup of organisms, previously members of the Protista, and including *Giardia,* which causes the intestinal illness giardiasis. *Trypanosoma brucei,* which causes sleeping sickness, and *Trichomonas,* which causes trichomoniasis.

Exon An expressed nucleotide sequence in a gene that is transcribed into messenger RNA and spliced together with the transcribed sequences of other exons from the same gene. Exons are separated from one another by intervening non-translated sequences.

Exon shuffling The recombination or exclusion of exons such that they remain active as sources of genetic information.

Expanding universe The theory that the space-time between galaxies is expanding so that the universe is continuing to increase.

Extinction The disappearance of a species or higher taxon.

Extranuclear inheritance Patterns of heredity in which the transmission is not via nuclear genes, for example transmission by mitochondrial genes.

F

F (inbreeding coefficient) *See* **Inbreeding coefficient.**

Family A taxonomic category that stands between order and genus; an order may comprise a number of families, each of which contains a number of genera.

Fauna All animals of a particular region or time period.

Fecundity A measure of the potential production of offspring by an individual.

Ferns (Pterophyta) Spore-bearing plants that carry their sporangia on the fronds.

Fertility A trait measured by the number of viable offspring produced.

Fertilization In multicellular organisms, the fusion of a sperm with an egg that allows sperm and egg nucleus to fuse as a zygote nucleus and development to be initiated.

Fitness Central to evolutionary theory evaluating genotypes and populations, fitness has had many definitions, ranging from comparing growth rates to comparing long-term survival rates. The basic fitness concept that population geneticists commonly use is relative reproductive success, as governed by selection in a particular environment.

Fixity of species A theory held by Linnaeus and others that members of a species could only produce progeny like themselves, and therefore each species was fixed in its particular form(s) at the time of its creation.

Flora All plants of a particular region or time period.

Flower (botany) The reproductive structure of many seed-producing plants; often brightly colored and elaborate in form.

Fossils The geological remains, impressions, or traces of organisms that existed in the past.

Founder effect The effect caused when a few individuals ("founders") derived from a large population begin a new colony. Since these founders carry only a small fraction of the parental population's genetic variability, different gene frequencies can become established in the new colony.

Frequency-dependent selection Instances where the effect of selection on a phenotype or genotype depends on its frequency (for example, a genotype that is rare may have a higher adaptive value than when it is common).

Frozen accident The hypothesis that an accidental event in the distant past was responsible for the presence of a defining feature in living organisms. Such events may include an accident in which the present genetic code was used by a group of early organisms that managed to survive some population bottleneck, thereby conferring this particular code on later organisms.

Fungi More closely related to animals than to plants, fungi are a lineage of multicellular organisms that lack chlorophyll, obtaining nutrition from parasitizing live hosts or digesting dead organic matter.

G

Galaxy A system of numerous stars, held together by mutual gravitational effects, and often spiral or elliptical in shape.

Gamete A germ cell (eggs in females, sperm in males) that is usually haploid and that fuses with a germ cell of the opposite sex to form a zygote (usually diploid) at fertilization.

Gametophyte The haploid life cycle phase in plants.

Gastrula A (typically) cuplike embryonic stage in animal development that follows the blastula stage. Its hollow cavity (archenteron) is lined with endoderm and opens to the outside through a blastopore.

Gene A unit of hereditary genetic material composed of a segment of DNA (sequence of nucleotides), usually with a specific function — coding for a protein or sometimes coding for RNA.

Genealogy A record of familial ties and ancestral connections among members of a group.

Gene duplication *See* **Duplication.**

Gene family Two or more gene loci in an organism whose similarities in nucleotide sequences indicate they have been derived by duplication from a common ancestral gene.

Gene flow The migration of genes into a population from other populations by interbreeding.

Gene frequency The proportion of a particular allele among all alleles at a gene locus (also called allele or allelic frequency).

Gene locus The chromosomal position (nucleotide sequence) occupied by a particular gene.

Gene pool All the genes present in the gametes of individuals in a sexually reproducing population.

Gene regulation The processes by which a gene is turned on or off or by which the level of activity of a gene is controlled.

Gene therapy Human-directed repair or replacement of genes that cause inherited diseases. When confined to somatic (body) cells rather than to sex cells (sperm or eggs), such gene repairs are not passed on to future generations.

Genetic assimilation The processes by which a trait elicited by an environmental stimulus can be genetically

incorporated and appear developmentally in later generations in the absence of the environmental stimulus.

Genetic code The sequences of nucleotide triplets (codons) on messenger RNA that specify each of the different kinds of amino acids positioned on polypeptides during the translation process.

Genetic distance A measure of the divergence among populations based on their differences in frequencies of given alleles.

Genetic drift Random change in the frequency of alleles in a population.

Genetic engineering Manipulation of genetic material from different sources to produce new combinations that are then introduced into organisms in which such genetic material does not normally occur.

Genetic load The loss in average fitness of individuals in a population because the population carries deleterious alleles or genotypes.

Genetic polymorphism The presence of two or more alleles at a gene locus over a succession of generations. (Called balanced polymorphism when the persistence of the different alleles cannot be accounted for by mutation alone.)

Genome The complete genetic constitution of a cell or individual.

Genotype The genetic constitution of an individual.

Genus (plural, **genera**) A taxonomic category that stands between family and species. In taxonomic binomial nomenclature, the genus is used as the first of two words in naming a species; for example, *Homo* (genus) *sapiens* (species).

Geographic isolation The separation between populations caused by geographic distance or geographic barriers.

Geological dating The determination of the ages of rocks and strata using a variety of methods.

Geological strata A series of discrete layers laid down on top of each other as rock by natural forces such as lava flows, siltation, infilling of a marsh, and so on.

Geological time scale The correlation between rocks (or the fossils contained in them) and time periods of the past.

Germ layers The primary layers from which multicellular embryos develop.

Germ line *See* **Germ Plasm.**

Germ plasm Cells in animals that are exclusively devoted to transmitting hereditary information to offspring, in contrast to somatic cells, which comprise all other tissues of the body.

Germ plasm theory The theory that, in some animals, gametes form from a special part of the egg (the germ plasm) and so cannot be influenced by environmental influences acting on the somatic tissues.

Gondwana The supercontinent in the Southern Hemisphere formed from the breakup of the larger Pangaea landmass about 180 Mya. Gondwana was composed of what are now South America, Africa, Antarctica, Australia and India.

Granite A coarse-grained igneous rock commonly intruded into the continental crust.

Gravitation The universal law discovered by Isaac Newton explaining how planets remain in their orbits because of a force (gravity).

Great Chain of Being The eighteenth-century hypothesis that instead of a static universe, there is a continuous progression of stages leading to a superior supernatural being; the transformation of the "Ladder of Nature" into a succession of moving platforms.

Greenhouse gas A gas such as carbon dioxide (CO_2) and methane (CH_4) that absorbs infrared radiation, reduces the loss of heat from Earth's surface and so raises the global temperature.

Group A population(s) in a species that shares a geographically and/or ecologically identifiable origin and has gene frequencies and phenotypic characters that distinguishes it from other groups.

Group selection A form of selection acting on the attributes of a group of related individuals in competition with other groups rather than only on the attributes of an individual in competition with other individuals.

Gymnosperms A group of vascular plants with seeds unenclosed in an ovary (naked); mainly cone-bearing trees.

H

Habitat The place and conditions in which an organism normally lives.

Haeckel's Gastraea theory The theory that animals developed from swimming hollow-balled colonies of flagellated protozoans.

Haldane's rule If one sex is absent, rare or sterile in a cross between two species with sex-determining chromosomes, that sex is the heterogametic one.

Half-life (radioactivity) The time required for the decay of one-half the original amount of a radioactive isotope. Each radioactive isotope has a distinctive and constant half-life.

Haplodiploidy A reproductive system found in some animals, such as bees and wasps, in which males develop from unfertilized eggs and are therefore haploid, while females develop from fertilized eggs and are therefore diploid.

Haploid Cells or organisms that have only one set ($1n$) of chromosomes.

Haplorhines The "dry-nosed primates" comprising New World monkeys, Old World monkeys and apes and human.

Hardy–Weinberg equilibrium (principle) The conservation of gene (allelic) and genotype frequencies in large populations under conditions of random mating and in the absence of evolutionary forces, such as selection, migration, and genetic drift, which act to change gene frequencies.

Heat shock proteins Molecular chaperones that help other proteins maintain their 3D conformation, prevent them from degrading, and mask changes mutations would otherwise have on the phenotype. An environmental shock can unmask these gene products creating new patterns and combinations of expressed proteins.

Hemoglobin A protein with as basic unit of an iron-containing porphyrin (heme) that reversibly binds oxygen attached to a globin polypeptide chain.

Herbivores Animals that feed mainly on plants.

Heritability The degree to which variations in the phenotype of a character are caused by genetic differences. Traits with high heritabilities can be more easily modified by selection than traits with low heritabilities. A measure of an organism's potential to respond to selection.

Hermaphrodite An individual possessing both male and female sexual reproductive systems.

Heterochrony A term Haeckel proposed to describe changes in timing of an organ's development during evolution. Its present usage varies but still hinges on a phylogenetic change in developmental timing, whether of one organ relative to other organs, or of one organ relative to the same ancestral organ.

Heterosis (Hybrid vigor) The increase in vigor and performance that can result when two different, often inbred strains are crossed.

Heterotopy A term Haeckel proposed to describe changes in the position of an organ during evolution.

Heterotroph An organism that cannot use inorganic materials to synthesize the organic compounds needed for growth but obtains them by feeding on other organisms or their products, such as a carnivore, herbivore, parasite, scavenger or saprophyte.

Heterozygote The situation in a genotype or an individual in which the two copies of a gene are different; having different alleles at a particular gene locus on homologous chromosomes (for example, *Aa* in a diploid).

Heterozygote advantage (superiority) The superior fitness of some heterozygotes relative to homozygotes.

Hierarchy A term used to designate an ordered grouping of the items within a system, often associated with increasing levels of complexity or organization.

Histones A family of small acid-soluble (basic) proteins that are tightly bound to eukaryotic nuclear DNA molecules and help fold DNA into thick chromosome filaments.

Hitchhiking When a gene persists in a population, not because of selection, but because of close linkage to one or more selected genes.

Homeobox A transcriptional sequence of 180 base pairs that unites genes known as homeotic, homeobox or Hox genes. The overall conservation of homeobox sequences, of the genes containing them, and of their linkage orders indicates common developmental functions in different phyla preserved for many hundreds of millions of years, extending back to Precambrian times.

Homeodomain A specific protein domain that is shared among a family of transcription factors (homeotic genes) and that is transcribed from a conserved DNA sequence known as the homeobox.

Homeosis *See* **Homeotic mutations**.

Homeostasis The tendency of a physiological system to react to an external disturbance so that the system is not displaced from normal values.

Homeotic genes Genes that share the homeobox of 180 base pairs.

Homeotic mutations (homeosis) Homeosis was defined by William Bateson as "something [that] has been changed into the likeness of something else." In modern genetic usage, homeotic mutations cause the development of tissue in an inappropriate position; for example, *bithorax* mutations in *Drosophila* produce an extra set of wings.

Hominid A member of the family Hominidae, which includes humans, whose earliest fossils can now be dated to about 4 Mya (genus *Australopithecus*). Previously used exclusively for humans and the fossil species most closely related to humans.

Hominin All taxa on the human lineage after separation from the common ancestor with chimpanzee; members of the tribe *hominini*. In recent publications hominin has the same meaning as hominid in older literature.

Hominoids A group (superfamily Hominoidea) that includes hominids (Hominidae), gibbons (Hylobatidae) and apes (Pongidae).

Homogametic sex The sex that produces only one kind of gamete for sex determination in offspring, thus causing sex differences among offspring to depend on the kind of gamete contributed by the heterogametic sex. The homogametic sex is the female in mammals and the male in birds.

Homologous *See* **Homology**.

Homologous chromosomes Chromosomes that pair during meiosis, each pair usually possessing a similar sequence of genes.

Homology The similarity of characters (genes, structures, behaviors) in different species or groups because of their descent from a common ancestor.

Homoplasy Character similarity that arises independently in different groups through parallelism or convergence.

Homozygote The situation in a genotype or an individual in which the two copies of a gene are the same; having the same alleles at a particular gene locus on homologous chromosomes (for example, *AA*).

Horizontal (Lateral) gene transfer Transmission of genes from one organism to another without reproduction.

Host-plant specificity A term for the species-specific interactions between a pollinator (usually an insect) and a species of plant.

Hox gene A homeobox gene in vertebrates, taking the gene name from the first two letters of the orthologous gene in *Drosophila* and adding an x.

Hunter-gatherer societies A form of human subsistence based on hunting animals for meat as well as foraging for other foods such as plants, insects and scavenged meat.

Hybrid breakdown (inviability, sterility) Hybrids that suffer from loss of fitness and reproductive failure.

Hybridization (genetics) Reproduction between individuals from two species to form a fertile offspring (the hybrid).

Hybridization (molecular biology) The formation of double-stranded molecules of DNA or RNA from a complementary single strand.

Hybrids (genetics) Offspring of a cross between genetically different parents or groups.

Hybrid vigor *See* **Heterosis**.

Hydrothermal vents Openings in the ocean floor from which water flows at temperatures up to 350°C and in and around which thermophilic organisms live.

I

Idealism The philosophy that all objects, including the universe, have no independent existence apart from the minds of those perceiving them; perceiving objects as ideal forms.

Igneous rock A rock such as basalt (fine-grained) and granite (coarse-grained), formed by the cooling of molten material from Earth's interior.

Imaginal disks Clusters of cells set aside in the larvae of insects such as *Drosophila* from which the structures of the adult fly are formed.

Imprinting (behavior) The learning process by which newborn/hatched organisms associate with an object (usually a parent or an adult of the same species, or an environment) and orient their behavior toward that object. Fish imprinting on a home river, newly hatched chicks imprinting on a parent are two well-studied examples.

Imprinting (genetics) The silencing of genes in germ cells by methylation of DNA. As a state, imprinting is usually sex-specific and is removed after development of a new individual begins from the imprinted gamete(s).

Inbreeding Mating between genetically related individuals, often resulting in increased homozygosity in their offspring.

Inbreeding depression Decrease in the average value of a character or in growth, vigor, fertility and survival, as a result of inbreeding.

Inclusive fitness The fitness of an allele or genotype measured not only by its effect on an individual but also by its effect on related individuals that also possess it (kin selection).

Independent assortment A basic principle of Mendelian genetics — that a gamete will contain a random assortment of alleles from different chromosomes because chromosome pairs orient randomly toward opposite poles during meiosis.

Individual A single entity such as a person or plant that has a life with a beginning and an end.

Industrial melanism The effect of soot and other dark-colored pollution in industrial areas in increasing the frequency of darkly pigmented (melanic) forms perhaps because of selection by predators against nonpigmented or lightly pigmented forms.

Inheritance The transmission of information from generation to generation.

Inheritance of acquired characters The theory used by Lamarck to explain evolutionary adaptations — that phenotypic characters acquired by interaction with the environment during the lifetime of an individual are transmitted to its offspring.

Innate Behavior *See* **Instinct**.

Instinct An inherited (innate), relatively inflexible behavior pattern that is often activated by one or several environmental factors (releasers).

Insular dwarfism The process of the reduction of body size in organisms isolated on islands.

Intrinsic rate of natural increase (r) The potential rate at which a population can increase in an environment free of limiting factors.

Intron A nucleotide sequence (region of DNA) within a gene that is transcribed to produce mRNA but the mRNA is not translated into protein.

Inversion An aberration in which a section of DNA or chromosome has been inverted 180 degrees, so that the sequence of nucleotides or genes within the inversion is now reversed with respect to its original order in the DNA or chromosome.

Isolating mechanisms Biological mechanisms that act as barriers to gene exchange or interbreeding between populations.

Isotope One of several forms of an element, with a distinctive mass based on the number of neutrons in the atomic nucleus. Radioactive isotopes decay at a rate that is constant for each isotope and release ionizing radiation as they decay.

J

"Junk" DNA *See* **Selfish DNA.**

K

Kelvin scale (K) A temperature scale in which absolute zero (the point at which molecules oscillate at their lowest possible frequency, $-273°C$) is designated as 0K, and the boiling point of water as 373K.

Kingdom The highest inclusive category of taxonomic classification. Each kingdom includes phyla or subkingdoms.

Kin selection Processes that influence the survival and reproductive success of genetically related individuals (kin). This contrasts with selection confined solely to an individual and its own offspring.

K-T extinction The disappearance of many species, including all non-avian dinosaurs and pterosaurs at a time that corresponds to the Cretaceous (*Kreidezeit* in German)–Tertiary (K-T) boundary.

L

Ladder of Nature A concept based on Aristotle's view (the Scale of Nature) that nature can be represented as a succession of stages or ranks that leads from inanimate matter through plants, lower animals, higher animals and finally to the level of humans.

Lamarckian inheritance The concept that the phenotype of an organism is itself hereditary: that characters acquired or lost during life experience, as well as characters that organisms acquire in order to meet environmental needs, can be transmitted to offspring.

Language A structured system of communication among individuals using vocal, visual or tactile signs to describe thoughts, feelings, concepts and observations.

Larva (plural, **larvae**) A sexually immature stage in various animal groups, often with a form and diet distinct from those of the adult.

Lateral transmission *See* **Horizontal transmission.**

Laurasia The supercontinent in the Northern Hemisphere (comprising what is now North America, Greenland, Europe and parts of Asia) formed from the breakup of Pangaea about 180 Mya.

Law of superposition For any given series of sedimentary rocks, the oldest layers (strata) lie at the bottom and the youngest layers at the top.

Learning Acquisition of a behavior through experience.

Life The capability of performing various organismal functions such as metabolism, growth and reproduction of genetic material.

Life cycle The series of stages that takes place between the formation of zygotes in one generation of a species and the formation of zygotes in the next generation.

Life history All the stages of an individual life from its beginning to death.

Life-history trait Characters associated with the survival and reproduction of an individual such as number of offspring produced, age at reproduction, growth rate.

Light-year The distance traveled by light, moving at 186,000 miles a second, in a solar year; approximately 6×10^{12} miles or 9.5×10^{12} kilometers.

Lineage An evolutionary sequence, arranged in linear order from an ancestral (stem) group or species to a descendant (crown) group or species (or *vice versa*).

Linkage (genetics) The occurrence of two or more gene loci on the same chromosome.

Linkage equilibrium The attainment of genotypic frequencies in a population indicating that recombination between two or more gene loci has reached the point at which their alleles are now found in random genotypic combinations.

Linkage map The linear sequence of known genes on a chromosome obtained from recombination data.

Lithosphere (Earth sciences) The term for the crust plus the uppermost portion of the mantle of Earth.

Living *See* **Life.**

Living fossil An existing species whose similarity to a fossil taxon indicates that very few morphological changes have occurred over a long period of geological time.

Locus (plural, **loci**) The site (nucleotide sequence) on a chromosome occupied by a specific gene. Some researchers use it more broadly as a synonym of gene.

Logistic growth curve Population growth that follows a sigmoid (S-shaped) curve in which numbers increase slowly at first, then rapidly, and finally level off as the population reaches its maximum size or carrying capacity for a particular environment.

Longevity The average life span of individuals in a population.

Lophotrochozoa One of two groups (the other being Ecdysoza) of protostomes, Lophotrochozoa being animals such as mollusks, annelids, brachiopods with tentacles.

M

Macroevolution The pattern of evolution at and above the level of the species. Fossils are the chief evidence for macroevolution.

Macromutation The origin of a new species or higher taxonomic category by a single large mutation rather than by selection acting on many mutations.

Mammals Homeothermic, vertebrates that suckle their offspring with milk produced in the mammary glands, have hair, three middle ear ossicles and a neocortex region in the brain.

Mammary glands One or more pairs of ventrally placed glands used by mammalian females for nursing offspring.

Mantle (Earth sciences) A layer of partly plastic or ductile rock 2,900 km thick, that has experienced repeated melting and crystallization, and which constituting four-fifths of Earth's volume.

Marsupials Mammals of the infraclass Metatheria possessing, among other characters, a reproductive process in which tiny live young at an early stage of development are born and then nursed in a female pouch.

Maternal inheritance Transmission of heredity information from the female parent by deposition of gene products into the egg cytoplasm.

Meiosis The two eukaryotic cell (maturation) divisions that produce haploid gametes (animals) or spores (plants) from a diploid cell. One is a reduction division that ensures that each gamete or spore contains one representative of each pair of homologous chromosomes in the parental cell.

Mendelian population A group of interbreeding, diploid individuals that exchange genes through sexually reproduction.

Mendel's laws *See* **Independent assortment, Segregation.**

Mesenchyme One of the two fundamental cell types in animals, consisting of loosely connected cells in an extracellular matrix that may be "solid" as in bone or fluid as in blood.

Mesoderm The embryonic tissue layer between ectoderm and endoderm in triploblastic animals that gives rise to muscle tissue, kidneys, blood, internal cavity linings and so on.

Mesozoic Era The middle era of the Phanerozoic Eon, covering an approximately 220-million-year interval between the Paleozoic (ending about 248 Mya) and the Cenozoic (beginning about 65 Mya). It is marked by the origin of mammals in the earliest period of the era (Triassic), the dominance of dinosaurs throughout the last two periods of the era (Jurassic and Cretaceous) and the origin of angiosperms.

Messenger RNA (mRNA) An RNA molecule produced by transcription from a DNA template, bearing a sequence of triplet codons used to specify the sequence of amino acids in a polypeptide.

Metabolic pathway A sequence of enzyme-catalyzed reactions that convert a precursor substance to one or more end products.

Metabolism A network of enzyme-catalyzed reactions used by living organisms to maintain themselves.

Metamorphic rock Rock that has been subjected to high but non-melting temperatures and pressures, causing chemical and physical changes.

Metamorphosis (zoology) The transition from one form into another during the life cycle, for example, a larva into an adult.

Metaphyta The Kingdom containing the green plants; also known as Embryophyta, Plantae.

Metatheria *See* **Marsupials.**

Metazoa Multicellular animals.

Methanogens Single-celled, methane-generating organisms levels that only survive in the absence of oxygen and that are known to have existed at least 3.8 Bya.

Microevolution Evolutionary changes within populations of a species.

Micro RNA (miRNA) Short (18–25 nucleotide) sequences of non-coding, single-stranded RNA that regulate the translation of proteins in plants and animals by binding to matching target mRNAs leading to destruction of the mRNA.

Microsatellites Tandem repeats of short di-, tri-, and tetranucleotide sequences. Such loci are abundant and mutate at a relatively high rate.

Microspheres Microscopic membrane-bound spheres formed when proteinoids are boiled in water and allowed to cool. Some cell-like properties, such as osmosis, growth in size and selective absorption of chemicals, have been ascribed to them.

Migration (ecology) Movement of a population to a different geographical area or its periodic passage from one region to another.

Migration (genetics) The transfer of genes from one population into another by interbreeding (gene flow).

Milky Way (cosmology) A large spiral galaxy containing more than 200 billion stars and their planets, including our own solar system, and surrounded by more than a dozen much smaller galaxies.

Mimicry Resemblance of individuals in one species (mimics) to individuals in another (models) as a result of selection.

Missing link A specimen or species of an extinct organism thought to be intermediate between two clades and so to provide a link between them.

Mitochondria Organelles in eukaryotic cells that use an oxygen-requiring electron transport system to transfer chemical energy. Mitochondria have their own genetic material (circular mtDNA without histones) and

generate some mitochondrial proteins by using their own protein-synthesizing apparatus.

Mitochondrial DNA (mtDNA) The DNA found within the intracellular organelles known as mitochondria.

Mitochondrial Eve The name given to the common mitochondrial DNA human ancestor estimated from mtDNA sequences to have existed 140,000 years ago.

Mitosis The mode of eukaryotic cell division that produces two daughter cells possessing the same chromosome complement as the parent cell.

Modern synthesis (evolutionary theory) The union of Darwin's theory of evolution by natural selection with population genetics.

Modularity The concept that units of life, such as gene networks, aggregations of cells and organ primordia, develop and evolve as units (modules) that interact with other modules.

Molecular clock The rate at which nucleotides are substituted over evolutionary time.

Monocotyledons Flowering plants (angiosperms) in which the embryo bears one seed leaf (cotyledon).

Monophyletic A taxonomic group united by having arisen from a single ancestral lineage.

Monotremes Egg-laying mammals, presently restricted to Australasia; echidnas (*Tachyglossus, Zaglossus*) and the platypus (*Ornithorhynchus*).

Morphogenesis Development of the form (morphology) of an organism or part of an organism.

Morphological species (morphospecies) Assemblages of individuals with shared structural characters that allow them to be separated from other assemblages.

Morphology The study (science) of the anatomical form and structure of organisms.

Morphospace A three-dimensional representation of the morphological characters of an organisms or groups of organisms used to show how much of the possible range of morphologies is expressed.

Mosaic evolution The independent evolution of different parts of an organism.

Mousterian tools Stone tools, including spear points and scrapers, made and used by *Homo neanderthalensis* and *Homo sapiens*.

Mullerian mimicry Sharing of a common warning coloration or pattern among a number of species that are all dangerous or toxic to predators; resemblances maintained because of common selective advantage.

Muller's ratchet The generalization that because of sampling errors, populations more easily lose that class of individuals bearing the fewest harmful mutations, so that classes with increasing numbers of such mutations tend to increase with time.

Multicellularity The condition of having many cells as opposed to unicellularity.

Multigene family *See* **Gene family**.

Mutagenesis Production of mutations by chemical treatment or radiation.

Mutation A change in gene structure and often function; change in the nucleotide sequence of genetic material whether by substitution, duplication, insertion, deletion or inversion.

Mutational load That portion of the genetic load caused by production of deleterious genes through recurrent mutation.

Mutualism A relationship among different species in which the participants benefit.

N

Natural scientist (natural historian) The term used before the twentieth century when referring to individuals we would now refer to as biologists, botanists, zoologists, physicists and so forth.

Natural selection Differential reproduction or survival of replicating organisms caused by agencies other than humans. Since such differential selective effects are widely prevalent, and often act on hereditary (genetic) variations, natural selection is a major cause for a change in the gene frequencies of a population that leads to a new distinctive genetic constitution.

Nebular hypothesis Originally proposed by Emanuel Kant and Pierre-Simon Laplace in the eighteenth century to explain the origin of the universe; a cloud of gas and dust (a nebula) collapses as gravitational force overcome the pressure of the gas.

Negative eugenics Proposals to eliminate deleterious genes from the human gene pool by identifying their carriers and restraining or discouraging their reproduction.

Neo-Darwinism The theory of evolution as a change in the frequencies of genes introduced by mutation, with natural selection as the most important, although not the only, cause for such changes.

Neoteny The retention of juvenile morphological traits in the sexually mature adult.

Neural crest A fourth germ layer found in vertebrates and from which arise skeletal tissues, cells that form the dentine of teeth, pigment cells, peripheral nerves and ganglia, some hormone-synthesizing cells, valves and septa of the heart and other cell types.

Neutral mutation A mutation that does not affect the fitness of an organism in a particular environment.

Neutral theory of molecular evolution The theory that most mutations that contribute to genetic variability (genetic polymorphism on the molecular level) consist of alleles that are neutral in respect to the fitness of the organism and that their frequencies can be explained in terms of mutation rate and genetic drift.

Neutron stars (pulsars) The collapsed core of an exploding star or supernova Type II.

Niche The environmental habitat of a population or species, including the resources it uses and its interactions with other organisms.

Normalizing selection *See* **Stabilizing selection.**

Notochord The dorsal axial supporting rod found in all chordates.

Nucleic acid An organic acid polymer, such as DNA or RNA, composed of a sequence of nucleotides.

Nucleotide A molecular unit consisting of a purine or pyrimidine base, a ribose (RNA) or deoxyribose (DNA) sugar, and one or more phosphate groups.

Nucleus A membrane-enclosed eukaryotic organelle that contains all the histone-bound DNA in the cell (that is, practically all the genetic material).

Numerical taxonomy A statistical method for classifying organisms by comparing them on the basis of measurable phenotypic characters and giving each character equal weight. The degree of overall similarity between individuals or groups is then calculated, and a decision is made as to their classification.

O

Oldowan tools Stone tools made with a single blow of one stone by another, made and used by *Homo habilis* between 2.5 and 1 Mya.

Ontogeny The development of an individual from zygote to maturity.

Operon A cluster of coordinately regulated structural genes.

Opisthokonts A subgroup of unikonts containing some parasitic protists, choanoflagellates, fungi and animals.

Order A taxonomic category between class and family. A class may contain a number of orders, each of which contains a number of families.

Organelles Functional intracellular membrane enclosed bodies such as nuclei, mitochondria and chloroplasts in eukaryotic cells.

Organic Carbon-containing compounds. Also refers to features or products characteristic of organisms or life.

Organism A living entity.

Origination In evolution, the first appearance or origin of a character, feature or organism.

Orthogenesis The concept that evolution proceeds in a particular direction because of internal or vitalistic causes.

Orthologous genes Gene loci in different species that are sufficiently similar in their nucleotide sequences (or amino acid sequences of their protein products) to suggest they originated from a common ancestral gene.

Outgroup A taxon that diverged from an ingroup before the members of the ingroup diverged from one another.

Ozone (O_3) A molecule consisting of three atoms of oxygen and that forms the ozone layer of the upper atmosphere.

P

Paedomorphosis The incorporation of adult sexual characters into immature developmental stages.

Paleobiology The study of the biology of extinct organisms and their ecosystems.

Paleomagnetism The magnetic fields of ferrous (iron-containing) materials in ancient rocks.

Paleontology The study of extinct fossil organisms or traces of organisms.

Paleozoic (Palaeozoic) Era The first era of the Phanerozoic Eon, extending from 545 to about 248 Mya.

Pangaea A very large supercontinent formed about 250 Mya comprising most or all of the present continental landmasses.

Pangenesis The theory of heredity, held by Darwin and others, that small, particulate "gemmules," or "pangenes" are produced by each of the various tissues of an organism and sent to the gonads where they are incorporated into gametes.

Parallel evolution The evolution of similar characters in related lineages that do not share a recent common ancestor.

Paralogous genes Two or more different gene loci in the same organism that are sufficiently similar in their nucleotide sequences (or in the amino acid sequences of their protein products) to indicate they originated from one or more duplications of a common ancestral gene.

Parapatric Geographically adjacent, non-overlapping species or populations that at the zone of contact do not interbreed.

Parapatric speciation A population at the periphery of a species adapts to different environments but remains contiguous with its parent so that gene flow is possible between them.

Paraphyletic A taxonomic grouping which includes some descendants of a single common ancestor, but not all.

Parasitism An association between species in which individuals of one (the parasite) obtain their nutrients by living on or in the tissues of the other species (the host), often with harmful effects to the host.

Parental investment Parental provision of resources to offspring with the effect of increasing the offspring's reproductive success.

Parsimony method Choice of a phylogenetic tree that minimizes the number of evolutionary changes necessary to explain species divergence.

Parthenogenesis Development of an egg without fertilization.

P-D axis A major axis of symmetry of bilateral animals where distal (D) is furthest from the midline and proximal (P) is closest to the midline.

Peptide (polypeptide) An organic molecule composed of a sequence of amino acids covalently linked by peptide bonds (a bond formed between the amino group of one amino acid and the carboxyl group of another through the elimination of a water molecule).

Period (geological) A major subdivision of an era of geological time distinguished by a particular system of rocks and associated fossils.

Peripatric speciation When a population divides and becomes genetically isolated because of the budding off of a small completely isolated founder colony from a larger population.

Phagocytic Cellular engulfment of external material.

Phanerozoic eon A major division of the geological time scale marked by the relatively abundant appearance of fossilized skeletons of multicellular organisms, dating from about 545 Mya to the present.

Pharyngeal jaws An additional set of jaws that develop from posterior pharyngeal arches in some teleost fish.

Phenetics *See* **Cladistics.**

Phenotype The characters that constitute the structural, functional and behavioral; properties of an organism.

Phenotypic plasticity Variation in the phenotype expressed in response to environmental changes and indicative of underlying genotypic plasticity.

Phloem The conducting tissue of green land plants that takes nutrient to all parts of the plant.

Photosynthesis The synthesis of organic compounds from carbon dioxide and water through a process that begins with the capture of light energy by chlorophyll.

Phyletic evolution Evolutionary changes within a single non-branching lineage. Although new species are produced by this lineage over time (chronospecies), there is no increase in the number of species existing at any one time.

Phylogenetic branching *See* **Cladogenesis, Phylogenetic evolution.**

Phylogenetic evolution (also called branching evolution) Evolutionary changes producing two or more lineages that diverged from a single ancestral lineage.

Phylogenetic systematics *See* **Cladistics.**

Phylogenetic tree A branching diagram showing the relationships and evolutionary lineages of one or more groups of organisms.

Phylogeny The evolutionary history of a species or group of species in terms of their derivations and connections. A phylogenetic tree is a schematic diagram designed to represent that evolution—ideally, a portrait of genetic relationships.

Phylogeography The study of the evolutionary processes regulating the geographic distributions of lineages/groups by reconstructing genealogies of individual genes, groups of genes or populations.

Phylum (plural, **phyla**) The major taxonomic category below the level of kingdom, used to include classes of organisms that may be phenotypically quite different but share some general characters or body plan.

Piltdown man A fossil thought to be of an early hominin and discovered in Piltdown England in 1910. Subsequently shown to be a human cranium and the lower jaw of a female orangutan made to appear to be a fossil.

Placentals Mammals of the infraclass Eutheria, possessing, among other characters, a reproductive process that uses a placenta to nourish their young until a relatively advanced stage of development compared to other mammalian groups (monotremes and marsupials).

Planetesimals (cosmology) Aggregations of dust and gas that form once aggregates reach one kilometer in diameter, allowing gravity to influence their formation. As they expand planetesimals can become protoplanets and finally planets.

Plasmid A self-replicating, circular DNA element that can exist outside the host chromosome. There are various kinds, some maintaining more than one copy per cell.

Plasticity. *See* **Phenotypic plasticity.**

Plate tectonics The geological mechanism by which the continental plates on Earth's crust move.

Pleiotropy Phenotypic effects of a single gene on more than one character.

Plesiomorphy Instances when a species character is similar to that character in an ancestral species.

Pollen (botany) Small grains composed of protein, produced by the male organs (anthers) of seed plants (flowers, trees, grasses, weeds) and containing the male DNA.

Polygene A gene that interacts with other genes to produce an aspect of the phenotype.

Polymerase chain reaction (PCR) A laboratory technique that can replicate a sequence of DNA nucleotides into millions of copies in a very short time.

Polymorphism The presence of two or more genetic or phenotypic variants in a population.

Polypeptide *See* **Peptide.**

Polyphyletic The presumed derivation of a single taxonomic group from two or more different ancestral lineages through convergent or parallel evolution.

Polyploidy When the number of chromosome sets (n) is greater than the diploid number ($2n$).

Polytene chromosomes Giant chromosomes that replicate many times without separating and show detailed

banding patterns when stained. Especially studied in salivary glands of species of *Drosophila* and other similar insects.

Polytypic species A species consisting of individuals of two or more forms which may be varieties, races or subspecies.

Population A group of conspecific organisms occupying a more or less well-defined geographical region and exhibiting reproductive continuity from generation to generation.

Population genetics The study of evolution as represented by change in gene (now allele) frequencies because of the action of mutation, selection and genetic drift, a view of evolution that became known as the neo-Darwinian theory.

Population explosion A term used, usually for human populations, for rapid increases in numbers of individuals, with the implication that the rate of increase is unsustainable.

Positive eugenics Proposals to increase the frequency of beneficial genes in the human gene pool by identifying their carriers and encouraging/permitting their reproduction.

Postzygotic mating barrier (isolating mechanism) The situation in which mating occurs between individuals of two species but hybridization is prevented because embryos fail to develop or if they begin to develop fail to survive.

Preadaptation A character that was adaptive under a prior set of conditions and later provides the initial stage (is "co-opted") for the evolution of a new adaptation under a different set of conditions.

Precambrian Eon A major division of the geological time scale that includes all eras from Earth's origin about 4.5 Bya to the beginning of the Phanerozoic Eon, about 545 Mya. The Precambrian (also known as the Cryptozoic) is marked biologically by the appearance of prokaryotes about 3.5 Bya and small, non-skeletonized, multicellular organisms in the Ediacarian Period about 50 or 60 My before the Phanerozoic.

Predation The killing and consumption of one living organism — the prey — by another, the predator.

Preformationism The theory that an organism is preformed at conception in the form of a miniature adult and that embryonic development consists of enlargement of preformed structures.

Prezygotic mating barrier (isolating mechanism) The situation in which hybridization fails to occur because gametes fail to form if individuals from two species are crossed.

Primary atmosphere The first gaseous envelope surrounding Earth, which arose between 4.6 and 4.2 Bya, was composed of hydrogen and helium but not oxygen.

Primary endosymbiosis The formation of mitochondria and chloroplasts through the engulfment by a unicells of an aerobic bacterium.

Principle of divergence Hypothesis developed by Charles Darwin that competition between subpopulations favors specialization and separation of the populations to the point of speciation.

Prion disease A state resulting from a change in the 3D structure of a prior.

Prions The smallest agent of infection, composed of a hydrophobic protein but neither DNA nor RNA.

Progenote The hypothetical ancestral cellular form that gave rise to archaebacteria, eubacteria and eukaryotes.

Progress (evolution) A controversial concept in evolutionary biology that evolution has been accompanied by change reflected in increasing complexity.

Prokaryotic cells Single cells that lack histone-bound DNA, endoplasmic reticulum, a membrane-enclosed nucleus and other cellular organelles found in eukaryotic cells.

Promoter A DNA nucleotide sequence that enables transcription (RNA synthesis) by serving as the starting point for transcription and as the site for binding the enzyme RNA polymerase.

Protein A macromolecule composed of one or more polypeptide chains of amino acids, coiled and folded into specific shapes based on its amino acid sequences.

Protein world The hypothesis that the first chemicals of life to evolve were DNA.

Proteinoids Synthetic polymers produced by heating a mixture of amino acids. Some show protein-like properties in respect to enzyme activity, color test reactions, hormonal activity and so on.

Protista One of the four eukaryotic kingdoms; includes protozoa, algae, slime molds and some other groups. (Called "Protoctista" by some biologists.)

Protocells A membrane-bounded system containing molecules considered an early step in the origin of cells.

Protogalaxy (cosmology) A galaxy in the process of formation.

Protoplanet (cosmology) A forming planet that arises from the aggregation of planetesimals in a protoplanetary disc.

Protoplasm Cellular material within the plasma membrane but outside the nucleus.

Prototheria *See* **Monotremes.**

Pulsars *See* **Neutron stars**

Punctuated equilibrium The view that evolution of a lineage follows a pattern of long intervals in which there is relatively little change (stasis or equilibrium), punctuated by short bursts of speciation during which new taxa arise.

Q

Quantitative character A character whose phenotype can be numerically measured or evaluated; a character displaying continuous variation.

Quantitative trait loci (QTLs) Regions of a chromosome containing genes (alleles) that influence a quantitative trait such as height or weight.

Quantum evolution A rapid increase in the rate of evolution over a relatively short period of time.

R

Radiation (phylogenetic) *See* **Adaptive radiation.**

Radioactivity Emission of radiation by certain elements as their atomic nuclei undergo changes.

Radiometric dating (Earth sciences) The dating of rocks by measuring the proportions present of a radioactive isotope and the stable products of its decay.

Random genetic drift *See* **Genetic drift.**

Random mating Sexual reproduction within a population regardless of the phenotype or genotype of the sexual partner.

Range The geographical limits of the region habitually traversed by an individual or occupied by a population or species.

Rangeomorphs A group of organisms with a frond-like organization that existed in complex ecological communities within the Ediacaran biota/fauna 570 to 575 Mya. One species, *Charnia wardi,* grew to heights of two meters.

Reaction norm A measure of the responsiveness of a character of the phenotype to a range of levels or concentrations of an environmental factor, temperature or predator abundance for example. A reaction norm is plotted as the phenotypic response (size, shape, number of elements and so forth) against the environmental parameter (temperatures, predator abundance and so forth).

Receptor A protein in the cell membrane that binds to a ligand to allow gene action to continue.

Recessive allele An allele without phenotypic effect in a heterozygote.

Recessive lethal An allele whose presence in homozygous condition causes lethality.

Reciprocal altruism A mutually beneficial exchange of altruistic behavioral acts between individuals.

Recombinant DNA A DNA molecule composed of nucleotide sequences from different sources.

Recombination (genetics) A chromosomal exchange process that produces offspring that have gene combinations different from those of their parents. (Also used by some geneticists to describe the results of independent assortment.)

Red giant (cosmology) A star that has exhausted its core hydrogen fuel so that only the center is hot enough to burn and in which hydrogen fusion continues in a shell around the core.

Red Queen hypothesis The hypothesis that adaptive evolution in one species of a community causes a deterioration of the environment of other species. As a consequence, each species must evolve as fast as it can in order "to stay in the same place" (to survive).

Redshift (cosmology) The degree to which the photons reaching Earth have been stretched so that their wavelength moves towards the red end of the spectrum.

Reductionism The theory that explanations for events at one level of complexity can or should be reduced to explanations at a more basic level. For example, that all biological events should be explained in the form of chemical reactions.

Regulation (developmental biology) The ability of an embryo or part of a developing embryo to compensate for the loss of parts.

Regulator gene A gene that controls other gene, either by turning them on or off or by regulating their rate.

Relative fitness The relative reproductive success of an allele or genotype as compared to other alleles or genotypes.

Repetitive DNA Nucleotide sequences of DNA that are repeated many times in the genome.

Replication Doubling of DNA.

Repressor A regulator gene that produces a repressor (usually a protein) that binds to a particular nucleotide sequence and prevents transcription.

Reproductive barriers Any mechanisms that prevents individuals from two populations or species from breeding.

Reproductive incompatibility Failure of two individuals to reproduce because of isolation by season, habitat or behavior; death of gametes, zygote or early embryo; formation of a hybrid with low viability; or hybrid sterility.

Reproductive isolation The absence of gene exchange between populations.

Reproductive success The proportion of reproductively fertile offspring produced by a genotype relative to other genotypes.

Reproductive technology Improvement of the human condition through the use of such medical advances as genetic screening or *in vitro* fertilization.

Resource partitioning The situation in which competing groups of organisms minimize the harmful effects of direct competition by using different aspects of their common environmental resources.

Restriction enzymes Enzymes that recognize particular nucleotide sequences and cut DNA molecules at or near those sequences.

Restriction fragment length polymorphisms (RFLPs) Differences between individuals in the size of DNA fragments for a particular DNA section cut by restriction enzymes. These are inherited in Mendelian fashion and furnish a basis for estimating genetic variation. They also provide linkage markers used to track mutant genes between generations.

Rhizaria A supergroup of more than 4000 species of eukaryotic organisms including foraminiferans and radiolarians.

Ribosomal RNA (rRNA) RNA sequences that are incorporated into the structure of ribosomes.

Ribosomes Intracellular particles composed of ribosomal RNA and proteins that furnish the site at which messenger RNA molecules are translated into polypeptides.

Ribozymes Sequences of RNA nucleotides that can perform catalytic roles.

RNA (ribonucleic acid) A typically single-strand nucleic acid, characterized by the presence of a ribose sugar in each nucleotide, whose sequences serve either as messenger RNA, ribosomal RNA, or transfer RNA in cells, or as genetic material in some viruses. In contrast to the base composition of DNA, RNA usually bears uracil instead of thymine.

RNA editing Information changes in RNA molecules by the addition, deletion or transformation of ribonucleotide bases after these molecules have been transcribed from their DNA templates.

RNA interference (RNAi) The process of regulating gene transcription using small interference RNA (siRNA). A class of short (18 to 25 nucleotide) sequences of non-coding RNA that repress the translation of proteins by binding to matching target mRNAs leading to degradation of the mRNA.

RNA splicing The joining of exons by the excision of introns.

RNA virus A small intracellular infectious agent, often parasitic, composed of RNA in a protein coat.

RNA world The hypothesis that the first chemicals of life to evolve were DNA.

S

Saltation The hypothesis that new species or higher taxa originate abruptly.

Saprophyte An organism that feeds on decomposing organic material.

Scientific method A universal means of proposing a hypothesis, designing experiments, collecting data to test the hypothesis, interpreting the data in the context of past knowledge and accepting or rejecting the hypothesis.

Sea-floor spreading Expansion of oceanic crust through the deposition of mantle material along oceanic ridges.

Seasonal isolation *See* **Reproductive incompatibility**.

Seasonal polymorphism The presence of different morphological types of a species in different seasons.

Secondary atmosphere The second gaseous envelope surrounding Earth, which arose between 4.2 and 3.5 Bya, was composed of water vapor and carbon dioxide but not oxygen.

Secondary endosymbiosis The acquisition of chloroplasts through engulfment by eukaryotes of another eukaryotic alga.

Sedimentary rock Rock formed by the hardening of accumulated particles (sediments) that had been transported by agents such as wind and water. Sedimentary rocks are the prime source of fossils.

Seed (botany) A complex structure of plants containing the embryo along with parental diploid and haploid tissues.

Segmentation The repetition of body structures along an animal's anterior–posterior axis, as found generally in annelids, arthropods and chordates.

Segregation The Mendelian principle that the two different alleles of a gene pair in a heterozygote segregate from each other during meiosis to produce two kinds of gametes in equal ratios, each with a different allele.

Selection A composite of all the forces that cause differential survival and differential reproduction among genetic variants. When the selective agencies are primarily those of human choice, the process is called artificial selection; when the selective agencies are not those of human choice, it is called natural selection.

Selection coefficient (s) A relative measure of the effect of selection, usually in terms of the loss of fitness endured by a genotype, given that the genotype with greatest fitness has a value of 1.

Self-assembly The spontaneous aggregation of macromolecules into biological configurations that can have functional value.

Self-incompatibility Mechanisms by which plants avoid inbreeding, usually by preventing their own pollen from reaching the ovules.

Selfish DNA The hypothesis that the persistence of DNA sequences with no discernible cellular function (for example, various repetitive DNA sequences) arises from the likelihood that, once present in the genome, they are impossible to remove without the death of the organism — that is, they act as "selfish,"

or "junk" DNA, which the cell has no choice but to replicate along with functional DNA.

Serial ("iterative") homology Similarities between parts of the same organism, such as the vertebrae of a vertebrate or the different kinds of hemoglobin molecules produced by a mammal. The genetic basis for such homology can often be ascribed to gene duplications that have diverged over time but still produce somewhat similar effects.

Sex chromosomes Chromosomes associated with determining the difference in sex. These chromosomes are alike in the homogametic sex (for example, XX) but differ in the heterogametic sex (for example, XY).

Sex determination The process by which the gender of an individual is determined, either by genes on sex chromosomes or environmentally.

Sex linkage Genes linked on a sex chromosome.

Sex ratio The relative proportions of males and females in a population.

Sexual dimorphism When males and females of a species have distinctive phenotypes.

Sexual reproduction Zygotes produced by the union of genetic material from different sexes through gametic fertilization.

Sexual selection Selection that acts directly on mating success through direct competition between members of one sex for mates (intrasexual selection), or through choices made between them by the opposite sex (epigamic selection), or through a combination of both selective modes. In any of these cases, sexual selection may cause exaggerated phenotypes to appear in the sex on which it is acting (large antlers, striking colors, and so on).

Sibling species Species so similar to each other morphologically that they are difficult to distinguish but that are reproductively isolated.

Sickle cell anemia Destruction of red blood cells in humans homozygous for the autosomal sickle cell gene, *Hs*; clinically presents as a usually fatal form of hemolytic anemia.

Signal transduction The process whereby a receptor is activated inside a cell to regulate gene activity.

Sister group (sister taxon) A group (taxon) that is the closest relative of another group (taxon). Derives from the concept that each significant evolutionary step marks a dichotomous split that produces two sister taxa equal to each other in rank.

Small interference RNA (siRNA) Twenty-twenty-five-nucleotide RNA sequences that assemble into RNA-induced silencing complexes (RISCs) that they guide to complementary RNA sequences, which they cleave and destroy.

Snowball Earth The theory that for 10 My about 600 Mya when the average temperature was around −40°C, Earth was enveloped in a blanket of ice as much as 1 km thick, and that four such episodes may have occurred between 750 and 600 Mya.

Social Darwinism The theory that social and cultural differences in human societies (political, economic, military, religious and so on) arise through processes of natural selection, similar to those that account for biological differences among populations and species.

Social organization The situation when there is a division of labor among different members of a species as seen in termites and many bees, ants and wasps.

Sociobiology The study of the biological basis of social behavior.

Solvent A liquid that can dissolve molecules or compounds to form a solution. Water is regarded as a universal solvent.

Somatic cells (or tissues) All body cells (also known as soma) other than those that produce sperm or eggs.

Space-time (cosmology) The four-dimensional construct proposed to contain all celestial bodies (stars, planets, galaxies, black holes). Because it continues to inflate, space-time is responsible for the ongoing expansion of the universe.

Speciation The splitting of one species into two or more new species or the transformation of one species into a new species over time.

Species A basic taxonomic category for which there are various definitions. Among these are an interbreeding or potentially interbreeding group of populations reproductively isolated from other groups (the biological species concept) and a lineage evolving separately from others with its own unitary evolutionary role and tendencies (the evolutionary species concept).

Split gene A gene whose nucleotide sequence is divided into exons and introns.

Spontaneous generation The theory that complex organisms can appear spontaneously from inert materials without biological parentage; life without parents.

Sporophyte The diploid spore-producing stage of plants.

Stabilizing selection Selection that favors the survival of organisms in a population that are at an intermediate phenotypic value for a particular character, thus eliminating extreme phenotypes. (Also called normalizing selection.)

Stasis A period of time (usually geological) without evident evolutionary change.

Stem cells Cells of multicellular organisms capable of producing more than one different cell type.

Stem group The group of extinct organisms considered closest to and more basal than the most basal members of a clade.

Stone tools Stones that have been modified (usually by humans) for use in food capture/preparation or defense.

Stop codon One of the three messenger RNA codons (UAA, UAG, UGA) that terminates the translation of a polypeptide. (Also called chain-termination codon or non-sense codon.)

Strata *See* **Geological strata.**

Strepsirrhines The "wet-nosed" primates comprising lemurs and lorises.

Stromatolites Laminated rocks produced by layered accretions of benthic microorganisms (mainly filamentous cyanobacteria) that trap or precipitate sediments.

Structural gene A DNA nucleotide sequence that codes for RNA or protein. Some definitions restrict this term to a protein-coding gene.

Subduction (Earth sciences) The process by which a tectonic plate descends (is subducted) beneath the edge of another plate into the mantle, often giving rise to earthquakes and/or volcanic activity.

Subspecies A taxonomic subdivision of a species often distinguished by special phenotypic characters and by its origin or localization in a given geographical region. Like other species subdivisions (*see* **Group**), a subspecies can still interbreed with the remainder of the species.

Supergroups Major groupings of eukaryotes at levels that replace kingdoms in some classifications.

Supernova *See* **Type Ia supernova, Type IIa supernova.**

Survival of the fittest The phrase coined by Herbert Spencer to describe the result of the operation of natural selection.

Survivorship The proportion of individuals born at a given time (cohort) who survive to a given age.

Symbiont A participant in the interactive association (symbiosis) between two individuals or two species. This term is often restricted to mutually beneficial associations.

Sympatric Species or populations whose geographical distributions coincide or overlap.

Sympatric speciation Speciation that occurs between populations occupying the same geographic range.

Symplesiomorphy A trait (feature, character) shared between two or more taxa and also shared with other taxa with which those two or more taxa share a last common ancestor; a shared or ancestral state of a character.

Synapomorphy The possession by two or more related lineages of the same phenotypic character derived from a different but homologous character in the ancestral lineage.

Synthetic theory of evolution *See* **Modern synthesis.**

Systematics Although defined by G. G. Simpson as the study of the diversity of organisms and all their comparative and evolutionary relationships, it is often used interchangeably with the terms classification and taxonomy.

T

Taxon (plural, **taxa**) A taxonomic unit at any level of classification.

Taxonomy The principles and procedures used in classifying organisms.

Tectonic plates The fairly rigid plates composing Earth's crust whose boundaries are marked by earthquake belts and volcanic chains. Continental masses ride on some of these plates, accounting for continental drift and such processes as the mountain building that occurs when these plates collide.

Teleology The theory that natural processes such as development or evolution are guided by their final stage (*telos*) or for some particular purpose, for example, "the reason plants engage in photosynthesis and animals seek food is for survival, and the ultimate purpose of survival is for reproductive success."

Telome hypothesis Thin branches (*telomes*) evolved toward greater complexity and vascularization to produce the leaves and branches of ferns and vascular plants, or regressed toward a single unbranched form, producing bryophytes.

Tetrapods Literal meaning, "four-footed." Commonly used to specify a member of the land-evolved vertebrates: amphibians, reptiles and mammals.

Thermophilic organisms (thermophiles) Organisms that thrive at temperatures as high as 350°C as found in hydrothermal vents.

Tinkering (bricolage) The hypothesis that evolutionary change consists of minor modifications of existing genes, pathways and processes.

Tissue A group of cells all performing a similar function in a multicellular organism.

"Tit for tat" An evolutionarily stable strategies in which an individual behaves cooperatively as the first move in an interaction, and then repeats its opponent's next move.

Tools *See* **Stone tools.**

Trait *See* **Character.**

Transcription The process by which the synthesis of an RNA molecule (for example, messenger RNA) is initiated and completed on a DNA template by RNA polymerase enzyme.

Transcription factors Gene products that bind to specific (regulatory) sequences of other genes to control their activity.

Transfer RNA (tRNA) Relatively small RNA molecules (about 80 nucleotides long) that carry specific amino

acids to the ribosome for polypeptide synthesis. Each kind of tRNA has a unique anticodon complementary to messenger RNA codons that specify the placement of particular amino acids in the polypeptide chain.

Transgenic An organism (sometimes a cell) containing a gene from another organism (or cell) that has been incorporated into its genome.

Translation The protein-synthesizing process that takes place on the ribosome, linking together a particular sequence of amino acids (polypeptide) on the basis of information received from a particular sequence of codons on messenger RNA.

Transposable elements *See* **Transposons.**

Transposons (transposable elements) Nucleotide sequences that produce enzymes to promote their own movement from one chromosomal site to another and that may carry additional genes such as those for antibiotic resistance.

Tree of Life A phylogenetic tree of all the organisms on Earth. Based on the proposal by Charles Darwin that the similarities and differences between organisms reflected a single, branched and hierarchical tree of nature.

Triploblastic An animal that produces all three major types of cell layers during development — ectoderm, endoderm and mesoderm.

Type specimen An individual of a species that defines a species and is housed in a museum where it may be examined.

Typology The study of organic diversity based on the principle that all members of a taxonomic group conform to a basic plan, and variation among them is of little or no significance.

U

Ultraviolet (UV) radiation Electromagnetic radiation at wavelengths between about 4 and 400 nanometers, shorter than visible light but longer than X-rays. UV radiation is absorbed by purine and pyrimidine ring structures and is therefore quite damaging to nucleic acid genetic material.

Unequal crossing over The result of improper pairing between chromatids, causing their crossover products to differ from each other in the amounts of genetic material.

Uniformitarianism The theory in earth sciences, popularized by Charles Lyell, that none of the forces active in past Earth history were different from those active today.

Unikonta A supergroup of organisms including slime molds, many types of amoebae, some parasitic protists, choanoflagellates, fungi and animals.

Uniparental inheritance Transmission of heritable characters through a single parent, for example,

mitochondria and chloroplasts through the female parent.

Universal genetic code The use of the same genetic code in virtually all living organisms.

Universal tree of life (UToL) The proposal that all organisms can be represented by branches and twigs on a single phylogenetic tree.

Upper Paleolithic tools Hooks, barbed harpoons, needles, ivory beads and carved figurines made and used by *Homo sapiens,* some as early as 90,000 years ago.

Upstream gene A regulatory gene(s), that controls (is upstream of) another gene or genes.

Use and disuse A theory used by Lamarck to explain evolution as resulting from the transmission of characters that became enhanced or diminished because of their use or disuse, respectively, during the life experience of individuals.

V

Variability The propensity of genotypes or phenotypes to vary.

Variation A term commonly used to indicate differences in the qualitative or quantitative values of a character among individual members of a population, whether molecules, cells or organisms.

Vascular plants Land plants that have special water- and food-conducting vessels and tissues (xylem and phloem).

Vertical transmission Transmission of heredity from parent to offspring.

Vestiges *See* **Vestigial organs.**

Vestigial organs Organs or structures that appear to be small and functionless but can be shown to be homologous with ancestral organs and structures that were larger and functional.

Virus A small intracellular infectious agent, often parasitic, often composed of DNA or RNA in a protein coat, and that depends on the host cell to replicate its genetic material and to synthesize its proteins.

Vitalism The concept that the activities of living organisms cannot be explained by any underlying physical or chemical principles but arise from unknowable internal or supernatural causes.

W

White dwarf (cosmology) The name for the final dying stage in the life of a planet, visible as a faint object in the sky.

Wild type The most commonly observed phenotype or genotype for a particular character. Variations from wild type are considered mutants.

X

X chromosome The name given in various groups to a sex chromosome usually present twice in the homogametic sex (XX) and only once in the heterogametic sex (XY or XO).

X-linked genes Genes present on the X chromosome.

Xylem The conducting tissue of green land plants that takes water to all parts of the plant.

Y

Y chromosome A sex chromosome present only in the heterogametic (XY) sex.

Y-chromosome Adam The name given to the common Y-chromosome human ancestor estimated from Y-chromosome nucleotide sequences to have existed 150,000 years ago.

Z

Zoogeography The science of the study of the distribution of animals in a particular region.

Zygote The cell formed by the union of male and female gametes.

Lexicon of Key Terms, Concepts, Principles, Processes and Groups of Organisms

This cross-list of terms allows you to access similar or related (or opposite) terms and concepts (for example, **Abiotic** links to Biotic and to Life; **Adaptive radiation** links to Biogeography, Speciation, Zoogeography). All terms are in the text and in the glossary. This lexicon will be especially useful for those who want to investigate or study related topics and for those who want to assign projects/presentations on related topics.

A

Abiotic — Biotic, Life

Acheulean tools — Mousterian tools, Oldowan tools, Stone tools

Adaptation — Environment, Exaptation, Inheritance of acquired characters, Preadaptation

Adaptive landscape — Fitness, Genetic load, Heterozygote advantage (superiority), Hybrid breakdown (inviability, sterility), Inclusive fitness, Natural selection, Population genetics, Relative fitness, Selection, Selection coefficient (s)

Adaptive radiation — Biogeography, Speciation, Zoogeography

Algae — Archaeplastida, Chromalveolata, Cyanobacteria, Photosynthesis, Protista

Allele — Allele frequency, Dominant allele, Gene, Gene pool, Genetic polymorphism, Heterozygote, Homozygote, Independent assortment, Linkage, Linkage equilibrium, Neutral theory of molecular evolution, Population genetics, Recessive allele, Recessive lethal, Relative fitness, Segregation

Allele frequency — Allele, Dominant allele, Gene, Gene pool, Genetic polymorphism, Heterozygote, Homozygote, Independent assortment, Linkage, Linkage equilibrium, Neutral theory of molecular evolution, Population genetics, Recessive allele, Recessive lethal, Relative fitness, Segregation

Allopatric speciation — Parapatric speciation, Peripatric speciation, Speciation, Sympatric speciation

Alternation of generations — Alternation of generations, Diploid, Haploid, Meiosis

Altruism — Evolutionary stable strategies, Mutualism, Reciprocal altruism, "Tit for tat"

Analogy — Convergence, Homology

Anatomically modern humans — Fossils, Hominid, Mousterian tools, Upper Paleolithic tools

Angiosperms — Biota, Dicotyledons, Double fertilization, Flora, Flower, Gymnosperms, Mesozoic Era, Monocotyledons, Pollen, Seeds, Telome hypothesis, Vascular plants

Antennapedia complex — Bithorax complex, Homeotic mutations

A-P axis — Axes of symmetry, P-D axis

Apomorphic character — Character, Character state, Convergence, Homology, Homoplasy, Parallel evolution, Phenotype, Plesiomorphy, Symplesiomorphy, Synapomorphy

Archaebacteria — Greenhouse gas, Methanogens

Arms race — Parasitism

Artificial selection — Canalizing selection, Constraint, Darwinism, Directional selection, Disruptive selection, Frequency-dependent selection, Group selection, Kin selection, Natural selection, Selection, Selection coefficient (s), Sexual selection, Stabilizing selection

Asexual reproduction — Clone, Sexual reproduction

Axes of symmetry — A-P axis, Bilateral symmetry, P-D axis

B

Balanced polymorphism — Genetic polymorphism

Baryonic matter — Cosmic microwave background, Dark ages (cosmology), Dark matter

Base (chemistry) — Base pairs, Base substitutions, Complementary base pairs, DNA (deoxyribonucleic acid), Nucleic acid, Nucleotide, RNA (ribonucleic acid)

Base substitutions — Base, Base pairs, Complementary base pairs

Batesian mimicry — Mimicry, Mullerian mimicry

Bauplan — Archetype, Phylum

Behavior — Altruism, Dominance (social), Homology, Imprinting, Instinct, Learning, Phenotype, Reciprocal altruism, Sociobiology

Term	Related terms
Big Bang theory	Evolutionary universe
Binomial nomenclature	Classification, Genus, Species
Biogenetic law	Development, Evolutionary developmental biology, Heterochrony, Ontogeny, Paedomorphosis, Parthenogenesis, Phylogeny
Biogeography	Endemic, Phylogenetic evolution, Phylogeography, Zoogeography
Biological species concept	Evolutionary species, Fixity of species, Morphological species, Polytypic species, Species
Biota	Abiotic, Angiosperms, Biotic, Life, Burgess shale fossils, Chengjiang fauna, Ediacaran fauna/biota, Fauna, Organism, Rangiomorphs
Biotic	Abiotic, Biotic, Life, Organism
Bithorax complex	Antennapedia complex, Homeotic mutations
Blending inheritance	Extranuclear inheritance, Inheritance, Inheritance of acquired characters, Lamarckian inheritance, Uniparental inheritance
Bottleneck effect	Founder effect, Population, Peripatric speciation, Population genetics
Box (Boxes)	Homeobox, Homeodomain, Homeotic Genes, Hox gene
Burgess shale fossils	Biota, Cambrian explosion, Cambrian period, Chengjiang fauna, Ediacaran fauna/biota, Fauna, Fossils, Macroevolution, Paleontology, Rangiomorphs

C

Term	Related terms
Cambrian explosion	Burgess shale fossils, Cambrian Period, Chengjiang fauna
Cambrian Period	Burgess shale fossils, Cambrian explosion, Chengjiang fauna, Geological strata, Geological time scale, Period (geological)
Canalization	Canalizing selection, Constraint, Variation
Canalizing selection	Artificial selection, Canalization, Constraint, Darwinism, Directional selection, Disruptive selection, Fitness, Frequency-dependent selection, Group selection, Inclusive fitness, Kin selection, Natural selection, Neo-Darwinism, Population genetics, Relative fitness, Selection, Selection coefficient (s), Sexual selection, Stabilizing selection, Survival of the fittest

Term	Related terms
Carrying capacity (K)	Ecological footprint, Environment, Intrinsic rate of natural increase, Logistic growth curve, Niche, Population, Population explosion, Range
Catastrophism (Earth sciences)	Geological dating, Geological strata, Geological strata, Law of superposition, Paleontology, Sedimentary rock, Snowball, Earth, Uniformitarianism
Cenozoic (Caenozoic) Era	Dinosaurs, Era, Mammals, Mesozoic Era, Paleozoic (Palaeozoic) Era, Phanerozoic Eon
Character	Analogy, Apomorphic character, Character, Character state, Convergence, Homology, Homoplasy, Inheritance of acquired characters, Morphology, Parallel evolution, Phenotype, Pleiotropy, Plesiomorphy, Quantitative character, Symplesiomorphy, Synapomorphy
Character displacement	Character state, Competition, Niche
Character state	Analogy, Apomorphic character, Character, Convergence, Homology, Homoplasy, Parallel evolution, Phenotype, Pleiotropy, Plesiomorphy, Quantitative character, Symplesiomorphy, Synapomorphy
Chengjiang fauna	Biota, Burgess shale fossils, Cambrian explosion, Cambrian Period, Ediacaran fauna/biota, Fauna, Fossils, Paleontology, Rangiomorphs
Chlorophyll	Chloroplast, Cyanobacteria, Photosynthesis
Chloroplast	Algae, Chlorophyll, Cyanobacteria, Endosymbiosis, Photosynthesis, Protista
Choanoflagellates	Opisthokonts, Unikonta
Chordates	Notochord, Phylum, Vertebrates
Chromosomes	Autosomes, Deletion, Gene locus, Heterozygote, Homozygote, Independent assortment, Locus, Quantitative trait loci (QTLs), Polytene chromosomes
Cis-regulation	Developmental control gene, Enhancer, Messenger RNA, Promoter, Repressor, Transcription
Clade	Crown Group, Missing link
Cladistics	Cladogram, Classification, Phylogenetic method
Cladogenesis	Lineages, Monophyletic, Paraphyletic, Phyletic evolution, Phylogenetic evolution, Principle of divergence

Cladogram	Cladogenesis, Phyletic evolution, Phylogenetic evolution, Principle of divergence
Class	Classification, Family, Order, Phylum, Systematics, Taxon, Taxonomy
Classification	Binomial nomenclature, Cladistics, Class, Family, Kingdom, Numerical Taxonomy, Order, Phylum, Supergroups, Systematics, Taxon, Taxonomy
Clone	Asexual reproduction, Cloning
Cloning (genetics)	Clone
Coacervate	Microspheres, Protocells
Codon	Degenerate (redundant) code, Genetic code, Messenger RNA (mRNA), Peptide (polypeptide), Stop codon, Translation, Universal genetic code
Coextinction	Extinction, Lineages
Collision theory	Condensation theory, Cosmology, Dwarf planets, Gravitation, Nebular hypothesis, Planetesimals, Protoplanet, White dwarf
Commensalism	Parasitism, RNA virus, Virus
Competition	Competitive exclusion, Environment, Group selection, Niche, Principle of divergence, Resource partitioning, Sexual selection
Competitive exclusion	Competition, Environment, Group selection, Niche, Principle of divergence, Resource partitioning, Sexual selection
Complementary base pairs	Base, Base pairs, DNA (deoxyribonucleic acid), Homeobox, Homeotic genes, RNA (ribonucleic acid)
Complexity	Design, Hierarchy, Progress, Reductionism
Condensation theory	Collision theory, Cosmology, Dwarf planets, Gravitation, Nebular hypothesis, Planetesimals, Protoplanet, White dwarf
Constraint	Canalization, Canalizing selection, Variation
Continental drift	Continents, Crust, Gondwana, Laurasia, Paleomagnetism, Pangaea, Plate tectonics, Sea-floor spreading, Tectonic plates
Continuous variation	Canalization, Character, Character state, Darwinism, Quantitative character, Variation
Convergence (Convergent evolution)	Analogy, Apomorphic character, Character, Character state, Homology, Homoplasy, Parallel evolution, Phenotype, Pleiotropy, Plesiomorphy, Symplesiomorphy, Synapomorphy
Core (Earth sciences)	Crust, Earth, Lithosphere, Mantle
Cosmic microwave background (CMB)	Baryonic matter, Dark ages (cosmology), Dark matter
Cosmology	Collision theory, Condensation theory, Cosmological constant, Cosmological redshift, Critical density, Dwarf planets, Expanding universe, Galaxy, Gravitation, Nebular hypothesis, Planetesimals, Protoplanet, Redshift, Space-time, White dwarf
Crown Group	Classification, Stem group, Taxonomy
Crust (Earth sciences)	Continental drift, Continents, Core, Earth, Igneous rock, Lithosphere, Mantle (Earth sciences), Metamorphic rock, Sedimentary rock
Culture	Altruism, Dominance (social), Homology, Imprinting (behavior), Instinct, Learning, Phenotype, Reciprocal altruism, Sociobiology
Cyanobacteria	Algae, Chlorophyll, Chloroplast, Photosynthesis
Cytoplasm	Eukaryotic cells, Extranuclear inheritance, Fertilization, Histones, Maternal inheritance, Nucleus, Organelles, Prokaryotic cells, Protoplasm, Uniparental inheritance

D

Dark ages (cosmology)	Baryonic matter, Collision theory, Cosmic microwave background (CMB), Dark matter, Galaxy, Milky Way, Neutron stars (pulsars), Red giant
Dark energy	Baryonic matter, Dark matter
Dark matter	Cosmic microwave background (CMB), Dark ages (cosmology), Dark energy
Darwinism	Continuous variation, Discontinuous variation, Domestication, Homology, Natural selection, Neo-Darwinism, Population genetics, Selection, Sexual selection, Social Darwinism, Survival of the fittest, Tinkering, Variation
Degenerate (redundant) code	Codon, Genetic code, Messenger RNA (mRNA), Peptide (polypeptide), Stop codon, Translation, Universal genetic code

Deleterious allele	Allele, Allele frequency, Heterozygote, Homozygote, Recessive lethal, Relative fitness, Segregation
Deletion	Chromosome, Mutation, RNA editing
Deme	Gene flow, Group, Isolating mechanisms, Mendelian population, Migration (genetics), Parapatric, Population, Species, Subspecies
Descent with modification	Darwinism, Domestication, Homology, Tinkering
Design	Complexity, Fixity of species, Hierarchy, Great chain of being, Ladder of Nature, Progress, Reductionism
Development	Biogenetic law, Developmental control gene, Embryology, Fertilization, Epigenesis, Evolutionary developmental biology, Heterochrony, Homeotic mutations (homeosis), Morphogenesis, Ontogeny, Paedomorphosis, Parthenogenesis, Preformationism, Regulation
Developmental control gene	*Cis*-regulation, Downstream gene, Enhancer, Gene regulation, Genetic assimilation, Heat shock proteins, Hierarchy, Homeotic mutations (homeosis), Modularity, Promoter, Regulator gene, Repressor, Signal transduction, Transcription, Upstream gene
Dicotyledons	Angiosperms, Double fertilization, Flower, Mesozoic Era, Monocotyledons, Pollen, Seeds
Dinosaurs	K-T extinction, Mammals, Mesozoic Era, Phanerozoic Eon
Diploblastic	Ectoderm, Endoderm, Germ layers, Triploblastic
Diploid	Alternation of generations, Double fertilization, Gamete, Haploid, Haplodiploidy, Meiosis, Polyploidy, Seed, Sporophyte
Directional selection	Artificial selection, Canalizing selection, Constraint, Darwinism, Disruptive selection, Fitness, Frequency-dependent selection, Group selection, Inclusive fitness, Kin selection, Natural selection, Neo-Darwinism, Population genetics, Relative fitness, Selection, Selection coefficient (s), Sexual selection, Stabilizing selection, Survival of the fittest
Discontinuous variation	Canalization, Constraint, Continuous variation, Quantitative character, Variation
Disruptive selection	Artificial selection, Canalizing selection, Constraint, Darwinism, Directional selection, Disruptive selection, Fitness, Frequency-dependent selection, Group selection, Inclusive fitness, Kin selection, Natural selection, Neo-Darwinism, Population genetics, Relative fitness, Selection, Selection coefficient (s), Sexual selection, Stabilizing selection, Survival of the fittest
DNA (deoxyribonucleic acid)	Base, Base Pairs, Base substitutions, Complementary base pairs, Nucleic acid, Nucleotide, RNA (ribonucleic acid)
DNA virus	RNA virus, Virus
DNA world	Protein world, RNA world
Domestication	Artificial selection, Darwinism
Dominance (social)	Altruism, Culture, Imprinting (behavior), Instinct, Learning, Reciprocal altruism, Sociobiology
Dominant allele	Allele, Allele frequency, Gene, Gene pool, Genetic polymorphism, Heterozygote, Homozygote, Independent assortment, Linkage, Linkage equilibrium, Neutral theory of molecular evolution, Population genetics, Recessive allele, Recessive lethal, Relative fitness, Segregation
Dosage compensation	Haldane's rule, Homogametic sex, Sex chromosomes, Sex-linked genes, X-chromosome, Y-chromosome
Double fertilization	Angiosperms, Dicotyledons, Diploid, Flower, Mesozoic Era, Monocotyledons, Nucleus, Pollen, Polyploidy, Seeds, Triploid
Downstream gene	*Cis*-regulation, Developmental control gene, Gene regulation, Heat shock proteins, Hierarchy, Homeotic mutations (homeosis), Modularity, Upstream gene
Duplication	Gene family, Genome, Mutation, Paralogous genes, Repetitive DNA, Selfish DNA, Serial ("iterative") homology
Dwarf planets	Condensation theory, Cosmology, Gravitation, Planetesimals, Protoplanet, White dwarf

E

Ecdysoza	Arthropods, Lophotrochoza
Ecological footprint	Carrying Capacity, Ecological footprint, Environment, Intrinsic rate of natural increase, Logistic growth curve, Niche, Population, Population explosion, Range

Microspheres	Coacervates, Enzyme, Metabolic pathway, Metabolism, Microspheres, Proteinoids, Protocells
Migration (ecology)	Ecology, Gene flow, Group, Population
Migration (genetics)	Deme, Gene flow, Group, Isolating mechanisms, Mendelian population, Parapatric, Population, Species, Subspecies
Mimicry	Batesian mimicry, Mullerian mimicry
Mitochondria	Chloroplasts, Endosymbiosis, Eukaryotic cells, Histones
Mitochondrial DNA (mtDNA)	Mitochondria, Mitochondrial Eve
Mitochondrial Eve	Mitochondrial DNA, Y-chromosome Adam
Mitosis	Chromosomes, Homologous chromosomes, Independent assortment, Meiosis, Segregation
Modularity	Developmental control gene, Gene regulation, Mosaic evolution
Molecular clock	Geological time scale, Speciation, Stasis
Monocotyledons	Angiosperms, Dicotyledons, Double fertilization, Flower, Mesozoic Era, Pollen, Seeds
Monophyletic	Cladogenesis, Coextinction, Lineage, Parallel evolution, Phyletic evolution, Phylogenetic evolution, Phylogenetic tree, Phylogeography, Polyphyletic, Punctuated equilibrium, Species, Synapomorphy
Monotremes	Lactation, Mammals, Mammary glands, Mesozoic Era, Monotremes, Placentals, Tetrapods, Cenozoic (Caenozoic) Era
Morphogenesis	Development, Embryology, Epigenesis, Evolutionary developmental biology, Homeotic mutations (homeosis), Morphology, Ontogeny, Regulation
Morphological species	Biological species concept, Evolutionary species, Fixity of species, Morphology, Polytypic species, Species
Morphology	Character, Morphogenesis, Morphological species, Morphospace, Neoteny, Paedomorphosis, Seasonal polymorphism, Sibling species
Morphospace	Character, Morphogenesis, Morphology
Mosaic evolution	Modularity
Mousterian tools	Acheulean tools, Anatomically modern humans, Oldowan tools, Stone tools, Upper Paleolithic tools
Mullerian mimicry	Batesian mimicry, Mimicry
Muller's ratchet	Genetic load, Mutation, Mutational load, Population
Mutagenesis	Mutation, Radioactivity
Mutation	Chromosome, Deletion, Duplication, Inversion, Mutagenesis, RNA editing
Mutational load	Genetic load, Muller's ratchet, Mutation, Population
Mutualism	Altruism, Evolutionary stable strategies, Reciprocal altruism, "Tit for tat"

N

Natural selection	Adaptation, Adaptive landscape, Artificial selection, Canalizing selection, Continuous variation, Darwinism, Fitness, Frequency-dependent selection, Group selection, Inclusive fitness, Kin selection, Neo-Darwinism, Population genetics, Relative fitness, Reproductive success, Selection, Selection coefficient (s), Sexual selection, Stabilizing selection, Survival of the fittest, Variation
Nebular hypothesis	Collision theory, Condensation theory, Cosmology, Gravitation, Planetesimals
Neo-Darwinism	Fitness, Natural selection, Population genetics
Neoteny	Heterochrony, Morphology, Paedomorphosis
Neural crest	Ectoderm, Germ layers, Triploblastic
Neutral mutation	Fitness, Inclusive fitness, Natural selection, Neutral theory of molecular evolution, Population genetics, Relative fitness, Selection, Selection coefficient (s)
Neutral theory of molecular evolution	Allele, Allele frequency, Fitness, Gene, Genetic polymorphism, Heterozygote, Homozygote, Independent assortment, Linkage equilibrium, Neutral mutation, Population genetics, Recessive allele, Recessive lethal, Relative fitness, Segregation
Neutron stars (pulsars)	Collision theory, Dark ages (cosmology), Galaxy, Milky Way, Red giant (cosmology)

Niche	Carrying Capacity, Character displacement, Competition, Competitive exclusion, Environment, Habitat, Intrinsic rate of natural increase, Logistic growth curve, Population, Population explosion, Range, Resource partitioning
Notochord	Chordates
Nucleic acid	Base, Base Pairs, Base substitutions, Complementary base pairs, DNA (deoxyribonucleic acid), Nucleotide, RNA (ribonucleic acid)
Nucleotide	Base, Exon, Gene, Intron, RNA splicing, Split gene, Transcription, Translation
Nucleus	Cytoplasm, Double fertilization, Eukaryotic cells, Fertilization, Histones, Organelles, Prokaryotic cells, Protoplasm
Numerical taxonomy	Binomial nomenclature, Cladistics, Classification, Kingdom, Phylum, Supergroups, Systematics, Taxon, Taxonomy

O

Oldowan tools	Acheulean tools, Mousterian tools, Stone tools, Upper Paleolithic tools
Ontogeny	Biogenetic law, Development, Developmental control gene, Embryology, Fertilization, Epigenesis, Evolutionary developmental biology, Heterochrony, Homeotic mutations (homeosis), Life cycle, Morphogenesis, Paedomorphosis, Phylogeny, Preformationism, Zygote
Operon	Gene regulation, Structural gene, Transcription
Opisthokonts	Fungi, Protista, Unikonta
Order	Class, Classification, Family, Systematics, Taxon, Taxonomy
Organelles	Chloroplasts, Cytoplasm, Eukaryotic cells, Mitochondria, Nucleus, Prokaryotic cells, Protoplasm, Ribosomes
Organic	Amino acids, Life, Metabolism, Peptide (polypeptide), Photosynthesis, Saprophyte
Orthogenesis	Progress, Vitalism
Orthologous genes	Homology, Hox gene, Paralogous genes
Outgroup	Ingroup, Taxon

P

Paedomorphosis	Development, Embryology, Evolutionary developmental biology, Heterochrony, Heterotopy, Ontogeny, Preformationism
Paleomagnetism	Continental drift
Paleontology	Ancient DNA, *Archaeopteryx*, Burgess shale fossils, Chengjiang fauna, Ediacaran fauna/biota, Fossils, Hominid, Living fossil, Macroevolution, Period (geological), Phanerozoic Eon, Piltdown man, Rangeomorphs, Sedimentary rock
Paleozoic (Palaeozoic) era	Cenozoic (Caenozoic) Era, Dinosaurs, Era, Mesozoic Era, Phanerozoic Eon
Pangaea	Continental drift, Continents, Gondwana, Laurasia, Tectonic plates
Pangenesis	Extranuclear inheritance, Inheritance, Maternal inheritance, Uniparental inheritance Vertical transmission
Parallel evolution	Analogy, Character, Character state, Convergence, Homology, Homoplasy, Lineage, Phenotype, Plesiomorphy, Symplesiomorphy, Synapomorphy
Paralogous genes	Duplication, Gene family, Mutation, Orthologous genes, Serial ("iterative") homology
Parapatric	Deme, Gene flow, Group, Isolating mechanisms, Mendelian population, Migration (genetics), Population, Species, Subspecies
Parapatric speciation	Allopatric speciation, Peripatric speciation, Sympatric speciation
Paraphyletic	Cladogenesis, Lineage, Monophyletic, Phyletic evolution, Polyphyletic
Parasitism	Arms race, Commensalism
Parsimony method	Cladistics, Phylogenetic tree
Parthenogenesis	Asexual reproduction, Development, Embryology, Fertility, Fertilization, Epigenesis, Evolutionary developmental biology, Ontogeny, Preformationism, Sexual reproduction
P-D axis	Axes of symmetry, A-P axis
Peptide (polypeptide)	Amino acids, Organic
Period (geological)	Fossils, Geological strata, Geological time scale, Macroevolution, Paleontology
Peripatric speciation	Allopatric speciation, Bottleneck effect, Founder effect, Parapatric speciation, Population genetics, Sympatric speciation

Phanerozoic eon	Archean eon, Cenozoic (Caenozoic) Era, Dinosaurs, Eon, Geological time scale, Mammals, Mesozoic Era, Paleozoic (Palaeozoic) Era, Paleontology, Precambrian Eon
Phenotype	Analogy, Character, Character state, Convergence, Genotype, Homology, Homoplasy, Parallel evolution, Plesiomorphy, Quantitative character, Symplesiomorphy, Synapomorphy
Phenotypic plasticity	Canalization, Constraint, Discontinuous variation, Genetic assimilation, Natural selection, Reaction norm, Variation
Photosynthesis	Algae, Chlorophyll, Chloroplast, Cyanobacteria, Metabolism, Organic
Phyletic	Cladogenesis, Geological time scale, Lineage, Monophyletic, Parallel evolution, Paraphyletic, Phylogenetic evolution, Phylogenetic tree, Phylogeography, Polyphyletic, Principle of divergence, Punctuated equilibrium, Principle of divergence, Speciation
Phylogenetic evolution	Cladogenesis, Lineage, Monophyletic, Phyletic evolution, Principle of divergence
Phylogenetic tree	Evolutionary trees, Lineages, Parsimony, Phylogeny, Tree of life, Universal Tree of Life
Phylogeny	Biogenetic law, Ontogeny, Phylogenetic tree
Phylogeography	Biogeography, Endemic, Lineage, Phylogenetic evolution, Zoogeography
Phylum	Arthropods, *Bauplan*, Chordates, Class, Classification, Kingdom, Numerical Taxonomy, Systematics, Taxon, Taxonomy
Piltdown man	Fossils, Hominid, Paleontology
Placentals	Lactation, Mammals, Mammary glands, Marsupials, Mesozoic Era, Monotremes, Tetrapods, Cenozoic (Caenozoic) Era
Plate tectonics	Continental drift, Crust, Earth, Sea-floor spreading, Subduction (Earth sciences), Tectonic plates
Pleiotropy	Character, Convergence, Gene
Plesiomorphy	Apomorphic character, Character, Character state, Convergence, Homology, Homoplasy, Parallel evolution, Symplesiomorphy, Synapomorphy
Pollen (botany)	Angiosperms, Dicotyledons, Double fertilization, Flower, Gamete, Mesozoic Era, Monocotyledons, Seeds
Polymorphism	Balanced polymorphism, Genetic polymorphism, Seasonal polymorphism
Polyphyletic	Cladogenesis, Convergence, Lineage, Monophyletic, Parallel evolution, Paraphyletic, Phyletic evolution, Phylogenetic evolution
Polyploidy	Diploid, Haploid, Meiosis
Polytene chromosomes	Autosomes, Chromosomes
Polytypic species	Biological species concept, Evolutionary species, Fixity of species, Morphological species, Polytypic species, Species
Population	Bottleneck effect, Founder effect, Carrying Capacity, Deme, Environment, Gene flow, Group, Intrinsic rate of natural increase, Isolating mechanisms, Logistic growth curve, Mendelian population, Migration (ecology), Migration (genetics), Niche, Population, explosion, Range
Population genetics	Adaptive landscape, Allele, Allele frequency, Bottleneck effect, Fitness, Founder effect, Gene, Genetic load, Genetic polymorphism, Heterozygote, Heterozygote advantage (superiority), Homozygote, Inclusive fitness, Independent assortment, Linkage, Linkage equilibrium, Mutation, Mutational load, Natural selection, Neo-Darwinism, Neutral mutation, Neutral theory of molecular evolution, Recessive allele, Recessive lethal, Relative fitness, Reproductive success, Segregation, Selection, Selection coefficient (*s*)
Population explosion	Carrying Capacity, Environment, Intrinsic rate of natural increase, Logistic growth curve, Population, Range
Postzygotic mating barrier (isolating mechanism)	Geographical isolation, Isolating mechanisms, Prezygotic mating barrier, Reproductive barriers, Reproductive isolation
Preadaptation	Adaptation, Environment, Exaptation
Precambrian eon	Archean eon, Ediacaran fauna/biota, Eon, Geological time scale, Phanerozoic Eon, Precambrian Eon, Rangiomorphs
Preformationism	Development, Embryology, Fertilization, Epigenesis, Evolutionary developmental biology, Morphogenesis, Ontogeny, Paedomorphosis

Prezygotic mating barrier (isolating mechanism) Geographical isolation, Isolating mechanisms, Postzygotic mating barrier, Reproductive barriers, Reproductive isolation

Primary atmosphere Atmosphere, Secondary atmosphere

Primary Endosymbiosis Chloroplasts, Endocytosis, Endosymbiosis, Mitochondria, Secondary endosymbiosis

Principle of divergence Cladogenesis, Competition, Phyletic evolution, Principle of divergence

Prions Prion disease

Progress (evolution) Complexity, Design, Hierarchy, Orthogenesis, Reductionism, Vitalism

Prokaryotic cells Cytoplasm, Eukaryotic cells, Histones, Mitochondria, Nucleus, Protoplasm

Promoter Cis-regulation, Enhancer, Gene regulation, Messenger RNA, Repressor, Signal transduction, Transcription

Protein world DNA world, RNA world

Proteinoids Enzyme, Metabolic pathway, Metabolism, Microspheres

Protista Chloroplast, Excavata, Opisthokonts, Unikonta

Protocells Coacervates, Microspheres

Protogalaxy Galaxy, Gravitation, Milky Way

Protoplanet Condensation theory, Cosmology, Dwarf planets, Gravitation, Planetesimals, White dwarf

Protoplasm Cytoplasm, Eukaryotic cells, Nucleus, Organelles, Prokaryotic cells

Punctuated equilibrium Lineage, Phyletic evolution, Phylogenetic evolution, Phylogenetic tree, Species

Q

Quantitative character Character, Character state, Variation

Quantitative trait loci (QTLs) Allele, Chromosomes, Gene, Gene frequency, Gene locus, Independent assortment, Linkage, Linkage equilibrium, Locus, Population genetics, Segregation

R

Radiometric dating Geological dating, Half-life (radioactivity), Isotope

Random mating Genotype, Hardy–Weinberg equilibrium (principle), Inclusive fitness, Phenotype, Relative fitness, Reproductive success, Selection coefficient (s), Variability

Range Carrying Capacity, Environment, Intrinsic rate of natural increase, Logistic growth curve, Population, Population explosion

Rangeomorphs Biota, Burgess shale fossils, Chengjiang fauna, Ediacaran fauna/biota, Fauna, Fossils, Macroevolution, Paleontology, Precambrian Eon

Reaction norm Genetic assimilation, Phenotypic plasticity, Variation

Recessive allele Allele, Allele frequency, Gene, Genetic polymorphism, Heterozygote, Homozygote, Independent assortment, Linkage, Linkage equilibrium, Neutral theory of molecular evolution, Population genetics, Relative fitness, Segregation

Recessive lethal Allele, Allele frequency, Dominant allele, Gene, Genetic polymorphism, Homozygote, Independent assortment, Lethal allele, Recessive allele, Relative fitness, Segregation

Reciprocal altruism Altruism, Evolutionary stable strategies, "Tit for tat"

Recombination (genetics) Exon shuffling, Linkage equilibrium, Independent assortment

Red giant (cosmology) Collision theory, Dark ages (cosmology), Galaxy, Milky Way, Neutron stars (pulsars)

Reductionism Complexity, Design, Hierarchy, Progress

Regulation (developmental biology) Development, Developmental control gene, Embryology, Epigenesis, Evolutionary developmental biology, Morphogenesis, Ontogeny, Preformationism

Regulator gene Developmental control gene, Enhancer, Gene regulation, Promoter, Repressor, Signal transduction, Transcription, Upstream gene

Relative fitness Adaptive landscape, Allele, Allele frequency, Fitness, Gene, Genetic load, Genetic polymorphism, Genotype, Heterozygote, Heterozygote advantage (superiority), Homozygote, Hybrid breakdown (inviability, sterility), Inclusive fitness, Independent assortment, Linkage, Linkage equilibrium, Natural selection, Neutral theory of molecular evolution, Population genetics, Recessive allele, Recessive lethal, Reproductive success, Segregation, Selection, Selection coefficient (s)

Repetitive DNA	Duplication, Genome, Selfish DNA
Repressor	Enhancer, Promoter, Regulator gene, Repressor, Signal transduction, Transcription
Reproductive barriers	Geographical isolation, Isolating mechanisms, Postzygotic mating barrier, Prezygotic mating barrier, Reproductive isolation
Reproductive Incompatibility	Geographical isolation, Habitat, Isolating mechanisms, Niche, Postzygotic mating barrier, Prezygotic mating barrier, Reproductive barriers, Reproductive isolation
Reproductive isolation	Geographical isolation, Isolating mechanisms, Postzygotic mating barrier, Prezygotic mating barrier, Reproductive barriers, Speciation
Reproductive success	Fecundity, Fertility, Fitness, Heterozygote advantage (superiority), Genotype, Hybrid breakdown (inviability, sterility), Hybridization (genetics), Inbreeding, Inbreeding depression, Population genetics, Random mating, Relative fitness, Selection, Selection coefficient (s), Sexual reproduction
Reproductive technology	Eugenics, Gene therapy, Genetic engineering, Transgenic
Resource partitioning	Competition, Competitive exclusion, Environment, Group selection, Niche, Principle of divergence, Sexual selection
Restriction enzymes	Enzyme, Restriction fragment length polymorphisms (RFLPs)
Restriction fragment length polymorphisms (RFLPs)	Gene locus, Genetic polymorphism, Linkage, Linkage equilibrium, Linkage map, Locus, Microsatellites, Quantitative trait loci (QTLs)
Ribosomal RNA (rRNA)	Messenger RNA (mRNA), Micro RNA (miRNA), Nucleotide, ribosomes, RNA (ribonucleic acid), RNA interference (RNAi), Small interference RNA (siRNA), Transfer RNA (tRNA)
Ribosomes	Messenger RNA (mRNA), Organelles, Peptide (polypeptide), Ribosomal RNA (rRNA), Transfer RNA (tRNA), Translation
RNA (ribonucleic acid)	Base, Base Pairs, Base substitutions, Complementary base pairs, DNA (deoxyribonucleic acid), Nucleic acid, Nucleotide
RNA editing	Chromosome, Deletion, Mutation
RNA interference (RNAi)	Messenger RNA (mRNA), Micro RNA (miRNA), Nucleotide, Ribosomal RNA (rRNA), RNA (ribonucleic acid), Small interference RNA (siRNA), Transcription, Transfer RNA (tRNA)
RNA splicing	Exon, Exon splicing, Intron, Nucleotide, Split gene, Transcription, Translation
RNA world	Protein world, RNA world

S

Saprophyte	Fungi, Heterotroph, Inorganic, Life, Metabolism, Organic
Sea-floor spreading	Continental drift, Continents, Crust, Subduction (Earth sciences), Plate tectonics, Tectonic plates
Seasonal polymorphism	Morphology, Polymorphism
Secondary atmosphere	Atmosphere, Primary atmosphere
Secondary endosymbiosis	Endocytosis, Endosymbiosis, Primary endosymbiosis
Sedimentary rock	Catastrophism (Earth sciences), Crust (Earth sciences), Geological dating, Geological strata, Igneous rock, Law of superposition, Paleontology, Metamorphic rock, Uniformationarism
Seed (botany)	Angiosperms, Dicotyledons, Diploid, Double fertilization, Flower, Gamete, Gametophyte, Haploid, Meiosis, Mesozoic era, Monocotyledons, Pollen, Seeds
Segregation	Allele, Allele frequency, Chromosomes, Gene, Genetic polymorphism, Heterozygote, Homologous chromosomes, Homozygote, Independent assortment, Linkage, Linkage equilibrium, Meiosis, Neutral theory of molecular evolution, Population genetics, Recessive allele, Recessive lethal, Relative fitness
Selection	Adaptive landscape, Artificial selection, Canalizing selection, Constraint, Darwinism, Directional selection, Disruptive selection, Fitness, Frequency-dependent selection, Genetic load, Group selection, Heterozygote advantage (superiority), Hybrid breakdown (inviability, sterility), Inclusive fitness, Kin selection, Natural selection, Neo-Darwinism, Neutral mutation, Neutral theory of molecular evolution, Population genetics, Relative fitness, Reproductive success, Selection coefficient (s), Sexual selection, Stabilizing selection, Survival of the fittest

Selection coefficient (s)	Adaptive landscape, Fitness, Genetic load, Genotype, Heterozygote advantage (superiority), Hybrid breakdown (inviability, sterility), Inclusive fitness, Natural selection, Population genetics, Relative fitness, Reproductive success, Selection
Selfish DNA	Duplication, Genome, Repetitive DNA
Serial ("iterative") homology	Duplication, Gene family, Mutation, Paralogous genes
Sex chromosomes	Autosomes, Dosage compensation, Haldane's rule, Homogametic sex, Sex-linked genes, X-chromosome, Y-chromosome
Sex linkage	Chromosomes, Gene locus, Independent assortment, Linkage, Linkage equilibrium, Linkage map, Locus
Sexual reproduction	Fertilization, Gamete, Hermaphrodite, Life cycle, Ontogeny, Reproductive Incompatibility, Zygote
Sexual selection	Competition, Environment, Principle of divergence, Resource partitioning
Sibling species	Morphology, Reproductive isolation, Speciation
Signal transduction	Developmental control gene, Enhancer, Gene regulation, Promoter, Regulator gene, Repressor, Transcription, Upstream gene
Small interference RNA (siRNA)	Codon, Messenger RNA (mRNA), Micro RNA (miRNA), Nucleotide, Ribosomal RNA (rRNA), RNA (ribonucleic acid), RNA interference (RNAi), Transfer RNA (tRNA), Translation
Social Darwinism	Neo-Darwinism, Population genetics
Sociobiology	Altruism, Culture, Dominance (social), Homology, Imprinting (behavior), Instinct, Learning, Phenotype, Reciprocal altruism
Somatic cells (or tissues)	Gamete, Germ plasm, Germ plasm theory, Imprinting (genetics)
Space-time (cosmology)	Cosmological constant, Cosmology, Expanding universe, Galaxy, Gravitation
Speciation	Adaptive radiation, Anagenesis, Allopatric speciation, Principle of divergence, Parapatric speciation, Peripatric speciation, Reproductive isolation, Phyletic evolution, Sibling species, Stasis, Sympatric speciation
Species	Binomial nomenclature, Deme, Gene flow, Genus, Group, Isolating mechanisms, Lineages, Mendelian population, Migration (genetics), Parapatric, Polytypic species, Population, Sibling species, Species, Subspecies
Split gene	Exon, Intron, Nucleotide, RNA splicing, Transcription, Translation
Sporophyte	Alternation of generations, Diploid, Meiosis, Seed
Stabilizing selection	Artificial selection, Canalizing selection, Directional selection, Disruptive selection, Fitness, Frequency-dependent selection, Group selection, Natural selection, Neo-Darwinism, Population genetics, Relative fitness, Selection, Selection coefficient (s), Sexual selection, Survival of the fittest
Stasis	Geological time scale, Molecular clock, Stabilizing selection
Stem group	Clade, Crown group
Stone tools	Acheulean tools, Mousterian tools, Oldowan tools, Stone tools, Upper Paleolithic tools
Stop codon	Codon, Degenerate (redundant) code, Genetic code, messenger RNA (mRNA), Peptide (polypeptide), Translation, Universal genetic code
Subduction (Earth sciences)	Continental drift, Plate tectonics, Sea-floor spreading, Tectonic plates
Subspecies	Deme, Gene flow, Group, Isolating mechanisms, Mendelian population, Migration (genetics), Parapatric, Population, Species, Subspecies
Supergroups	Binomial nomenclature, Cladistics, Classification, Kingdom, Numerical Taxonomy, Systematics, Taxon, Taxonomy
Survival of the fittest	Artificial selection, Canalizing selection, Constraint, Darwinism, Directional selection, Disruptive selection, Fitness, Frequency-dependent selection, Group selection, Inclusive fitness, Kin selection, Natural selection, Neo-Darwinism, Population genetics, Relative fitness, Selection, Selection coefficient (s), Sexual selection, Stabilizing selection
Symbiont	Altruism, Evolutionary stable strategies, Mutualism, Reciprocal altruism, "Tit for tat"

Sympatric	Allopatric speciation, Parapatric speciation, Peripatric speciation, Speciation, Sympatric speciation
Sympatric speciation	Allopatric speciation, Parapatric speciation, Peripatric speciation, Speciation, Sympatric
Symplesiomorphy	Apomorphic character, Character, Character state, Homology, Plesiomorphy, Synapomorphy, Plesiomorphy
Synapomorphy	Apomorphic character, Character, Character state, Homology, Lineages, Plesiomorphy, Symplesiomorphy
Systematics	Binomial nomenclature, Cladistics, Class, Classification, Family, Kingdom, Numerical Taxonomy, Order, Phylum, Supergroups, Taxon, Taxonomy

T

Taxon	Binomial nomenclature, Cladistics, Class, Classification, Family, Kingdom, Numerical Taxonomy, Order, Phylum, Sister group (sister taxon), Supergroups, Systematics, Taxonomy
Taxonomy	Binomial nomenclature, Cladistics, Class, Classification, Family, Kingdom, Numerical Taxonomy, Order, Phylum, Supergroups, Systematics, Taxon, Type specimen
Tectonic plates	Continents, Continental drift, Continents, Crust, Gondwana, Laurasia, Pangaea, Plate tectonics, Subduction
Telome hypothesis	Angiosperms, Dicotyledons, Enation hypothesis, Ferns, Gymnosperms, Mesozoic Era, Monocotyledons, Vascular plants
Thermophilic organisms (thermophiles)	Hydrothermal vents
Tinkering (bricolage)	Darwinism, Descent with modification Domestication, Homology
"Tit for tat"	Altruism, Evolutionary stable strategies, Mutualism, Reciprocal altruism
Transcription	*Cis*-regulation, Enhancer, Exon, Intron, Messenger RNA (mRNA), Promoter, RNA interference (RNAi), RNA splicing, Repressor
Transcription factors	*Cis*-regulation, Enhancer, Promoter, RNA interference (RNAi), RNA splicing, Repressor, Transcription

Transfer RNA (tRNA)	Amino acids, Codon, Messenger RNA (mRNA), Micro RNA (miRNA), Nucleotide, Peptide (polypeptide), Ribosomal RNA (rRNA), RNA (ribonucleic acid), RNA interference (RNAi), Small interference RNA (siRNA), Translation
Transgenic	Gene therapy, Genetic engineering, Horizontal (Lateral) gene transfer, Reproductive technology, Transposons (transposable elements)
Translation	Amino acid, Codon, Degenerate (redundant) code, Genetic code, Messenger RNA (mRNA), Peptide (polypeptide), Ribosomes, Stop codon, Universal genetic code
Transposons (transposable elements)	Enzyme, Horizontal (Lateral) gene transfer, Transgenic
Tree of Life	Evolutionary trees, Phylogenetic tree, Universal Tree of Life
Triploblastic	Diploblastic, Ectoderm, Endoderm, Germ layers, Mesoderm, Neural crest

U

Unequal crossing over	Chromosomes, Homologous chromosomes, Independent assortment, Meiosis, Mitosis, Segregation
Uniformitarianism	Catastrophism (Earth sciences), Law of superposition
Unikonta	Amoebozoans, Fungi, Opisthokonts, Protists
Uniparental inheritance	Extranuclear inheritance, Inheritance, Maternal inheritance
Universal genetic code	Codon, Degenerate (redundant) code, Frozen accident, Genetic code, Messenger RNA (mRNA), Peptide (polypeptide), Stop codon, Translation
Universal Tree of Life (UToL)	Evolutionary trees, Phylogenetic tree, Tree of Life
Upper Paleolithic tools	Acheulean tools, Anatomically modern humans, Mousterian tools, Oldowan tools, Stone tools
Upstream gene	Downstream gene, Enhancer, Gene regulation, Homeotic mutations (homeosis), Regulator gene, Repressor, Signal transduction, Transcription
Use and disuse	Lamarckian inheritance

V

Variability	Environment, Evolvability, Genotype, Variation

Variation	Canalization, Constraint, Continuous variation, Darwinism, Discontinuous variation, Natural selection, Phenotypic plasticity, Quantitative character, Reaction norm, Variability
Vascular plants	Angiosperms, Gymnosperms, Mesozoic Era, Phloem, Seed, Telome hypothesis, Xylem
Vertical transmission	Extranuclear inheritance, Inheritance, Maternal inheritance, Pangenesis, Uniparental inheritance
Virus	DNA virus, RNA virus
Vitalism	Life, Metabolism, Organism, Orthogenesis, Progress

W

White dwarf (cosmology)	Condensation theory, Cosmology, Dwarf planets, Gravitation, Planetesimals, Protoplanet
Wild type	Genetic load, Genotype, Hardy–Weinberg equilibrium (principle), Mutation, Phenotype, Relative fitness, Selection Variability, Variation

X

X-chromosome	Dosage compensation, Haldane's rule, Homogametic sex, Sex chromosomes, Sex-linked genes, Y-chromosome

Y

Y-chromosome	Dosage compensation, Haldane's rule, Homogametic sex, Sex chromosomes, Sex-linked genes, X-chromosome
Y-chromosome Adam	Mitochondrial DNA, Mitochondrial Eve

Z

Zoogeography	Biogeography, Endemic, Phylogenetic evolution, Phylogeography
Zygote	Fertilization, Gamete, Life cycle, Ontogeny, Reproductive Incompatibility, Sexual reproduction

ADDITIONAL READING

The following **encyclopedias**, **texts** and **collections of essays** are recommended, either because they cover evolution in its entirety or because they cover a major aspect of evolution.

Appleman, P, (ed.), 2001. *A Norton Critical Edition. Darwin: Texts, Commentary, Third Edition*, selected and edited by Philip Appleman. Norton, New York.

Bowler, P. J., 2003. *Evolution: The History of an Idea. Third Edition*. University of California Press, Berkeley, CA.

Browne, E. J., 1995. *Charles Darwin: Voyaging*. Alfred A. Knopf, New York.

Browne, E. J., 2002 *Charles Darwin: The Power of Place*. Volume II of a Biography. Alfred A. Knopf, New York.

Coyne, J. A., and H. A. Orr, 2004. Speciation. Sinauer, Sunderland, MA.

Cracraft, J., and R. W. Bybee (eds.), 2005. *Evolutionary Science and Society: Educating a New Generation*. Biological Sciences Curriculum Study, Washington, DC.

Diamond, J. M., 1997. *Guns, Germs, and Steel: The Fates of Human Societies*. Norton, New York.

Diamond, J. M., 2005. *Collapse: How Societies Choose to Fail or Succeed*. Viking Books, New York.

Erwin, D. H., 2006. *Extinction: How Life on Earth Nearly Ended 250 Million Years Ago*. Princeton University Press, Princeton, NJ.

Evolution, 2006. A Scientific American Reader. The University of Chicago Press, Chicago.

Grant, P. R., and B. R. Grant, 2008. *How and Why Species Multiply: The Radiation of Darwin's Finches*. Princeton University Press, Princeton, NJ.

Hall, B. K., and B. Hallgrímsson, 2008. *Strickberger's Evolution, Fourth Edition*. Jones and Bartlett, Sudbury, MA.

Hall, B. K., and W. M. Olson (eds), 2003. *Keywords & Concepts in Evolutionary Developmental Biology*. Harvard University Press, Cambridge MA.

Hallgrímsson, B., and B. K. Hall, 2005. *Variation: A Central Concept in Biology*. Elsevier/Academic Press, New York.

Knoll, A. H., 2003. *Life on a Young Planet: The First Three Billion years of Evolution on Earth*. Princeton University Press, Princeton, NJ.

Mayr, E., 2001. *What Evolution Is. With a Foreword by Jared Diamond*. Basic Books, New York

Pagel, M., (ed. in chief), 2002. *Encyclopedia of Evolution*, Two Volumes. Oxford University Press, New York.

Quammen, D., 2006. *The Reluctant Mr. Darwin*. Norton, New York.

Scott, E. C., 2005. *Evolution vs. Creationism: An Introduction*. University of California Press, Berkeley, CA.

Wood, B., 2005. *Human Evolution. A Very Short Introduction*. Oxford University Press, Oxford, England.

Index

Note: Page numbers in the Glossary and in the Lexicon are not included in this index.

first on Earth, 48–52
hydrothermal vents and, 44
in meteorites, 44–45
and origin of life, 52
prerequisites for, 41–43
proteins first, 48
replicating in laboratory, 43
RNA first, 48
sites of formation of, 43–44
volcanoes and, 44–45
Origin of species, 112–114, 325–344
in absence of geographical isolation,
 329, 331–336
adaptation and differentiation, 327
change without speciation, 326–327
genes associated with, 342–344
by geographical isolation, 89,
 114–116, 327–330
hybridization and, 340–342
mechanisms of reproductive isolation,
 336–338
reproductive incompatibility,
 338–340
sexual isolation, 339–340
Origination, defined, 11
Owen, Richard, 320, 435. *See also*
 Homology
Oxygen
and animal origins, 198
in atmosphere, 55–56
and cyanobacteria, 27, 55–56, 137
Ozone layer, 19, 27, 198

P
Paleolithic stone age, 391–392
Paley, William, 438. *See also* Design
Pangaea, 35
fragmentation, 352
mammalian radiations, 352–354
Pangenesis, 117, 120–122; *See also*
 Darwin, Charles
Paramecium
competitive exclusion between
 species of, 265–266
symbiosis with algae, 159.
Parallelism, 318, 319, 323
anteaters, 317–318
definition, 320
of "wolf" phenotype, 317–318, 323
Parasitism, 260, 267–268
and complexity, 372
myxoma virus and rabbits, 268

Schistosoma, 372
viruses and, 268
Pasteur, Louis, 91, 129
Pax-6 gene, 204
PCR. *See* Polymerase chain reaction
Pesticides, evolution of resistance to,
 237–238, 240–241, 441
Phenotype, 61
relation to genotype, 208, 213, 325.
 See also Genotype
Phenotypic plasticity, 260, 269–271,
 286. *See also* Canalization;
 Reaction norms
birds, 270–271
cichlid fishes, 331–333, 441
Daphnia pulex (water flea), 269–270
ecomorphs, 261
life history theory and, 269
morphs, 269
moths, 270
predator-prey induced, 260, 269–271
rotifers, 269–270
speciation and, 331
sticklebacks, 270
sunfish, 270
tadpoles, 270
Phillips, John, 27, 29
Phospholipids and cell membranes,
 56–57
Photons, 20
Photosynthesis, 55
cyanobacteria and, 56, 159
Phyla, 108, 152, 192
Phyletic evolution, 114–116
Phylogenetic trees, 106, 143, 144, 301.
 See also Classification
of birds, 361
of cichlid fishes, 335
of dinosaurs, 361
Ernst Haeckel and, 156, 313
hominoids, 379
human populations, 394
humans and apes, 377
of life, 153
from molecular data, 147, 153
Phylogeny, classification and,
 303–305, 307–308
Phylogeography, 254–256
and the spread of wheat, 254–256
Piltdown man, 382–383. *See also*
 Australopithecines
Pithecanthropus erectus, 385

Piwi-interacting RNAs (piRNAs).
 See Gene regulation
Planets orbits, 4
Plants. *See also* Coevolution; Insects;
 Pollination
algal ancestry, 168–169, 176
alternation of generation, 166,
 171–173, 177
Amborella trichopoda as most basal
 extant, 184, 186
angiosperms, 166, 167, 179–181, 184
bryophytes, 166, 170–174
coevolution with insects, 179, 180,
 182–184, 267–269, 365
convergent evolution, 179, 180
Devonian land, 173–175
early vascular, 173–176
evolution of spores, 174, 175
ferns, 174, 176–178
flowers as reproductive organs,
 179–185
frequency of hybridization, 341–342
gametophyte generation, 166,
 171–173, 177
gymnosperms, 166, 167, 179, 180
horsetails, 174, 176
Knox homeobox genes, 176
MADS box genes, 184
move to land, 177–181
mutualism, 269
origin, an abominable mystery, 166
origin of leaves, 176–178
origination of, 153, 158
phloem, 169
phylogenetic tree, 167
relationships of flowering, 167,
 181, 184
speciation by hybridization, 340–342
sporophyte generation, 166,
 171–174, 177
xylem, 169
Plasmodium falciparum, 154, 293. *See
 also* Malaria
Plato, 75, 76, 302, 418. *See also*
 Eugenics; Idealism
Pleiotropy, 223
Pollination
by birds, 166, 185
by butterflies, 180–181, 185
host-plant specificity, 181–185
by insects, 166, 185
Polygenes, 96

Photo Acknowledgments

Table of Contents

Page vii © Misulka/Dreamstime.com; **page viii** © tom67/ShutterStock, Inc.; **page xi** © Ismael Montero Verdu/ShutterStock, Inc.; **page xiii** © ClimberJAK/ShutterStock, Inc.; **page xiv** © Mccls1030/Dreamstime.com; **page xv** © Ecophoto/Dreamstime.com; **page xvi** © Daniel Bellhouse/Dreamstime.com.

All Part Openers © AbleStock.

Chapter 1

Opener © Katrina Brown/ShutterStock, Inc.; **1.2a** Courtesy of Richard Borowsky, New York University; **1.2b** © Domen Lombergar/ShutterStock, Inc.; **1.2c** © Misulka/Dreamstime.com; **1.3b** Courtesy of National Library of Medicine; **Box 1.1** © Popovici loan/ShutterStock, Inc.

Chapter 2

Opener © Jhaz Photography/ShutterStock, Inc.; **2.1** Courtesy of NASA/WMAP Science Team; **2.6** © maurizio grimaldi/age footstock.

Chapter 3

Opener © Pichugin Dmitry/ShutterStock, Inc.

Chapter 4

Opener © rebvt/ShutterStock, Inc.; **4.6a** © David McCarthy/Photo Researchers, Inc.; **4.6b** © Science VU/Dr. Sidney Fox/Visuals Unlimited; **4.7 insert** (embryo) © Joe Mercier/Dreamstime.com; **4.7 insert** (cells) Courtesy of Janice Carr/CDC; **4.7 insert** (sperm) © Sebastian Kaulitzki/ShutterStock, Inc.; **4.7 insert** (egg) © Sebastian Kaulitzki/ShutterStock, Inc.; **4.8** © AbleStock/Alamy Images.

Chapter 5

Opener © nouseforname/ShutterStock, Inc.; **5.1a** © English Heritage Photo Library/By kind permission of Darwin Heirlooms Trust; **5.1b-c** © National Library of Medicine; **5.2a** © SPL/Photo Researchers, Inc.; **5.2b** © Mary Evans Picture Library/Alamy Images; **5.4** © Photos.com; **Box 5.5** © Photos.com.

Chapter 6

Opener © javarman/ShutterStock, Inc.; **6.1a** Courtesy National Library of Medicine; **6.2** © Lordprice Collection/Alamy Images; **6.3a** © Eric Isselée/ShutterStock, Inc.; **6.3b** © Utekhina Anna/ShutterStock, Inc.; **6.3c** © Utekhina Anna/ShutterStock, Inc.; **6.3d** © Eric Isselée/ShutterStock, Inc.; **6.4a-b** Images courtesy of Tamara Franz-Odendaal and Megan Dufton, Mount Saint Vincent University; **Box 6.6** Photo courtesy of Amy Tsuneyoshi.

Chapter 7

Opener © Andrew Corney/Dreamstime.com; **Box 7.2** © Kåre Telnes/seawater.no; **Box 7.3a-b** Reprinted from *Deep Sea Research Part II: Topical Studies in Oceanography*, vol. 56, Robert C. Vrijenhoek, Cryptic species, phenotypic plasticity . . . , pp. 1713–1723. Copyright 2009, with permission from Elsevier. [http://www.sciencedirect.com/science/journal/09670645]. Photo courtesy of Robert C. Vrijenhoek; **Box 7.3c** Courtesy of Greg Rouse.

Chapter 8

Opener © Carlos Caetano/ShutterStock, Inc.; **8.4a** © Chung Ooi Tan/ShutterStock, Inc.; **8.4b** © Marli Miller/Visuals Unlimited; **8.4c-d** © Sinclair Stammers/Photo Researchers, Inc.; **8.6a-b** Courtesy of J. William Schopf, Professor of Paleobiology & Director of IGPP CSEOL; **8.7a-b** Reproduced from Schopf, J.W., *Science* 260 (1993): 640–646. Reprinted with permission from AAAS. Courtesy of J. William Schopf, Professor of Paleobiology & Director of IGPP CSEOL.

Chapter 9

Opener Courtesy of Dr. Stan Erlandsen/CDC; **9.1a-c** Reproduced from Schopf, J.W., *Scientific American* 239 (1978): 111–138. Courtesy of J. William Schopf, Professor of Paleobiology & Director of IGPP CSEOL; **9.3** Courtesy of Anthony L. Swinehart, Hillsdale College.

Chapter 10

Opener © Brian Maudsley/ShutterStock, Inc.; **10.1** Courtesy of U.S. Botanic Garden; **10.2** Courtesy of Dr. Robert Ricker, NOAA/NOS/ORR; **10.3a** © Todd Boland/ShutterStock, Inc.; **10.3b** © tom67/ShutterStock, Inc.; **10.3c** Courtesy of James Mauseth, University of Texas at Austin; **10.5c** Courtesy of Francisco Javier Yeste Garcia, www.flickr.com/photos.fryega/; **10.8b** © Ismael Montero Verdu/ShutterStock, Inc.; **10.13a** © Alex Neauville/ShutterStock, Inc.; **10.13b** © Photodisc; **10.13c** © Micw/Dreamstime.com; **10.13d** © Steve Allen Travel Photography/Alamy Images; **Box 10.3** © Sami Sarkis Studio/Alamy Images; **Box 10.4a** Courtesy of Dave Knepper, www.flickr.com/photos/outside/; **Box 10.4b** © The Natural History Museum/Alamy Images; **Box 10.5** © Nathan Muchhala; **10.16** Photo by Thomas J Lemieux, University of Colorado; **10.17a-b** Reproduced from Taylor, T. *Paleobotany: An Introduction to Fossil Plant Biology.* McGraw-Hill, 1981. Used with permission of The McGraw-Hill Companies.

Chapter 11

Opener © Photos.com; **11.2** Reproduced from Guy Narbonne, 2004. *Science* 305: 1141–1144. Reprinted

with permission from AAAS; **Box 11.1** Courtesy of Dr. Elizabeth C. Raff; **11.3 insert** (jellyfish) © Dwight Smith/ShutterStock, Inc.; **11.3 insert** (worm) Courtesy of NOAA, National Estuarine Research Reserve; **11.3 insert** (bird) © AbleStock; **11.3 insert** (giraffe) © AbleStock; **11.10b** © David Scharf/Peter Arnold, Inc.

Chapter 12

Opener © twobluedogs/ShutterStock, Inc.; **12.1** © R0b/Dreamstime.com; **12.5** (top) Courtesy of Chuck Dresner/Saint Louis Zoo; **12.5** (bottom) Courtesy of the Saint Louis Zoo; **12.9b** © Eye of Science/Photo Researchers, Inc.; **12.10a** © Nature's Images, Inc.; **12.10b** Courtesy of Dr. Simon Walker, Department of Zoology, University of Oxford; **Box 12.2a** © Mike Flippo/ShutterStock, Inc.; **Box 12.2b** © Greg30127/Dreamstime.com.

Chapter 13

Opener © Srcromer/Dreamstime.com; **13.1 insert** © Frank B. Yuwono/ShutterStock, Inc.; **13.6** Reproduced from *Trends in Ecology & Evolution*, 19(9), Foster, S. A, and Baker, J. A., Evolution in parallel . . . , pp. 456–459, copyright 2004, with permission from Elsevier. Photographs courtesy of Dr. W. A. Cresko, University of Oregon; **Box 13.1b** © Sergey Chushkin/ShutterStock, Inc.; **Box 13.1c** © Brzostowska/ShutterStock, Inc.

Chapter 14

Opener © Rich Lindie/ShutterStock, Inc.; **14.1b** (left) © Philip Delos/Dreamstime.com; **14.1b** (middle) © ClimberJAK/ShutterStock, Inc.; **14.1b** (right) © Bob Blanchard/ShutterStock, Inc.; **14.4a** © Nigel Cattlin/Visuals Unlimited, Inc.; **14.5a** Photos courtesy of Susan Bragg, Thomas Massie and Gregor Fussmann; **14.5b** © Christian Laforsch/Photo Researchers, Inc.; **14.6a** © Ronald Altig; **14.6b** © Gary Nafis; **14.8 insert** © Sebastian Knight/ShutterStock, Inc.

Chapter 15

Opener © Chris Lofty/Dreamstime.com; **15.3** © Dmitry Maslov/Dreamstime.com; **15.7a** © Tlarsen/Dreamstime.com; **15.7b** © Florian Andronache/ShutterStock, Inc.; **15.7c** © Steffen Schellhorn/age fotostock; **15.7d** © Sbotas/Dreamstime.com; **15.8** Seals photographed on St. Paul Island in the Bering Sea by Dr. Sara Iverson, who kindly supplied the image, taken under MMPA Permit No.782-1708-05; **Box 15.1a** © Milan Vasicek/ShutterStock, Inc.; **Box 15.1b** © Mccls1030/Dreamstime.com; **15.12a-b** © Science VU/Visuals Unlimited; **15.13a-b** © R B Forbes and the Mammal Images Library of the American Society of Mammalogists.

Chapter 16

Opener © rebvt/ShutterStock, Inc.; **Box 16.5a** Courtesy of Erin Green, Crown Copyright, Department of Conservation, N.Z., 2004; **Box 16.5b-c** Courtesy of Dr. J. G. M. Thewissen, Northeastern Ohio Universities Colleges of Medicine and Pharmacy; **Box 16.6c** © 3drenderings/ShutterStock, Inc.; **16.3a** © Brian Arbuthnot/Dreamstime.com; **16.3b** © Ainars Aunins/ShutterStock, Inc.

Chapter 17

Opener © LubaShi/ShutterStock, Inc.; **Box 17.1a-c** Modified from Edwards K.A, Doescher L.T., Kaneshiro K.Y., Yamamoto D., (2007) A Database of Wing Diversity in the Hawaiian Drosophila. *PLoS ONE* 2(5): e487. doi:10.1371/journal.pone.0000487. Courtesy of Kevin Edwards, Illinois State University; **17.2** (top left) © Wawritto/Dreamstime.com; **17.2** (top right) © Savone/Dreamstime.com; **17.2** (middle left) © Showkontor/Dreamstime.com; **17.2** (middle right) © Moori/Dreamstime.com; **17.2** (bottom left) © Schoor/Dreamstime.com; **17.2** (bottom right) © Ecophoto/Dreamstime.com; **Box 17.4** Photos **a, b** and **e**, courtesy of Dr. Axel Meyer, University of Konstanz and photos **c, d** and **f**, courtesy of courtesy of Dr. Ann Huysseune, Ghent University; **17.6a** © Colin D. Young/ShutterStock, Inc.; **17.6b** © Cousin Avi/ShutterStock, Inc.; **17.6c** Courtesy of Clarence A. Rechenthin and USDA NRCS Texas State Office; **17.7a insert** Photo by John Doebley; **17.7b insert** © ailenn/ShutterStock, Inc.; **17.8a-d** Photos courtesy Toby Bradshaw (University of Washington) and Douglas Schemske (Michigan State University).

Chapter 18

Opener © Styve Reineck/ShutterStock, Inc.; **18.3** Courtesy of Shannon Rankin, NMFS, SWFSC/NOAA; **18.5** © Paul B. Moore/ShutterStock, Inc.; **18.11** © Natural History Museum, London; **Box 18.6** Courtesy of Brian Hall; **18.12a** © Joy Stein/ShutterStock, Inc.; **18.12b** © Imagex/Dreamstime.com.

Chapter 19

Opener © loriklaszio/ShutterStock, Inc.; **Box 19.2a** © Antonio Petrone/ShutterStock, Inc.; **Box 19.2b** © Daniel Bellhouse/Dreamstime.com; **Box 19.3a** © Equinox Graphics/Photo Researchers, Inc.; **Box 19.3b** © Mark Boulton/Alamy Images.

Chapter 20

Opener © George Lamson/ShutterStock, Inc.; **20.1** © Behavioural Ecology Research Group, University of Oxford; **20.2** © attem/ShutterStock, Inc.; **20.3a** © Monkey Business Images/ShutterStock, Inc.; **20.3b** © Joe Mercier/Dreamstime.com; **20.4** © Antonio Petrone/ShutterStock, Inc.

Chapter 21

Opener © silver-john/ShutterStock, Inc.; **Box 21.1** © iBird/ShutterStock, Inc.; **Box 21.2** © sgame/ShutterStock, Inc.; **21.1** © FAO/Giampiero Diana; **21.2a** Courtesy of National Library of Medicine; **21.2b** © Photos.com; **21.3** © Smithsonian Institution Archives, SIA2007-0124.

Unless otherwise indicated, all photographs and illustrations are under copyright of Jones and Bartlett Publishers, LLC, or have been provided by the author.